Alkaloids: Chemical and Biological Perspectives

Alkaloids: Chemical and Biological Perspectives

Volume Two

Edited by

S. WILLIAM PELLETIER
Institute for Natural Products Research

and

The Department of Chemistry
University of Georgia, Athens

A Wiley-Interscience Publication

JOHN WILEY & SONS

New York   Chichester   Brisbane   Toronto   Singapore

CHEMISTRY

7205-850X

Library of Congress Cataloging in Publication Data:

(Revised for volume 2)
Main entry under title:

Alkaloids: chemical and biological perspectives.

"A Wiley-Interscience publication."
Includes bibliographies and indexes.
1. Alkaloids.   I. Pelletier, S. W., 1924–

[DNLM: 1 Alkaloids.   QD 421 A4156]
QD421.A56   1983   574.19′242   82-11071
ISBN 0-471-08811-0 (v. 1)
ISBN 0-471-89299-8 (v. 2)

Printed in the United States of America

10 9 8 7 6 5 4 3 2 1

Dedicated to
the memory of

Pierre Joseph Pelletier
(1788–1842)

and

Joseph Bienaimé Caventou
(1795–1877)

who, while working at the Faculty of Pharmacy in Paris, were the first to isolate the important alkaloids emetine (1817), brucine (1819), strychnine (1819), caffeine (1820), colchicine (1820), quinine (1820), cinchonine (1820), piperine (1821), coniine (1827), and thebaine (1835). These alkaloids have served as cornerstones for subsequent advancements in alkaloid chemistry over the past century and a half.

Contributors

Edward Arnold, Department of Biological Sciences, Purdue University, West Lafayette, Indiana

Manuel F. Balandrin, Plant Resources Institute, University Research Park, Salt Lake City, Utah

Jon Clardy, Department of Chemistry, Cornell University, Ithaca, New York

Janet Finer-Moore, Department of Biochemistry and Biophysics, University of California, School of Medicine, San Francisco, California

Richard K. Hill, Department of Chemistry, University of Georgia, Athens, Georgia

Balawant S. Joshi, Institute for Natural Products Research, University of Georgia, Athens, Georgia

A. Douglas Kinghorn, Department of Medicinal Chemistry and Pharmacognosy, College of Pharmacy, University of Illinois at Chicago, Chicago, Illinois

Naresh V. Mody, Institute for Natural Products Research, University of Georgia, Athens, Georgia

S. William Pelletier, Institute for Natural Products Research and Department of Chemistry, University of Georgia, Athens, Georgia

Richard G. Powell, Northern Regional Research Center, Agricultural Research Service, U.S. Department of Agriculture, Peoria, Illinois

Lee C. Schramm, Institute for Natural Products Research, University of Georgia, Athens, Georgia

Cecil R. Smith, Jr., Northern Regional Research Center, Agricultural Research Service, U.S. Department of Agriculture, Peoria, Illinois

Volume 2 of *Alkaloids: Chemical and Biological Perspectives* presents timely reviews of several important topics on alkaloids.

Chapter 1 is an excellent review of the applications of single-crystal X-ray diffraction to the elucidation of structures of alkaloids. Because single-crystal X-ray analysis has become the most powerful and reliable method for elucidation of organic structures, and because alkaloids have proved a fertile field for the application of this technique, a chapter reviewing methods of structure determination and illustrative examples of the determination of alkaloid structures seems appropriate.

Since the last comprehensive review of imidazole alkaloids, several novel imidazoles have been discovered. Chapter 2 is a concise review of this class of rare alkaloids.

Chapter 3 presents a comprehensive review of quinolizidine alkaloids occurring in Leguminosae. Plants of this family are of great importance for human foodstuffs, forage, and fodder, to fix nitrogen, and as sources of timber, fuel, oils, gums, and resins. In keeping with our objective of reviewing topics of biological interest, as well as chemistry, this chapter treats structure, methods of analysis, chemotaxonomy, and biological activities of these quinolizidine alkaloids.

The term *alkaloid* was redefined along modern, functional lines in Volume 1 of this series. Thus a review of the important maytansinoids appears appropriate. Many of these ansamacrolide alkaloids manifest potent anti-tumor activity and certain of them show insect antifeedant properties. Chapter 4 surveys the chemistry, isolation techniques, synthesis, and pharmacology of these interesting alkaloids.

Chapter 5 presents an extensive review of ^{13}C and proton NMR shift assignments and physical constants of C_{19}-diterpenoid alkaloids. Valuable features of this chapter are tables that correlate proton and ^{13}C shifts with functional groups.

Each chapter in this volume has been reviewed by an expert in the field. Indexes for both subjects and organisms are provided.

The editor invites prospective contributors to write him about topics for review in future volumes of this series.

S. WILLIAM PELLETIER

Athens, Georgia
August 1984

Contents

Alkaloids: Chemical
and Biological
Perspectives

Chapter One

Some Uses of X-ray Diffraction in Alkaloid Chemistry

Janet Finer-Moore
Department of Biochemistry and Biophysics
University of California
School of Medicine
San Francisco, California

Edward Arnold
Department of Biological Sciences
Purdue University
West LaFayette, Indiana

Jon Clardy
Department of Chemistry
Cornell University
Ithaca, New York

CONTENTS

1. INTRODUCTION

In the relatively recent past, single-crystal X-ray diffraction has become one of the most powerful, if not the most powerful, methods for elucidating organic structures. Alkaloids are not the exception and have proved a fertile field for the application of this technique. The Cambridge Crystallographic Data Center [1] has been adding structures classified as alkaloids to their data base at the rate of roughly one a week during the past few years. Some of the reasons for this rapid growth are improved experimental equipment and computational advances. The collection of intensity data used to be a tedious chore that required weeks of labor-intensive work. With a modern computer-controlled diffractometer, this process has been reduced to hours or days of experimental time on a unit that requires little human intervention. The computational situation has also improved. There are a number of "user friendly" but very powerful program libraries that allow a relatively unsophisticated investigator to work successfully on complex crystal structure determinations. These analyses would have been the province of an expert only a few years ago. Thus the time, cost, and knowledge required to carry out a diffraction experiment have all been reduced and are roughly comparable to those of other physical methods.

The other reason for the growth of X-ray diffraction is found in the nature of the diffraction experiment itself. The diffraction experiment is fundamentally different from other physical methods. When a structure is elucidated by chemical or spectroscopic methods, all structures consistent with the known data must be considered and further experiments must be devised to pare this list of candidate structures to a single solution. The analysis of diffraction data does not require this list of candidate structures. The final structure appears without it. Occasionally in our laboratory the original molecular formula we assumed was wrong and this has hampered us little if at all. Diffraction analysis arrives at a structure in a prejudice-free fashion and this is an enormous advantage when new chemotypes are being explored. Another advantage of the X-ray diffraction experiment is the lavish amount of experimental data available. While there will be variations, a typical case might involve 50 X-ray reflections for every nonhydrogen atom in a structure. The use of each of these reflections is, in principle, straightforward. The X-ray experiment is done with modest amounts of material. A typical crystal weighs tens of micrograms and is not destroyed during the course of an experiment.

There are also problems associated with the X-ray diffraction experiment. The most vexatious of these is the absolute requirement for single crystals. In many investigations the rate-limiting step is obtaining usable crystals. As far as the authors know there are no generally applicable tricks for guaranteeing the growth of good crystals. The old virtues of cleanliness and patience are still useful today. There is also the problem that the crystal structure may not be able to be solved. This is not a common occurrence but is extremely frustrating when it happens. With other physical methods even if a unique structure does not appear, something is learned. Partial information such as the number of methyl carbons, the existence of carbonyl functions, or the molecular weight can be gained from nuclear magnetic resonance, infrared, or mass spectra. Unfortunately partial analysis of a structure is not possible in a diffraction experiment. If the structure is not solved, little of chemical value is learned.

In the remainder of this chapter we will discuss the principles involved in X-ray diffraction, illustrate them with some examples drawn from the alkaloid field, and finally indicate that there is much more information obtained in a crystallographic analysis than is generally used. This chapter is not intended as a short primer on how to do X-ray crystallography, but rather as an invitation to begin exploring this powerful technique.

1.1. The X-Ray Equipment

X-rays are electromagnetic radiation with typical wavelengths of approximately 1 Å or 10^{-8} cm. In conventional sources X-rays are produced by electrons that have been accelerated by a potential field of tens of kilovolts slamming into a target material. One of the processes that occurs when these high-energy electrons collide with the target is the ejection of an electron from a low-lying energy level of the target material. This electron hole is filled by an electron dropping from a higher energy level with the simultaneous emission of radiation. This emitted radiation is the radiation used in the diffraction experiment. It is essentially monochromatic by virtue of its production by transitions from atomic energy levels. Since the overall process is rather inefficient, much energy is dissipated as heat, and suitable target materials are high melting metals which must be cooled. The most common targets are Cu with a characteristic wavelength of 1.54178 Å and Mo with a wavelength of 0.71069 Å. Any unwanted radiation is removed by a suitable filter or monochromator and the X-rays continue their journey to the sample.

The sample is made up of a collection of charged particles—electrons and protons. When a charged particle is bathed by electromagnetic radiation, it begins to oscillate with the same frequency as the incoming radiation. As a charged particle oscillates, it becomes a new source of electromagnetic radiation. This process is called elastic scattering. Since the intensity of this secondary radiation is proportional to the amplitude of the oscillation and the amplitude is inversely proportional to the square of the mass, essentially all of the scattering is done by electrons. It is the scattered X-rays that form the raw data for the experiment. Since only electrons give rise to scattered X-rays, the experiment reports only the distribution of electrons.

The scattered X-rays do not bear a readily perceived relation to the scattering object. Whenever an object is illuminated by light waves with roughly the same dimension as the object, the resulting pattern is not a simple picture of the object. It is a related figure called the diffracted image. A simple example might be recalled from high school physics where a narrow slit was illuminated with light and the resulting pattern was a series of lines. Since X-rays and atoms are roughly the same size, the patterns formed by the scattered X-rays are diffraction patterns of the original object. An object and its diffracted image are mathematically related by the Fourier transform. A diffraction pattern may be calculated for any given electron density distribution using the Fourier transform, and an electron density distribution can similarly be derived from its diffracted image. The electron density distributions computed with crystallographic data are sometimes called Fourier syntheses.

In principle the X-ray experiment could be done on a single molecule but X-ray sources are not bright enough to give an observable diffraction pattern and there would be substantial technical difficulties in orienting a single molecule when X-rays are diffracted from it. What is needed, then, is a device that will hold large numbers of molecules in an ordered fashion while X-rays are diffracted from them. Crystals are such devices. The use of crystals adds another complication. The diffraction pattern of a crystal is not a continuous distribution of intensities but a series of discrete points, since the scattering from a periodic array will constructively add together only when certain geometrical constraints are satisfied. The diffraction pattern of a crystal is a series of regularly spaced, variable intensity spots. The regular spacing is related to the periodicity of the crystalline array and measurement of this spacing leads in a straightforward way to the dimensions of the crystalline unit cell. The intensity of the radiation at any of these points is a function of the electron distribution of the scattering object. The diffraction experiment can, then, be divided into several phases. The repeating distances and symmetry of the diffraction pattern are first analyzed. This analysis leads to the unit cell parameters—the lengths and angles between the repeating vectors that build up the crystal—and the symmetry relations between the objects making up the unit cell. The next phase consists of measuring the intensities of each of the diffracted beams. The third phase is the analysis of the diffraction data to get a rough molecular picture and the last phase is the refinement of the molecular model to get the best agreement between calculated and experimental values.

Once a suitable crystal is available for diffraction work the first measurements are usually done by examining the diffraction pattern photographically. By analysis of these photographs, the crystallographer can determine the point group symmetry of the crystalline array and define a unit cell. While there are several ways to break a lattice into repeating units, the proper unit cell is the smallest repeating unit which incorporates the symmetry elements of the crystal's point group. In general, a unit cell is characterized by the lengths of three vectors and the angles between them. By convention the vectors are called a, b, and c, and represent the edges of the unit cell. The angles are given such that α is the angle between b and c, β is the angle between a and c, and γ is the angle between a and b. Often not all of these need be specified since some symmetry of the crystal will restrict the values. For example, suppose that a unit cell has a twofold rotation axis of symmetry. If this twofold axis were along b then the a and c axes would have to be at right angles to b and the angles α and γ would be 90° by symmetry. The 32 possible point groups for three-dimensional lattices can be divided into seven crystal classes, each class with a different set of restrictions on its cell parameters. Table 1 summarizes the available crystal classes and the restrictions on their cell parameters.

There is rarely any difficulty in assigning a crystal to a class and the first parameters usually given in an experimental discussion are the unit cell parameters. From a knowledge of the size and symmetry of a unit cell some useful information can be extracted. If the density of the crystal is known it is possible to calculate the molecular weight of the repeating unit. Even if the density is not known or not known accurately, it is still possible to estimate the contents of the unit cell. A rough rule for organic structures is that the number of nonhydrogen atoms in the unit cell is equal to the volume of the cell (in $Å^3$) divided by 18.

Once the unit cell is characterized, the next experimental task is to record the intensity of each diffracted beam. This is occasionally done by photographic techniques but is more commonly done with a scintillation counter. The scintillation counter is part of a piece of equipment called a diffractometer. The diffractometer has several angular degrees of freedom which are used to orient the crystal so that a diffracted beam is directed toward the scintillation counter. In modern instruments the diffractometer is computer controlled to scan diffracted beams in succession and typical rates for data collection fall in the

Table 1.   The Crystal Classes

Class	Independent Parameters	General Parameter Values		Diffraction Symmetry
Triclinic	6	$a \neq b \neq c$;	$\alpha \neq \beta \neq \gamma$	Ci
Monoclinic	4	$a \neq b \neq c$;	$\alpha = \gamma = 90°$, $\beta \neq 90°$	C2h
Orthorhombic	3	$a \neq b \neq c$;	$\alpha = \beta = \gamma = 90°$	D2h
Tetragonal	2	$a = b \neq c$;	$\alpha = \beta = \gamma = 90°$	D4h
Rhombohedral	2	$a = b = c$;	$\alpha = \beta = \gamma \neq 90°$	D3d
Hexagonal	2	$a = b \neq c$;	$\alpha = \beta = 90°$, $\gamma = 120°$	D6h
Cubic	1	$a = b = c$;	$\alpha = \beta = \gamma = 90°$	Oh

range of 500–1500 reflections per day. As a rough calibration, an alkaloid with 30 nonhydrogen atoms in the asymmetric unit would give rise to approximately 1500 useful reflections and data collection would take two days. This guideline will vary from laboratory to laboratory and from sample to sample.

When all of the intensities have been recorded the problem of analysis begins. Each of the diffracted beams can be thought of as a wave. The amplitude and wavelength of each of these waves is known experimentally and if all of these waves were added according to the Fourier transform relation, the resulting pattern would be the electron density in the crystal. Unfortunately, in order to add up waves one needs to know their amplitude, wavelength, *and phase*. To understand the importance of the phase of a wave, consider the simple case consisting of two cosine waves with identical wavelength and amplitude. If we matched these waves so that crest coincided with crest and trough with trough (phase difference 0°), the resulting sum would be a wave with double the amplitude of the original wave and the same period. If, on the other hand, we matched them crest to trough, (phase difference 180°), the resulting wave would be a straight line. Each different value of the phase difference would lead to a different resultant wave. Unfortunately, while the diffraction experiment can record the amplitude of the diffracted waves, the phases of these waves are not known. In our hypothetical alkaloid with 30 nonhydrogen atoms and 1500 diffracted waves we are faced with a problem of staggering proportions—deducing the phases of 1500 waves! To judge the magnitude of this problem suppose we required the phase of each wave to be correct to within 30° in order to get a recognizable picture of the molecule. This means that each reflection could have 12 different phases or that the picture of the resultant of all these waves could take on about 12^{1500} possibilities. This is a ridiculously large number. If we restricted our attentions to the 90 largest waves and considered only four phases for each one we would have 4^{90} possibilities which is still ridiculously large. Even 2^{20} is over 1 million. Clearly we cannot analyze diffraction data by considering all possibilities!

In the language of crystallography, a scattered beam is described by a complex number called a structure factor: $F = |F| \exp(i\phi)$. The magnitude of the structure factor, $|F|$, is the amplitude of the scattered beam. It is essentially the square root of the measured intensity. The phase of the scattered beam, ϕ, on the other hand, cannot be calculated from the measured intensities, which are real numbers. Determining the phases of the structure factors is the major problem in solving a structure. Crystallographers have devoted much time and effort to solving this phase problem and several reliable methods of attack have resulted. The most generally useful for alkaloid research are the heavy atom method and the direct method. Other methods are useful in special cases and will also be discussed in this chapter.

1.2. Finishing the Structure

A crystallographic experiment is not concluded when a recognizable electron density map has been calculated. The final stage of the analysis is to systematically vary the molecular parameters such as atom location and atomic thermal vibrations to produce the best fit between the experimental and calculated data. In general, each atom is specified by an x, y, and z coordinate and these are varied. Thermal motions of the atoms are incorporated into the calculations by describing the atoms as electron density spheres (isotropic motion) or ellipses (anisotropic motion) of variable dimensions. It should be pointed out that while the adjustable parameters defining the ellipses or spheres are called thermal parameters, they are also sensitive to a variety of other effects. For example, if a sample strongly absorbs the X-ray beam this affects anisotropic thermal parameters. Any imperfections in the way crystals form also affect the thermal parameters. There are many excellent studies where thermal motions have been carefully examined for chemical information, but in the usual crystal structure they may not contain much useful information, but rather serve as adjustable parameters which mimic a variety of imperfections.

The systematic variation of adjustable parameters is accomplished by a least-squares procedure [2]. In our hypothetical example of a 30 nonhydrogen atom alkaloid, we assume that we might have 1500 measured reflections. Each of these 30 atoms can have a variable x, y, and z coordinate (90 parameters) and let us assume that we describe each with anisotropic thermal motion which requires 6 more parameters per atom (180 parameters). This alkaloid might contain 50 hydrogen atoms and each of these would have an adjustable position (150 parameters). Let us further assume that we describe the hydrogens with isotropic thermal parameters (50 parameters). One more adjustable parameter is the scale factor which puts the calculated and experimental numbers on a common basis. In all, then, there are 471 adjustable parameters and 1500 pieces of data. Thus the problem is overdetermined by about 3 to 1. This figure is on the low side even for routine work and examples with 10 to 1 ratios or better are not uncommon.

At the end of the fitting procedure a crystallographer has generated an impressive set of numbers. In our example we have a list of 471 parameters and associated errors and a list of 1500 observed and calculated structure factors. The concluding sections of the chapter will deal with what the molecular parameters can be used for. The 1500 observed and calculated structure factors are important in the sense that the correctness of the analysis rests on their close correspondence. Crystallographers normally report this agreement in a highly compressed form called the R factor. The R factor is the average percent error between observed and calculated $|F|$'s or F^2's. An R factor of 0.08, for example, means that the average difference between observed and calculated values is 8%. It is a handy number to deal with but it should not be taken as an absolute indicator of the quality of an X-ray structure.

2. THE HEAVY ATOM METHOD

The assignment of phases to structure factors may be simplified if a compound contains a heavy atom or if it can be derivatized by heavy atom containing groups. A heavy atom is an atom containing many more electrons than the remaining, or light atoms, in a structure. A somewhat conservative rule is that

$$\frac{Z^2 \text{ (heavy atom)}}{\Sigma \ Z^2 \text{ (light atoms)}},$$

where Z = atomic number, should be about unity, although substantially lower values have been solved successfully by this method. For example, if our hypothetical 30-atom alkaloid were crystallized as the hydrochloride, $Z^2(\text{Cl})/(30 Z^2(\text{C})) = 0.27$. The corresponding figures for the hydrobromide and hydro-iodide are 1.13 and 2.60. Following our rough guide, the hydrobromide would be ideal for use in the heavy atom method.

It is usually possible to locate the positions of the heavy atoms in the unit cell by a straightforward calculation of the Patterson function [3]. The Patterson function is a Fourier series using F^2 as coefficients; thus it does not require knowledge of the phases. Calculation of the Patterson function results in a three-dimensional map with peaks at the tips of vectors from its origin which are equivalent to the vectors between atoms in the unit cell. The magnitude of a Patterson peak is proportional to the product of the atomic numbers of the two atoms in the corresponding vector. The largest peaks in the map will therefore occur at the tips of vectors corresponding to vectors between two heavy atoms. Simple algebra can be used to derive heavy atom positions from these peaks. The limited structural information thus derived is sufficient to calculate an approximate set of phases for the structure factors and, subsequently, an electron density map.

Even if the errors in the initial set of phases are large, the electron density map will usually show positions of additional atoms in the structure. This new information is then used to calculate new phases and an improved map in which more details of the structure are revealed. This procedure of map computation, selection of new atom positions, phase improvement, and recomputation of the electron density map is called Fourier refinement and is repeated until all of the nonhydrogen atoms in the structure are located.

Nitrogens do not have sufficient electron density to assist in phase determination by the method described. However, their basicity facilitates formation of heavy atom derivatives and salts. As a result, some of the most complex natural product structures studied before the automation of the direct methods of phase determination were alkaloids. The structure of lycoctonine 1, for example, was determined from the X-ray structure of des-(oxymethylene)-lycoctonine hydridiode 2 in 1956 [4]. The results were significant since they finally established the complicated skeleton of the C_{19}-diterpenoid alkaloids after years of chemical degradation experiments. [The assignment of stereochemistry to C(1) of lycoctonine has recently been revised [5] but the structure presented in the 1956 paper is in all other respects correct.] The discussion of the mitomycins below is exemplary of the heavy atom method.

There are disadvantages to using heavy atom derivatives in crystallography. The structures of heavy atom derivatives are frequently less precise than structures with no heavy atoms (see Section 5.1). Furthermore, as illustrated with heterophylloidine below, in preparing a heavy atom derivative, a chemist may introduce unexpected changes in the molecule.

2.1. Mitomycin A

The mitomycins are antibiotics from *Streptomyces verticillatus* [6]. Their chemical composition and atom connectivity were deduced from noncrystallographic studies, and the crystal structure of the N-brosyl derivative of mitomycin A was determined in order to define crucial stereochemical details of these compounds [7].

The molecular formula of the antibiotic derivative was $C_{22}H_{22}N_3OSBr$. The symmetry of its crystals was identified by preliminary observations of diffraction patterns as monoclinic, space group $C2$, and the dimensions of the unit cell were measured as $a = 19.70$, $b = 8.24$, and $c = 16.05$ Å, and $\beta = 95.80°$. Density measurements as well as chemical analysis of the crystals indicated that the unit cell contained four molecules of antibiotic and two benzenes of crystallization.

Therefore, the asymmetric unit, the smallest region with which by symmetry the unit cell could be generated, contained 1 molecule of antibiotic and 0.5 molecule of benzene. Diffraction data were collected on a diffractometer using CuK-α radiation.

Fortunately both sulfur and bromine are heavy enough with respect to carbon, nitrogen, and oxygen to be seen clearly in Patterson maps. Bromine is significantly heavier than sulfur so that bromine–bromine vectors in a Patterson map can also be distinguished from sulfur–sulfur vectors. Both the sulfur and the bromine positions in the unit cell of the mitomycin A derivative were unambiguously determined by Patterson methods. In addition, positions of the two brosyl carbons to which the heavy atoms were attached were calculated by assuming they were colinear with the heavy atoms, and by assuming standard C—Br and C—S bond distances.

Probably this structure could have been solved by locating only the bromine by Patterson methods, then using Fourier refinement to locate the remaining atoms. However, calculating phases with a single atom in certain space groups results in an ambiguous electron density map with false symmetry. In the case of space group $C2$, such a map would generally appear centrosymmetric, with peaks from both enantiomers of the molecule present. Only the peaks from a single enantiomer could be correctly included in refinement (either enantiomer could be chosen), and sorting out the two sets of peaks might have been difficult. Using the four-atom fragment of the brosyl group circumvented this problem and increased the accuracy of the initial phases.

An electron density map calculated with the first estimates of the phases had peaks for every atom in the molecule, as well as several spurious peaks. The structural information available from chemical degradation studies facilitated interpretation of the map, especially the assignment of the chemical identities of the peaks. To test the assignments, new structure factors were calculated from both structure **3a,** which was consistent with the results of noncrystallographic studies, and from the same model with all nitrogens and oxygens included as carbons. The agreement of these two sets of structure factors with observed data was $R = 22\%$ and 28%, respectively, indicating that the assignments were probably correct.

In a second electron density map, calculated with phases from the new model, most of the spurious peaks had disappeared. Four strong peaks near and on the twofold rotation axis of the unit cell were identified as half of a benzene of crystallization. A twofold rotation axis of the benzene coincided with the twofold axis in the unit cell. This fact explained how half the expected number of benzene molecules crystallized in the unit cell without violating the symmetry requirements of the space group.

The bromine and sulfur of the mitomycin derivative allowed determination of its absolute configuration. To a first approximation, X-rays are scattered from a crystal elastically and a pair of reflections $F(hkl)$ and $F(\overline{hkl})$ (called a Friedel pair) are of equal intensity. However, when a crystal contains heavy atoms, X-rays are frequently appreciably absorbed and this approximation breaks down. Absorption causes a slight shift in the phases of the diffracted X-rays. As a result of this anomalous scattering, the intensities of $F(hkl)$ reflections are measurably different from those of their Friedel mates for noncentrosymmetric structures. Calculated structure factors can be corrected for the anomalous scattering (see Section 5.6). Commonly, corrected structure factors are calculated for both enantiomers, and the set that best agrees with observed data is judged to correspond with the correct enantiomer. For N-brosylmitomycin A, structure factors calculated for enantiomer **3b** were in better agreement with observed data ($R = 0.087$) than were structure factors for the opposite enantiomer ($R = 0.094$). There was an even greater difference between agreement factors when only those 67 reflections that were most sensitive to the effects of anomalous scattering were used in the comparisons: $R = 0.107$ versus $R = 0.152$.

The crystal structure of N-brosylmitomycin A provided not only a complete stereochemical description of the parent compound, but also conformational details important for explaining earlier experimental results. Studies had shown mitomycin A to have one amide-type nitrogen, presumably N(2) of **3a** conjugated with the quinone system. However, the ultraviolet spectrum of the compound was not compatible with the hypothesized conjugated system [8]. Other unusual features of the compound were the nonbasicity of the aziridine nitrogen and the large barrier to its inversion, as demonstrated by nuclear quadrupole measurements.

Examination of the refined crystal structure of the mitomycin A derivative showed that the lone pair orbitals on N(1) and N(2) were in close proximity. In fact, deformation of the five-membered ring B, which brought N(2) out of the plane of the quinone ring, had been necessary to prevent serious overlap of the two lone pairs. Apparently N(2) was prevented from fully participating in the conjugated system by its position above the plane of the quinone ring, and steric interference from N(2) affected both the basicity of the aziridine nitrogen and its ability to invert. The bond between N(2) and C(7), with its short bond length of 1.36 Å, certainly had double-bond character, but the angles around N(2) were more characteristic of a tetrahedral than a trigonal atom and the C(6)–C(7) double bond was not lengthened as would be expected if N(2) were fully conjugated with the quinone system.

2.2. Heterophylloidine

One danger of making a derivative of a compound is the possibility for an unexpected modification of the molecule. If the molecule is altered, spectroscopic and chemical studies may play a crucial role in deducing the structure of the natural product from the crystallographic results. The importance of considering structural information from several sources was recently demonstrated by the X-ray diffraction study of the C_{20}-diterpenoid alkaloid heterophylloidine [9]. A ^{13}C NMR spectrum of the alkaloid had peaks characteristic of the atisine carbon skeleton **4**, and also indicated the presence of a C(4)-methyl group, two ketone groups, an N-methyl group, an exocyclic methylene, and an

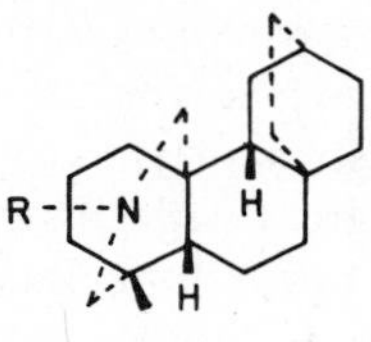

4

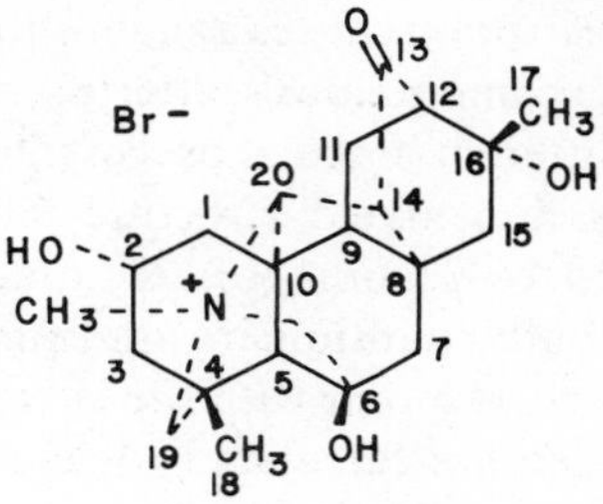

5

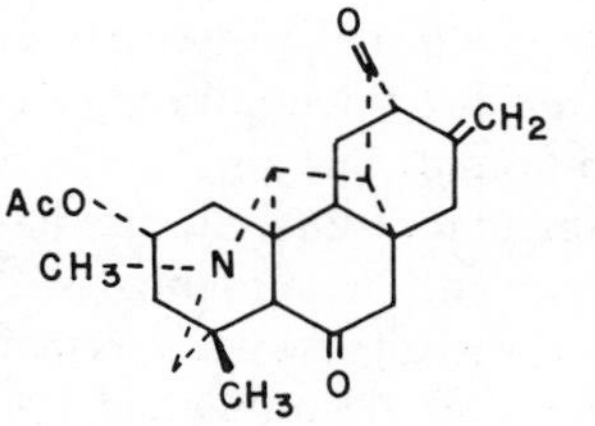

6

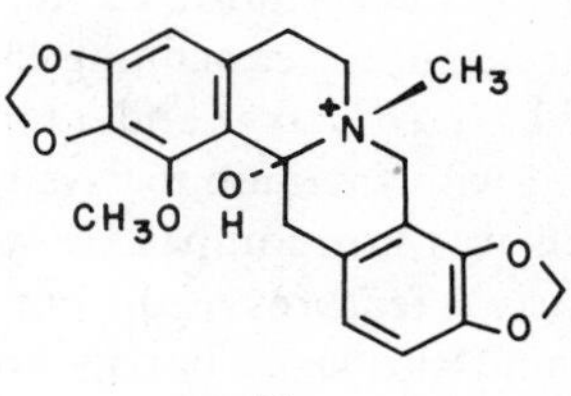

7

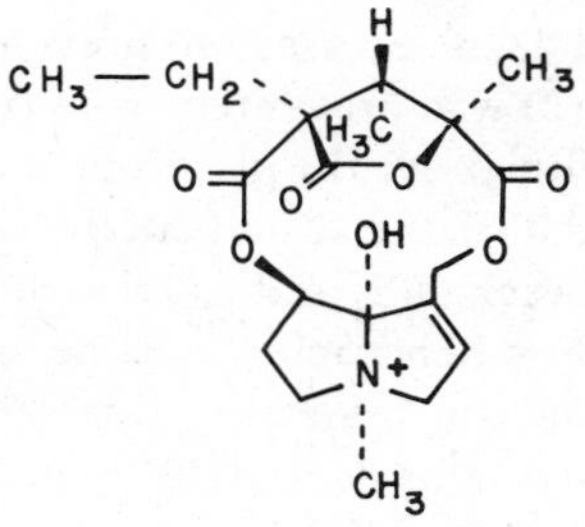

8

acetoxyl group. Supporting evidence for some of these features came from IR and ^{1}H NMR data.

The spectroscopic results for heterophylloidine, though extremely informative, were not sufficient to determine its structure, and the small amount of the compound available precluded additional experiments. At this point, the amorphous alkaloid was treated with 49% HBr in water in hopes of obtaining a crystalline bromine derivative suitable for X-ray diffraction analysis. The two crystalline forms resulting from this treatment, one colorless and the other dark red–brown, both had melting points in excess of 300° C, confirming that a bromine salt of the alkaloid had formed. X-ray photographs of the brown crystals showed no diffraction spots, even though the crystals were quite large. The thin, platelike colorless crystals, however, diffracted to moderate resolution. These crystals were orthorhombic, with cell constants $a = 11.738(3)$, $b = 12.155(4)$, and $c = 13.548(3)$ Å, and belonged to space group $P2_12_12_1$. Data were collected for a colorless crystal on a diffractometer using MoKα radiation.

The position of the bromine atom in the crystal structure was determined from a Patterson map and used to calculate an initial set of phases. The electron density map calculated with these phases showed one large peak for the bromine, and a complex assortment of smaller peaks, most of which were spurious. A chemically sensible 10-atom fragment chosen from these peaks was successfully elaborated during several cycles of Fourier refinement. The complexity of the first electron density map, and the difficulty in locating the light atoms in the structure may be attributed partly to the fact that a single atom was used to obtain the initial phases, and partly to the lack of good high resolution data (only 767 of the 1934 reflections measured were significantly above background).

All light atoms were included as carbons in the first model of the structure and their positions and temperature factors were refined by the method of least squares. Usually the temperature factors of oxygens and nitrogens that are refined as carbons will be significantly smaller than the temperature factors of true carbons with similar constraints on their movement. Therefore, they are a very sensitive indicator of chemical identity. The oxygens and the nitrogen of the heterophylloidine derivative were assigned on the basis of their refined isotropic temperature factors, their bond distances to neighboring atoms, and on the compatibility of the assignments with the atisine skeleton **4**.

The locations of the hydrogen atoms were deduced from the model of the structure and their coordinates were calculated. Though included in subsequent refinements, they were not themselves refined. Anisotropic refinements of the nonhydrogens eventually converged with a final R factor of 0.066. Structure factors corrected for the effects of anomalous scattering were then calculated for both enantiomers and compared to the observed data. Structure **5** was clearly distinguished as the correct enantiomer by the better agreement of its structure factors with those observed: $R = 0.063$ versus $R = 0.072$.

Structure **5** did not have all of the features predicted for heterophylloidine from the spectroscopic data. Puzzling were the absence of both an exocyclic methylene and an acetyl group, functionalities clearly indicated by both the ^{1}H NMR and ^{13}C NMR spectra. To reconcile spectroscopic evidence with crystallographic results, it was necessary to postulate the hydrolysis of the acetyl group and hydration of the exocyclic methylene during formation of the bromine derivative. The two functional groups would then have to be at C(2) and C(16), respectively, in the natural product. Also one of the ketone groups in the natural product appeared to have been converted to an hydroxyl by formation of a carbinolamine linkage. Therefore structure **6,** which is compatible with both spectroscopic and crystallographic results, was proposed for heterophylloidine.

The formation of a carbinolamine linkage during acid salt preparation is typical behavior for C_{20}-diterpenoid alkaloids with ketone groups at C(6), and also for the protopine and pyrrolizidine-type alkaloids [10, 11]. The carbinolamine linkage in **5** has a normal bond distance of 1.52(2) Å. This distance may be compared, for example, to the value of 1.49 Å for N—C(4) in *N*-brosylmitomycin A **3b** [7]. In the structure of coulterpine hydrobromide **7**, the length of the carbinolamine bond was 1.58 Å [10]. The unusual length of this nitrogen-carbon bond was attributed to crowding by the aromatic ring methoxy substituent. However, the carbinolamine bond in the salt of retusamine **8** is 1.64 Å, about six standard deviations greater than its expected value, in spite of no obvious steric constraints [11].

The crystal structures of some free bases of protopine and pyrrolizidine alkaloids have short transannular distances between the same nitrogens and carbonyl carbons that are covalently bonded in the acid salts [12–14]. These distances range from 1.99 to 2.58 Å and are much smaller than the sum of the van der Waals radii for nitrogen and oxygen, about 3 Å. These short distances may be considered incipient carbinolamine linkages frozen at various stages of bond formation [15]. There is an inverse correlation of the N—C distance with the C=O distance as well as with the planarity of the carbonyl system in the structures. Using the geometrical parameters from crystal structures of both free bases and salts of protopine and pyrrolizidine alkaloids, the stereochemical changes that occur during nucleophilic addition to a carbonyl group can be mapped [15].

3. THE DIRECT METHOD

The direct method of analyzing diffraction data is becoming increasingly powerful and with the proliferation of easily run programs to carry it out, more popular. The direct method provides a means of studying the structures of compounds whose atoms all have approximately the same number of electrons; thus it is particularly well-suited to the study of organic compounds. However, the effectiveness of the direct method decreases in proportion to the number of atoms in the structure studied and application of the direct method to large structures (75 or more atoms) often proves to be exceedingly difficult.

To understand the basis of the direct method, we have to consider the phase problem again. Our earlier discussion of the magnitude of the phase problem was predicated on the assumption that all of the phases were independent. In other words, we assumed that knowledge of any one phase did not tell us anything about possible values of other phases. This would be true if the object giving rise to the scattering were completely general, but for real scattering objects we can begin to rule out certain phase combinations. The most generally useful assumption is that electron density is not negative anywhere in the unit cell. Starting from this plausible assumption it can be shown that there are relations between the phases, each with a certain probability of being correct [16, 17, 18, 19].

A useful formula for predicting phases from these probable relationships is called the "tangent formula" and is incorporated into many automated procedures [20]. The data used in the tangent formula are known phases, and structure factor amplitudes scaled in such a way that they are no longer a function of scattering angle. Structure factors decrease with increasing scattering angle, whereas the scaled values used in the direct method, called E's, do not. E's are the structure factors that would result if each atom in the crystal structure were replaced by a stationary point atom. E's are used not only for phase prediction, but also in the initial Fourier synthesis using the predicted phases.

In practice, a small number of phases, the "starting set," are arbitrarily assigned values and then the phases of other reflections are worked out using probability relations [21, 22]. Since the number of phases in the starting set is usually small, the values of the phases can be independently varied in small increments, and the phase extension process repeated for each new combination of values. The result is a relatively modest number of possible solutions. The solutions are ranked using several figures of merit (see Section 3.3). Fourier syntheses are computed for the most promising solutions and the resulting E maps, which are similar to electron density maps, can be examined for plausible molecular fragments or possibly the entire molecule. If only a portion of the molecule is visible, it can be used to generate phases for a subset of strong reflections and these phases can be used for phase extension and refinement [23]. This procedure for combining partial structural information with phase probability relationships is usually described by the succinct phrase "tangent formula recycling."

The structures of veatchine, rugulovasine, and clavicipitic acid were all obtained by the direct method and are discussed in this section.

3.1. Veatchine

Veatchine (**9**) is a C_{20}-diterpenoid alkaloid containing an oxazolidine ring which isomerizes in the presence of protic solvents (Fig. 1). Doubling of the C(4) methyl peak in the ^1H NMR spectra of veatchine and the related alkaloid atisine (**10**), and coalescence of the peak at high temperatures were noted in 1968 [24]. Later, scientists were able to detect, in addition to the doubled methyl signal, a doubled peak for the C(20) hydrogen in the ^1H NMR spectrum of atisine which broadened with an increase in temperature [25]. They also observed broadening of the C(20) peaks in protic deuterated solvents, a phenomenon they attributed to deuterium exchange at C(20). On the basis of these results, they postulated the existence of C(20) epimers which rapidly interconverted via a zwitterion intermediate (Fig. 2).

Doubling of the ^{13}C NMR peaks for the carbons in the piperidine and oxazolidine rings of veatchine, atisine, and atisinone, reinforced the proposal that these compounds existed as C(20) epimers [26]. However, the persistence of the doubled peaks at high temperatures in nonionic solvents, and the fact that the ratio of peak sizes in the doublets remained at about 7:4 over the range of temperatures 25–90° C, suggested that the epimers were stable, noninterconverting mixtures. The temperature and solvent dependence of the ^{1}H NMR spectrum of atisine was attributed to isomerization of atisine to isoatisine by the mechanism shown in Fig. 1 [26].

To clarify the results of the spectroscopic experiments, the crystal structure of veatchine was studied. The heavy atom salt of veatchine is the open form (**11**) whose structure would not answer crucial questions about the stereochemistry of the oxazolidine ring *F*. Therefore, the free base was crystallized from acetone and its structure determined by the direct method [27].

Large clear crystals of veatchine belonged to space group $P2_12_12_1$ with one

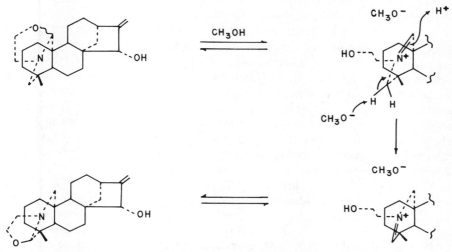

Figure 1. Isomerization of the oxazolidine ring in veatchine.

molecule in the asymmetric unit and cell constants $a = 10.034(3)$, $b = 21.581(8)$, and $c = 9.674(2)$ Å. Data for one of the crystals were collected using MoKα radiation. Direct phasing of the data followed by an E map calculation located all but one nonhydrogen atom. The remaining atom was found in a subsequent difference electron density map, and the entire model was refined by least squares.

A difference map calculated after refinement revealed alternate positions for atoms C(22) and O(22). After refinement of a model of veatchine with C(22) and O(22) in the new positions, a difference map showed strong peaks at the old positions. The authors concluded that both sets of peaks were real. The structure seen in the electron density map, which was a statistical average of all molecules in the unit cell, contained contributions from two C(20) epimers which had co-crystallized. The structure was disordered. That is, the regular pattern of repeating structural units in the crystal was interrupted by random replacement of a unit by another that was slightly different. In order for this unusual type of disorder to occur, the crystal packing forces holding together epimers must have been very similar to those holding together identical molecules. In fact, the disordered atoms C(22) and O(22) were not involved in any intermolecular contacts

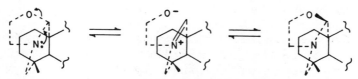

Figure 2. Postulated interconversion of epimers via a zwitterion intermediate.

and the nitrogen, which with OH(15) formed the only hydrogen bond in the structure, was equally accessible to OH(15) in both epimers.

Initially each epimer was assumed to contribute equally to the crystal structure. Then a more accurate estimate of the relative contributions of the two epimers was obtained by least-squares refinement of the occupancy factors of the disordered atoms. The occupancy factors are multiplicative weights which must sum to 1.0 over all possible positions of an atom. The occupancy factors for C(22) and O(22) of veatchine were 0.6 and 0.4 for the 20(R) and 20(S) epimers, respectively, indicating that the two epimers of veatchine were present in the crystals in a 3:2 ratio. That the two C(20) epimers were randomly distributed in the crystal was verified by observation of the same proportions of epimers in a disordered relationship in three other veatchine crystals [27].

The results of the X-ray diffraction analysis verified the existence of two C(20) epimers of veatchine and identified the major epimer as **12**. Furthermore, the co-crystallization of the two epimers in roughly the same proportion as was observed in solution over the range of temperatures 25–90° [26] was interpreted as additional evidence that the C(20) epimers of veatchine were not an equilibrium mixture. The authors proposed that the epimers of veatchine and atisine arose during regeneration of the free bases after their isolation as the ternary imminium salts **11** and **13**. The detailed geometrical parameters of the crystal structure of atisinium chloride **13** showed that attack on either side of the N—C(20) double bond would be sterically hindered and preference for either β- or α-closure of the oxazolidine ring could not easily be predicted on the basis of steric effects [27].

The ^{13}C NMR spectrum of cuachichicine **14** does not show doubling of the peaks for the oxazolidine and piperidine carbons in spite of the fact that cuachichicine has almost the same structure as veatchine. The differences between the two compounds are in substituents on the D ring, and the effect of the substituents on the closure of the oxazolidine ring is not obvious from molecular models. The crystal structure of the free base of cuachichicine was determined by the direct method and compared with the structure of the major epimer of veatchine [28]. One C(14) hydrogen, which had earlier been identified as a source of steric hindrance to α ring closure in atisine [27], was significantly closer to C(20) in cuachichicine than in veatchine (2.65 compared with 2.96 Å). Perhaps it was this subtle difference in conformation which prevented forma-tion of a second cuachichicine epimer during isolation of the compound via its ternary imminium salt. Other veatchine-type diterpenoid alkaloids with β-substituents at C(16) also appear to exist as single C(20) epimers [29].

3.2. Rugulovasine-A

The epimers of veatchine are so similar chemically, as evidenced by their cocrys-tallization, that it is unlikely that a method for their separation will be devised. In contrast, two epimeric fungal metabolites, rugulovasine-A and -B, were easily separated by chromatographic methods [30]. Structures **15** and **16** were proposed as structures for the epimers, though which structure corresponded to which compounds was not determined [31].

The rugulovasines interconvert on warming in polar solvents. A mechanism involving intermediate **17,** which was similar to that proposed by Pradhan and Girijavallabhan [25] for interconversion of atisine epimers, was suggested for this reaction. Reformation of the six-membered ring from the planar zwitterion would be expected to give not only structures **15** and **16,** but also their enantiomers. The X-ray study of rugulovasine-A [32] was initiated primarily to define the stereochemistry of this alkaloid, but the results of the study also provide indirect evidence of the proposed mechanism of epimer formation.

Rugulovasine-A, $C_{16}H_{16}N_2O_2$, crystallized in the triclinic crystal system with four molecules of alkaloid and four molecules of H_2O in the unit cell. Unit cell parameters for the crystals were $a = 8.526(2)$, $b = 11.213(3)$, and $c = 16.296(3)$ Å and $\alpha = 98.94(5)$, $\beta = 93.84(5)$, and $\gamma = 108.96(5)°$. Intensities of 3283 reflections were measured with $CuK\alpha$ radiation, of which 2248 were significantly above background. The distribution of the measured intensities was characteristic of a centrosymmetric structure, indicating that the space group was probably $P1$ with two molecules of the alkaloid in the asymmetric unit.

The crystallization of two molecules in the asymmetric unit required the location of two alkaloid molecules as well as two waters of crystallization in electron density maps. A crystal structure of this complexity might have been difficult to solve in a noncentrosymmetric space group. However, in $P1$, where phases are restricted to one of two values (0 or 180°), direct phasing of the data proceeded easily. The structure was refined by least squares to an R factor of 0.038.

Rugulovasine-A proved to be diastereomer **15.** Each asymmetric unit of the crystal structure contained one molecule of **15** and one molecule of its enantiomer. The observation of both enantiomers supported the proposed mechanism of interconversion, and suggested that the existence of epimers for rugulovasine was an artifact of the isolation process.

3.3. Clavicipitic Acid

During the course of classic studies of ergot alkaloid biosynthesis, scientists reported the isolation of a new metabolite from *Claviceps* strain SD 58 [33]. In order to inhibit the normal biosynthetic pathway leading to the tetracyclic ergolines at the N-methylation step, thereby increasing the production of trace metabolites, they had treated their culture medium with ethionine, a known methionine antagonist [34]. Structure **18** shows the normal tetracyclic ergoline ring system with the conventional numbering scheme. Structure **19** is the well-known *d*-lysergic acid, a derivative of the 6,8-dimethyl-δ9,10-ergolene system. Instead of the standard tetracyclic ergolines, the scientists isolated primarily the indole containing metabolite 4-dimethylallyltryptophan and, in small amounts, the new metabolite clavicipitic acid from their ethionine-treated cultures.

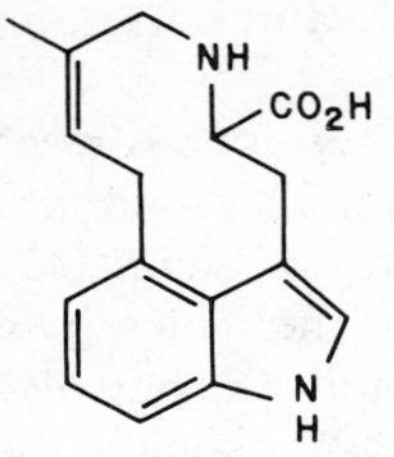

20

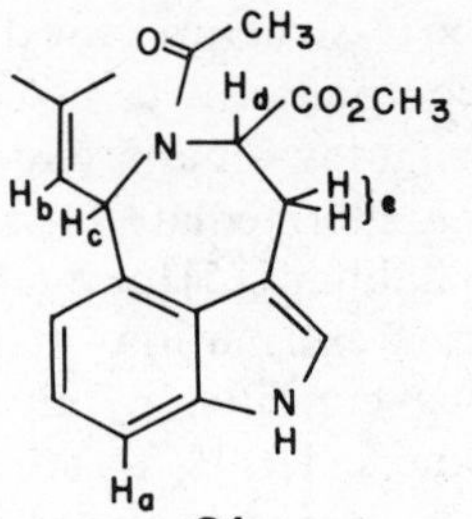

21

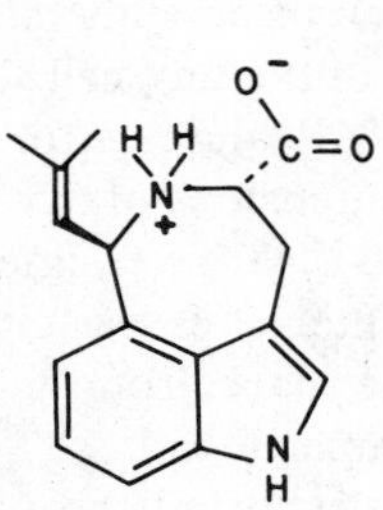

22

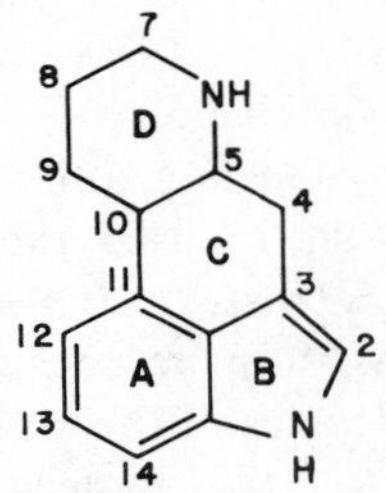

18

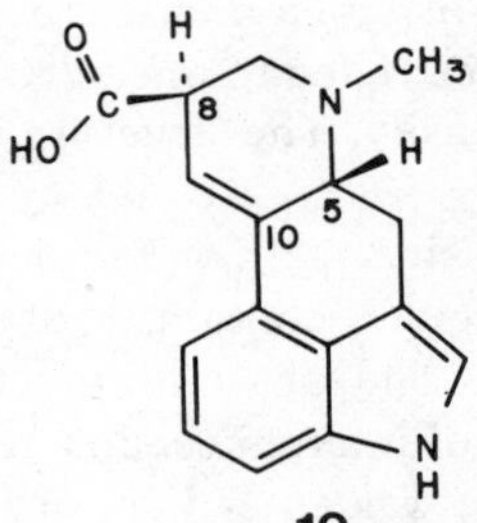

19

A high-resolution mass spectrum indicated a molecular formula of $C_{16}H_{18}$-N_2O_2 for clavicipitic acid. A UV spectrum for the alkaloid (λ = 288 nm, log ε = 3.81; λ = 225 nm, log ε = 4.58, EtOH) showed the presence of an indole ring and an isolated double bond. A ^{1}H NMR spectrum from a trimethyl derivative of clavicipitic acid showed the presence of a single vinyl proton (CDCl₃; δ 5.7, 1H, s; the rest of spectrum not reported) which was assigned to the unconjugated double bond. The results from tritium retention experiments using DL-tryptophan tritiated in the α position and DL-mevalonic acid (2-^{14}C, 5-^{3}H) indicated that there was no ring closure from C(5) to C(10) as in the path leading to the normal tetracyclic ergolines. On the basis of the aforementioned spectral properties and the results of the labeling studies, structure **20** was proposed for clavicipitic acid.

Another research group obtained clavicipitic acid from a strain of *Claviceps fusiformis* regardless of whether the culture medium had been treated with ethionine [35]. They had difficulty preparing the trimethyl derivative of the compound using diazomethane. They therefore treated the compound with acetic anhydride in methanol at room temperature to obtain the *N*-acetyl methyl ester derivative, which was soluble in chloroform. Based on an interpretation of the ^{1}H NMR spectrum measured at 100 MHz [CDCl$_3$; $\delta 8.2$(1H, broad singlet), $\delta 6.8$–7.2 (3H, mult), $\delta 6.7$ (1H, *d, a*), δ-5.74 (1H, *d, J* = 7 Hz, *b*), $\delta 5.14$ (1H, *d, J* = 7 Hz, *c*), $\delta 4.30$ (1H, mult, *d,* coupled to *e* at $\delta 3.2$–3.7), $\delta 3.2$–3.7 (2H, mult, *e,* coupled to *d* at δ-4.30), $\delta 3.64$ (3H, *s*), $\delta 2.06$ (3H, *s*), $\delta 1.82$ (3H, *s*), $\delta 1.68$ (3H, *s*)], structure **21** was proposed for this derivative.

The investigation of clavicipitic acid was renewed in 1979 with the X-ray diffraction analysis of a single crystal of the alkaloid [36]. A cylindrical crystal, which had been obtained from slow evaporation of an aqueous ethanol solution, was shown to have a diffraction pattern of sufficient intensity to be useful. The diffraction symmetry was triclinic. Unit cell lengths *a* = 5.667(1), *b* = 8.282(2), *c* = 15.183(3) Å and interaxial angles α = 99.18(2), β = 100.47(1), and γ = 86.45(2)° were determined from diffractometer measurements. The intensities of a total of 2200 independent diffraction maxima were measured using MoKα radiation on a computer-controlled diffractometer. These data were converted to structure factor amplitudes. A total of 1999 amplitudes were considered statistically significant measurements and used for subsequent structure solution and refinement. Assuming a plausible density of 1.3 g/cm^3 the unit cell volume of 691.6(2) Å^3 indicated two molecules of composition C$_{16}$H$_{18}$N$_2$ per unit cell. Since clavicipitic acid was known to be chiral, the observations were uniquely accommodated by the space group *P*1 with two independent molecules forming the asymmetric unit.

A successful application of the direct method of phasing in the lowest symmetry space group *P*1 can sometimes be elusive. In this and other space groups with no translational symmetry, the set of probability relations that are used in the tangent formula to assign phases to the X-ray data can be satisfied by a trivial solution in which all of the phases have a value of zero. The determination of the correct phases of several hundred X-ray data from a handful of starting set reflections may be thwarted by the tendency of the phases to drift toward this trivial solution. Figures of merit will misleadingly rank trivial solutions highest if they merely evaluate the consistency of the phases with respect to the probability relations used in the tangent formula. Fortunately, more subtle figures of merit [37, 38] have been developed which do not give high rank to the trivial solution and these often have been used successfully in space group *P*1. It is important to stress, however, that none of the methods for evaluating phase solutions are fully reliable, and correct (or acceptable) solutions are sometimes overlooked because of undistinguished figures of merit.

In spite of the problems discussed above, the structure of clavicipitic acid was solved without difficulty. Direct phasing of a subset of the X-ray diffraction data led to a Fourier synthesis which revealed the positions of 16 of the 18 nonhydrogen atoms in one of the two molecules. The best set of phases had been chosen from a large number of possible solutions by an outstanding value of a figure of merit known to be reliable in space group $P1$. The entire nonhydrogen framework of both molecules was completed by Fourier refinement and the results of the analysis indicated structure **22** for clavicipitic acid. Least-squares refinement clearly showed that both molecules in the asymmetric unit were in the zwitterionic form. The positions of all the hydrogen atoms in both molecules were determined from a difference electron density map calculated after the refinement. Further least-squares refinement of the X-ray model of clavicipitic acid yielded an R factor of 0.042 for all of the observed data.

Given the special problems of using the direct method in space group $P1$, the ease of solution of the clavicipitic acid structure was surprising. An inspection of the crystal structure shows that the indole rings of the two independent molecules in the unit cell are very cleanly related by an inversion center. As it must, this relationship breaks down for the saturated portion of the seven-membered ring and its substituents. Nevertheless, the symmetry of the crystal structure can be approximated by the centrosymmetric space group $P\bar{1}$, in which the phases of all reflections are restricted to values of 0 or 180°. Although this pseudosymmetry was not explicitly used in the X-ray study of clavicipitic acid, it did constrain the phases of many of the low-angle reflections to have values close to 0 or 180°. Thus it may have indirectly facilitated the solution of the structure.

4.   OTHER METHODS

When heavy atom derivatives are not available and the direct method does not work, one is of necessity forced to consider other approaches to the phase problem. One possibility is that a similar structure, similar not only in the molecular but also in the crystallographic sense, has been solved. If this is the case it is possible to borrow phase information from the known structure and use it in the unknown structure. This technique is of limited usefulness in small molecule crystallography but an interesting example is provided by paspalinine and is discussed below (Section 4.1).

There are also hybrid combinations of the heavy atom and direct method and the solution of the structure of the antitumor antibiotic luzopeptin A is provided to illustrate these (Section 4.2).

If a reasonably large portion of a molecule is known there is a possibility of searching for this known piece in a crystal structure using the same function used for the location of heavy atoms. This technique is called the Patterson search method and is more fully discussed later in this section.

4.1.   Paspalinine

It is sometimes mistakenly assumed that when the molecular structure of a compound is known, solving its crystal structure is a trivial matter. The reason this assumption is not true is that solving the phase problem for a crystal structure requires knowledge not only of molecular structure, but also of the packing arrangement of molecules in the unit cell. In special cases the packing arrangement can be predicted from molecular structure. Otherwise, it is possible to search the considerable number of possible molecular packing schemes for the arrangement that best reproduces the observed intensities (see Section 4.3), but it is usually easier simply to solve the phase problem using traditional heavy atom or direct methods.

Intuitively one would expect that two molecules closely related in structure would also have similar crystal structures, so that the molecular packing in crystals of one compound could be predicted from the crystal structure of the other. Unfortunately, even this is not generally true. Small changes in substitution pattern, and even changes in the conditions for crystallization can result in crystals with completely different symmetry.

This point is illustrated by the variety of crystal lattices for the structurally related fungal metabolites paspaline (**23**), paxilline (**24**), and paspalicine (**25**). Parameters for the lattices are summarized in Table 2. The compounds all have the same hexacyclic skeleton and differ in, at most, three substitution sites in the two right-most rings. Paspalicine crystallizes in two space groups, depending on the solvent used for crystallization. The crystal structures of all three compounds were solved independently by the direct method [39, 40].

Given these results, it seemed fortuitous that a fourth related alkaloid, paspalinine, should crystallize from a dioxane solution in space group $P2_12_12_1$ with lattice parameters nearly identical to those of paspalicine: $a = 9.801(4), b = 10.555(3)$, and $c = 21.605(7)$ Å [41]. Paspalinine's molecular formula was determined from its mass spectrum to be $C_{27}H_{31}NO_4$, compared to $C_{27}H_{31}NO_3$

23

24

25

26

Table 2. Solvent of Crystallization and Lattice Parameters for Paspaline, Paxilline, and Paspalicine

	Paspaline	Paxilline	Paspalicine 1	Paspalicine 2
Solvent	MeOH	Heptane–Acetone	MeOH	Acetone
Asymmetric unit	$C_{28}H_{39}NO_2 \cdot$ $(CH_3OH)_{1/2}$	$C_{27}H_{33}NO_4 \cdot$ C_3H_6O	$2(C_{27}H_{31}NO_3)$	$C_{27}H_{31}NO_3$
Space group	$P2_12_12_1$	$P2_12_12_1$	$P2_1$	$P2_12_12_1$
Cell constants				
a	49.388(5)	7.707(1)	13.439(2)	9.706(1)
b	6.527(1)	11.522(1)	11.740(1)	10.670(1)
c	7.891(1)	31.009(3)	15.295(1)	21.775(2)
β			97.38(1)	

for paspalicine. Its ^{13}C NMR spectrum was identical with that of paspalicine except for the absence of the doublet assigned to C(13) and the presence of a new singlet at 76.0 ppm, suggesting that it differed from paspalicine only by substitution with a hydroxyl group at C(13). This hypothesis was supported by evidence from IR, UV, and ^{1}H NMR spectroscopy. The apparent structural similarity of paspalinine and paspalicine and the correspondence of the unit cell parameters and symmetry of their crystals suggested that their crystal structures were isomorphous, that is, they had the same molecular packing scheme.

The crystal structure of paspalinine was studied to verify its suspected molecular structure **26**. Crystals of the compound were very small and diffracted only out to low resolution. Only 645 reflections with intensities significantly above background were measured—about 20 reflections per independent nonhydrogen atom. The small number of observed data complicated solution of the crystal structure. Whereas the direct method had routinely solved structures **23–25**, they failed to give an immediate solution to the crystal structure of paspalinine.

Rather than struggle with conventional methods of phase determination, the crystallographers exploited the suspected isomorphism of paspalicine and paspalinine. The small differences in cell parameters for the two compounds meant the fractional coordinates for the atoms they had in common would not be precisely the same. Therefore, the phases calculated from the paspalicine structure would not be strictly correct for paspalinine. Before assigning these phases to the paspalinine data, they were first modified by tangent formula recycling. The skeletal carbons of paspalicine were used to calculate phases for 100 reflections. These 100 phases were used in probability relationships to generate 100 more phases. The entire set of 200 phases was then refined to give those phases most consistent with all of the probability relationships among them.

An E map calculated with these 200 phased reflections clearly showed peaks for every atom in structure **26**. Least-squares refinement of this solution to an R

27

factor of 0.084 using all the data confirmed that it was correct. This experiment, then, was one of the rare instances when the structure of one natural product could be solved directly and simply from the structure of another. The tangent formula recycling technique proved to be a very efficient method for using information from the paspalicine crystal structure.

The story of paspalinine-type metabolites continues to unfold. The latest chapter is the structure of the fungal tremorgen penitrem A (27) [42]. Part of this interesting metabolite is clearly related to paspalinine but a new piece has been added to the indole ring.

4.2. Luzopeptin A

In the course of screening for microbial fermentation products which showed antitumor activity, workers at the Bristol-Banyu Research Institute in Tokyo isolated three potent antitumor antibiotics, which they named luzopeptin A, B, and C, from an actinomycete strain [43]. The actinomycete strain was characterized [44] and found to be the genus *Actinomadura*, and was hence designated *Actinomadura luzonensis*, nov. sp. Luzopeptins A, B, and C were obtained in crystalline form using a solvent extraction procedure for isolation followed by countercurrent distribution for purification. The Tokyo workers reported [43] a yield of 988, 420, and 848 mg for luzopeptins A, B, and C, respectively, from a harvested broth (pH 8.5) of 170 L.

From a series of exhaustive acetylation experiments, luzopeptins A and B were shown to be di- and monoacetylated derivatives, respectively, of luzopeptin C. The high-resolution field desorption mass spectrum indicated a molecular formula of $C_{64}H_{78}N_{14}O_{24}$ for luzopeptin A, and the presence of molecular two-fold symmetry was evident from the ^{13}C and ^1H NMR spectra for the luzopeptin complex. The ultraviolet spectrum indicated the presence of two heteroaromatic chromophores which were shown by chemical experiments to be linked to a depsipeptide cycle. These data suggested that the compounds might be similar to members of the quinoxaline series of antibiotics, which includes echinomycin [45], actinoleukin [46], the quinomycins [47], and the triostins [48]. However, each of the quinoxaline antibiotics has a sulfur-containing bridge across the depsipeptide cycle which is not present in the luzopeptins.

28

36

The major component of the complex, luzopeptin A (**28**), was shown [43] to possess significant antitumor activity in a number of experimental tumor systems in mice, including leukemia P388, leukemia L1210, melanoma B16, Lewis lung carcinoma, and sarcoma 180. The antitumor activities of luzopeptins A, B, and C, echinomycin, and mitomycin C were comparatively determined against leukemia P388. In anti-P388 activity, luzopeptin A was approximately three times as potent as both luzopeptin B and echinomycin and was about 100 times as potent as mitomycin C. Luzopeptin C showed no significant antitumor activity in any of the systems studied.

Luzopeptin A crystallizes from toluene with toluenes of crystallization. This fact was exploited by crystallizing the compound from a similar, heavy atom containing solvent, bromobenzene, thus facilitating X-ray analysis by means of the heavy atom technique [49]. Large single crystals that exhibited excellent diffracting power were grown from layered solutions of CH_2Cl_2/C_6H_5Br. Both the toluene solvate and bromobenzene solvate crystals of luzopeptin A belonged to space group $P2_1$, and the latter had cell constants of $a = 19.881(5)$, $b = 12.303(2)$, and $c = 22.919(3)$ Å and $\beta = 100.32(2)°$. The crystals decomposed rapidly when removed from the mother liquor, as manifested by cracking in the previously smooth crystal faces and rapid deterioration of the X-ray pattern. The decomposition presumably involved loss or rearrangement of the solvent molecules in the crystal and subsequent collapse of the crystal lattice. Since stability of the crystals over time is important in an X-ray diffraction experiment, the crystals were bathed in mother liquor at all times. For data collection a single crystal was drawn into a thin-walled glass capillary via suction accomplished by adapting a gas-tight syringe to one end of the capillary. A specimen prepared in this manner remained stable for the duration of X-ray intensity data collection (10 days). Of a total of 5995 unique X-ray reflections with $2\theta \leq 100°$, more than 4600 were considered observed and used for structure solution and refinement.

The number of independent atoms in the crystal structure of luzopeptin A was well over 100, so it probably would not have been easily solved by the direct method. The bromines in the new crystals made the heavy atom method a possible approach to solving the crystal structure. Three bromines were located in a Patterson map calculated with the intensities measured from the new crystals. For three bromobenzenes and one depsipeptide in the unit cell,

$$\frac{\Sigma \left[Z^2(\text{Br}) \right]}{\Sigma \left[Z^2(\text{light atoms}) \right]} = 0.7$$

which is considerably lower than the optimum ratio of 1.0 given in Section 2. This calculation actually overestimates the phasing power of the three bromines because it underestimates the contents of the unit cell (12 more partially ordered solvent molecules, including a fourth bromobenzene were included in the final model), and because the three located bromine sites were not fully occupied. Not surprisingly, then, the heavy atom method alone was not succesful. Only when phase information obtained with the bromine sites was combined with information from the phase probability relationships of the direct method could the structure be solved. Details of the structure solution are discussed below.

A sharpened Patterson synthesis was calculated using E^2 as the coefficients. The bromine–bromine interactions stood out from what was otherwise a farily featureless map. The crystal positions of three solvent bromine atoms were obtained by interpretation of the Patterson map. Tangent formula recycling, in which the most reliable phases calculated from the three-bromine model were input as starting phases to the tangent formula for phase extension and refinement, enabled, after much toil, the positioning of a chemically sensible fragment of 22 atoms, including what appeared to be a quinaldamide system. Attempts to extend this model using either Fourier or tangent formula refinements were at first unsuccessful. The tangent formula recycling procedure was then modified in one important way. The small set of reliable input phases from the model were used to phase other reflections as before, but were kept invariant during phase refinement. This modification prevented the phases from drifting back to the phases of the initial three-bromine model [50]. An E map calculated after this modification clearly showed both the input fragment and a chemically identical portion. After one more cycle of tangent formula recycling, which used 1378 of 5995 E's, 101 of the 102 atoms of the depsipeptide were clearly visible in an E map. The structure was completed by Fourier refinement.

Alternating block-diagonal least-squares refinements of the model with Fourier syntheses succeeded in locating 4 bromobenzene molecules and 11 other solvent molecules, many of which were disordered. The bromines and 101 atoms in luzopeptin A were modeled using anisotropic thermal parameters; the rest of the 141 nonhydrogen atoms in the structure, including one disordered atom in the depsipeptide, were given isotropic thermal parameters [51]. Positions for all nonhydroxylic hydrogens of luzopeptin A, and hydrogens of the two bromo-benzenes of highest occupancy, were calculated and included in the model. The R factor for the model with hydrogens included was 0.13.

The crystallographic results for luzopeptin A conclusively established the chemical connectivity of the molecule, which had been simultaneously deter-mined by an independent chemical study [52]. Two views of luzopeptin A are shown in Fig. 3. The molecule consists of two identical regions, each composed of

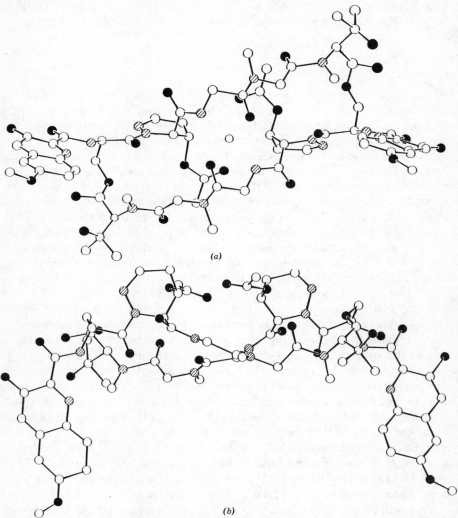

(a)

(b)

Figure 3. (*a*) Computer-generated perspective drawing of luzopeptin A viewed down the twofold axis. (*b*) Computer-generated perspective drawing of luzopeptin A viewed at right angles to the twofold axis. Reprinted with permission from reference 49. Copyright 1981, American Chemical Society.

a 3-hydroxy-6-methoxyquinaldic acid, the amino acids glycine, $R(\text{D})$-serine, N-methyl-β-hydroxy-$S(\text{L})$-valine, and sarcosine, and a very unusual cyclic imino acid, $trans$-$(3S,4S)$-4-acetoxy-2,3,5-tetrahydropyridazine-3-carboxylic acid. In the crystal structure the decadepsipeptide cycle has a rectangular shape with its two identical regions related by a noncrystallographic twofold axis. The depsipeptide cycle is perhaps better described as having a propellor twist shape, as is evident when the rectangle is viewed on edge as in Fig. 3b. Two hydrogen bonds bridge this rectangle, and several other intramolecular, but no intermolecular hydrogen bonds were identified in the structure.

Luzopeptin A is thought to function as a bisintercalator [53–55]. A model for intercalation that is consistent with the three-dimensional structure of the antibiotic has the quinoline rings effectively sandwiching two base pairs in double-helical DNA [49]. In this model, each quinoline moiety overlaps almost completely with the sandwiched base pair adjacent to it, and the DNA axis coincides with the twofold axis in luzopeptin A. This model assumes that there is not radical alteration of the geometry of the DNA, apart from the unwinding of the helix which normally accompanies intercalation [56]. Other models for intercalation are possible if the depsipeptide is not constrained to its crystal structure conformation.

A space-filling model of the luzopeptin A–DNA bisintercalation complex was constructed and showed that the depsipeptide cycle can be very snugly accommodated in the narrow groove of DNA [51]. The functional groups on the buried face of the luzopeptin A molecule in the complex are arranged so as to provide preferential stabilization with an included dGdC base sequence. These studies also suggest that the base sequence specificity of luzopeptin A could extend beyond the two sandwiched base pairs: the lactone carbonyl group is in a favorable position to act as a hydrogen bond acceptor with the outer neighboring base pairs. A highly preferred base sequence might thus be dCdGdCdG. Although luzopeptin A would also bisintercalate at other sequences, the lifetime, and hence the biological significance of the resulting complexes would probably be less.

4.3. The Patterson Search Method

The Patterson function is used most often in small molecule crystallography to determine the location of heavy atoms. Yet this function provides information about the relative positions of all of the atoms in a structure. Whereas the phases of structure factors are necessary to calculate an electron density map from a diffraction pattern, a similar map may be calculated with the Patterson function directly from measured intensities. General features of atomic arrangement in the crystal, such as the orientation of planar groups, can sometimes be recognized by simple inspection of this map. However, such approximate information is usually tantalizingly incomplete in terms of allowing a complete structure determination.

An important use of the Patterson function in X-ray crystallography is the location of known rigid fragments of a crystal structure. If a model of a substantial portion of a molecule in a crystal is available, it may be used to provide approximate phases for diffraction data once it is correctly positioned in the unit cell. For the phase information to be useful, the model should be an accurate representation of about 20% or more of the structure. Ideally, the atoms of the optimally positioned model should deviate from the positions of the corresponding atoms in the crystal structure by 0.3 Å or less, although random errors of up to 0.7 Å can sometimes be tolerated.

The number of possible arrangements of a fragment in a crystal may be astronomical, so it is unrealistic to locate the substructure by trial and error. The Patterson function can be used to locate the fragment in a systematic procedure called the Patterson search technique [57]. Given a model structure, the set of vectors that relate each atom in the model to every other atom in the model can be calculated. From the description of the Patterson function given in Section 2, it is apparent that if the model structure is present in the crystal, the calculated vectors between its atoms can be placed with tails at the origin of the Patterson map and an appropriate rotation will bring their tips into coincidence with a set of peaks in the Patterson map. The rotation that brings the vector tips into coincidence with the Patterson peaks is the rotation that will properly orient the fragment. Once a fragment has been oriented it is translated along the three crystal axes until it is properly positioned with respect to space group symmetry elements of the unit cell.

Patterson search techniques have been used to solve a number of complex crystal structures in recent years. Two examples, marcfortine C and alstonisidine will be discussed.

4.3.1. Marcfortine C.

Scientists recently reported the isolation of a number of alkaloids from an extract (crude ~ 380 mg) of lyophilized mycelium (196 g) from the common fungus *Penicillium rogueforti* strain B26 [58, 59]. The structure of marcfortine A (29A), a major component, was elucidated by a single-crystal X-ray diffraction analysis [59]. The structure of a second major component, roquefortine, had been previously determined by spectroscopic methods [58, 60]. Two minor components that exhibited spectral properties similar to marcfortine A were designated marcfortine B and C [61]. Comparison of the mass spectra, and the ^{1}H and ^{13}C NMR spectra of marcfortines A and B suggested that A was an N-methylated derivative of B. Marcfortine B was therefore assigned structure 29B. The ^{1}H NMR and mass spectra of marcfortine C indicated that it had a structure similar to B, but with a singly oxygenated ring fused to the tryptophan moiety. The complete structure of marcfortine C was provided by single-crystal X-ray diffraction [61].

Suitable crystals of marcfortine C, $C_{27}H_{33}N_3O_3$, were obtained from ethyl acetate as colorless prisms. The crystals belonged to space group $P2_1$ with two independent molecules in the asymmetric unit and unit cell parameters $a = 25.274(4)$, $b = 8.238(3)$, and $c = 13.084(3)$ Å and $\beta = 91.56°$. Of 4377 reflections scanned on a four-circle diffractometer using CuKα radiation, only 2396 were

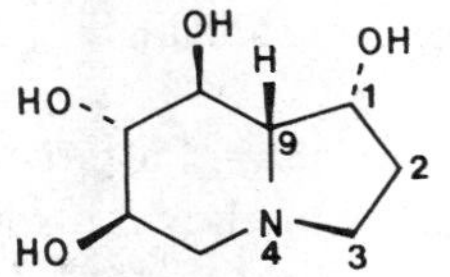

37

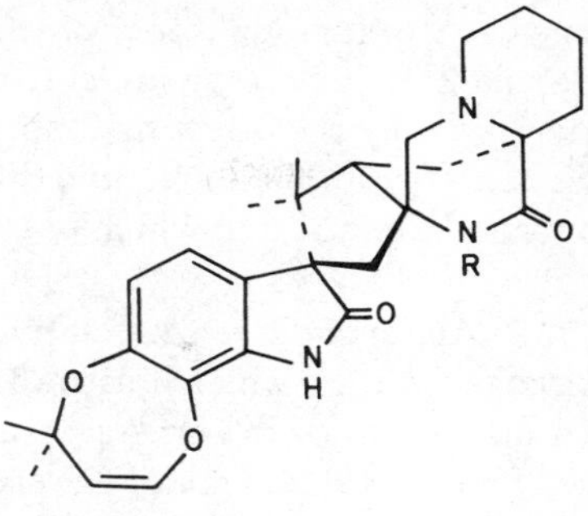

$R_A = CH_3$

$R_B = H$

29

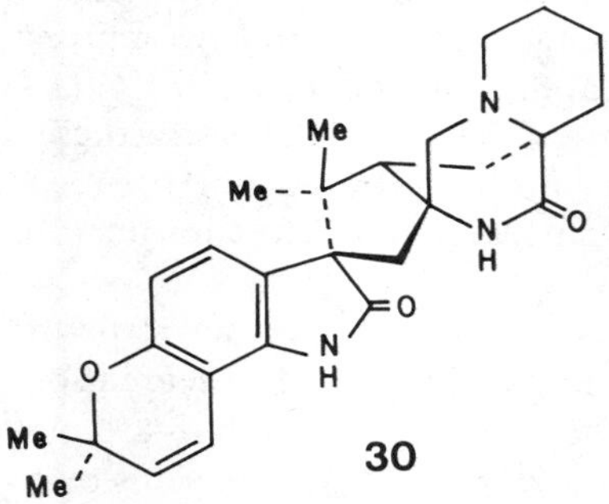

30

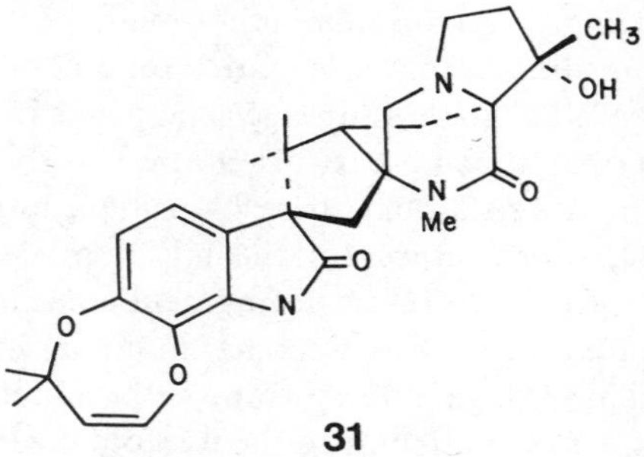

31

considered statistically significant. The crystal structure was solved by locating a rigid 21-atom fragment from the accurate model of marcfortine A in the unit cell of C, using an automated Patterson search algorithm [62]. The fragment was first oriented by rotating it until its calculated vector set overlapped vectors in the Patterson map, then it was positioned with respect to the twofold screw axis of the unit cell by a translational search procedure. The structure, which consisted of 66 independent nonhydrogen atoms, was completed by Fourier refinement. Considering the paucity of X-ray data, the structure of marcfortine C (**30**), may not have been easily solved by the direct method.

A toxic metabolite from *Penicillium paraherque*, paraherquamide **31**, whose structure is quite similar to the marcfortines, has also been reported [63]. Crystals of this alkaloid were obtained from ethyl acetate and the crystal structure was determined by the direct method. Paraherquamide has a seven-membered ring fused to a tryptophan unit, and differs from marcfortine A only in the nature of its second amino acid, which is 2-hydroxyproline instead of pipecolic acid.

4.3.2. Alstonisidine. A relatively detailed account of the use of Patterson search techniques which illustrates some of the difficulties of the method is given in a discussion of the crystal structure determination of alstonisidine [64]. In 1972 two bis-indole alkaloids were isolated from the bark of the Australian tree *Alstonia muelleriana* [65]. The major alkaloid isolated was villastonine (**32**), whose structure had previously been determined by single crystal X-ray analysis [66] and, concurrently, by chemical and spectral studies [67, 68]. Villastonine may be thought of as a combination of the two indole "monomers" macroline (**33**) and pleiocarpamine. The second bis-indole aklaloid, alstonisidine, $C_{42}H_{48}N_4O_4$, was shown by spectral and chemical evidence to be composed of "monomeric" units resembling macroline and quebrachidine (**34**) [69]. This result was confirmed by the biomimetic synthesis of alstonisidine using these two indole alkaloids [70]. However, the stereochemistry of the junction between the two "monomeric" units remained undefined, and an X-ray diffraction analysis of the structure was undertaken.

Since a good model for macroline was available from the crystal structure of villastonine, the crystallographers decided to use this model to search for the macroline substructure of alstonisidine. They initially used an 18-atom rigid model which constituted the greater part of the macroline unit, omitting only the atoms linked to the second "monomer." The rotational search using the vector set of this model showed broad peaks in the overlap function, with several local maxima, and no best orientation of the fragment was obvious. In light of the difficulties that can arise by using a large, possibly distorted, search fragment [71], substructures of the initial model were tried in the search procedure. Eventually an 11-atom fragment containing an indole moiety was found whose vector set gave a distinct maximum in the orientation search. The location of the oriented fragment in the unit cell was determined by a translational search in the plane perpendicular to the twofold screw axis. Attempts to develop this partial structure using tangent formula recycling were not successful, and the fragment was too small (~25% of the atoms in the asymmetric unit) to be used in Fourier refinement. The positioned 11-atom fragment was expanded to the original 18-atom model, but the vectors calculated from this model did not match the Patterson data. Apparently the search procedure had located not the indole portion of the macroline unit, but a structurally similar portion of the bis-indole alkaloid.

Using a routine to adjust a model to fit the Patterson data, the scientists succeeded in extending the 11-atom fragment to 16 atoms. These atoms were used in tangent formula recycling and the resulting E map showed the main features of the crystal structure. The search fragment was identified as the in-

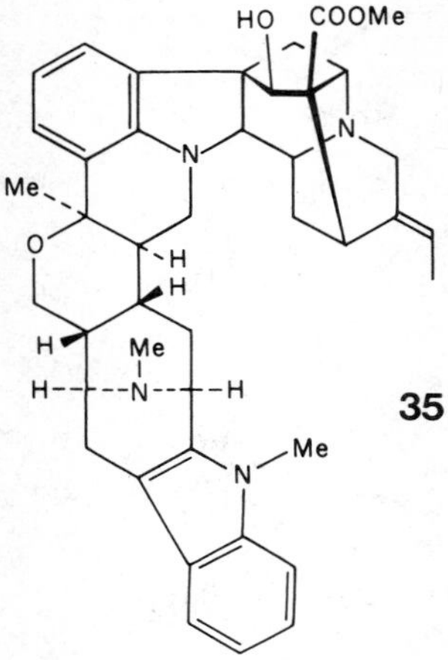

35

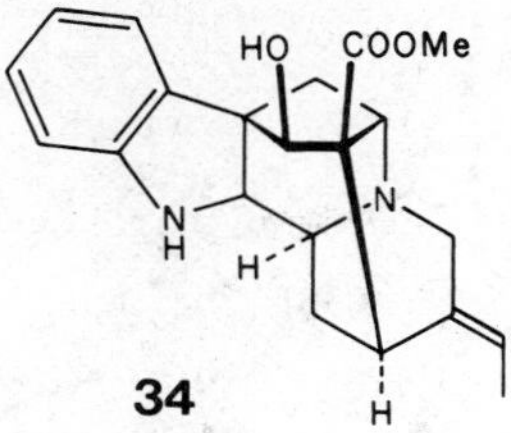

34

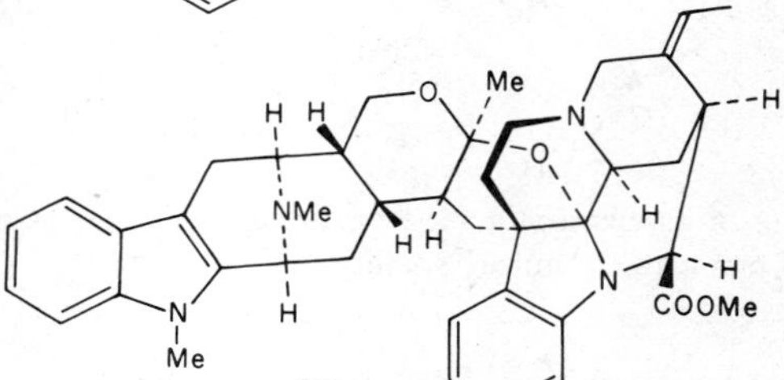

32

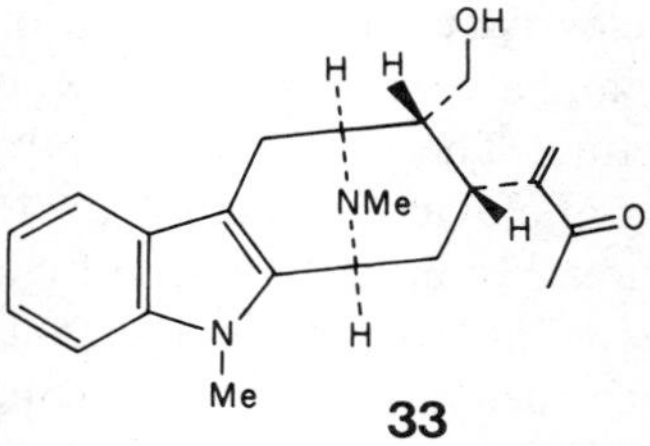

33

doline region of the quebrachidine unit. Further formula refinement located all nonhydrogen atoms in the structure and showed alstonisidine to be **35.**

5. DERIVED RESULTS

Since electrons scatter X-rays, we learn something about the arrangement of the electrons in a crystal from an X-ray diffraction experiment. The results that are reported for a single-crystal X-ray analysis are usually obtained from an interpretation of the electron density map of the unit cell. These results are presented as a list of numbers which represent the positional and vibrational characteristics of the atoms that are used to model the crystal structure.

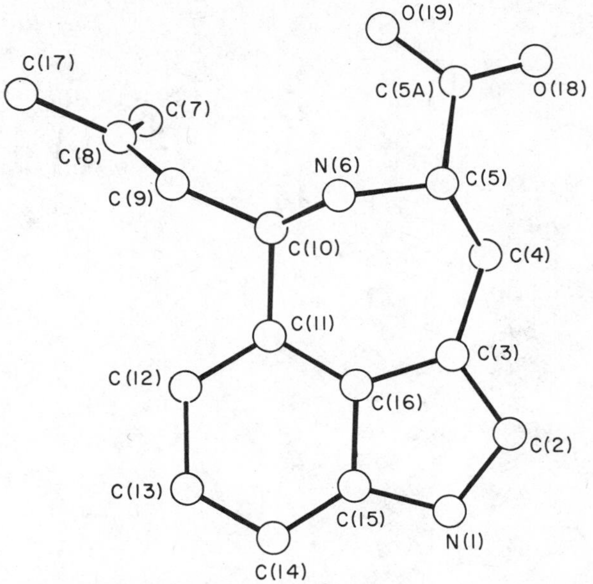

Figure 4. ORTEP drawing of clavicipitic acid. Reprinted with permission from reference 36. Copyright 1980, American Chemical Society.

The computer-generated perspective drawings that usually accompany the report of an X-ray analysis are a familiar and accessible representation of crystallographic results. Such drawings are closely related to conventional chemical representations, such as line drawings, and are derived from the positional and thermal parameters of a crystal structure. The illusion of depth that can be created in a figure such as the drawing of clavicipitic acid (Fig. 4) allows easy recognition of stereochemical information. For instance, the configuration at the chiral centers can be readily assigned from an inspection of such drawings.

Although positional and thermal parameters give a highly compressed and somewhat less informative picture of a structure than the electron density map, chemically interesting information may be calculated from them in a straightforward manner. This information includes bond distances, bond angles, torsion angles, and other metric details of the X-ray model. Calculations of molecular geometry are most conveniently done in Cartesian, or rectangular, coordinates. General computer programs [72] which transform fractional crystal coordinates to rectangular coordinates and calculate all derived quantities of interest are used in virtually every modern crystallographic laboratory.

5.1. Precision and Accuracy of Results

Uncertainties in the positions of atoms in a crystal structure determine the precision of all geometrical parameters calculated from them. Commonly, un-

certainties in fractional coordinates are calculated in the least-squares procedures which refine the structure. Statistical arguments for propagation of errors
[73] can then be used to estimate the errors in the geometrical parameters.
Factors that affect the uncertainties in the position of an atom include amount
and quality of the data collected in the X-ray experiment, thermal vibration, and
scattering power of the atom relative to other atoms in the structure. Oxygens,
nitrogens, and carbons may be more precisely located than hydrogens, for
example, because they have more electrons than hydrogen; hence a carbon–
carbon bond distance is much more precise than a carbon–hydrogen distance. In
a structure with a very heavy atom, such as uranium, carbons contribute
relatively little to X-ray scattering, and their geometry may be as imprecisely
defined as the geometry of hydrogens in an average organic structure.

The accuracy of X-ray results is limited by the nature of the diffraction experiment. Positional parameters from an X-ray crystal structure represent only the
center of the time-averaged electron density distribution about the atoms. In
bonded atoms the electrons are not necessarily symmetrically distributed around
the nuclei. Therefore, atomic positions from X-ray crystallography will differ
slightly from the parameters determined by other methods, such as neutron
diffraction, which locate atomic nuclei. For example, the electron cloud of a
hydrogen bonded to an oxygen is polarized toward the oxygen, and the position
of the hydrogen, as determined by X-ray diffraction, will be displaced from the
position of its nucleus by a small amount. The O—H bond distance determined
by X-ray diffraction will then be shorter than that determined by neutron diffraction, or any other method that measures nuclear–nuclear distances.

Furthermore, the time-averaged picture of the electron density distribution
seen in a crystal structure is smeared out because of thermal vibration. The time-
averaged distance between two atoms in a structure may be quite different from
the center-to-center distance of the ellipses used to describe the atoms in the
X-ray model. The center-to-center distances are systematically shorter than the
time-averaged bond lengths, often by amounts much greater than their estimated
standard deviations [74]. If the vibrational modes of the atoms in the crystal can
be deduced, it is possible to better estimate atomic arrangement at any one point
in time and calculate improved bond distances and angles [75], but rigorous corrections are not possible.

5.2. Bond Distances and Angles

The reliability of calculated bond distances and angles may be evaluated by comparing the X-ray diffraction results for two analyses of the same structure, or by
comparing geometrical parameters of two independent molecules of the same
compound in a single-crystal structure. If we focus our attention on the bond
lengths of clavicipitic acid **22,** shown in Fig. 5, we see that the results for the two
independent molecules in the asymmetric unit in most cases show no significant

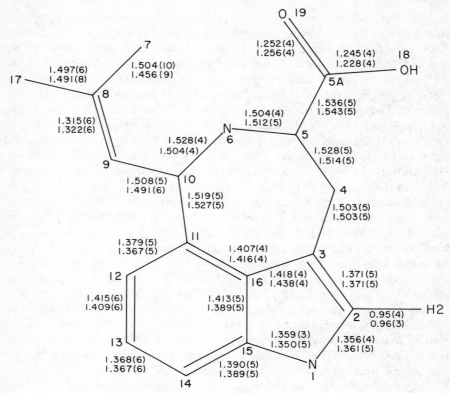

Figure 5. Bond distances of clavicipitic acid.

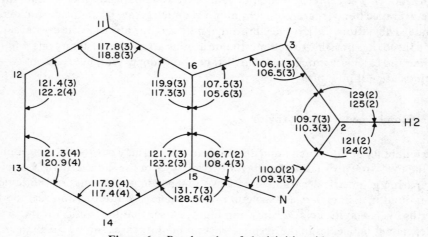

Figure 6. Bond angles of clavicipitic acid.

differences relative to the estimated standard deviations, which average about 0.005 Å. The agreement is particularly striking for bond lengths of the indole moiety which do not involve an atom of the seven-membered ring. The few bond length differences greater than three times their standard deviations involve bonds of the seven-membered ring, or of ring substituents, and could result from either an underestimation of error, or significant differences in true bond lengths. The seven-membered rings are in different conformations in the two molecules; thus it does not seem unreasonable for their bond distances to be observably different.

Bond angles of the indole portions of the two independent molecules are compared in Fig. 6. Only the bond angles at the C(15)—C(16) ring junction differ in the two molecules by a significant amount. Again, these may indicate real differences in the two structures as they are packed in the crystal. As shown in Figs. 5 and 6, the standard deviations of bond lengths and angles involving hydrogens are in general approximately an order of magnitude larger than for bond lengths and angles involving only nonhydrogens.

5.3. Torsion Angles and Conformation

Bond lengths and angles are not by themselves sufficient to construct a three-dimensional model of a molecule which has conformational flexibility. Molecular shape is defined largely by the values of the torsion angles [76] and these are required for a complete description of a structure. The definition and sign for the torsion angle are shown in Fig. 7a. The sign for the torsion angle ω is considered positive if a clockwise rotation of bond 1–2 brings it into a plane defined by atoms 2, 3, and 4, and is considered negative if this rotation is counterclockwise. The Newman projection in Fig. 7b for the staggered form of ethane shows the values of the torsion angles for hydrogens attached to the rear carbon with respect to the hydrogen designated H1.

Conformations of small rings (3–8 atoms) are conveniently classified by symmetry group. Each symmetry group is associated with a characteristic pattern of symmetry-related torsion angles. Figure 8 illustrates the torsion angles for the three standard conformations of five membered rings. The planar form has D5h symmetry, with all torsion angles equal to zero, the Cs, or envelope conformation, has mirror symmetry, with two sets of twofold-related torsion angles, and the C_2 conformation has twofold symmetry, with two sets of mirror-

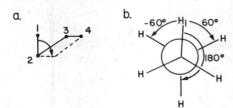

Figure 7. (a) Definition of a positive torsion angle. (b) Torsion angles for the hydrogens in ethane.

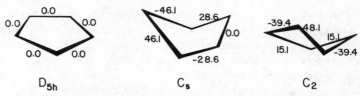

Figure 8. Standard conformations with their characteristic torsion angles for five membered rings.

related torsion angles. The symmetry classes of small rings have been discussed in the literature by several authors [77–79].

The conformations of rings in most structures deviate somewhat from ideal symmetry. Unequal bond lengths or steric strain imposed by bonded groups perturb the symmetry of a ring. The deviation of a ring conformation from any particular symmetry group can be quantitated using the following equation [79]:

$$\Delta = \left(\sum_{i=1}^{m} \frac{(\phi(i) \pm \phi'(i))^2}{m} \right)^{1/2} \tag{1}$$

where $\phi(i)$ and $\phi'(i)$ are two torsion angles related by symmetry in the ideal case, and the sum is over all such symmetry-related angles. The $+$ sign is used if the torsion angles should be related by twofold symmetry, and $-$ is used if they should be related by a mirror plane. In an ideal conformation, Δ would be zero.

As a simple example of conformational analysis, we may examine the crystal structure of sesbanine (**36**), a cytotoxic alkaloid from the seeds of *Sesbania drummondii* [80]. This compound crystallizes with two molecules in an asymmetric unit. The two molecules are nearly related by the symmetry operation $x' \sim \frac{3}{2} - x$, $y' \sim 1 - y$, $z' \sim 1 - z$, and the crystal structure of sesbanine was actually solved by direct phasing of the data in the centro-symmetric space group $P2_1 a$ to obtain the achiral portion of the molecule. The structures of the two molecules in the asymmetric unit were then completed and independently refined in the true space group $P2_1$. The two six-membered rings in sesbanine are aromatic, therefore constrained to planarity, and only the five-membered ring conformation must be defined.

The torsion angles of the hydroxycyclopentane rings for the two sesbanine molecules are shown in Fig. 9. A comparison of the set of torsion angles for the ring in molecule one (upper numbers) with the ideal torsion angles in Fig. 8 suggests that the ring has nearly ideal C_2 symmetry with a twofold axis bisecting the C(4)—C(9) bond. On the other hand, the ring in molecule two (lower numbers), does not seem to have any of the standard conformations shown in Fig. 8. The conformation of this ring appears to be somewhere between the conformation of the ring in molecule one, and a Cs conformation with a mirror plane bisecting the C(10)—C(11) bond. From Equation (1),

$$\Delta(C_2) = [\tfrac{1}{2}((-27.2 + 16.7)^2 + (-34.3 + 38.9)^2)]^{1/2} = 8.11$$

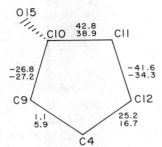

Figure 9. Torsion angles for the five-membered ring in sesbanine.

and

$$\Delta(Cs) = [\tfrac{1}{2}((-27.2 + 34.3)^2 + (5.9 - 16.7)^2)]^{1/2} = 9.14$$

indicating almost equal deviation from both symmetries. Analysis of errors in the torsion angle calculations is required before concluding that the apparent conformational differences between the two chemically identical sesbanine molecules are significant.

5.4.   Torsion Angles and $^3Jhh'$ Couplings

Approximate formulas, such as the Karplus equation [81, 82] which correlate vicinal proton–proton coupling constants with torsion angles may be used in conjunction with crystal structure data to help interpret complex high-resolution ^{1}H NMR spectra, or to compare an averaged solution structure with that found in the solid state. A careful single-crystal X-ray diffraction study of an organic molecule usually provides the investigator with the approximate location of the hydrogen atoms, and even more accurate hydrogen positions may be obtained by neutron diffraction. Hydrogen positions poorly defined by experiments can be reliably estimated from the positions and hybridizations of the heavy atoms to which they are bonded. Therefore, torsion angles between hydrogens are readily derived from most crystal structures. Further, bond distances and angles necessary to adjust the parameters of the Karplus equation [82] are standard output of crystallographic studies. The following example illustrates the usefulness of the Karplus equation in the interpretation of crystallographic results.

Castanospermine (37) was isolated from a 2 M NH₃ wash of the ethanol/water extract from toxic seeds of the Australian legume *Castanospermum australe*. The connectivity of this alkaloid was determined by ^{13}C and ^{1}H NMR analyses, including extensive double resonance experiments at 360 MHz which completely defined the pattern of vicinal proton–proton coupling constants. The relative stereochemistry of castanospermine was determined by a single-crystal X-ray analysis. The crystal structure provided all of the data necessary to estimate the vicinal proton–proton coupling constants which would be observed if the alkaloid conformation in the crystal were also present in solution.

To simplify the example, the original equation and parameters for ethane reported by Karplus will be used for the coupling constant calculation:

$$Jhh' = A + B \cos (\phi) + C \cos (2\phi)$$

where ϕ is the torsion angle, and the parameters are $A = 4.22$, $B = 0.5$, and $C = 4.5$ cps. The observed and calculated coupling constants are presented in Table 3. Comparison of the data reveals at least qualitative agreement for most of the coupling constants. However, the predictions break down for the hydrogens attached to carbons 3 and 5, both of which are adjacent to the nitrogen position. A correction to the calculation based on the electronegativity difference between nitrogen and hydrogen [82] only makes the agreement worse.

In the crystal, the conformation of the five-membered ring has C_2 symmetry with a twofold axis bisecting the N(4)—C(9) bond and containing atom C(2). Inspection of a molecular model indicates that another C_2 conformation in which the twofold axis would bisect the C(3)—C(4) bond and contain C(1) is readily accessible from the crystal conformation via a pseudorotation which would involve minimal distortion and would not appear to introduce any severe steric interactions. Coupling constants calculated from this conformation are in closer agreement with the observed values than are those from the crystal structure. Therefore, this conformation is more likely to represent the structure of castanospermine in solution. Intermolecular hydrogen bonds or other packing forces may affect the conformation of the molecule in the crystal.

5.5. Hydrogen Bonding

Nonbonded distances between all atoms in a crystal structure are routinely calculated and examined for unusually short values, that is, distances less than the sum of the van der Waals radii of the atoms involved. A short nonbonded distance between two electronegative atoms, at least one of which is bonded to a hydrogen, may signal the existence of a hydrogen bond. Ranges of hydrogen bond distances reported in the literature have been tabulated in the *International Tables for X-ray Crystallography* [83] and assistance in the interpretation of close contacts.

It is generally safe to assume that two electronegative atoms which are in close contact and are on different molecules in the unit cell are hydrogen bonded. If the two atoms are in the same molecule, however, care must be taken to ensure that the close contact is not the result of steric constraints. The only way to conclusively identify a hydrogen bond is to locate the hydrogen involved. The hydrogen should be close to the nonbonded electronegative atom (less than the sum of their van der Waals radii) and the $X - H \cdots Y$ angle should be close to 180°. However, it should be noted that hydrogen bonds that deviate from linearity by as much as 30° have been observed [84].

A hydrogen bond between atoms in the same molecule is called an intramolecular hydrogen bond, and between atoms in different molecules, an inter-

Table 3. Observed and Predicted Vicinal Proton–Proton Coupling Constants for Castanospermine

Vicinal Hydrogens	Torsion Angle(°)	J_{vic} Obs. in D_2O	J_{vic} Calc. Simple[a]	J_{vic} Calc. Corrected[b]	ΔJ Obs.	ΔJ Calc.
1–2A	−111	2.0	1.0	0.9	1.0	1.1
1–2B	7	7.0	8.1	7.3	−1.1	−0.3
1–9	−30	4.4	6.0	5.1	−1.6	−0.7
2A–3A	17	10.0	7.5	7.0	2.5	3.0 (see text)
2A–3B	143	10.0	5.9	5.5	4.1	4.5 (see text)
2B–3A	−86	0.0	−0.3	−0.3	0.3	0.3
2B–3B	41	2.0	4.4	4.1	−2.4	−2.1
5A–6	−55	5.0	2.4	2.2	2.6	2.8
5B–6	−163	10.0	8.4	7.6	1.6	2.4
6–7	178	8.5	9.2	7.5	−0.7	0.7
7–8	−172	8.5	9.0	7.3	−0.5	1.2
8–9	174	10.0	9.1	8.2	0.9	1.8

[a] Calculated from the simple Karplus equation $J_{vic} = A + B\cos\phi + C\cos 2\phi$ where ϕ is the dihedral angle. For this calculation $A = 4.2$, $B = -0.5$, and $C = 4.5$ Hz.

[b] A correction for substituents different in electronegativity from a hydrogen $[J_{vic} = J_{vic}(1.0 - 0.07\Delta\chi$ where $\Delta\chi = \chi_{hetero} - \chi_H$ and the following values were assumed: $\chi_H = 2.1$, $\chi_N = 3.0$, and $\chi_O = 3.5]$.

molecular hydrogen bond. A hydrogen bond to two acceptor atoms is a bifurcated hydrogen bond. Bifurcated hydrogen bonds are rare, and definitely require location of the hydrogen for identification [84]. The hydrogen in a bifurcated hydrogen bond may be equidistant from both acceptors (a symmetric bifurcated hydrogen bond), or closer to one than the other (an asymmetric bifurcated hydrogen bond), but must be within hydrogen bonding distance of both. The $X - H \cdots Y$ angles for bifurcated hydrogen bonds deviate significantly from 180°, and are in fact closer to 120° in the ideal, symmetric case.

5.6. Absolute Configuration

The phenomenon of anomalous scattering of X-rays has been routinely used to determine the absolute configuration of chiral molecules from their crystal structures. Anomalous scattering occurs when atoms in a crystal absorb incident radiation used in a diffraction experiment. While all atoms absorb the radiation to some extent, it is the heavy atoms, particularly those whose atomic numbers are close the the atomic number of the elements used to generate the X-rays, which absorb the radiation most strongly. Therefore, the effect of anomalous scattering on X-ray diffraction data are most pronounced for crystals that contain heavy atoms.

The effects of anomalous scattering on the intensities of reflections can be calculated using theoretical scattering factor corrections [85]. Scattering factors indicate the theoretical scattering of X-rays from nonvibrating atoms. To the first approximation, they are real numbers calculated with the assumption that the atoms do not absorb X-rays. Structure factors calculated with these scattering factors will be identical for two reflections (hkl) and $(\overline{hkl})$, called a Friedel pair. When the scattering factors are corrected for the effects of anomalous scattering they become complex numbers. For a noncentrosymmetric structure, intensities calculated with the corrected scattering factors will be slightly different for two Friedel mates. Some reflections are more sensitive to anomalous scattering effects than others, and their intensities may differ from those of their Friedel mates by several percent.

Transformation of a crystal structure model to its enantiomorph interchanges the calculated structure factor amplitudes of Friedel mates. That is, the calculated amplitude of any reflection $| F_c(hkl) |$ for one enantiomer will equal $| F_c(\overline{hkl}) |$ for the opposite enantiomer. For the correct enantiomer, the relationship of $F_c(hkl)$ to $F_c(\overline{hkl})$ (i.e., whether it is greater or less than the latter) will be the same as that of $I(hkl)$ to $I(\overline{hkl})$. Only 10 to 20 of the Friedel pairs most sensitive to anomalous scattering effects need to be compared. For a successful experiment a clear majority should give the same prediction of absolute configuration.

Alternatively, the two entire calculated data sets may be compared to observed data and the one in best agreement, as measured by the R factor, is judged to correspond to the correct enantiomer. This method is convenient because the R factor is standard output of the computer programs used by crystallographers to calculate structure factors, but it is not as sensitive as the first method. Differences in the R factors for the two enantiomers may be as small as 0.1%, and the significance of these differences must be judged on the basis of statistics [86].

In recent years, crystallographers have succeeded in establishing the absolute configurations of structures containing no atoms heavier than oxygen. The anomalous scattering corrections for oxygen and other light atoms are extremely small. The detection of anomalous scattering effects in the data sets of light atom structures is possible only when great care has been taken to minimize errors in the measurements. Optimal experimental procedures for determining the absolute configurations of light atom structures have been discussed in the literature [87, 88]. The recent success in predicting with 99.9% probability the absolute configuration of the achiral compound 4,4'-dimethylchalcone, $C_{17}H_{16}O$, from the anomalous scattering effects of its single oxygen [88] suggests that it may be possible to determine the absolute configuration of many highly oxygenated alkaloids without preparing their heavy atom derivatives.

The assignment of R or S designations to chiral centers is a distinct procedure from the determination of absolute configuration. The R and S nomenclature was developed for describing the known configuration of asymmetric atoms [89]. The nomenclature allows symbolic, nonpictorial representation of molecular structure which is unambiguous except for conformational details. Information on the configuration of asymmetric atoms is inherent in the atomic coordinates of a crystal structure [90]. In principle, then, one may design an algorithm to automatically assign configurations to asymmetric atoms of a molecule given only the coordinates of its crystal structure. Alternatively, one may create a perspective plot or a molecular model of the final crystal structure and determine the correct configuration label for each chiral center by inspection.

6. CONCLUDING REMARKS

The vast number of published crystal structures cannot practically be reviewed in this introduction to the technique of X-ray crystallography. We conclude instead with a few remarks on the impact of these structures on natural products research. The results of alkaloid X-ray diffraction analyses have contributed to a large data base of structural parameters, useful in predicting bond distances and angles for common groups of atoms in organic structures. Correlation of X-ray results with spectroscopic information has improved the predictive ability of spectroscopic methods and provided a means of comparing solution and solid-state conformations of molecules. Such correlations have become easier and more meaningful with the development of techniques, such as ^{13}C magic angle spinning, for recording solid-state spectra.

Alkaloid crystallographic studies have stimulated research in the areas of biosynthesis and total synthesis. Crystal structure data on important classes of biologically active alkaloids, such as the analgesics and the neurotransmitters, are helping to define specific structural requirements for activity, thus providing a basis for drug design. A few structures, such as luzopeptin A, even suggest detailed, testable models for biological activity and encourage further research in both chemistry and biology.

Advances in direct methods of crystal structure analysis may soon allow the small-molecule crystallographer to solve noncentrosymmetric structures with hundreds of atoms. Fascinating biological molecules, which cannot now be easily studied by either small molecule or protein crystallographic techniques, will then become tractible structural problems. The results of these studies will undoubtedly be relevant to several areas of science.

ACKNOWLEDGMENTS

The authors wish to thank Prof. C. E. Nordman for providing a copy of the section of L. Hoard's thesis which describes the structure determination of alstonisidine.

REFERENCES

1. O. Kennard, F. H. Allen, and D. G. Watson, Eds., *Molecular Structure and Dimensions, Guide to the Literature 1935–1976, Organic and Organometallic Crystal Structures,* Bohn, Scheltema and Holkema, Utrecht, The Netherlands, 1977.
2. E. W. Hughes, *J. Am. Chem. Soc.* **63,** 1737 (1941).
3. G. H. Stout and L. H. Jensen, *X-ray Structure Determination,* Macmillan, London, 1968, pp. 270–288, and references cited therein.
4. M. Przybylska and L. Marion, *Can. J. Chem.* **34,** 185 (1956).
5. S. W. Pelletier, N. V. Mody, K. I. Varughese, J. A. Maddry, and H. K. Desai, *J. Am. Chem. Soc.* **103,** 6536 (1981).
6. D. V. Lefemine, M. Dann, F. Barbatschi, W. K. Hausmann, V. Zbinovsky, P. Monnikendam, J. Adam, and N. Bohonos, *J. Chem. Soc.* **84,** 3184 (1962).
7. A. Tulinsky and J. H. van den Hende, *J. Am. Chem. Soc.* **89,** 2905 (1967).
8. J. S. Webb, D. B. Cosulich, J. H. Mowat, J. B. Patrick, R. W. Broschard, W. E. Meyer, R. P. Williams, C. F. Wolf, W. Fulmor, C. Pidacks, and J. E. Lancaster, *J. Am. Chem. Soc.* **84,** 3185 (1962).
9. S. W. Pelletier, N. V. Mody, J. Finer-Moore, H. K. Desai, and H. S. Puri, *Tetrahedron Lett.* **22,** 313 (1981).
10. F. R. Stermitz, R. M. Coomes, and D. R. Harris, *Tetrahedron Lett.,* 3915 (1968).
11. J. A. Wunderlich, *Acta Crystallogr.* **B23,** 846 (1967).
12. S. R. Hall and F. R. Ahmed, *Acta Crystallogr.* **B24,** 337 (1968).
13. S. R. Hall and F. R. Ahmed, *Acta Crystallogr.* **B24,** 346 (1968).
14. K. B. Birnbaum, *Acta Crystallogr.* **B28,** 2825 (1972).
15. H. B. Burgi, J. D. Dunitz, and Eli Shefter, *J. Am. Chem. Soc.* **95,** 5065 (1973).

16. D. Sayre, *Acta Crystallogr.* **5,** 60 (1952).
17. W. Cochran, *Acta Crystallogr.* **5,** 65 (1952).
18. W. H. Zachariasen, *Acta Crystallogr.* **5,** 68 (1952).
19. W. Cochran, *Acta Crystallogr.* **8,** 473 (1955).
20. G. Germain, P. Main, and M. M. Woolfson, *Acta Crystallogr.* **A27,** 368 (1971).
21. J. Karle and I. L. Karle, *Acta Crystallogr.* **21,** 849 (1966).
22. G. Germain and M. M. Woolfson, *Acta Crystallogr.* **B24,** 91 (1968).
23. J. Karle, *Acta Crystallogr.* **B24,** 182 (1968).
24. S. W. Pelletier and T. N. Oeltmann, *Tetrahedron* **24,** 2019 (1968).
25. S. K. Pradhan and V. M. Girijavallabhan, *J.C.S. Chem. Commun.*, 644 (1970).
26. S. W. Pelletier and N. V. Mody, *Tetrahedron Lett.,* 1477 (1977).
27. S. W. Pelletier, W. H. DeCamp, and N. V. Mody, *J. Am. Chem. Soc.* **100,** 7976 (1978).
28. S. W. Pelletier, H. K. Desai, J. Finer-Moore, and N. V. Mody, *J. Am. Chem. Soc.* **101,** 6741 (1979).
29. S. W. Pelletier, N. V. Mody, H. K. Desai, J. Finer-Moore, and J. Nowacki, *J. Org. Chem.*, **48,** 1787 (1983).
30. M. Abe, S. Ohmomo, T. Ohashi, and T. Tabuchi, *Agr. Biol. Chem. (Tokyo)* **33,** 469 (1969).
31. S. Yamatodani, Y. Asahi, A. Matsukura, S. Ohmomo, and M. Abe, *Agr. Biol. Chem. (Tokyo)* **34,** 485 (1970).
32. R. J. Cole, J. W. Kirksey, J. Clardy, N. Eickman, S. M. Weinreb, P. Singh, and D. Kim, *Tetrahedron Lett.,* 3849 (1976).
33. J. E. Robbers and H. G. Floss, *Tetrahedron Lett.,* 1857 (1969).
34. S. Agurell and J. Lindgren, *Tetrahedron Lett.,* 5127 (1968).
35. G. S. King, P. G. Mantle, C. A. Szczyrbak, and E. S. Waight, *Tetrahedron Lett.,* 215 (1973).
36. J. E. Robbers, H. Otsuka, H. G. Floss, E. V. Arnold, and J. Clardy, *J. Org. Chem.* **45,** 1117 (1980).
37. W. Cochran and A. S. Douglas, *Proc. Roy. Soc. London* **A243,** 281 (1957).
38. G. T. DeTitta, J. W. Edmonds, D. A. Langs, and H. Hauptman, *Acta Crystallogr.* **A31,** 472 (1975).
39. J. P. Springer and J. Clardy, *Tetrahedron Lett.* **21,** 231 (1980).
40. J. P. Springer, J. Clardy, J. M. Wells, R. J. Cole, and J. W. Kirksey, *Tetrahedron Lett.,* 2531 (1975).
41. R. T. Gallagher, J. Finer, and J. Clardy, *Tetrahedron Lett.* **21,** 235 (1980).
42. A. E. DeJesus, P. S. Steyn, F. R. Van Heerden, R. Vleggaar, P. L. Wessels, and W. E. Hull, *J.C.S. Chem. Commun.*, 289 (1981).
43. H. Ohkuma, F. Sakai, Y. Nishiyama, M. Ohbayashi, H. Imanishi, M. Konishi, T. Miyaki, H. Koshiyama, and H. Kawaguchi, *J. Antibiotics* **33,** 1087 (1980).
44. K. Tomita, Y. Hoshino, T. Sasahira, and H. Kawaguchi, *J. Antibiotics* **33,** 1098 (1980).
45. A. Dell, D. H. Williams, H. R. Morris, G. A. Smith, J. Feeney, and G. C. K. Roberts, *J. Am. Chem. Soc.* **97,** 2497 (1975).
46. M. Ueda, Y. Tanigawa, Y. Okani, and H. Umezawa, *J. Antibiotics Ser. A.* **7,** 125 (1954).
47. H. Otsuka and J. Shoji, *Tetrahedron* **23,** 1535 (1967).
48. J. Shoji and K. Katagiri, *J. Antibiotics Ser. A.* **14,** 325 (1961).
49. E. Arnold and J. Clardy, *J. Am. Chem. Soc.* **103,** 1243 (1981).
50. I. Karle, *J. Am. Chem. Soc.* **96,** 4000 (1974).
51. E. Arnold, Ph.D. Thesis, Department of Chemistry, Cornell University, Ithaca, N.Y., 1982.

52. M. Konishi, H. Ohkuma, F. Sakai, T. Tsuno, H. Koshiyama, T. Naito, and H. Kawaguchi, *J. Am. Chem. Soc.* **103**, 1241 (1981).

53. J. S. Lee and M. J. Waring, *Biochem. J.* **173**, 115 (1978).

54. J. S. Lee and M. J. Waring, *Biochem. J.* **173**, 129 (1978).

55. C. H. Huang, S. Mong, and S. T. Crooke, *Biochemistry* **19**, 5537 (1980).

56. J. C. Wang, *J. Mol. Biol.* **89**, 783 (1974).

57. C. E. Nordman and K. Nakatsu, *J. Am. Chem. Soc.* **85**, 353 (1963).

58. P. M. Scott, M. A. Merrien, and J. Polonsky, *Experientia* **32**, 140 (1976).

59. J. Polonsky, M. A. Merrien, T. Prange, C. Pascard, and S. Moreau, *J.C.S. Chem. Commun.*, 601 (1980).

60. P. M. Scott, J. Polonsky, and M. Merrien, *J. Agric. Food Chem.* **27**, 201 (1979).

61. T. Prange, M. Billion, M. Vuilhorgne, C. Pascard, J. Polonsky, and S. Moreau, *Tetrahedron Lett.*, **22**, 1977 (1981).

62. B. P. Braun, J. Hornstra, and J. L. Leenhouts, *Philips Research Report* **24**, 85 (1969).

63. M. Yamazaki, B. Okuyama, M. Kobayashi, and H. Inoue, *Tetrahedron Lett.*, **22**, 135 (1981).

64. L. G. Hoard, Ph.D. Thesis, Department of Chemistry, University of Michigan, 1977, pp. 22–79.

65. R. C. Elderfield and R. E. Gilman, *Phytochemistry* **11**, 339. (1972).

66. C. E. Nordman and S. K. Kumra, *J. Am. Chem. Soc.* **87**, 2059 (1965).

67. M. Hesse, H. Hurzeler, C. W. Gemenden, B. S. Joshi, W. I. Taylor, and H. Schmid, *Helv. Chim. Acta* **48**, 689 (1965).

68. M. Hesse, F. Bodmer, C. W. Gemenden, B. S. Joshi, W. I. Taylor, and H. Schmid, *Helv. Chim. Acta.* **49**, 1173 (1966).

69. J. M. Cook and P. W. LeQuesne, *J. Org. Chem.* **36**, 582 (1971).

70. D. E. Burke, J. M. Cook, and P. W. LeQuesne, *J.C.S. Chem. Commun.*, 697 (1972).

71. R. Hoge and C. E. Nordman, *Acta Crystollogr.* **B30**, 1435 (1974).

72. J. D. Dunitz, *X-ray Analysis and the Structure of Organic Molecules,* Cornell University Press, Ithaca, 1980, pp. 423–424.

73. P. R. Bevington, *Data Reduction and Error Analysis for the Physical Sciences,* McGraw-Hill, New York, 1969, pp 56–65.

74. R. E. Marsh and J. Donohue, *Advances in Protein Chemistry* **22**, 235 (1967).

74. R. E. Marsh and J. Donohue, *Advances in Protein Chemistry,* **22**, 235 (1967).

75. W. R. Busing and H. A. Levy, *Acta Crystallogr.* **17**, 142 (1964).

76. E. L. Eliel, N. L. Allinger, S. J. Angyal, and G. A. Morrison, *Conformational Analysis,* Wiley-Interscience New York, 1965, p. 6.

77. J. B. Hendrickson, *J. Am. Chem. Soc.* **83**, 4537 (1961).

78. K. S. Pitzer and W. E. Donath, *J. Am. Chem. Soc.* **81**, 3213 (1959).

79. W. L. Duax, C. M. Weeks, and D. C. Rohrer, "Crystal Structures of Steroids," in N. L. Allinger and E. L. Eliel, Eds., *Topics in Stereochemistry,* Vol. 9, Wiley, New York, 1976, pp. 271–383.

80. R. G. Powell, C. R. Smith, Jr., D. Weisleder, D. A. Muthard, and J. Clardy, *J. Am. Chem. Soc.* **101**, 2784 (1979).

81. M. Karplus, *J. Chem. Phys.* **30**, 11 (1959).

82. M. Karplus, *J. Am. Chem. Soc.* **85**, 2870 (1963).

83. K. Lonsdale, Ed., *International Tables for X-ray Crystallography,* Vol. 3, The Kynoch Press, Birmingham, England, 1968, p. 273.

84. J. Donohue, "Selected Topics in Hydrogen Bonding," in A. Rich and N. Davidson, Eds., *Structural Chemistry and Molecular Biology,* Freeman, San Francisco, 1968, pp. 443–465.

85. D. T. Cromer, *Acta Crystallogr.* **18,** 17 (1965).

86. W. C. Hamilton, *Acta Crystallogr.* **18,** 502 (1965).

87. D. W. Engel, *Acta Crystallogr.* **B28,** 1496 (1972).

88. D. Rabinovich and H. Hope, *Acta Crystallogr.* **A36,** 670 (1980).

89. R. S. Cahn and C. K. Ingold, *J. Chem. Soc.* 612 (1951).

90. V. Prelog and G. Helmchen, *Angew. Chem. Int. Ed.* **21,** 567 (1982).

Chapter Two

The Imidazole Alkaloids

Richard K. Hill
Department of Chemistry
University of Georgia
Athens, Georgia

CONTENTS

1. INTRODUCTION

The amino acid histidine shares with phenylalanine, tyrosine, and tryptophan the structural features of a reactive aromatic ring connected to a 2-aminoethyl side-chain. These two units comprise two of the three functional groups required to form a new ring by an intramolecular Mannich (Pictet–Spengler) reaction, and it is because this reaction type apparently occurs so readily in plants that the isoquinoline alkaloids and indole alkaloids formed in this way constitute the two dominant families of alkaloids. It is consequently surprising that relatively few alkaloids containing an imidazole ring occur in nature, and very few of these appear to be biosynthesized by Mannich reactions. A number of alkaloids are simple derivatives of histidine and its metabolic products, but at the time of the last comprehensive review [1] of the imidazole alkaloids in 1953 the alkaloids of *Pilocarpus* were the only more complex members known. In the intervening years several novel imidazoles have been discovered, but the total is still comparatively small, and nature does not seem to employ histidine as a major source of alkaloids.

2. SIMPLE HISTIDINE DERIVATIVES

2.1. Histamine and *N*-Methyl Derivatives of Histamine and Histidine

Histamine (**1**), the enzymatic decarboxylation product of histidine, occurs mainly in animal organisms and plays an extremely important physiological role in humans. It has been found in the venom of wasps, hornets, and honeybees and in the tentacles of sea anemones. It is ubiquitous in nature and has been isolated from or detected in a variety of terrestrial plants and marine organisms. Examples include perennial rye grass (*Lolium perenne*). [2], the foliage of shepherd's purse (*Capsella bursapastoris*) [3], and the leather coral (*Alcyonium digitatum*) [4]. Histamine, *N*-methylhistamine (**2**), and *N*,*N*-dimethylhistamine (**3**) have been isolated from the skin of South American amphibians of the genus *Leptodactylus* [5], and **1**, **3**, and *N*-acetylhistamine were found in spinach, *Spinacea oleracea* [6]. Histamine and *N*,*N*-dimethylhistamine co-occur in the mushroom *Coprinus comatus* [7, 8]. In a search for the hypotensive agent present in the seeds of *Casimiroa edulis*, a tree of Mexico and Central America whose extracts have been used in native medicine as a sedative and hypnotic, *N*,*N*-dimethylhistamine was isolated as the picrate in 0.05% yield from methanol extracts of the seeds [9]. Histamine was isolated in surprising amounts (100 mg/kg) from the giant sponge, *Geodia gigas* [10].

A ring *N*-methylhistidine, identified as 1-methylhistidine (**4**) occurs in the brown seaweed *Phyllospora courosa*, and the 1,3-dimethyl quaternary salt (**5**) has been isolated from the Australian red seaweed *Gracilaria secundata* [11]. This compound (**5**) has been synthesized by methylating *N*-phthaloylhistidine with dimethyl sulfate and removing the protecting group. It is toxic to first instar mosquito larvae.

The isomeric *N*-methylhistidine (**6**) has been found in human urine [12],

$CH_2CH_2NR_1R_2$ (imidazole structure, HN)

1 $R_1 = R_2 = H$ HISTAMINE
2 $R_1 = CH_3$, $R_2 = H$
3 $R_1 = R_2 = CH_3$

$CH_2CHCOOH$, NH_2 (imidazole, N–CH_3)

4

CH_3-N^+ ... $CH_2CHCOOH$, NH_2 (N–CH_3)

5

CH_3-N ... $CH_2CHCOOH$, NH_2

6

$COOCH_3$, NH, O (cyclic urea structure)

7

CH_2CHCOO^-, $^+N(CH_3)_3$ (HN, R)

8 R = H   HERCYNINE
9 R = SH ERGOTHIONEINE

$CH_2CHCO_2CH_3$, NH_2 (HN)

10

$CH_2CHCOOR$, Cl (HN)

11 R = H
29 R = CH_2CH_2Cl

muscle proteins [13], and as a dipeptide component in whalemeat extracts [14]. Noordam et al. [15] have devised a clever regioselective preparation of this *N*-methylated isomer, based on forming the cyclic urea (7) by treating histidine methyl ester with *N,N*-carbonyldiimidazole in DMF. Since the only basic nitrogen in 7 is the one not involved in urea formation, methylation occurs specifically to give 6 after acid hydrolysis.

2.2. Hercynine

Hercynine (8) and ergothioneine (9) are two related betaines first isolated from fungi. Hercynine was initially obtained by Kutscher [16] from an aqueous extract of the mushroom *Agaricus campestris* which was sold under the name "Hercynia" as a patent medicine and appetite stimulant. Hercynine is a dextro-rotatory base, $[\alpha]_D + 41.1°$ (dil. HCl). Kutscher established the formula $C_9H_{15}N_3O_2$, observed a strong Pauly test for imidazoles (colored product from coupling with diazobenzene-*p*-sulfonic acid [17]), and noted that the formula agreed with a trimethylhistidine. Kutscher proposed the betaine structure 8 and confirmed this by synthesis from histidine [18]. Attempted methylation of histidine led to a pentamethyl derivative, but this problem was circumvented by diazotizing histidine methyl ester (10) in the presence of excess HCL and treating the resulting chloroacid **11** with trimethylamine to give hercynine, isolated as the chloroaurate, mp 183°.

Hercynine was subsequently found in other mushrooms, [19], including *Amanita muscaria* [20], *Coprinus comatus* [8], *Limulus polyphemus* [21], and *Polyporus sulphureus* [22]. More recently it has been detected in animal organisms; it is found in the red slug *Arion empiricorum* [23] and has been detected by paper chromatography and paper electrophoresis, along with ergothioneine, in boar ejaculate and the blood corpuscles of cattle [24].

2.3. Ergothioneine

Ergothioneine (**9**) is a crystalline base, $C_9H_{15}N_3O_2S$, $[\alpha]_D + 110°$, first isolated from ergot fungus by Tanret [25]. Although it forms a series of crystalline salts, it is a very weak monoacidic base. Its high nitrogen content and the formation of mercury and silver salts suggested to Barger and Ewins [26] a relationship to histidine, confirmed by a positive Pauly test. When boiled with 50% KOH, a Hofmann elimination took place with evolution of trimethylamine and formation of a yellow acid, $C_6H_6N_2O_2S$ (**12**), which was desulfurized by boiling with nitric acid to afford the known urocanic acid (**13**). Reduction of **13** with sodium and alcohol gave imidazole-4-propionic acid (**14**), identical with a synthetic sample.

The location of the sulfhydryl group was not established by this degradation, but its removal by oxidizing agents, its stability to KOH, and the weak basicity of ergothioneine all pointed to its attachment at C(2) of the imidazole ring. Desulfurization of ergothioneine by ferric chloride oxidation gave a trimethylhistidine betaine shown to be identical with hercynine [27, 28]. The location of the sulfur atom was proved conclusively by Akabori [29], who synthesized the ergothioneine degradation product **17**. Sodium amalgam reduction of diethyl glutamate (**15**) hydrochloride gave aldehyde **16,** which, in a general imidazole synthesis designed by Wohl and Marckwald, was immediately condensed with ammonium thiocyanate to form the thiolimidazole ring of **17**. This synthetic acid was identical with the acid formed by reducing **12** with sodium amalgam.

Ergothioneine itself was synthesized by Heath et al. [30], taking advantage of an earlier synthesis [31] of 2-histidinethiol (**20**) which had been achieved by opening and then reclosing the imidazole ring of histidine. Benzoylation of histidine methyl ester occurs with ring opening to give the tribenzoyl compound **18.** Hydrolysis of the enamide led to the aminoketone **19,** which underwent Wohl–Marckwald cyclization to **20** on treatment with sodium thiocyanate. In order to selectively methylate the side-chain nitrogen, the thiol group was first protected as the carbonate ester **21**; methylation with methyl iodide and silver oxide then gave ergothioneine, optically active although partially racemized.

Ergothioneine is a widespread fungal product and was found to be produced by at least nine common fungi, though it does not appear to be synthesized by bacteria [32]. It has been isolated from the mushrooms *Coprinus comatus* [8], *C. micaceus* [33], and *Limulus polyphemus* [21]. Moreover, it seems to be widespread in animal tissues and blood. Crystalline materials isolated from

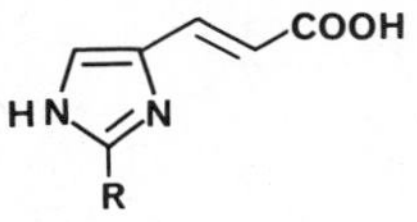

12 R = SH

13 R = H UROCANIC ACID

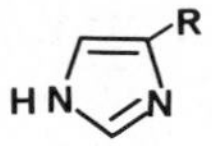

14 R = CH_2CH_2COOH

22 R = CH_2COOH

23 R = $CHOH-COOH$

24 R = CH_2CH_2OH

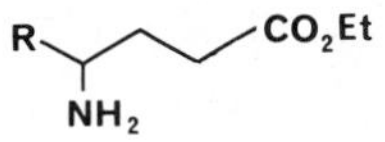

15 R = CO_2Et

16 R = CHO

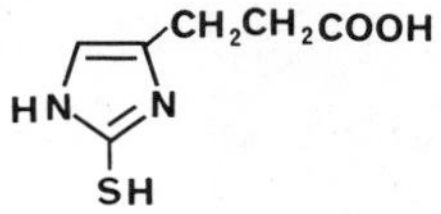

17

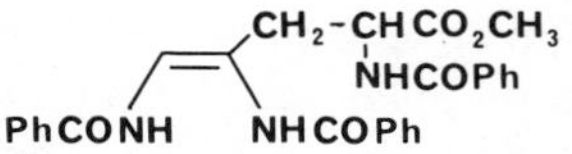

18

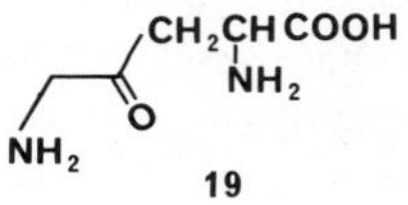

19

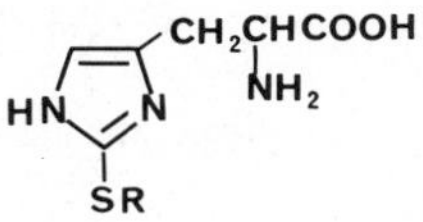

20 R = H

21 R = CO_2Et

pig and human blood and originally named "sympectothion" and "thiasine" [34] were shown [35, 36] to be identical with ergothioneine. It was detected in cattle blood and boar ejaculate [24] and is present in most tissues of rats, particularly the liver [37].

The biosynthesis of ergothioneine has been studied in the fungi *Claviceps purpurea* and *Neurospora crassa* with the use of radioactive precursors. Initial studies with [2-^{14}C]acetate gave a labeling pattern which indicated a biosynthetic pathway from histidine rather than glutamate [38]; the methyl group of acetic acid was primarily incorporated into C(2) of the imidazole ring as well as the *N*-methyl groups. Indeed, although histamine is not a precursor [39], histidine is incorporated with the imidazole ring and side-chain intact [40, 41]. 2-Histidinethiol is a poor precursor, but the sulfur can be provided by sulfate [32], thiosulfate [40], methionine [39], or cysteine [40]; the latter amino acid gives the highest incorporation and seems to be the immediate sulfur donor. Finally, the *N*-methyl groups originate from methionine [40]; use of methionine with the *S*-methyl doubly labeled with ^{14}C and ^{2}H shows that the methyl groups are transferred intact [42]. Though the incorporation of hercynine has not yet been tested, a plausible biosynthetic sequence is histidine → hercynine → ergothioneine.

Despite its wide occurrence, ergothioneine has no marked physiological action and no biological role has been postulated for it.

2.4. Zooanemonin (26)

This base was initially isolated from the "Noah's Ark" mussel, *Arca Noae* [43]. It was later found in the sea anemone *Anemonia sulcata* [44], the horny coral *Plexaura flexuosa* [45], two other anemones *Metridium dianthus* and *Condylactis gigantea* [45], and the sponge *Hippospongia equina* [46], where it co-occurs with hercynine and the acids **13**, **14**, **22** and **23**. Ackermann [44] established the empirical formula $C_7H_{10}N_2O_2$, prepared a series of crystalline salts, named it [47], and suggested that it was a dimethylbetaine of imidazole-acetic acid. This was confirmed by synthesis by methylating imidazole acetic acid **22** with dimethyl sulfate [48]. Ackermann and Janka proposed structure **25** for zooanemonin [48], but revised this to **26** when R. B. Woodward pointed out [45, 49] that methylation of **22** should occur 1,3 rather than 1,1 in order to preserve the aromatic character of the imidazole ring in the quaternary cation.

Zooanemonin has no physiological action when tested on crab legs [45].

2.5. Norzooanemonin (27)

This betanine alkaloid was isolated from the Caribbean gorgonian *Pseudoptero-gorgia americana*, found on the Florida Keys, by extracting the ground cortex with 95% ethanol [50]. It is an optically inactive colorless solid, mp 260–263° C, analyzing for $C_6H_8N_2O_2$, and forms a crystalline hydrochloride and chloro-aurate. The infrared spectrum showed the carboxylate ion at 1640 cm^{-1}, shifted to 1710 cm^{-1} upon conversion to the hydrochloride. The NMR spectrum showed two *N*-methyl groups at δ 3.87 and 3.99 and two aromatic protons at δ 7.69 and 8.63. The latter signal disappeared rapidly when a D_2O solution was made slightly basic with ammonia vapor, typical of the rapid base-catalyzed proton exchange at C(2) of imidazolium salts.

This information was sufficient to assign structure **27** to norzooanemonin, and this assignment was confirmed by the ready synthesis by methylation of imidazole-4-carboxylic acid with dimethyl sulfate below pH 9.

2.6. Murexine and Imidazole Carboxylic Acids

An early step in the metabolism of histidine is the loss of ammonia, catalyzed by the enzyme histidine ammonia lyase, to give the unsaturated acid urocanic acid (**13**). Several related acids have been found in nature. Imidazolepropionic acid (**14**) is a bacterial degradation product of histidine [51], and **13** and **14**, as well as the related imidazolylacetic acid **22**, imidazolyllactic acid **23**, and imidazolylethanol **24** occur in mushrooms [22, 52] and the snail *Arion empiricorum* [23]. Histidine metabolites have also been found in sponges [53].

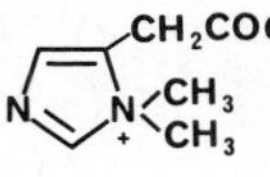

25

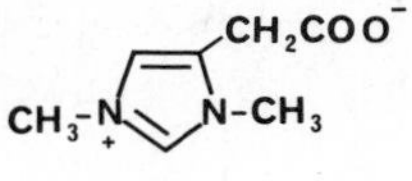

26 ZOOANEMONIN

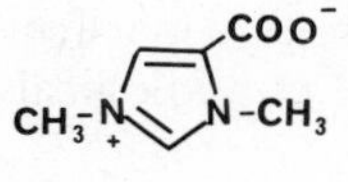

27 NORZOOANEMONIN

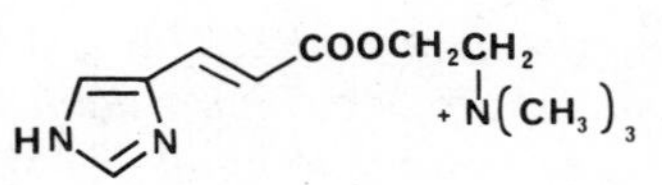

28 MUREXINE
30 DIHYDROMUREXINE
  (SATURATED SIDECHAIN)

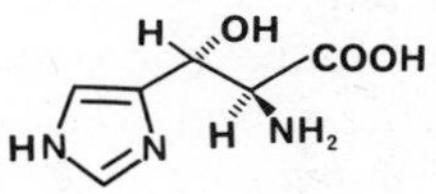

31

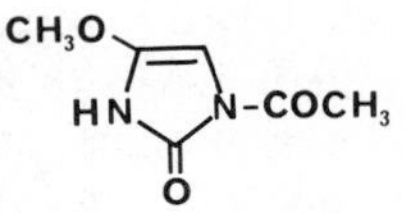

32 MONOSPERMINE

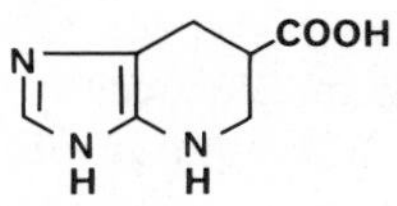

33

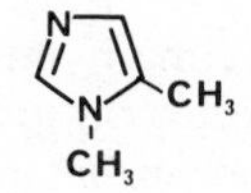

34

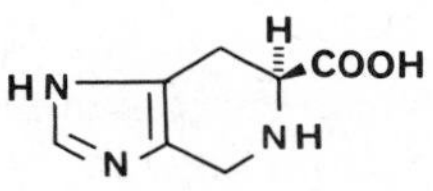

35 SPINACINE

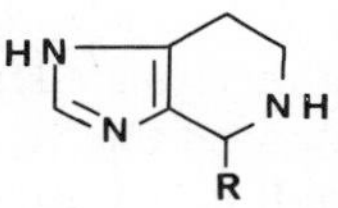

36 R = H   SPINACEAMINE
37 R = CH₃ 6-METHYL-SPINACEAMINE

Members of the family *Muricidae* (murex snails) are venomous predatory molluscs. Their hypobranchial glands have attracted the interest of zoologists for many years, for they secrete a chromogen (a sulfur-containing conjugate of 6-bromoindoxyl) which, in the presence of light and air, is oxidized to an intense purple dye, well known to the ancient world as Tryian purple. These glands also secrete a complex mixture of substances used in relaxing the prey (e.g. barnacles and bivalve molluscs) [54]. In an investigation of the hypobranchial glands of the Adriatic snail *Murex trunculus* [55], Erspamer and Benati found that an active compound, murexine (**28**), was present in surprisingly large amounts. Murexine is a quaternary ammonium salt, isolated as the picrate in amounts of 2–4 g/100 g tissue. It gives the Feigl hydroxylamine test for an ester, and was hydrolyzed by dilute acid or base to urocanic acid (**13**) and choline. The urocanylcholine structure **28** was confirmed by synthesis [56]: the chloroacid **11** was esterified with ethylene chlorohydrin to give ester **29,** and heating with trimethylamine effected both displacement and elimination to give murexine chloride.

Murexine has subsequently been identified in other members of the *Muricidae* [57]: *Urosalpinx cinereus* [58], *Murex fulvescens* [58], *M. brandaris* [59], *Thais lapillus* [58], *T. haemastoma* [60], *Concholepas concholepas* [59], and *Acanthina spirata* [61], accompanied in several cases [59, 60] by free urocanic acid, methyl urocanate and **14**. As a choline derivative, murexine exhibits intense nicotinic and curariform activity [55], and is an effective neuromuscular blocking agent [62, 63]. It has a marked hypertensive action and strongly stimulates respiration at low doses [63].

Dihydromurexine (**30**) occurs in large amounts (12–17 g/kg of tissue) in the hypobranchial gland of the marine snail *Thais haemastoma* [60, 64], and an *N*-methylmurexine was reported to occur in the north Pacific marine snail *Nucella emarginata* [61] and the Australian mollusc *Dicathais orbita* [65]. However, the compound from *D. orbita* was shown to be murexine, with the misassignment resulting from intermolecular methyl transfer in the mass spectrometer [66]. Unambiguous synthesis of the two possible *N*-methyl-murexines showed that neither was identical with the compound from *N. emarginata* [66]. Consequently the natural occurrence of an *N*-methylmurexine remains to be demonstrated.

2.7. L-*Erythro-β*-Hydroxyhistidine (31)

This amino acid occurs as a component of two peptide antibiotics, bleomycin A_2 [67] and phleomycin [68]. It was isolated as the crystalline hydrochloride, $[\alpha]_D^{28} + 40°$ (H_2O), by ion exchange chromatography of the hydrolysate. Analysis corresponded to a histidine possessing an extra oxygen atom. The NMR spectrum was consistent with the *β*-hydroxyhistidine structure, and the vicinal relationship of the hydroxyl and amino groups was proven by formation of an oxazolidone derivative upon treatment with phosgene in aqueous KOH. An X-ray single-crystal structure [69] on the hydrobromide salt revealed the *erythro* stereochemistry. The racemic form of the amino acid was synthesized by condensation of imidazolecarboxaldehyde with the copper salt of glycine, which gave a separable mixture of the two diastereomeric racemates [67].

2.8. Monospermine

The seeds of the leguminous plant *Butea monosperma* are reported to possess anthelmintic and antifertility properties. Extraction of the outer seed coat with alcohol, followed by methylation with diazomethane and chromatography, gave the crystalline alkaloid monospermine [70], $C_6H_8N_2O_3$, mp 161–163°C. The infrared spectrum showed the presence of N—H and *N*-acetyl groups along with another carbonyl function, and an *O*-methyl and vinyl proton were revealed by the NMR spectrum, leading to the imidazolidone structure **32**. Because of the unusual methylation step in the isolation procedure, it is not clear whether the *O*-methyl group is present in the natural product or introduced during isolation.

2.9. Spinacine and Spinaceamine

Spinacine is an optically active crystalline base, $C_7H_9N_3O_2$, mp 264° C, $[\alpha]_D - 169.9°$, first isolated from the liver of the shark, *Acanthias vulgaris* [71]; it was later found in the crab *Crangon vulgaris* [72]. Spinacine is resistant to boiling 20% sulfuric acid, but gives the Pauly diazonium test for imidazoles. From the molecular formula and the absence of an isolated double bond, Ackerman and Müller [73] deduced that a second ring must be present, and proposed the tentative structure **33**. Later Ackermann isolated 1,5-dimethyl-imidazole (**34**) by soda lime distillation of spinacine in a hydrogen atmosphere [74], and revised the structure [75] to **35**. This was confirmed by synthesis [75]; condensation of L-histidine with formaldehyde gave (−)-spinacine, with mp and optical rotation identical with those of the natural material. The structure has been confirmed by X-ray analysis [76, 77].

The related alkaloids spinaceamine (**36**) and 6-methyl spinaceamine (**37**) were isolated from skin glands of the South American frogs *Leptodactylus penta-dactylus labyrinthicus* and *Leptodactylus laticeps* [5, 78]. Structures were assigned [5] based on the resemblance to spinacine and confirmed by X-ray analysis [77]. Spinaceamine and its 6-methyl congener have been synthesized by Pictet–Spengler condensation of histidine with formaldehyde and acetaldehyde, respectively [79], and a number of analogs have been prepared by a similar route for pharmacological testing [80]. Both spinaceamine and 6-methyl-spinaceamine show antimicrobial activtiy against Gram-negative bacteria [81].

These three alkaloids, along with glochidicine (Section 3.3), are among the few examples of natural products derived by Pictet–Spengler-type condensations of histidine.

3. AMIDES OF HISTAMINE

3.1. Simple Amides of Histamine

A handful of simple amides of histamine occur naturally. *N*-Acetylhistamine (**38**) has been found in spinach leaves [6]. *N*-Cinnamoylhistamine (**39**) mp 178–179° C, was first isolated from the leaves of *Acacia argentea* and the bark of *A. polystacha*, both plants of the *Leguminosae* native to Queensland [82]; it also occurs in *A. spirorbis* [83]. The structure was quickly deduced from the NMR and mass spectra and by hydrolysis to histamine and cinnamic acid, and the alkaloid was synthesized from histamine and *trans*-cinnamoyl chloride. It was later found in the leaves and bark of the New Guinea tree *Glochidion philippicum* [84] and in the leaves and stems of *Argyrodendron peralatum*, a large tree of the family *Sterculiaceae* common to the tropical rain forests of northern Queensland [85].

3.2. Dolichotheline (40)

From the nonphenolic fraction of the chloroform extracts of *Dolichothele sphaerica*, a small cactus indigenous to southern Texas and northern Mexico, the alkaloid *dolichotheline*, $C_{10}H_{17}N_3O$, mp 130–131°C, was isolated in 0.7% yield. The spectroscopic properties and hydrolysis to histamine and isovaleric acid established the structure as *N*-isovalerylhistamine (40), confirmed by synthesis from histamine and isovaleric anhydride [86]. It has also been synthesized by aminolysis of 3-imidazolylthiazolidone-2-thione [87].

In studies of the biosynthesis [68, 69], it has been shown by specific labeling experiments that the plant incorporates either histamine or histidine into the imidazole portion and either isovaleric acid, valine, leucine, or (in poorer conversion) mevalonic acid into the acyl portion. The levels of incorporation suggest that leucine is the major source of the isovaleric acid. Rosenberg and his coworkers have shown that *D. sphaerica* will produce aberrant alkaloids when administered unnatural analogs of histamine and isovaleric acid [90–93]. 4(5)-Aminomethylimidazole (41) and the pyrazole 42 are incorporated in place of histamine, although the plant will not accept *N*-isopropylhistamine, while isocaproic and cinnamic acids, but not isobutyric and benzoic acids, can substitute for isovaleric acid. Histidine decarboxylase inhibitors suppress the incorporation of histidine in competition with 41 [93].

3.3. Glochidine and Glochidicine

Chromatography of the ethanol extracts of the leaves of *Glochidion philippicum*, a small tree of the family *Euphorbiaceae* growing in New Guinea and New Britain, yielded *N*-cinnamoylhistamine (39) and three interrelated imidazole alkaloids [94]. One of these, mp 115–117°C, has the formula $C_{15}H_{25}N_3O_2$ and is hydrolyzed by refluxing concentrated HCl to histamine and 4-oxodecanoic acid. Since the IR spectrum of this alkaloid shows both ketone (1710 cm^{-1}) and amide (1665 cm^{-1}) bands as well as NH absorption, it was considered to be the amide 43, N^α-4'-oxodecanoylhistamine.

The second alkaloid, glochidine, $C_{15}H_{23}N_3O$, mp 65–67°C, is an optically inactive base also hydrolyzed to histamine and 4-oxodecanoic acid by boiling conc. HCl. The IR spectrum of glochidine, however, shows a single carbonyl band at 1690 cm^{-1} characteristic of a five-membered lactam but no absorption due to N—H. The NMR spectrum shows two sharp singlets at δ 7.40 and 6.70 for the C—H protons of the imidazole ring along with the long side-chain containing a terminal methyl. Based on this evidence and the mass spectrum, structure 44 was assigned to glochidine.

The third alkaloid, glochidicine, mp 102–103°C, is isomeric with glochidine; it is also optically inactive, but unlike its two relatives is stable to refluxing HCl. The close relationship among the alkaloids was established by the formation of glochidicine when either glochidine or amide 43 was heated in dilute acetic acid. The infrared spectrum of glochidicine shows the five-membered lactam at 1670 cm^{-1} and an imidazole N—H at 3450 cm^{-1}. The

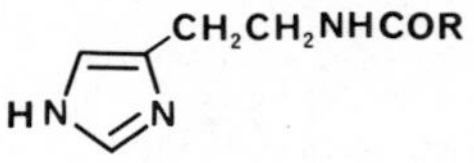

38 R = CH₃
39 R = TRANS -CH=CH-PH
40 R = -CH₂CH(CH₃)₂
          DOLICOTHELINE
43 R = -CH₂CH₂-CO-(CH₂)₅-CH₃

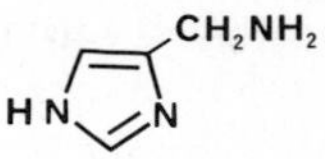

41

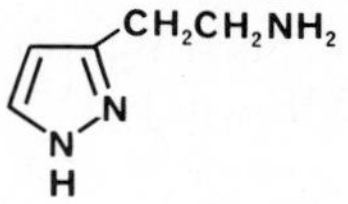

42

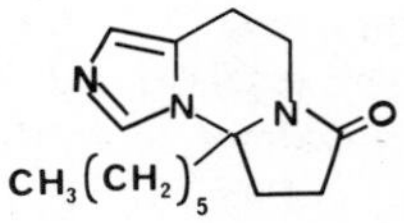

44 GLOCHIDINE 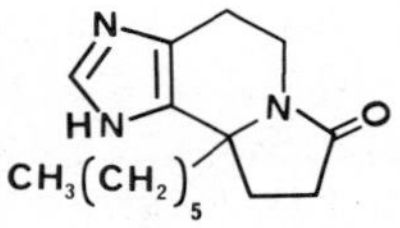   **45** GLOCHIDICINE 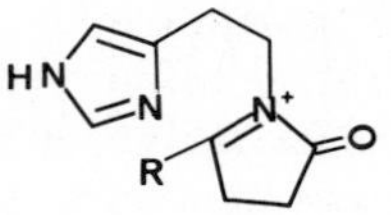   **46** R = -(CH₂)₅CH₃

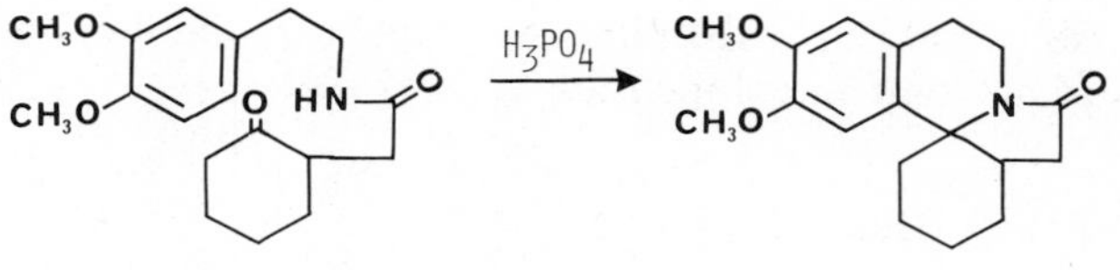

47   **48**

NMR spectrum shows only one imidazole C—H proton (δ 7.38), along with the imidazole N—H and the side-chain protons. This information establishes structure **45** for glochidicine.

It is remarkable that the attempted synthesis of amide **43** by treating histamine with 4-oxodecanoyl chloride in pyridine gives a mixture of all three alkaloids **43, 44** and **45**. Heating either **43** or **44** in aqueous acetic acid also gives the same mixture of all three, in which the proportion of **45** is increased by prolonged heating. It is clear that glochidine and glochidicine are alternate cyclization products of **43**, the former being an acid-labile nitrogen ketal structure and the latter the more stable Pictet–Spengler cyclization product. The cyclic bases are probably formed biosynthetically in just this way, and the ease with which equilibration takes place by ring opening and recyclization of intermediate **46** would account for the lack of optical activity in **44** and **45**. A good analogy for the cyclization of amide **43** to glochidicine is the acid-catalyzed cyclization [95] of amide **47** to the tetracyclic Erythrina skeleton **48**.

3.4.   Casimiroedine (49)

The seeds of *Casimiroa edulis*, a tree of the family *Rutaceae* found in Mexico and Central America, have long been used in native medicine as a hypnotic and sedative. In Mexico the tree is known as "Zapote blanco," and extracts have been used clinically for treatment of insomnia. The early history of chemical investigations has been summarized by Power and Callan [96], who were the first to isolate the alkaloid casimiroedine by extraction with hot alcohol, steam distillation of volatile material, and extraction of the aqueous residue with amyl alcohol. Casimiroedine, mp 226.5–228°C, is a levorotatory base to which the formula $C_{17}H_{24}N_2O_5$ was assigned. A series of investigations [97–99] in different laboratories in 1956 revised the empirical formula to $C_{21}H_{27}N_3O_6$ and established the presence of an N-methyl group, four active hydrogens belonging to four hydroxy groups which form a tetraacetate, but no C-methyl, O-methyl, or O-acetyl groups. The infrared spectrum showed absorption characteristic of an amide, and this result was confirmed by alkaline or acid hydrolysis to cinnamic acid and an amine, casimidine (**50**), $C_{12}H_{21}N_3O_5$. Casimiroedine could be hydrogenated to dihydrocasimiroedine, which was hydrolyzed to casimidine and hydrocinnamic acid.

The presence of an imidazole ring was first established by Djerassi et al. [100], by treating dihydrocasimiroedine with periodic acid followed by zinc. Hydrolysis of the resulting amide gave a base identified as N-methylhistamine (**2**). The nature of the presumed sugar residue was elucidated by liberating it from casimiroedine tetraacetate with HBr in acetic acid, then treating with silver carbonate followed by acetylation to afford β-D-glucose pentaacetate [101]. The final details of structure and stereochemistry were revealed by a single-crystal X-ray analysis [101, 102] of **50**. Casimiroedine was shown to be an imidazole nucleoside (**49**) with the imidazole ring in the β-configuration. The pure alkaloid has none of the spasmolytic, hypotensive, or hypnotic properties of the seeds of *C. edulis* [97], though it was subsequently shown to possess anticancer activity against lymphoid leukemia L1210 [103].

The total synthesis of casimiroedine was achieved by Panzica and Townsend [103]. Reaction of the chloromercury derivative (**51**) of 4-(2-chloroethyl) imidazole with 2,3,4,6-tetra-O-acetyl-α-D-glucosyl bromide (**52**) gave the nucleoside **53**. The crude product was treated directly with methanolic methylamine at 100° to replace the halogen by a methylamino group and deacetylate the glucose moiety; this step led to crystalline casimidine (**50**). The amide bond was then formed selectively by coupling **50** with *trans*-cinnamic acid in the presence of N-ethoxycarbonyl-2-ethoxy-1,2-dihydroquinoline; this reagent was chosen to avoid reaction with the sugar hydroxyls. The product was identical with natural casimiroedine and distinct from the coupling product with *cis*-cinnamic acid.

3.5. Zapotidine (54)

This sulfur-containing alkaloid was also isolated from the seeds of *Casimiroa edulis* by Syntex chemists [99] and assigned the formula $C_7H_9N_3S$. Zapotidine, mp 96–98°C, has an N-methyl group; the NMR spectrum also shows four aliphatic protons and singlets for two aromatic protons. The sulfur atom can be removed by boiling with ethanolic silver nitrate and sodium hydroxide

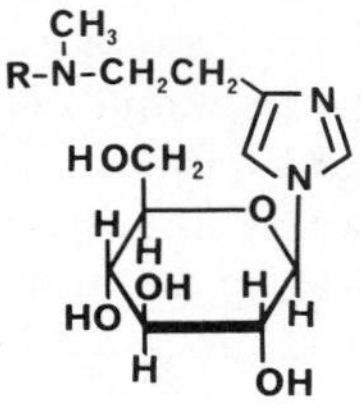

49 R = Ph⌒⌒CO—

CASIMIROEDINE

50 R = H

CASIMIDINE

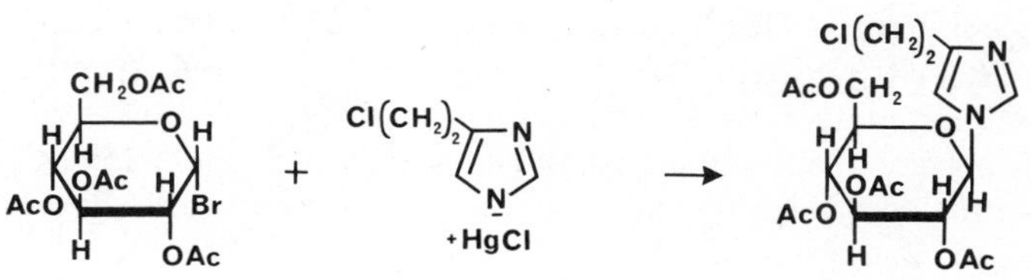

52   51   53

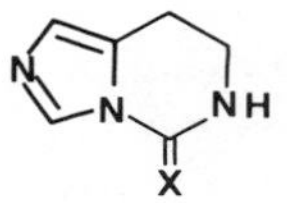

54   X = S   ZAPOTIDINE

55   X = O

to afford the urea **55**. LiAlH₄ reduction of zapotidine gave *N,N*-dimethyl-histamine (**3**), while hydrolysis with KOH gave *N*-methylhistamine (**2**). These results establish structure **54**, a rare example of a naturally occurring thiourea [104].

3.6. Anserine (60)

From an aqueous extract of goose muscles Ackermann et al. [105] isolated a dextrorotatory amino acid named anserine, $C_{10}H_{16}N_4O_3$. It formed a crystalline picrate, chloroaurate, and copper salt, and could be precipitated with mercuric salts, but was best isolated by precipitation with flavianic acid. Keil et al. [106–107] pointed out that the formula corresponded to a methyl derivative of carnosine (**56**), a well-known constituent of animal and human muscles, and identified β-alanine and a ring *N*-methylhistidine as the products of barium hydroxide hydrolysis of anserine. β-Alanine was also obtained in the form of its *N*-dinitrotolyl derivative (**57**) by treatment of anserine [107] (or carnosine [108]) with 2,4,5-trinitrotoluene (**58**) followed by acid hydrolysis. This nucleophilic aromatic substitution, in which the nitro group *ortho* and *para* to the two other nitro substituents is displaced by a free amino group [109], was a very early forerunner of the technique later developed by Sanger for peptide end-group analysis, and proved that the amide bond of anserine (and carnosine) is formed from the carboxyl group of β-alanine and the amino group of histidine. Anserine consequently must have either structure **59** or **60**.

Heating anserine with sodium carbonate afforded a dimethylimidazole. After some initial confusion this was identified [107, 110] as the 1,5-dimethyl compound **34,** identical with the dimethylimidazole obtained by Pyman from degradation of pilocarpine. Anserine must then be the 1,5-isomer **60**.

More recently Matthews and Rapoport [111] have provided an unambiguous NMR method for solution of a recurring problem in imidazole chemistry, the differentiation of 1,4- and 1,5-disubstituted isomers such as **59** and **60**; the method is based on the difference in cross-ring coupling constants of the aromatic protons in the two isomers.

Anserine is probably formed in nature by methylation of carnosine, and these two amides occur together in muscle of various animal and reptile species. The methyl group of anserine has been shown to be derived from the S-methyl group of methionine [112].

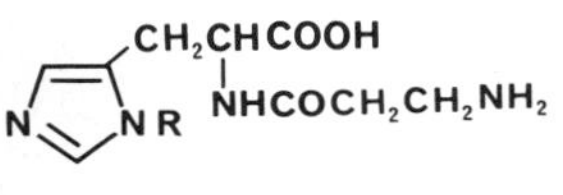

56 R = H CARNOSINE

60 R = CH₃ ANSERINE

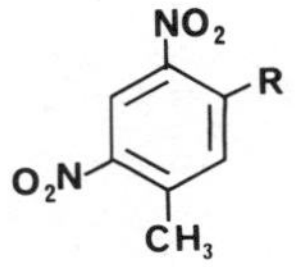

57 R = NHCH₂CH₂CO₂H

58 R = NO₂

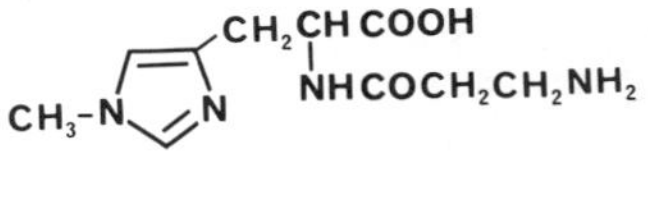

59

61 R = NH₂
62 R = NO₂ AZOMYCIN

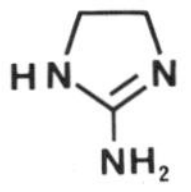

63

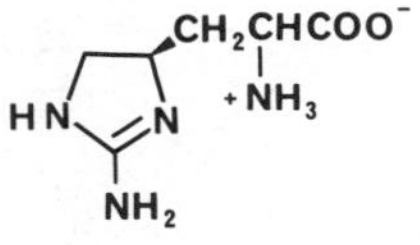

64 ENDURACIDIDINE

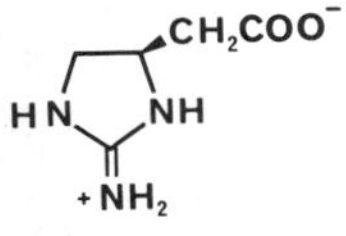

65

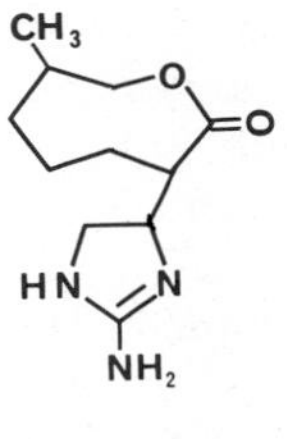

66 CHAKSINE

4. DERIVATIVES OF 2-AMINOIMIDAZOLE

4.1. 2-Aminoimidazole (61)

This simple imidazole is produced by *Streptomyces eurocidicus* [113], where it serves as a precursor of the antibiotic azomycin (2-nitroimidazole, 62). It has also been isolated from the marine sponge, *Reniera cratera* [114], where it constitutes 0.7% of the weight of the dry sponge. The first occurrence in higher plants was reported in methanol extracts of the seeds of the leguminous shrub, *Mundulea sericea*, a plant distributed throughout tropical Africa [115]. 2-Aminoimidazole is readily reduced to the dihydro derivative 63. It is resistant to hydrolysis by 6 *M* HCl at 100°C but is easily degraded by 1 *M* NaOH at room temperature. The reader should note that 2-aminoimidazole and most of the structures derived from it which are discussed in this section can exist in several tautomeric structures.

4.2. Enduracididine (64)

The seeds of another tropical legume, *Lonchocarpus sericeus*, which occurs mostly in the West Indies and tropical America, contain two derivatives of 4,5-dihydro-2-aminoimidazole [116]. One of these was identified as enduracididine (64), known earlier as a hydrolysis product of the polypeptide antibiotics enduracidin [117], from *Streptomyces fungicidicus*, and minosaminomycin [118]. In these cases the vigorous acid hydrolysis of the peptides epimerized the α-carbon, leading to mixtures of enduracididine and alloenduracididine; the absolute configuration at both asymmetric centers has been established [119]. It seems likely that enduracididine originates, not from histidine, but rather by cyclization of γ-hydroxyarginine, which also occurs in the seeds of *L. sericeus*. The second alkaloid in this species is 2-amino-4,5-dihydro-imidazolyl-4-acetic acid (65), the only report of its occurrence in nature.

The 2-amino-4,5-dihydroimidazole moiety also appears in the alkaloid chaksine (66) of *Cassia absus* [120]. In this case the skeleton seems to have been formed by combination of a monoterpene fragment with guanidine.

4.3. The *Alchornea* Alkaloids

4.3.1. Alchornine. A novel group of hexahydroimidazo[1,2-a]pyrimidine alkaloids has been isolated from the bark and leaves of *Alchornea javanensis*, a small tree of the family *Euphorbiaceae* growing in the New Guinea rain forest [121]. The major alkaloid, alchornine (67), mp 134–135°C, is a dextrorotatory base of the formula $C_{11}H_{17}N_3O$. The oxygen atom was shown to be present as an amide by the IR band at 1692 cm^{-1} and by LiAlH$_4$ reduction to the oxygen-free base 68. The structure was largely deduced from the NMR spectrum, which shows (1) a single exchangeable N—H proton, (2) two nonequivalent

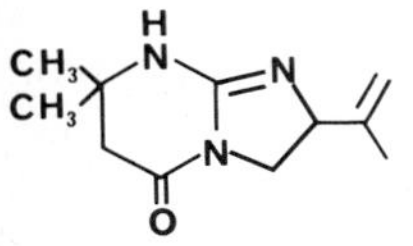

67 ALCHORNINE

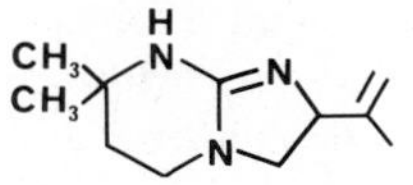

68

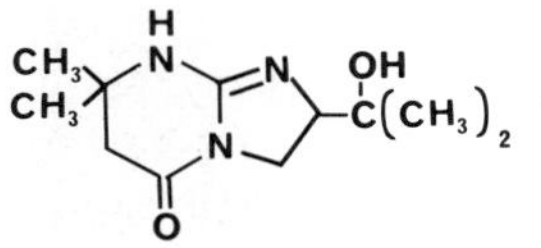

69

$(CH_3)_2C=CH-CO-X$

70a X = OH

70b X = NH$_2$

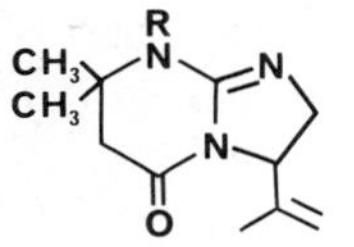

71 R = H
ISOALCHORNINE

72 R = $COCH=C(CH_3)_2$
ALCHORNIDINE

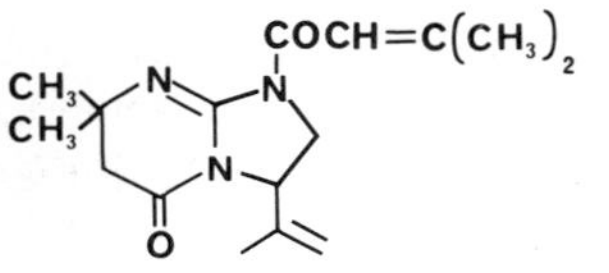

73

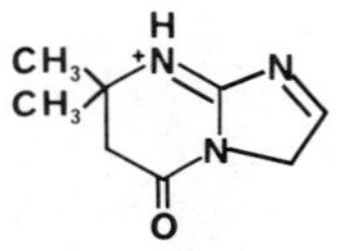

74

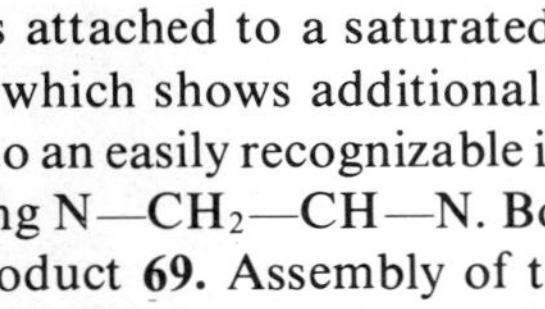

$(CH_3)_2C=CHCH_2$
$N-C-NH_2$
$(CH_3)_2C=CHCH_2$ $\overset{\|}{NH}$

75

$(CH_3)_2C=CHCH_2NH$
$C=N-CH_2$
$(CH_3)_2C=CHCH_2NH$ $CH=C(CH_3)_2$

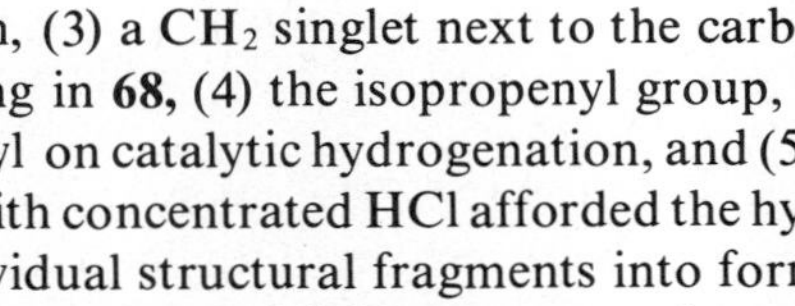

76

methyls attached to a saturated carbon, (3) a CH$_2$ singlet next to the carbonyl group, which shows additional coupling in **68**, (4) the isopropenyl group, converted to an easily recognizable isopropyl on catalytic hydrogenation, and (5) the grouping N—CH$_2$—CH—N. Boiling with concentrated HCl afforded the hydration product **69**. Assembly of the individual structural fragments into formula **67** was done on the basis of double resonance NMR spectra and analysis of the mass spectral fragmentation pattern. It is, of course, possible that alchornine has the tautomeric structure with the C=N double bond in the six-membered ring and the NH in the five-membered ring.

4.3.2. Alchornidine. A second alkaloid from *A. javanensis*, alchornidine, mp 96–97°C, has the formula C$_{16}$H$_{23}$N$_3$O$_2$. The IR spectrum shows the absence of NH and OH groups. Treatment with ethanolic KOH gives alchornine (**67**) and

2,2-dimethylacrylic acid (**70a**), but alchornidine is not simply the *N*-acyl derivative of alchornine, for mild hydrolysis with dilute acetic acid (under conditions to which alchornine is stable) gives isoalchornine, mp 137–138°C, a levorotatory isomer of **67**. Treatment of isoalchornine with methanolic KOH gives a solution from which nothing can be isolated by dilution with water and extraction with chloroform, but when the solution is acidified and warmed briefly, then neutralized, alchornine can be extracted in high yield with chloroform. The NMR spectra make it clear that alchornine (**67**) and isoalchornine (**71**) differ only in the location of the isopropenyl group; rearrangement of **71** to **67** involves hydrolysis of the lactam to a carboxylate salt which can reclose to either of two nitrogens, alchornine representing the thermodynamically more stable isomer in this equilibrium. Alchornidine is consequently either **72** or **73**, with **72** favored because of the chemical shift difference between the *gem*-dimethyls near the amide carbonyl in the NMR spectrum [121]. Additional support for the location of the isopropenyl sidechain comes from the mass spectra; the major ion (m/z 166) in the spectrum of **67** corresponds to structure **74**, derived by loss of the isopropenyl sidechain; this m/z 166 ion is relatively small in the mass spectrum of isoalchornine (**71**), reflecting the lower stabilization of a positive charge by an amide nitrogen.

A clue to the biosynthetic origin of alchornine and alchornidine comes from the presence of the two guanidine alkaloids **75** and **76** from *A. javanensis*. The alchornine family is apparently derived by condensation of isoprene units (from mevalonic acid) with guanidine. 3,3-Dimethylacrylamide (**70b**) is also found in the extracts, but this may be an artifact resulting from ammonolysis of **72** during neutralization with ammonia in the isolation procedure.

4.3.3. Alchorneine (77).

The related *Alchornea* species, *A. floribunda*, grows in the Congo and Cameroun regions of Africa, where it is prized for its aphrodisiac properties. The powdered root of *A. floribunda* has been classified as a sympathicosthenic drug because it increases the sensitivity of the sympathetic nervous system to adrenaline. Although the aphrodisiac action was initially attributed to the presence of yohimbine, subsequent studies showed the absence of yohimbine in *A. floribunda* bark. An extract of the alkaloids has an intense vagolytic action and strongly inhibits intestinal peristalsis in the anesthetized dog.

Alchorneine [122], the major alkaloid of *A. floribunda* and *A. hirtella*, is extracted as a colorless oil which crystallizes on standing several weeks, mp 43°C. Analysis shows the formula $C_{12}H_{19}N_3O$. The mass spectrum shows the molecular ion at m/z 221, the base peak at m/z 206 (loss of methyl), and other strong peaks at m/z 175 and 134, due to further loss of methoxyl and C_3H_5 (isopropenyl). The infrared spectrum shows a strong band at 1690 cm^{-1}, attributed to the C=N bond of a guanidine, and absorption characteristic of an N—OCH$_3$ group.

The NMR spectrum is particularly informative, clarifying the environment of all 19 protons. It shows an *O*-methyl, two nonequivalent methyls on a tetrasubstituted carbon, and a methyl attached to a C=C double bond. Four vinyl hydrogens are present; two of these show the characteristic AB pattern and coupling constant ($J = 7.5$ Hz) of cis hydrogens on a disubstituted double bond, and two multiplets at δ 5.03 and 5.16, which become doublets on irradiation of the olefinic methyl, indicate the methylene portion of an isopropenyl group. The remaining three protons between δ 2.93 and 4.23 are all adjacent to heteroatoms and form an AMX pattern at 300 MHz: two doublets of doublets at δ 3.13 and 4.05 and a triplet at δ 3.48. The downfield signal is ascribed to a methine proton located between the N—OCH$_3$ and isopropenyl groups.

Alchorneine titrates as a monoacidic base, but it is inert to acetic anhydride and NaBH$_4$. It forms a crystalline methiodide (**78**) which is transformed to a secondary amine (**80**), C$_{13}$H$_{23}$N$_3$O$_2$, on treatment with aqueous alkali. The change corresponds to the hydrolysis of an immonium salt: $R_2C{=}N^+R_2'$ + $H_2O \longrightarrow R_2C{=}O + HNR_2'$, in this case the opening of the pyrimidine ring. The new amino group is secondary, as evidenced by the presence of an exchangeable N—H and the formation of N-acetyl and N-methyl derivatives, but the latter does not form a quaternary salt due to the steric hindrance of the *gem*-dimethyl groups. These two methyls appear as a singlet in the NMR specrum of **80**, as does the new N-methyl. Finally, the carbonyl group of **80** absorbs at 1740 cm^{-1} in the IR, characteristic of an N-substituted 2-imidazolidone.

Acid hydrolysis of **80** hydrolyzed the enamine grouping to give the imidazolidone **81**, C$_7$H$_{12}$N$_2$O$_2$, a neutral liquid with a strong IR band at 1740 cm^{-1}. The NMR spectrum of **81** was similar to that of **80** except for the disappearance of the amine side-chain. Hydrogenation over Adams' catalyst gave the saturated 4-isopropyl-3-methoxy-2-imidazolidone (**82**), whose structure was confirmed by synthesis from ethyl 4-methyl-3-oxopentanoate (**84**). After protecting the ketone function of **84** as the dioxolane, the ester group was converted to the carbamate **85** by Curtius rearrangement. The ketone function was converted to the methoxime and reduced with diborane to give **86,** which cyclized to racemic **82** on heating with potassium ethoxide. This synthesis proved the presence of the unusual N—OCH$_3$ group in alchorneine, located the isopropenyl side-chain, and led to structure **77** for the alkaloid. This assignment was subsequently confirmed by X-ray crystal structure analysis [123] of the methobromide **79**.

Alchorneine and its degradation products are all optically active, and the absolute configuration at the asymmetric carbon was established by synthesis of the imidazolidone **83** from L-valine (**87**). Reduction of valinamide (**88**) with lithium aluminum hydride gave diamine **89**, which was cyclized to imidazolidone **83**, $[\alpha]_D - 22°$, by condensation with phosgene. On the other hand, reduction of degradation product **82** with lithium in ethylamine gave the enantiomer, $[\alpha]_D + 22°$. Since the imidazolidone synthesized from L-valine is (S), the natural alkaloid must have the (R) configuration.

4.3.4. Isoalchorneine (90).

Isoalchorneine, an isomer of alchorneine, is a minor alkaloid in the bark and roots of *A. floribunda*, but is the major alkaloid of the leaves [122]. Like alchorneine, it shows IR absorption peaks for C=N, C=CH$_2$, and N—OCH$_3$, but it completely lacks the large M-15 peak in the

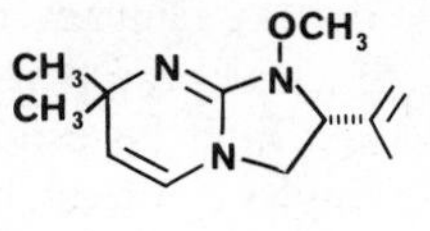

77 ALCHORNEINE

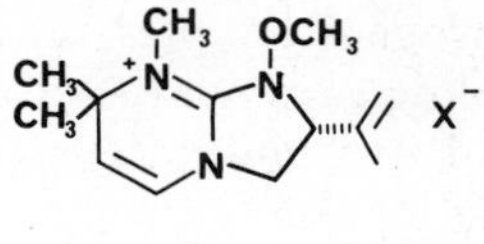

78 X = I
79 X = Br

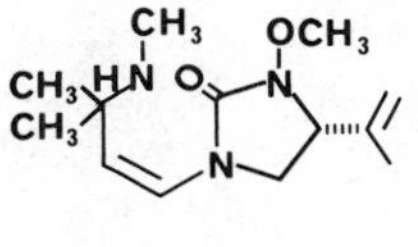

80 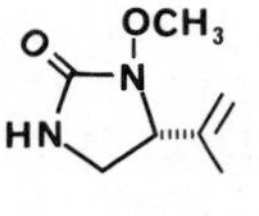   **81** 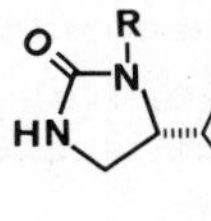   **82** R = OCH$_3$
                                                    83 R = H

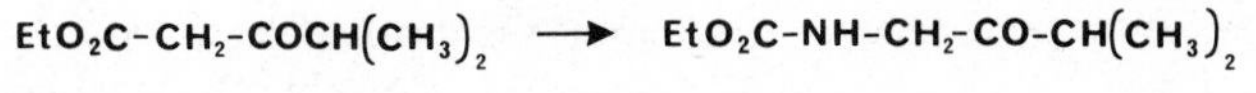

$$EtO_2C-CH_2-COCH(CH_3)_2 \longrightarrow EtO_2C-NH-CH_2-CO-CH(CH_3)_2$$

84   **85**

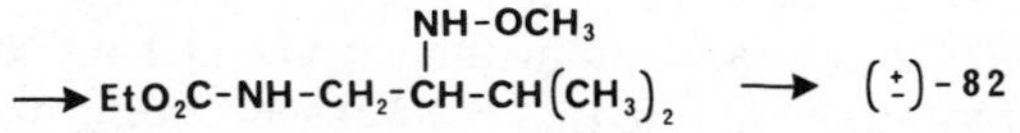

$$\longrightarrow EtO_2C-NH-CH_2-\overset{\overset{\textstyle NH-OCH_3}{|}}{CH}-CH(CH_3)_2 \longrightarrow (\pm)-82$$

86

$$H_2N\diagdown\underset{R\diagup\overset{|}{H}}{CH(CH_3)_2}$$

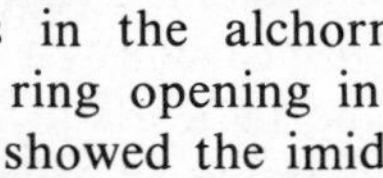

87 R = COOH   L-VALINE
88 R = CONH$_2$
89 R = CH$_2$NH$_2$

mass spectrum, and the NMR spectrum shows *two* isopropenyl groups and six hydrogens α to heteroatoms. Based on these data, structure **90,** a derivative of imidazo(1,2-*a*)imidazole, was suggested.

Paralleling reactions in the alchorneine series, the methiodide of iso-alchorneine underwent ring opening in alkali to give the secondary amine **91,** C$_{12}$H$_{23}$N$_3$O$_2$, which showed the imidazolidone carbonyl at 1720 cm^{-1} and an N—CH$_3$ in the NMR. Amine **91** underwent methylation to a quaternary salt, degraded by Hoffmann elimination to the conjugated diene **92.** Acid hydrolysis of the enamine gave **81,** which was reduced to (*R*)-**83;** these latter compounds were identical, including optical rotations, with those obtained by degradation of alchorneine.

The optical rotation of isoalchorneine is nearly zero and it shows no circular dichroism absorption at normal concentrations, although alchorneine exhibits strong positive *CD* peaks at 258 and 232 nm. Khuong-Huu et al. [122] argued that isoalchorneine must have the (2*R*,6*S*) configuration, and that the near sym-

90 ISOALCHORNEINE

91

92

93

94 ALCHORNEINONE

metry is responsible for the low rotation. In agreement with this conclusion, reductive cleavage of the N—O bond gave **93,** an optically inactive compound whose tautomeric forms are enantiomers.

4.3.5. Alchorneinone (94). A third alkaloid from the leaves of *A. floribunda* is alchorneinone [122], a neutral liquid, $C_{12}H_{20}N_2O_3$. The mass spectrum shows large fragmentation peaks at *m/z* 209 (M—OCH₃) and 169 [M-71; loss of (CH₃)₂CHCO]. Two carbonyl peaks are evident in the infrared spectrum, an imidazolidone at 1740 cm⁻¹ and a ketone at 1725 cm⁻¹. The NMR spectrum reveals isopropenyl *and* isopropyl groups, O—CH₃, three protons α to nitrogen, and a methylene singlet at δ 4.1. The presence of a ketone was confirmed by formation of an oxime and reduction to a secondary alcohol. Structure **94** was deduced from these data and by analogy with alchorneine.

These three alkaloids of *Alchornea fluoribunda* can also be considered to originate in nature from the union of guanidine with two isoprenoid units.

4.4. Oroidin, Sceptrin, and Dibromophakellin

This novel family of related 2-aminoimidazoles containing a dibromopyrrole unit has been isolated from marine sponges.

4.4.1. Oroidin (95). Methanol extraction of *Agelas oroides*, a sponge collected in the bay of Naples, gave oroidin (2.3%) along with three simple dibromopyrroles **96a–c** [124]; oroidin has also been isolated from two species of *Axinella, A. damicornis* and *A. verrucosa* [125]. The infrared absorption at 3360, 3240, 1685, and 1570 cm⁻¹ is characteristic of a secondary amide, and the NMR spectrum shows two aromatic protons at δ 6.84 and 6.45, as well as an ABX$_2$ system indicative of the partial structure —CH=CH—CH$_2$—N. Acetylation of oroidin afforded an *N*-acetyl derivative, C$_{13}$H$_{13}$Br$_2$N$_5$O$_2$, which on catalytic hydrogenation gave the dihydro derivative **97**. Hydrolysis of either *N*-acetyloroidin or **97** with 20% KOH gave dibromopyrroles **96a** and **96b,** while acid hydrolysis of **97** gave 2-amino-4(or 5)-(3-aminopropyl)-imidazole (**99**).

Oroidin is clearly an amide formed by linking acid **96a** with one of the amino groups of **98**. Since hydrolysis gave amide **96b**, Forenza et al. [124] argued that the nitrogen of **96b** must be joined to a hydrolyzable group, and consequently proposed structure **101** for oroidin.

This assignment was questioned by several authors [126, 127], who argued that the alternative amide structure **95** would bear a closer relationship to the structure of a related sponge product, dibromophakellin (**103**). Garcia et al. [126] confirmed this proposal by an unambiguous synthesis of *N*-acetyldi-hydrooroidin (**97**). The amide of imidazolepropionic acid (**14**) was dehydrated with POCl$_3$ to the nitrile; azo coupling with *p*-bromobenzenediazonium chloride, followed by reduction with stannous chloride, introduced the 2-amino group to give the key intermediate **100**. Acetylation, followed by reduction of the

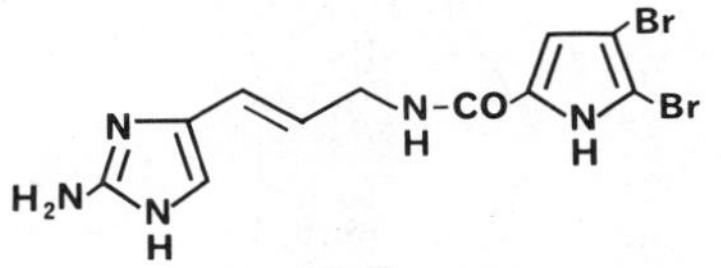

95 Oroidin

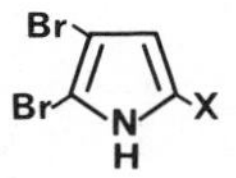

96a X = COOH
b X = CONH$_2$
c X = CN
d X = COCCL$_3$

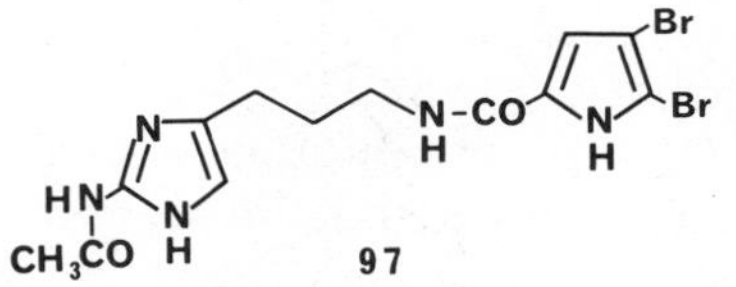

97

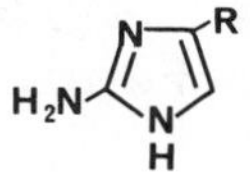

98 R = –CH=CH–CH$_2$NH$_2$
99 R = –(CH$_2$)$_3$NH$_2$
100 R = –(CH$_2$)$_2$CN

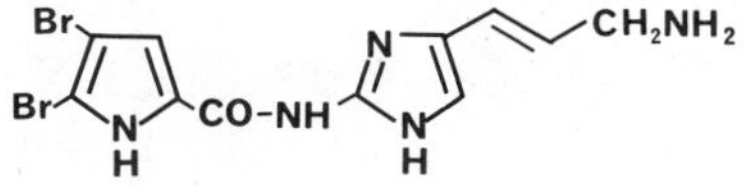

101

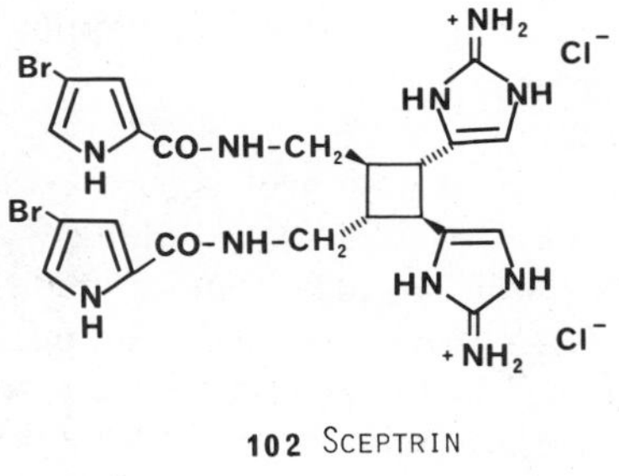

102 Sceptrin

nitrile group and reaction with the acid chloride of **96a,** provided **97.** Catalytic hydrogenation of **97** removed the bromine atoms, giving an amide identical with the hydrogenation product of *N*-acetyloroidin. Independent confirmation of structure **95** later came from an X-ray single-crystal structure analysis [128].

4.4.2. Sceptrin (102). From the methanol extracts of the Carribean sponge *Agelas sceptrum*, there was isolated oroidin (0.5%) and a new antimicrobial constituent, sceptrin (2.1%). Sceptrin, isolated as the bis-hydrochloride salt, $C_{22}H_{24}Br_2N_{10}O_2 \cdot 2HCl$, mp 215–225°d., exhibited spectral properties which indicated that it was a symmetrical dimer of 2-debromooroidin. The structure and relative configuration were determined to be **102** by single-crystal X-ray analysis [128]. Sceptrin is a head-to-head dimer of debromooroidin, but the biosynthesis cannot be regarded as simply a photodimerization process because (a) insufficient light is present at the 20–30 m depth where *A. sceptrum* was found, and (b) sceptrin is optically active, $[\alpha]_D - 7.4°$, while the debromo-oroidin precursor is achiral.

Sceptrin exhibited antimicrobial activity against a variety of microorganisms [128].

4.4.3. Dibromophakellin (103). This alkaloid was isolated from the marine sponge *Phakellia flabellata*, found on the Great Barrier Reef [129]. It is isomeric with oroidin, and is accompanied by the monobromo analog **104.** Both mono-bromo- and dibromophakellin show mild but broad spectrum antimicrobial activity [129, 130]. Catalytic hydrogenation of **103** and **104** removed the halogen atoms to afford the parent base phakellin (**105**). The infrared spectra of all three phakellins showed an amide band at 1675 cm^{-1} as well as the presence of NH and methylene groups and a heteroaromatic ring, and the UV spectra exhibited maxima around 233 and 281 nm suggestive of a pyrrole ring with a carbonyl group at the 2-position. The NMR spectrum of phakellin (**105**) showed three NH protons exchangeable in D_2O, a 1,2-disubstituted pyrrole, a highly deshielded methine proton, and a trimethylene chain linked to an amide nitrogen. The presence of a guanidine moiety was deduced from the mass spectral fragments, although dibromophakellin is a much weaker base (pK$_a$ 7.7) than most guanidines (pK$_a \sim$ 13.4). Heating dibromophakellin with dilute HCl afforded compound **106** with concomitant elimination of one end of the cyclic guanidine unit.

The spectroscopic data summarized above, along with the ^{13}C-NMR spectrum, led to structures **103–105** for the phakellins, and these were confirmed by single-crystal X-ray analysis of the *N*-acetyl derivative of dibromophakellin [129]. The X-ray structure revealed that the five-membered ring containing the guanidine group is almost perpendicular to the plane of the other three rings and, moreover, has a twisted conformation which resists protonation to a planar, resonance-stabilized, guanidinium ion, accounting for the unusually low basicity. The nitric acid oxidation product **107** has a planar imidazoline ring and has the basicity of a normal guanidine.

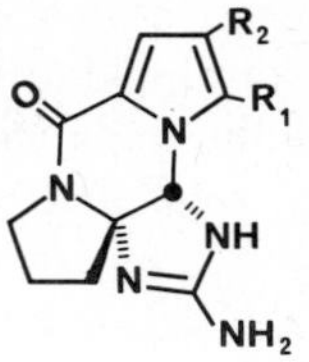

103 $R_1 = R_2 = Br$
         DIBROMOPHAKELLIN
104 $R_1 = H, R_2 = Br$
105 $R_1 = R_2 = H$

106

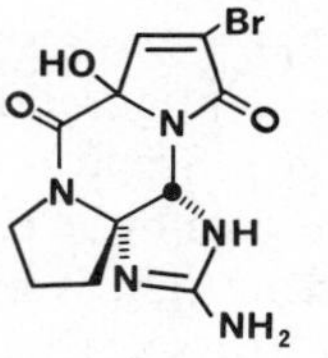

107

SCHEME I

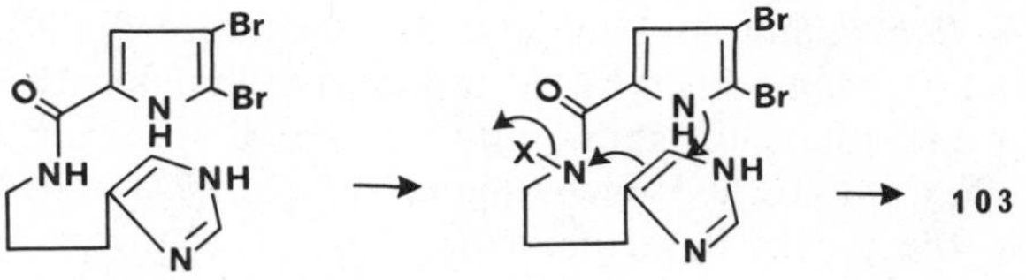

108 DIHYDROOROIDIN

Dibromophakellin is obviously biogenetically related to oroidin (**95**), and it has been suggested [129] that it may be formed in the sponge by oxidative cyclization of dihydrooroidin (**108**) as shown in Scheme 1. Foley and Büchi have accomplished an imaginative biomimetic synthesis of dibromophakellin modeled on this hypothesis [131]. These authors first developed an alternative synthesis of dihydrooroidin (Scheme 2), beginning with the commercially available amino acid L-citrulline (**109**), which was converted to aldehyde **110** by sodium amalgam reduction of the ethyl ester. The 2-aminoimidazole ring of **111** was constructed by condensation of **110** with cyanamide and cyclization with 15% HCl. Hydrolysis of the urea and acylation with **96d** then gave **108**.

Verifying the chemical plausibility of the biosynthetic hypothesis, treatment of dihydrooroidin with bromine in acetic acid precipitated an unstable hydrobromide salt, believed to have structure **112**. Reaction with potassium *t*-butoxide converted it quantitatively to racemic dibromophakellin. Interestingly, the intermediate salt **112** cyclizes in an alternative fashion when warmed in DMSO or DMF in the absence of strong base to give **113**, the product of electrophilic substitution on the *carbon* atom of the pyrrole ring.

SCHEME 2

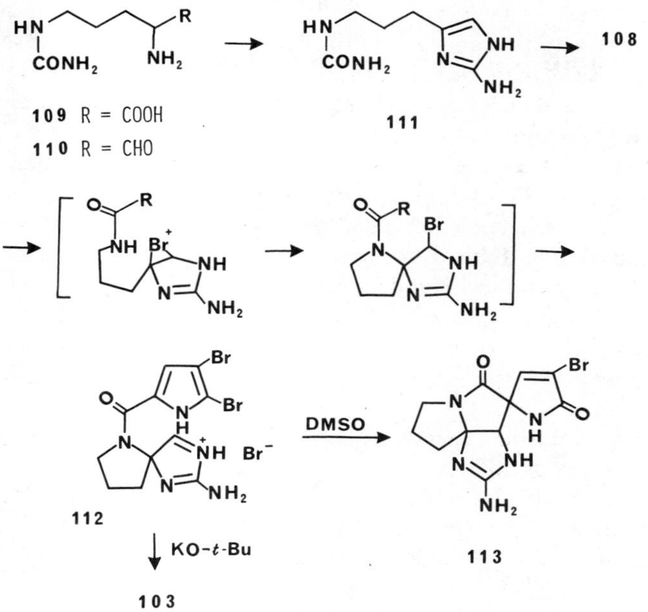

109 R = COOH
110 R = CHO

111

108

112

↓ KO-*t*-Bu

103

DMSO

113

4.5. Zoanthoxanthins

The zoanthoxanthins are a group of basic yellow-colored pigments, exhibiting a powerful fluorescence in ordinary light, which occur in marine coelenterates of the class *Anthozoa*, related to anemones and stony corals. They are characterized by possessing either of two novel isomeric tetraazacyclopentazulene ring systems. The properties of the known members are given in Table 1.

Table 1.   Zoanthoxanthins

			mp (°C)	UV (CH₃OH)	UV (HCl)
Zoanthoxanthin		$C_{13}H_{16}N_6$	275–276	427, 293	392, 293, 259
Parazoanthoxanthin	A	$C_{10}H_{10}N_6$	>310	404, 295	381, 284
	B	$C_{11}H_{12}N_6$		404, 294	384, 287
	C	$C_{12}H_{14}N_6$		412, 302	392, 297
	D	$C_{12}H_{14}N_6$	303–304 d.	415, 306	394, 300, 255
	E	$C_{14}H_{18}N_6$	>310	429, 306, 296	404, 309, 260
	F	$C_{14}H_{18}N_6$	>310	391, 309, 293	403, 306, 261, 245
Epizoanthoxanthin	A	$C_{13}H_{16}N_6$	191–192	419, 310, 295	396, 303
	B	$C_{14}H_{18}N_6$		412, 310	396, 308
Pseudozoanthoxanthin		$C_{12}H_{14}N_6$	>310	421, 367, 307, 281	399, 348, 335, 290
3-Norpseudozoanthoxanthin		$C_{11}H_{12}N_6$	>310	400, 360, 296, 253	398, 332, 287
Paragracine		$C_{13}H_{16}N_6$	258–262 d.	428(sh), 409, 373, 316, 305(sh), 252, 230(sh)	
Pseudozoanthoxanthin A		$C_{10}H_{10}N_6$	280 d.	402, 363, 296, 245	395, 357, 288, 230

4.5.1. Zoanthoxanthin (114). Zoanthoxanthin, $C_{13}H_{16}N_6$, was isolated by dipping the Mediterranean zoanthid *Parazoanthus axinellae* in ethanol and passing the extracts through a cationic ion exchange resin to absorb the pigments, which were then eluted with ammonia [132]. It crystallizes as yellow needles and shows a characteristic greenish-blue fluorescence in daylight.

The NMR spectrum shows four methyl groups, identified as N—CH_3 (δ 4.26), —$N(CH_3)_2$ (δ 3.62), and C—CH_3 (δ 3.42), the latter confirmed by Kuhn–Roth oxidation. The spectrum also shows two aromatic protons and a signal at δ 8.05 for two exchangeable protons, suggesting a primary amino group.

Zoanthoxanthin is stable to the action of acid, alkalai, and various oxidizing agents, but forms a monoacetyl derivative with acetic anhydride. Diazotization in HCl leads to a chloro compound $C_{13}H_{14}N_5Cl$ (**115**), which regenerates zoanthoxanthin on treatment with ammonia. Alkaline hydrolysis of **115** gives $C_{13}H_{15}N_5O$, which exists predominantly as the carbonyl tautomer **116**, IR 1722 cm^{-1}.

These data established that zoanthoxanthin has a tricyclic ring system composed of nine carbons and four nitrogens, substituted by CH_3, —$N(CH_3)_2$, and a primary amino group located next to a ring nitrogen. Its low solubility and unreactivity made further chemical investigation unrewarding, and the structure was elucidated by X-ray single-crystal analysis of the chloro derivative **115**. Zoanthoxanthin (**114**) contains the previously unknown 1,3,5,7-tetraazacyclopent[*f*]-azulene ring system, a nearly planar structure possessing two 2-amino-imidazole units [132].

4.5.2. Parazoanthoxanthins A, B, and C. Four additional pigments, differing only in the number and location of *N*-methyl groups, were also isolated from *P. axinellae* [133]. Parazoanthoxanthin A, $C_{10}H_{10}N_6$, shows only the *C*-methyl group in the NMR spectrum and can consequently be assigned structure **117**.

Parazoanthoxanthins B and C, whose molecular formulas indicate the presence of one and two *N*-methyls, respectively, were isolated in too small quantity to be identified [133]. Parazoanthoxanthin B is believed to be either the 1-methyl or 3-methyl compound.

4.5.3. Parazoanthoxanthin D. The NMR spectrum of this pigment, $C_{12}H_{14}N_6$, showed the —$N(CH_3)_2$ and C—CH_3 of zoanthoxanthin but no additional methyl groups. The two aromatic protons appeared as a pair of doublets centered at δ 8.82 and 8.91. When treated with sodium nitrite in HCl it was converted to a mixture of a desaminochloro compound, $C_{12}H_{12}N_5Cl$, and its hydrolysis product, $C_{12}H_{13}N_5O$, which also exists in solution as the ureido tautomer **118**, IR 1708 cm^{-1}. This information was sufficient to assign structure **119** to parazoanthoxanthin D. These assignments of structure to parazoanthoxanthins A and D were supported by isolating them from the mixture of demethylation products formed when zoanthoxanthin (**114**) was heated in boiling 40% hydrobromic acid [133].

4.5.4.   Parazoanthoxanthins E and F.   These two pigments are trace constituents of *P. axinellae*, isolated in too small amount to permit chemical degradation. Since the identified pigments differed only in the number and position of *N*-methyl groups, Cariello et al. [134] carried out methylation studies of the more available members in the hope of identifying the minor constituents. With parazoanthoxanthin D **(119)** as the substrate, methyl iodide in DMSO, or diazomethane in the presence of BF$_3$ etherate, effected methylation at only nitrogens 1 and/or 3, the ring nitrogens of the right-hand ring. Methylations of **114** or **119** initiated with NaNH$_2$ in liquid ammonia methylated only the 2-amino group. In all cases methylation was restricted to the nitrogens of the guanidine moiety in the right-hand ring (ring A), suggesting that rings B and C form an "aromatic" diazaazulene system.

Comparison with the various methylation products showed that parazoanthoxanthin E had structure **120** and that parazoanthoxanthin F was compound **121**.

4.5.5.   Epizoanthoxanthins A and B.   From *Epizoanthus arenaceus*, another zoanthid common in the Bay of Naples, was isolated a different set of four fluorescent pigments, two with the zoanthoxanthin skeleton and two with an isomeric skeleton [135]. Epizoanthoxanthin A, $C_{13}H_{16}N_6$, proved to be identical with **122,** one of the methylation products of **119.**

Epizoanthoxanthin B appeared, from its UV and NMR spectra, to be a ring *N*-methyl derivative of **122.** It was assigned structure **123** on the basis of (a) the chemical shift of the *C*-methyl group (δ 3.27) is characteristic of a zoanthoxanthin unsubstituted at N(3), and (b) methylation with CH$_3$I—NaNH$_2$ gave **124,** identical with a derivative obtained by methylation of **119.**

4.5.6.   Pseudozoanthoxanthin.   The principal nitrogenous metabolite of *E. arenaceus* is pseudozoanthoxanthin, $C_{12}H_{14}N_6$, yellow prisms that exhibit an intense blue fluorescence at 446 nm. The NMR spectrum shows two *N*-methyls at δ 4.05 and 4.61, a *C*-methyl at δ 3.03, two aromatic protons at δ 8.21, and three D$_2$O-exchangeable NH groups.

Refluxing 6 *N* HCl converted pseudozoanthoxanthin to the urea derivative **126,** $C_{12}H_{13}N_5O$, IR 1724 cm^{-1}, which was diazotized in aqueous acid to the hydroxyurea **127.** If the order of these steps was reversed, diazotization of pseudozoanthoxanthin gave a mixture of the chloro compound **128** and a hydrolysis product **129,** $C_{12}H_{13}N_5O$. Although **129** showed no carbonyl absorption in the IR, it was hydrolyzed by refluxing acid to the urea **127.** In the NMR spectrum of **127,** the aromatic protons appear as an AB quartet ($J = 11$ Hz), indicating that they are adjacent.

The UV spectrum of pseudozoanthoxanthin appeared to rule out the known zoanthoxanthin skeleton, and the specific possible structure **130** was eliminated when it was found that methylation products of urea **126** were not identical with known methylation products of zoanthoxanthin. Finally, since the *C*-methyl of pseudozoanthoxanthin did not experience a nuclear Overhauser

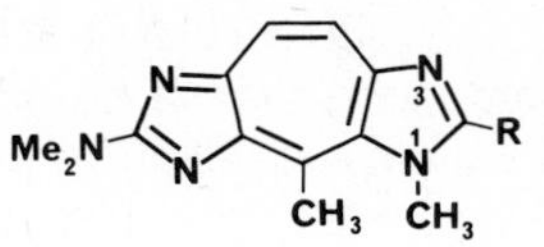

114 R = NH$_2$ ZOANTHOXANTHIN
115 R = CL
120 R = NHCH$_3$ PARAZOANTHOXANTHIN E

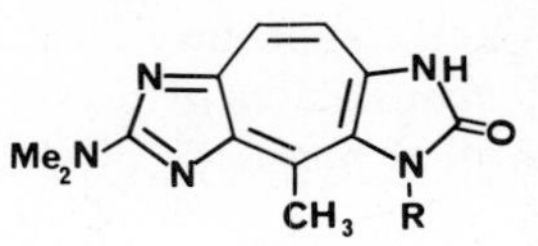

116 R = CH$_3$
118 R = H

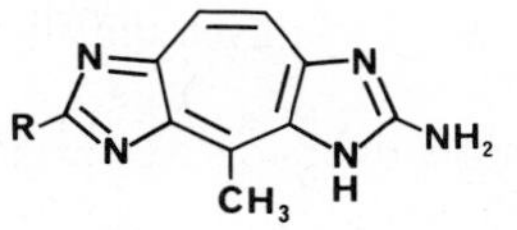

117 R = NH$_2$ PARAZOANTHOXANTHIN A

119 R = NME$_2$ PARAZOANTHOXANTHIN D

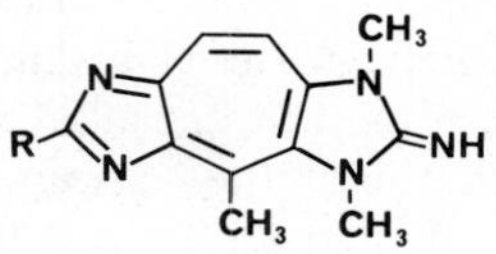

121 R = N(CH$_3$)$_2$
     PARAZOANTHOXANTHIN F
130 R = NH$_2$

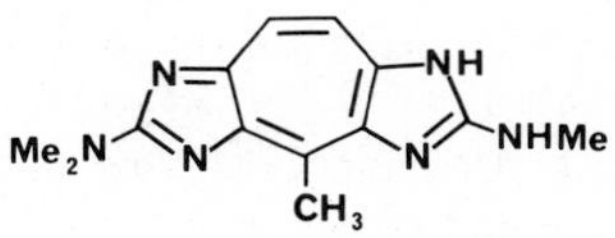

122 EPIZOANTHOXANTHIN A

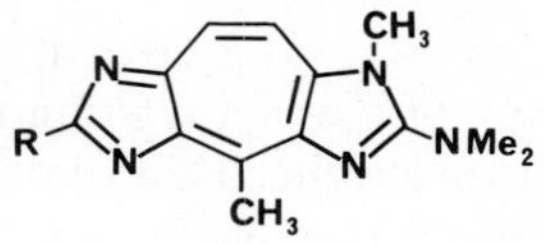

123 R = NHCH$_3$
     EPIZOANTHOXANTHIN B
124 R = NME$_2$

effect when the N^3-methyl was irradiated, formula **125** was left as the only reasonable structure for pseudozoanthoxanthin [135]. It contains the isomeric 1,3,7,9-tetraazacyclopent[*e*]azulene skeleton.

4.5.7. 3-Norpseudozoanthoxanthin.   The fourth metabolite of *E. arenaceus* has the formula $C_{11}H_{12}N$ and lacks the *N*-methyl signal at δ 4.05 in the NMR spectrum. Structure **131** was confirmed by formation from pseudozoanthoxanthin by demethylation in boiling HBr [135].

4.5.8. Paragracine.   Another member of the pseudozoanthoxanthin family was isolated from the methanol extracts of *Parazoanthus gracilis*, a zoanthid collected at Sagami Bay in Japan [136]. Paragracine, $C_{13}H_{10}N_6$, shows the characteristic UV spectrum of this group. The NMR spectrum reveals N—CH$_3$, —N(CH$_3$)$_2$, and C—CH$_3$ groups, along with two neighboring aromatic protons and two exchangeable NH groups. X-ray analysis of the hydrobromide elucidated the structure as **132**. Paragracine possesses papaverine-like activity, with antihistamine and antiacetylcholine action.

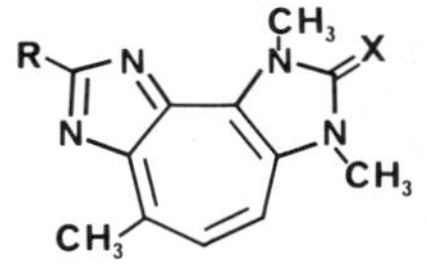

125 R = NH₂, X = NH
    PSEUDOZOANTHOXANTHIN
126 R = NH₂, X = O
127 R = OH, X = O
128 R = CL, X = NH
129 R = OH, X = NH

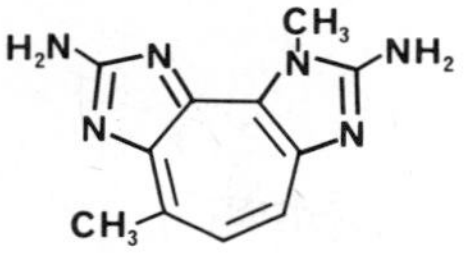

131 3-NORPSEUDOZOANTHOXANTHIN

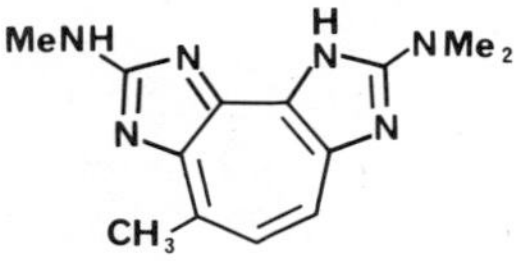

132 PARAGRACINE

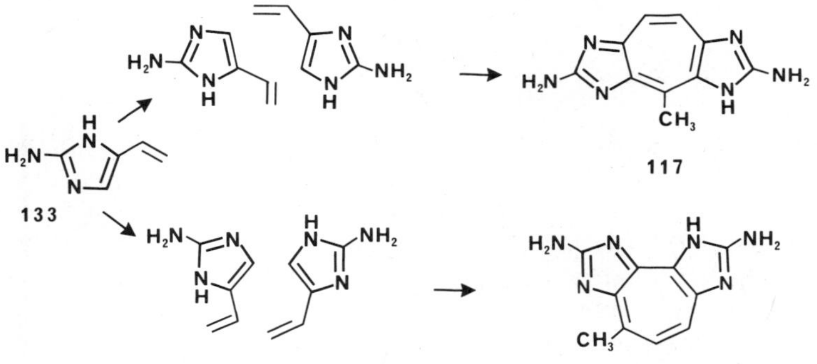

117

133

134

4.5.9. Synthesis. Cariello and coworkers [133] suggested that the basic zoanthoxanthin skeleton probably arises biogenetically from two C_5N_3 units derived from arginine. Büchi et al. devised an imaginative and short synthetic route [137] which embodies this principle of coupling two such C_5N_3 units. The scheme takes the form of generating 2-amino-4-vinylimidazole (133) or its equivalents in a strong acid medium which catalyzes the 6 + 4 cycloaddition of two units to give both zoanthoxanthin skeletons.

Four variations of the monomeric imidazole were prepared and dimerized: the alcohol 137, its *N*, *O*-dibenzoylderivative 138, the isomeric alcohol 141, and the vinyl derivative 145. Reduction of lactone 135 with sodium amalgam gave hemiacetal 136, which was condensed with cyanamide and cyclized to the 2-aminoimidazole 137. Condensation of threonine ethyl ester with benziminoethyl ether hydrochloride gave oxazoline 139, which was reduced to aldehyde 140 with diisobutylaluminum hydride. Hydrolysis and condensation with cyanamide gave the imidazole 141. Finally, the oxadiazole 142 was

135   136   137 R = H
                                138 R = COPH

139 R = CO$_2$ET
140 R = CHO

141

142 R = H
143 R = CH=CHCOCH$_3$

144   145

condensed with 4-methoxybut-3-en-2-one to give **143**, which underwent the oxadiazole–imidazole rearrangement on treatment with sodium hydride to afford **144**. Reduction with NaBH$_4$ and dehydration with *p*-toluenesulfonic acid gave the protected vinylimidazole **145**.

Each of the four monomers **137, 138, 141**, and **145**, when heated in conc. HCl or H$_2$SO$_4$, gave a mixture of two colored pigments identified as the naturally occurring parazoanthoxanthin A **(117)** and its isomer in the pseudo series, named pseudozoanthoxanthin A **(134)**. The product ratio depended on the precursor and acid used, and these oxidative dimerizations could be accelerated by added ferric chloride. The yield of **117** reached as high as 50% when **137** was coupled with FeCl$_3$ in HCl.

This remarkable synthesis is postulated to involve the 6 + 4 cycloaddition of two units of the protonated vinyl imidazole **133**. One molecule may exist in the diazafulvene form **146**, to which the second unit adds as a diene, paralleling the addition of 1-diethylaminobutadiene to 6-phenylfulvene [138]. This orientation would lead to the pseudozoanthoxanthin precursor **147**. Cycloaddition in the opposite orientation would afford the parazoanthoxanthin precursor **148**.

5. THE JABORANDI ALKALOIDS

The most important imidazole alkaloid is pilocarpine, which occurs together with a group of related alkaloids in the leaves of South American *Pilocarpus* species, a family of shrubs of the order *Rutaceae* popularly known as jaborandi.

SCHEME 3

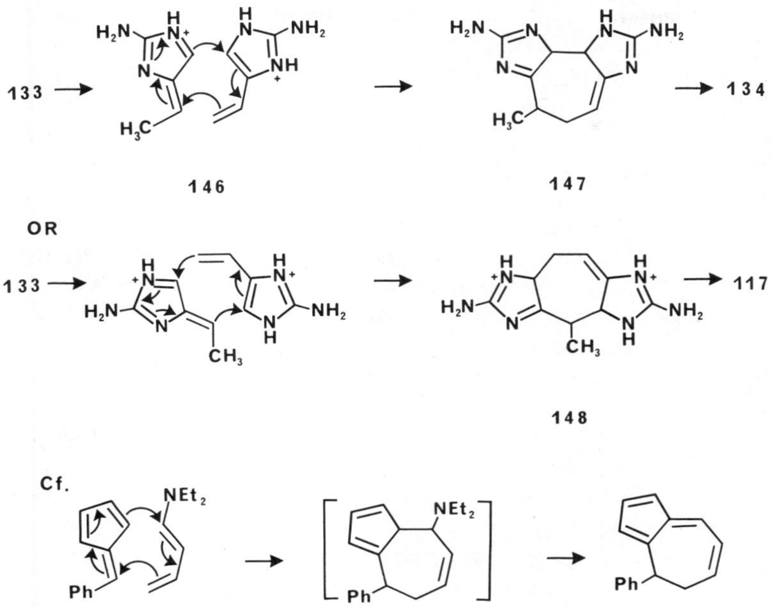

The jaborandi alkaloids have attracted great interest because of their valuable pharmacological properties. The drug first sent to Europe in 1874 was isolated from *P. jaborandi* and *P. pennatifolius*, but the most important source since 1895 has been *P. microphyllus*, the so-called Maranham jaborandi. The major alkaloid, which constitutes 0.8% of the dry weight of the plant, is pilocarpine, a widely studied peripheral stimulant of the parasympathetic system. The main use of pilocarpine today is topically in a 0.5–1.0% solution as a miotic in the treatment of glaucoma.

5.1. The Pilocarpine Group

5.1.1. Pilocarpine and Isopilocarpine.

Pilocarpine (**149a**) was isolated independently by Hardy [139] and Gerrard [140] in 1875; it is accompanied by a stereoisomer, isopilocarpine (**149b**) and the *N*-desmethyl compound, pilocarpidine (**150**). The easy isomerization of pilocarpine and the uncommonly difficult separation of the alkaloids led to conflicting claims of the presence of other alkaloids, most of which proved to be mixtures of **149** and **150**, as well as arguments over names and structures; the confused early history is summarized by Jowett [141] and also in the excellent 1953 review by Battersby and Openshaw [1]. Accurate descriptions of the three alkaloids were given in 1897 by Petit and Polonovski [142] and in 1900 by Jowett [141], who prepared

a number of crystalline salts and established the molecular formula of pilocarpine as $C_{11}H_{16}N_2O_2$. Pilocarpine is a colorless, monoacidic base, mp 34°C,
$[\alpha]_D$ + 100.5°. It is largely isomerized to isopilocarpine, $[\alpha]_D$ + 42.8°, when
distilled or warmed with bases, or when the acid salts are heated. Even neutralization of pilocarpine hydrochloride with silver oxide at 25°C results in substantial isomerization to isopilocarpine [172]. Both alkaloids dissolve in hot
aqueous alkali, giving isolable salts of a hydroxy acid, and the bases are regenerated when acidified, showing the presence of a lactone ring. The lactone
hydrolyzes slowly even in water, though hydrolysis is catalyzed by acid or
base [143, 144]. No reaction occurs with acetyl or benzoyl chloride, but alkylation with methyl or ethyl iodide gives quaternary salts, showing that both
nitrogens are tertiary. The lactone ring reacts normally with ammonia and
simple amines and is reduced by LiAlH₄ to a diol [145].

The most informative degradative experiments proved to be permanganate
oxidation and alkali fusion. Oxidation of pilocarpine and isopilocarpine,
studied by Jowett [146] and Pinner [147], gave acetic and propionic acids and
two homologous lactonic acids, isopilopic (151) and homoisopilopic (152)
acids; pilocarpine and isopilocarpine gave identical products in this oxidation
due to the alkali-induced isomerization. The carbon skeleton of homoisopilopic
acid was established by potassium hydroxide fusion to the racemic triacid
153, identical with a sample synthesized by Jowett [148]. This result allowed
only several plausible structures for the permanganate oxidation products,
and the observation that isopilopic acid is stable to heat, ruling out malonic acid
derivatives, left only structures 151 and 152. These have been confirmed by
synthesis (vide infra).

The nature of the nitrogen-containing residue was revealed soon thereafter.
Pinner and Schwarz [149], noting the similarity to known imidazole derivatives
in the formation and reactions of quaternary salts, the destruction of the ring
by oxidation, and behavior on bromination, proposed the correct imidazole
structure (149) for pilocarpine. Jowett confirmed this by isolating 1-methyl-
imidazole, a dimethylimidazole, and an N-methyl-C-amylimidazole from soda–
lime distillation of isopilocarpine [146, 150]. Pyman [151] proved by degradation
that the dimethylimidazole was the 1,5-isomer 34, and the structure of the
amyl derivative 154 was established by synthesis [152]. The identification of
these imidazoles confirmed structure 149 for pilocarpine.

Pinner [149] maintained for some time that pilocarpine and isopilocarpine
were structural isomers, differing in the location of the N-methyl group in
the imidazole ring, while Jowett argued that the alkaloids were stereoisomers,
supporting this view by showing that an equilibrium mixture, in which isopilocarpine predominates, could be reached by starting with either isomer [153].
Conclusive proof was provided by Langenbeck [154], who first showed that the
quaternary methosalts of pilocarpine and isopilocarpine were not identical, as
they must be if the alkaloids are 1,4- and 1,5-positional isomers. Then he provided the first case in which derivatives of the isomers lacking the imidazole ring
were still different, by using a degradation method avoiding epimerizing

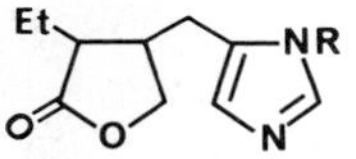

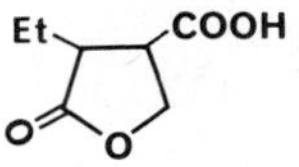

149 R = CH₃

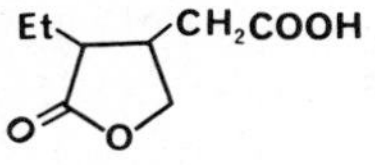

  a PILOCARPINE AND

  b ISOPILOCARPINE

150 R = H PILOCARPIDINE

151

152

Et-CH-CH-COOH
HOOC CH₂R

153 R = COOH
156 R = BR

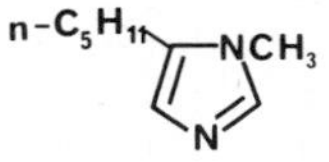

154

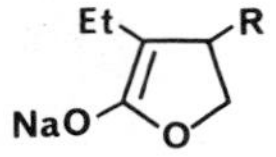

155

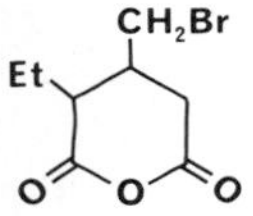

157 CIS
158 TRANS

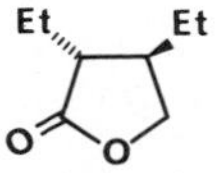

159

Et-CH-CH-R
HOCH₂ CH₂OH

160 R = ET
161 R = CH₂OH

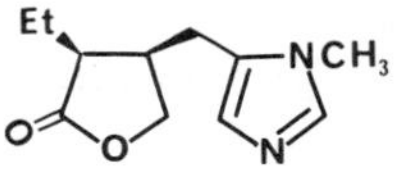

162 (2S, 3R) PILOCARPINE

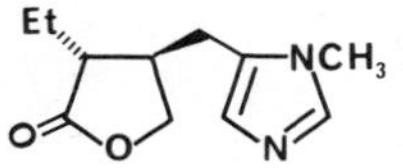

163 (2R, 3R) ISOPILOCARPINE

conditions. Ozonolysis of pilocarpine gave homopilopic acid *N*-methylamide, mp 104°C, while ozonolysis of isopilocarpine gave homoisopilopic acid *N*-methylamide, mp 53°C. These results proved that pilocarpine and isopilocarpine were stereoisomers. The base-catalyzed isomerization takes place via the enolate **155,** which can be isolated as a sodium salt whose hydrolysis leads to a mixture of the two epimers [155].

The direction of isomerization suggests that the more stable iso series has the *trans* configuration. Zavialov [156] induced further evidence; heating pilopic and isopilopic acids **(151)** with hydrobromic acid gave the stereoisomeric bromo-succinic acids **156. 156b** formed an anhydride much more easily than **156a,** suggesting that the anhydride from **156a** is the *cis* isomer and that from **156b** the *trans*. Moreover, Zavialov was able to convert homoisopilopic acid **(152)** to the less stable homopilopic acid by a similar technique, opening the lactone with HBr, forming the anhydride **157,** isomerizing it to the more stable *trans* **158,** and hydrolyzing this to homopilopic acid. The stereochemical assignment was put on an unambiguous basis by Hill and Barcza [157], who prepared lactone **159** by mixed Kolbe electrolysis of homoisopilopic and acetic acids.

Reduction with LiAlH$_4$ gave the dextrorotatory diol **160,** whose optical activity proves the trans configuration of **159** and homoisopilopic acid. A later X-ray analysis of pilocarpine hydrochloride confirmed that it is the *cis* isomer [158].

The absolute configuration was established by four independent correlations with compounds of known configuration: (a) synthesis of (+) diol **160** from (2*R*,3*R*)-2,3-diethylsuccinic acid [157]; (b) the synthesis [159] of a degradation product of strychnine from the enantiomer of lactone **159**; (c) preparation of the (−) triol **161,** the LiAlH$_4$ reduction product of (+)-isopilopic acid, from (*R*)-butane-1,1,2-tricarboxylic acid [157]; and (d) synthesis of **161** by a 13-step sequence from glucose [160]. These correlations all agree in assigning pilocarpine the (2*S*,3*R*) configuration **162** and isopilocarpine the (2*R*,3*R*) structure **163.**

5.1.2. Pilocarpidine (150).

Pilocarpidine was first isolated from jaborandi by Merck [161]; it is an oily base, C$_{10}$H$_{14}$N$_2$O$_2$, which forms crystalline salts. The formula and lack of an *N*-methyl group suggested that it was the lower homolog of pilocarpine, and this was eventually confirmed by methylating it to pilocarpine [162]. The methylation also gives an isomer, neopilocarpine, formed by methylation of the other imidazole nitrogen. Pilocarpidine is isomerized by sodium ethoxide to isopilocarpidine, and so belongs to the *cis* series.

5.1.3. Synthesis.

Synthetic efforts began with the pilopic acids **(151).** Preobrashenski [163] reduced diethyl α-ethyl-α′-formylsuccinate **(164)** to the alcohol **165;** the latter cyclized on heating, and hydrolysis gave a mixture of racemic pilopic and isopilopic acids. Other ways of introducing the —CH$_2$OH group include condensation of cyanoester **166** with formaldehyde or alkylation of the related malonic ester with methyl chloromethyl ether [164], followed by hydrolysis and cyclization; an alternative method involves reduction of the cyano group of **166** followed by diazotization [165]. Racemic isopilopic acid was resolved with strychinine [163], while pilopic acid could be resolved with brucine or cinchonine [166].

Synthesis of homoisopilopic acid **(152)** was initially achieved by a tedious homologation of isopilopic acid [167]; the more efficient Arndt–Eistert homologation was used to prepare homopilopic acid without isomerization [168]. Dey [169] published a more direct route which began with Michael addition of malonic ester to the unsaturated ester **167,** followed by ethylation and hydrolysis to diacid **168.** Refluxing **168** with HBr gave a mixture of racemic homopilopic and homoisopilopic acids.

The principal problem plaguing synthesis of pilocarpine is the easy isomerization of lactone intermediates to the iso series in the presence of alkaline or acid reagents or on heating, making separations necessary to realize even low yields of pilocarpine. One logical solution to this problem, developed by Chumachenko [170] and DeGraw [171], uses as the critical step the hydrogenation of the unsaturated lactone **171** to afford pure *cis* homopilopic acid.

The most practical route to **171** begins with air oxidation of an ethanol solution of furfural in sunlight to give the butenolide **169**. Michael addition of diethyl ethylmalonate to **169** gave diester **170** in 99% yield. Heating this compound in a mixture of hydrobromic and acetic acids gave **171** in 88% yield, and hydrogenation over a rhodium catalyst afforded (±)homopilopic acid which can be resolved with α-methylbenzylamine [172].

Several methods to construct the imidazole ring from the carboxyl of homopilopic acid were developed by Preobrashenski and his colleagues. In the first [173], (+)-homopilopic acid was converted to the diazoketone **172** and thence to chloroketone **173**. Gabriel synthesis gave the aminoketone **174**, which was cyclized with potassium thiocyanate (cf. Section 2.3) and desulfurized with ferric chloride to afford (+)-pilocarpidine (**150**). In an alternative approach

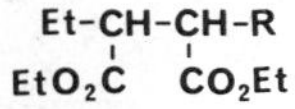

Et-CH-CH-R
EtO₂C   CO₂Et

164 R = CHO
165 R = CH₂OH
166 R = CN

ROCH₂CH=CH-CO₂Et

167 R = Et or Ph

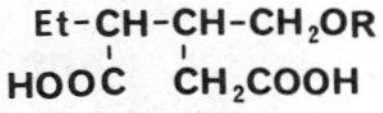

Et-CH-CH-CH₂OR
HOOC   CH₂COOH

168

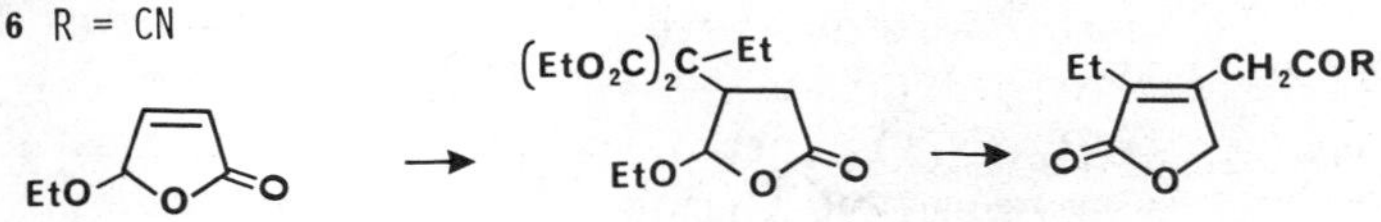

169 → **170** → **171** R = OH **178** R = CH₂PHTH

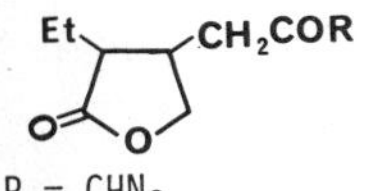

172 R = CHN₂
173 R = CH₂CL
174 R = CH₂NH₂
175 R = CH₂OH
176 R = CHO
177 R = CH=CH-PH

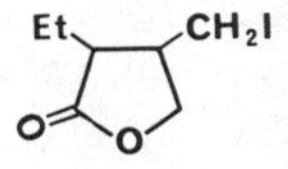

179

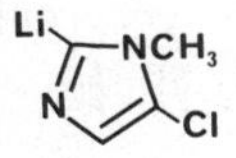

180

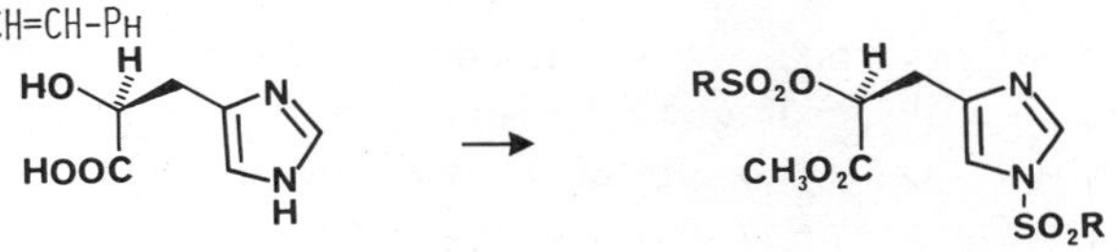

181 → **182**

183 → **184**

[147], diazoketone 172 was converted by treatment with acetic acid and hydrolysis to hydroxyketone 175. The latter was converted in one step to pilocarpidine by treatment with copper acetate, formaldehyde, and ammonia. The presumed intermediate ketoaldehyde 176 was prepared separately by Dey [169] by ozonolysis of the benzylidene derivative 177 and cyclized to pilocarpidine.

A practical improvement in the synthesis was the efficient preparation of 174 by treatment of homopilopyl chloride with acetamido-di-*t*-butyl malonate [171]. The problem of nonselective methylation of pilocarpidine, a drawback in all previous syntheses, was solved [175] by the use of methyl isothiocyanate in the Wohl–Marckwald synthesis with 174, leading unequivocally to 1-methyl-2-mercaptopilocarpine. DeGraw [171] found, however, that the desulfurization step with Raney nickel or even hydrogen peroxide caused some isomerization to isopilocarpine. Pilocarpine can be resolved with tartaric acid [169]. DeGraw [175] has used this general approach to prepare (+)-pilocarpine labeled with ^{14}C in the N-methyl group.

An approach to pilocarpine which postponed the hydrogenation step to a later stage failed when the phthalimide 178 could not be hydrolyzed [176]. An attempt to prepare pilocarpine analogs by coupling iodide 179 with lithium reagent 180 gave instead a product with the lactone ring opened [177]; other analogs were prepared [178] from 180.

A novel synthesis of (+)-pilocarpine from L-histidine was achieved by Beyerman and coworkers [179]. Diazotization converted histidine to the (*S*)-hydroxyacid 181, which was converted to the N,O-bis-*p*-nitrobenzenesulfonate 182. The sulfonate ester was displaced with lithium bromide, while the N-sulfonate allowed specific methylation on the basic nitrogen to afford 183. Displacement of halogen with sodium dibenzyl ethylmalonate, followed by hydrogenolysis and decarboxylation, gave 184. Selective reduction of the ester function with LiAlH₄ led to a 1:1 mixture of (+)pilocarpine and (+)isopilocarpine, both about 65% racemized.

5.1.4. Physiological Properties and Analysis.

Pilocarpine acts principally as a peripheral stimulant of the parasympathetic system; it is one of the oldest parasympathomimetic drugs. Like muscarine and acetylcholine it acts on the iris, salivary and sweat glands, and intestinal smooth muscle, but it has a dualistic activity and behaves like atropine on the frog heart [180]. It reduced or abolished inhibition of spinal motoneurons in the cat, apparently depleting the store of inhibitory transmitter at inhibitory synapses [181]. It is a strong diuretic, and was formerly used as a diaphoretic, that is, to induce sweating, especially in nephritis or dropsy, to relieve the kidneys and to remove toxic metabolites. It can act as an antidote to small doses of atropine. The main current application is in opthalmology as a miotic (contraction of the pupil). Isopilocarpine and pilocarpidine have much weaker activity [144].

A large number of methods have been described for the estimation of pilocarpine; the older methods are summarized in Ref. 1. Many involve titration with standard HCl after separation of the bases by chromatography [182].

Pilocarpine has a first basicity constant [143] pK_{b1} of 7.15, and NMR spectroscopic studies [183] have shown that the site of protonation is the nonmethylated nitrogen, similar to protonation of imidazole itself [184]. Recent assay methods include thin-layer electrophoresis [185] and thin-layer chromatography [186], potentiometry using ion selective membrane electrodes [187], complexation with an ion exchange resin [188], and a magnesium oxide extraction method [189]. Reaction of pilocarpine with hydroxylamine gives a hydroxamic acid [190] which is complexed with ferric ion [191]. Pilocarpine catalyzes the chemiluminescence of luminol, and can act as a model for peroxidase [192].

5.2. Pilosine, Isopilosine, and Epiisopilosine

Three stereoisomeric alkaloids which differ from pilocarpine by replacement of the ethyl group by an α-hydroxybenzyl substituent occur as minor constituents of *Pilocarpus species*.

5.2.1. Isopilosine (185b).

From the mother liquors remaining after separation of pilocarpine and isopilocarpine from extracts of *P. microphyllus*, Pyman [193] isolated in 0.07% yield a crystalline alkaloid which he named pilosine, later changed to isopilosine. The same alkaloid was isolated independently by Leger and Roques [194] and called carpiline. Isopilosine is a monoacidic dextrorotatory base of the formula $C_{16}H_{18}N_2O_3$, containing an *N*-methyl group. One oxygen is evidently present as a hydroxyl group, beause boiling with acetic anhydride effects dehydration to anhydropilosine (186), a crystalline dextrorotatory base, mp 133–134° C. The other two oxygens are contained in a lactone function, for isopilosine, though insoluble in cold alkali, dissolves in warm 5% KOH and can be recovered upon acidification. When isopilosine is boiled with 15–20% KOH, benzaldehyde is formed in a good yield along with a new base, pilosinine, $C_9H_{12}N_2O_2$, mp 78–79° C, a dextrorotatory lactone. The empirical formula and chemical properties suggested that pilosinine was 187, the parent lactone of the pilocarpine family. In order to account for the retroaldol reaction, Pyman placed the α-hydroxybenzyl group of isopilosine as shown in structure 185, and suggested that the stereochemical configuration was analogous to isopilocarpine.

5.2.2. Pilosine (185a).

In 1959, Voigtländer and Rosenberg [195] pointed out that, by analogy with the pilocarpine–isopilocarpine pair, an unstable pilosine stereoisomer should exist, and found it in *P. microphyllus* mother liquors. Their pilosine was also dehydrated to 186 and underwent the retroaldol conversion to 187. With dilute alkali or sodium ethoxide it was isomerized to Pyman's "pilosine." Voigtländer and Rosenberg thus considered their alkaloid the true "pilosine" (*cis*) and renamed Pyman's alkaloid isopilosine. They reported that of 13 specimens of *P. microphyllus* examined, most contained either only pilosine or only isopilosine; in only two batches were both alkaloids found together.

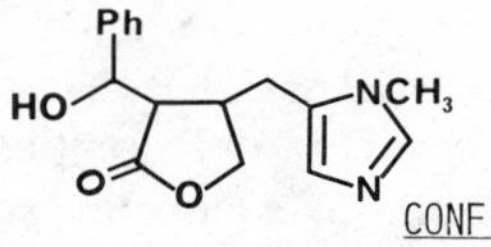

	CONFIGURATION	MP	$[\alpha]_D$
185a PILOSINE	2R,3R,6R	171-172°	+136.5°
185b ISOPILOSINE	2S,3R,6R	182-183°	+37.6°
185c EPIISOPILOSINE	2S,3R,6S	179-180°	-44.0°

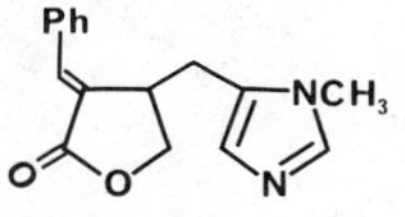

186 ANHYDROPILOSINE

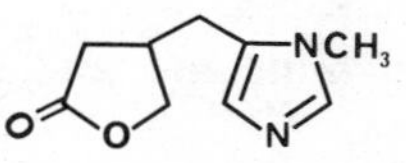

187 PILOSININE

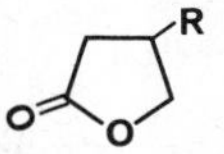

188 R = COOH
189 R = CH$_2$COOH
190 R = CH$_2$CO-CH$_2$NH$_2$

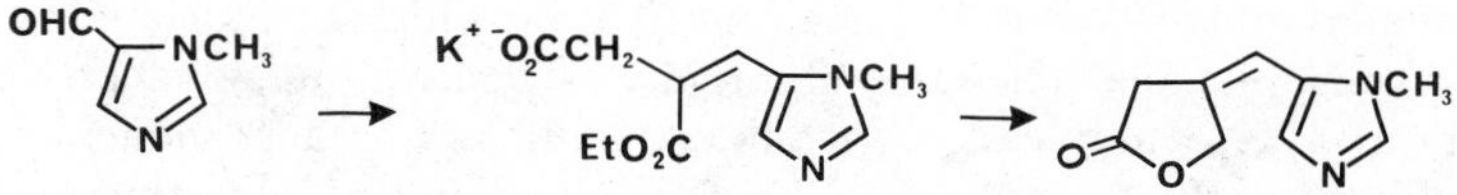

191   →   **192**   →   **193**

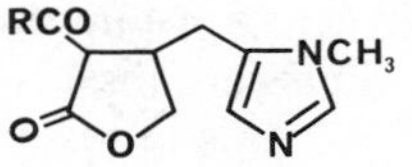

194 R = PH
195 R = CH$_3$

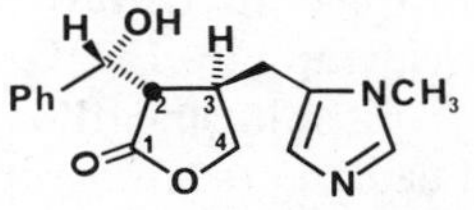

196 ISOPILOSINE

Later, however, two groups observed by TLC [196] and NMR [197] analysis that Voigtländer's "pilosine" was not homogeneous, but rather an approximately 1:1 mixture of pilosine and isopilosine. Pure pilosine was obtained by recrystallization of the salicylate [196].

5.2.3. Epiisopilosine (185c).

The third pilosine stereoisomer was isolated independently by Löwe and Pook [196] and by Sarel and coworkers [197] from *P. microphyllus*. It also underwent retroaldol cleavage to **187** and benzaldehyde. It was shown to be identical with a sample synthesized earlier by Link and Bernauer (vide infra). The melting points and optical rotations of the three stereoisomers are listed; NMR data are given in Ref. **196**.

Pilosine and isopilosine possess only weak pilocarpine-like pharmacological activity, but isopilosine has been patented for use in dermatological preparations for treatment of acne and seborrhea [198].

5.2.4.   Synthesis.   Several early syntheses of pilosinine (187), none particularly satisfactory, were developed by Soviet and Indian chemists. In the first, Preobrashenski and coworkers [199] followed the same methods they had used earlier to synthesize pilocarpine (Section 5.1.3), proceeding via acids **188** and **189** to pilosinine. In a modification by Mehrotra and Dey [200], Michael addition of ethyl 4-phthalimidoacetoacetate to 4-ethoxycrotonitrile gave the aminoketone **190** after hydrolysis; Sarel et al. [197] used crotonolactone as the Michael acceptor in a related approach.

The most efficient synthesis of pilosinine yet published is that due to Link and Bernauer [201] who, in contrast to the earlier schemes, began with the properly substituted imidazole and added the lactone ring last. 1-Methyl-5-formylimidazole **(191)** was converted to **192** by Stobbe condensation with ethyl succinate. Reduction of the ester group with $LiBH_4$ gave a mixture of **193** and **187** that was converted completely to (±) pilosine by hydrogenation. The racemic product was resolved with di-*p*-toluoyltartaric acid to give material identical with (+)-pilosinine.

Conversion of pilosinine to pilosine involves, in principle, aldol condensation with benzaldehyde. Sarel et al [197] have studied in detail the effect of base and solvent on the product distribution resulting from this mixed aldol reaction; isopilosine, epiisopilosine, and anhydropilosine were found in the product mixtures, but none of the unstable stereoisomer pilosine. In another approach using Claisen condensation of methyl benzoate with (+)-pilosinine, Link and Bernauer [201] obtained the benzoylated lactone **194,** shown by NMR to be the *trans* isomer. Catlytic hydrogenation gave a mixture of (+) isopilosine and (−)epiisopilosine.

5.2.5.   Stereochemistry.   With three asymmetric centers, pilosine may exist in eight stereoisomeric forms (four *dl* pairs). Three of these are represented by pilosine, isopilosine, and epiisopilosine. The chemical evidence presented above showed that pilosine has a *cis* relation of the substituents on the lactone ring, while isopilosine and epiisopilosine are *trans* isomers, but gave no information as to the configuration at the benzylic carbon.

The relative configuration of isopilosine was established by an X-ray crystal structure [202], shown in **196**. Epiisopilosine differs from isopilosine only at the hydroxylated carbon, and pilosine differs at the epimerizable C(2).

The absolute configuration was established by synthesis of pilocarpine from pilosinine [201]. Condensation of (+)-pilosinine with ethyl acetate and potassium *t*-butoxide gave ketolactone **195**. Reduction, dehydration, and catalytic hydrogenation gave a mixture of (+)pilocarpine and (+)isopilocarpine. Thus all three pilosine stereoisomers have the same (*R*) configuration at C(3) as pilocarpine.

The configurational assignments at C(6) were challenged [197] on the basis of tenuous circular dichroism correlations with aromatic amino acids, but the revisions were later withdrawn [203] and the original assignments upheld [204].

5.3.   Epiisopiloturine

From a Brazilian species of *P. microphyllus*, Voigtländer et al. [205] isolated a
new isomer of pilosine, mp 218–219°C. It underwent the characteristic pilosine
reactions, retroaldol cleavage to benzaldehyde and dehydration on heating with
acetic anhydride, but the products were isomeric, not identical, with pilosinine
(**187**) and anhydropilosine (**186**). The new alkaloid is thus not the missing fourth
stereoisomer of pilosine. The NMR spectra, particularly the carbon-13 spectrum, suggested that the source of the isomerism was the location of the *N*-methyl
group. Specifically, pilosinine and 1,5-dimethylimidazole show the imidazole
C(4) signal at 124.5–127 ppm and the C(5) signal at 128.9, while "neopilo-
sinine" and 1,4-dimethylimidazole exhibit these signals at 137 and 117 ppm,
respectively.

The conclusion that the new alkaloid is a 1,4- rather than 1,5-disubstituted

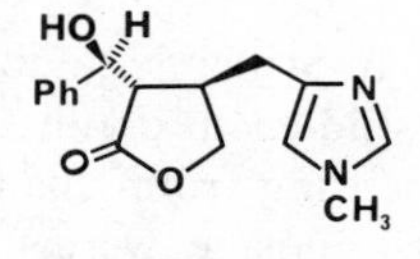

197   Epiisopiloturine

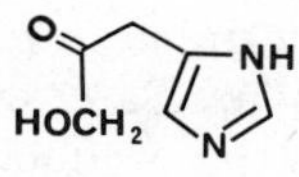

198

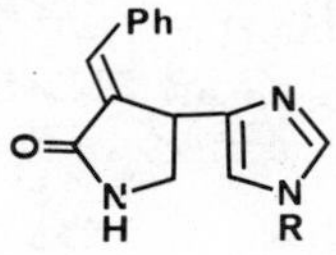

199   R = CH₃   Anantine
202   R = H   Noranantine

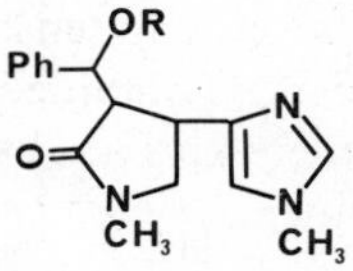

200   R = H   Cynometrine
201   R = COPh   Cynodine

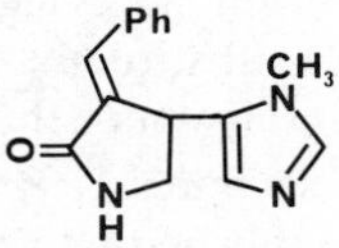

203   Isoanantine

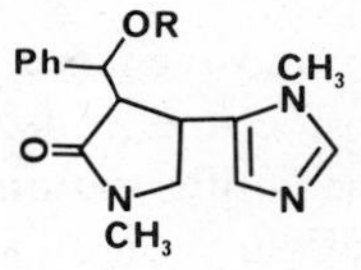

204   R = H   Isocynometrine
205   R = COPh   Isocynodine

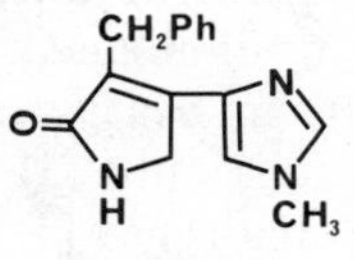

206

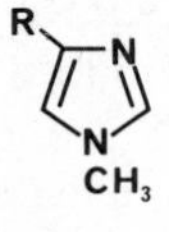

207   R = -CH=CH-CO₂Et
208   R =

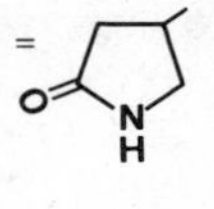

imidazole was proved in an elegant way, when methylation gave a quaternary salt identical with that formed by methylation of epiisopilosine. The new alkaloid, named epiisopiloturine, is consequently **197**. It is interesting that none of the pilosine stereoisomers nor any neopilocarpine was detected in this species.

5.4. Biosynthesis

Robinson [206] speculated that pilocarpine might originate by condensation of the known [207] histidine precursor **198** with a four-carbon acid (butyric or acetoacetic), while reaction with cinnamic or benzoylacetic acid could give pilosine. An alternative suggestion [208] is condensation of α-ketobutyrate (from threonine) with urocanic acid (**13**). In the only labeling experiments reported to date [208], however, neither radioactive acetate, threonine, nor histidine was incorporated significantly into the alkaloids of *P. pennatifolius;* only methionine was incorporated, the *S*-methyl providing the *N*-methyl of pilocarpine.

6. THE CYNOMETRA ALKALOIDS

A family of imidazole alkaloids somewhat related to pilosine occurs in two African *Cynometra* species; anantine (**199**), cynometrine (**200**), and cynodine (**201**) were isolated from *C. ananta* [209], while **199, 200** and noranantine (**202**) also occur in *C. lujae* [210] along with a set of alkaloids with the other imidazole nitrogen methylated, isoanantine (**203**), isocynometrine (**204**), and isocynodine (**205**).

6.1. Anantine (199)

Anantine [209], $C_{15}H_{15}N_3O$, mp 204°C, is a levorotatory base, $[\alpha]_D - 549°$. The ultraviolet spectrum (λ_{max} 218 and 277 nm) shows a conjugated system, confirmed by catalytic hydrogenation in a neutral solvent to a mixture of two isomeric dihydro derivatives which lack the maximum at 218 nm; the NMR spectrum of the dihydrocompound shows three new aliphatic protons, indicating the presence of a trisubstituted double bond. Permanganate oxidation affords benzoic acid. The IR spectrum of anantine shows a lactam carbonyl (1700 cm^{-1}), and anantine forms a monoacetyl derivative. The NMR spectrum shows an *N*-methyl, an AMX system corresponding to N—CH$_2$—CH—, a vinyl proton allylically coupled to the methine, and three protons in the region δ 7.0–7.7 in addition to the benzene protons. The double bond can be rearranged to the more stable tetrasubstituted isomer **206** by treating with palladium–charcoal in *p*-cymene; **206** is optically inactive, shows two —CH$_2$— singlets in the NMR at δ 3.91 and 4.36, and exhibits increased conjugation in the UV (λ_{max} 285 nm).

Both the double bond and the benzene ring are reduced by catalytic hydrogenation in acidic solution. The NMR spectrum of the crystalline octahydro derivative formed shows the N-methyl and two imidazole protons at very nearly the same chemical shift (δ 3.63, 6.58, 7.26) as they are in pilocarpine (3.56, 6.76, 7.38). These data lead to structure **199** for anantine.

This structure has been confirmed by synthesis [210]. 1-Methyl-4-formylimidazole was converted by a Wittig reaction to ester **207**. The lactam ring was constructed by Michael addition of nitromethane followed by catalytic reduction of the nitro group and cyclization to **208**. After protecting the lactam N—H as the N-acetyl derivative, condensation with benzaldehyde and sodium hydride, followed by dehydration with $POCl_3$, gave racemic anantine.

6.2. Cynometrine (200)

Cynometrine, $C_{16}H_{19}N_3O_2$, mp 213°, is a second levorotatory base from *C. ananta*. One oxygen is present as a lactam carbonyl (IR 1689 cm^{-1}) and the other as an alcohol (formation of an O-acetyl derivative). The NMR spectrum shows two N-methyl groups (δ 2.86, 3.43), a fragment N—CH$_2$—CH—, and a secondary benzylic alcohol, d ($J = 6$ Hz), δ 5.03, along with the aromatic protons. Dehydration with hot polyphosphoric acid gives a substance identified by its NMR spectrum as N-methylanantine, even though it was not possible to prepare it by methylation of anantine.

6.3. Cynodine (201)

The structure of cynodine, $C_{23}H_{23}N_3O_3$, $[\alpha]_D + 15°$, follows immediately from its saponification to benzoic acid and cynometrine [209].

6.4. Isoanantine (203)

The structure of this key alkaloid in the isomeric series from *C. lujae* was deduced from spectroscopic data and confirmed by synthesis [210], using the same set of reactions used for anantine but starting with 1-methyl-5-formylimidazole **(191)**.

7. THE MACRORUNGIA ALKALOIDS

From *Macrorungia longistrobus*, a shrub of the family *Acanthaceae* found in the Republic of South Africa, Arndt and his coworkers have isolated a group of seven new imidazole alkaloids. Four are the first members of a previously unknown family of imidazolyl quinoline alkaloids, while the other three appear to be closely related biogenetically.

7.1. Macrorine and Isomacrorine

Macrorine, mp 160°C, and isomacrorine, mp 110°C, are two isomeric, optically inactive, bases of the formula $C_{13}H_{11}N_3$ isolated by chromatography of the ethanol extracts of the leaves and branches [211]. The NMR spectra of both show only an N-methyl group and eight aromatic protons. Permanganate oxidation of macrorine afforded the amide (213) of 2-quinolinecarboxylic acid along with the N-acylurea 214, while similar oxidation of isomacrorine gave the N-methylamide 215. Hydrogenation of macrorine in acetic acid gave a tetrahydrocompound, shown by its UV spectrum and formation of an N-acetyl derivative to be the corresponding 1,2,3,4-tetrahydroquinoline. At this point, the $C_4H_3N_2$ fragment attached to the quinoline ring was suggested to be an N-methylimidazole, joined at C(4) in macrorine (209) and at C(5) in isomacrorine (210). In confirmation, macrorine and isomacrorine gave *identical* methiodides (216), mp 196°C, which underwent ring opening in alkali to 217. The latter could be oxidized by chromic acid to 218, which was hydrolyzed to 215.

7.2. Normacrorine

A third alkaloid, normacrorine [212], mp 156–157°C, has the formula $C_{12}H_9N_3$ and a UV spectrum similar to those of 209 and 210. Its NMR spectrum showed absorption only in the aromatic region, suggesting structure 211. As expected from this assignment, reaction with dimethyl sulfate gave a mixture of macrorine and isomacrorine, the former predominating.

Arndt et al. [212] synthesized normacrorine from 2-acetylquinoline, using the Wohl–Marckwald cyclization of the derived α-aminoketone with potassium isothiocyanate to construct the imidazole ring.

7.3. Macrorungine

Macrorungine [211], mp 267–270°C, has the formula $C_{13}H_{11}N_3O$. The oxygen was shown to be present in a carbonyl group by the infrared absorption at 1695 cm^{-1}. Catalytic hydrogenation resulted in the uptake of 6 mol of hydrogen to give a perhydro derivative, which was dehydrogenated by heating with palladium–charcoal to a mixture of quinoline, 2-methylquinoline, and 2-ethylquinoline. Chromic acid oxidation gave a ring-opened product identified as 219, which could be hydrolyzed by acid to 214; the latter was also obtained directly by oxidation of macrorungine with sodium periodate and osmium tetroxide. Oxidation with permanganate gave a hydantoin assigned structure 220. The key reaction of the structure proof was reduction with LiAlH$_4$ in refluxing dioxan to afford macrorine, confirming structure 212 for macrorungine.

The mass spectra of the imidazolylquinoline alkaloids have been analyzed in detail [213].

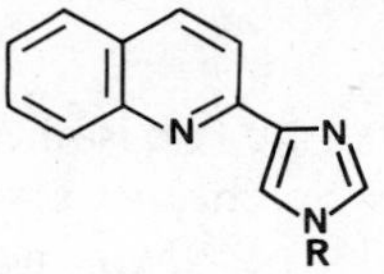

209 R = CH₃ MACRORINE
211 R = H   NORMACRORINE

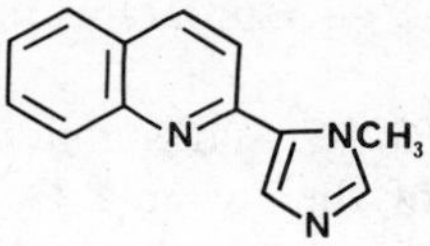

210 ISOMACRORINE

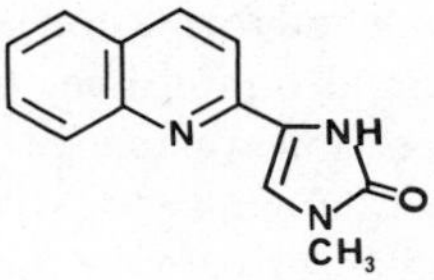

212 MACRORUNGINE

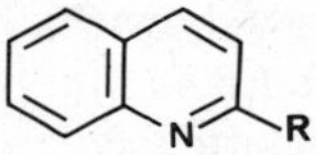

213 R = CONH₂
214 R = CO-NH-CO-NHCH₃
215 R = CONHCH₃
217 R = C=CH-NHCH₃
                N(CH₃)CHO

218 R = CO-N(CH₃)CHO
219 R = CONH-CO-N(CH₃)CHO
220 R =

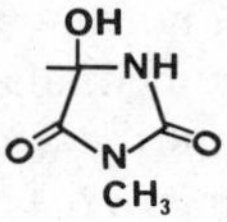

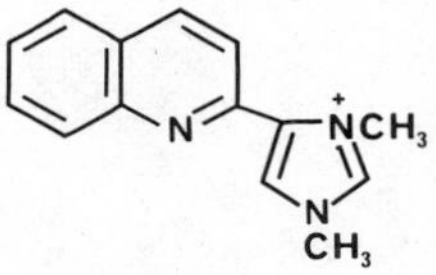

216

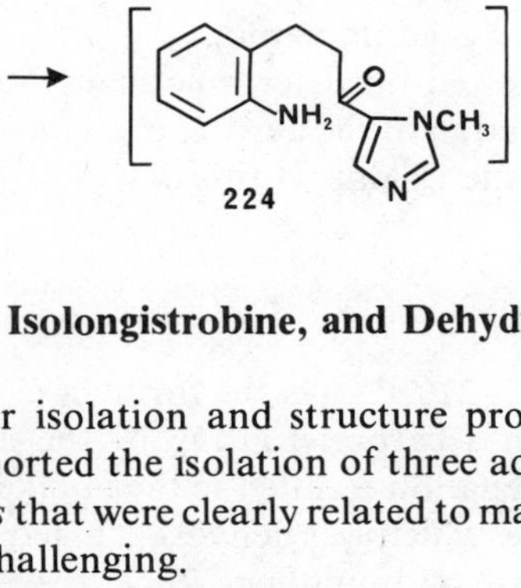

223 →

224

→ **210**

7.4. Longistrobine, Isolongistrobine, and Dehydroisolongistrobine

Five years after their isolation and structure proof of the macrorine family, Arndt et al. [214] reported the isolation of three additional imidazole alkaloids from *M. longistrobus* that were clearly related to macrorine but whose structures proved to be more challenging.

Longistrobine (**221**), mp 145–148°C, and isolongistrobine (**222**), mp 132–136°C, are optically inactive isomers of the formula $C_{17}H_{19}N_3O_3$, while dehydro-isolongistrobine (**223**), mp 131°C, has two fewer hydrogens. The latter alkaloid can be obtained quantitatively by chromic acid oxidation of isolongistrobine. The relationship of these new alkaloids to the macrorine group was shown immediately by zinc dust distillation of longistrobine to macrorine (**209**) and of isolongistrobine to isomacrorine (**210**).

All three alkaloids have similar ultraviolet spectra, with a maximum at 253–258 nm, shifted to 233–235 nm in acid solution; the spectra resemble those of tetrahydroquinoline (λ_{max} 250 nm) and 1-methyl-5-acetylimidazole (λ_{max} 255 nm, shifted to 235 nm in dilute HCl). The infrared spectra of longistrobine and its isomer show carbonyl absorption at 1660–1680 cm^{-1} and broad NH or OH absorption, while the spectrum of dehydroisolongistrobine was reported to show a carbonyl band at 1710 cm^{-1} but no absorption in the region 3100–4000 cm^{-1}. Isolongistrobine forms an *O*-acetate with acetic anhydride.

The NMR spectra of **221** and **222** are almost identical, showing (a) singlets characteristic of two nonadjacent protons on an imidazole ring, (b) four other aromatic protons, (c) an imidazole *N*-methyl, (d) a signal at δ 5.40–5.54 for a secondary carbinol C—H, (e) a broad one-proton signal at δ 6.70 or 5.92, exchangeable with D$_2$O, and (f) eight protons between δ 1.9 and 3.5, one of which also disappeared on D$_2$O exchange. The NMR spectrum of **223** was similar except for the absence of the carbinol C—H and the exchangeable proton at δ 5.9–6.7, indicating that CrO$_3$ had oxidized the secondary carbinol group of **222** to a carbonyl. The mass spectra of **221** and **222** were identical, and two fragment peaks at m/z 125 and 110 were taken to support the presence of an acylimidazole moiety.

Hydrolysis of dehydroisolongistrobine in refluxing acetic–hydrochloric acids afforded isomacrorine (**210**) and succinic acid, while similar hydrolysis of dehydrolongistrobine (which does not occur naturally but can be obtained by chromic acid oxidation of **221**) gave macrorine (**209**) and succinic acid. These two products account for all the carbon and nitrogen atoms, and made it

ARNDT ET AL [214]   WUONOLA AND WOODWARD [215]

221 LONGISTROBINE

$R_1 =$

222 ISOLONGISTROBINE

$R_2 =$

223 DEHYDROISOLONGISTROBINE

225 ANHYDROISOLONGISTROBINE
     (FROM HCL-HOAC)

226 ANHYDROISOLONGISTROBINE
     (FROM DMSO)

apparent that hydrolysis of the dehydrocompounds liberates a compound with a primary amino group (e.g., **224**) that cyclizes with the carbonyl group attached to the imidazole ring and then aromatizes to form the quinoline ring of **210**.

Heating isolongistrobine in HCl–AcOH afforded an anhydro compound $C_{17}H_{17}N_3O_2$ (**225**) that showed two carbonyl bands in the IR and a UV spectrum similar to the parent. An isomeric dehydration product (**226**) was obtained by heating **222** in dimethyl sulfoxide; it also had the UV spectrum of an acylimidazole, and the NMR spectrum showed the moiety —CH_2—CH=CH—.

Based on these data and the additional observations that isolongistrobine exchanged two hydrogens for deuterium when equilibrated with CH_3OH in the inlet of a mass spectrometer, but exchanged six protons when exposed to CH_3ONa—CH_3OD, Arndt et al. [214] proposed structures **221–223** for the three alkaloids and **225–226** for the anhydroisolongistrobines.

Wuonola and Woodward [215] called attention to a number of serious discrepancies between the data accumulated by Arndt et al. and the structures derived therefrom, including:

1. The carbinolamine structures proposed for **221** and **222** would be expected to open easily, dehydrate, and aromatize to quinolines.
2. The structures proposed are chiral but the alkaloids are optically inactive.
3. Structures **221, 222,** and **226** do not possess the acylimidazole system indicated by the ultraviolet spectra.
4. The ketone carbonyl of structures **221** and **222** would be expected to absorb in the IR around 1710 cm^{-1}, but the IR band reported for longistrobine and isolongistrobine is at 1660–1680 cm^{-1}.
5. The secondary lactam structure **223** cannot be correct for dehydroisolongistrobine, for its NMR and IR spectra show no evidence for an N—H bond. Moreover, the IR spectrum of this alkaloid was reported to show a single carbonyl band at 1710 cm^{-1}, whereas structure **223**, with three different carbonyl groups, would be expected to show IR bands at ~1710 (ketone), 1670 (acylimidazole), and 1680–1655 (amide) cm^{-1}.
6. It is difficult to rationalize the cleavage of a carbon–carbon bond in the formation of succinic acid by hydrolysis of **223**.

Wuonola and Woodward suggested that this last hydrolysis, as well as the spectroscopic data, would be better accommodated by a succinimide structure for dehydroisolongistrobine (**223**), and proved the correctness of this deduction by synthesis (Scheme 4).

Treatment of the imidazole ester **227** with the Grignard reagent derived from methyl phenyl sulfone gave sulfone **228,** the active methylene of which was alkylated with *o*-nitrobenzyl bromide to afford **229**. Reduction with aluminum amalgam gave a mixture of two amines **230** and **231,** the latter resulting from intramolecular cyclization with the ketone function before it could be reduced. Dehydrogenation of **231** with sulfur and palladium–charcoal provided a new

SCHEME 4

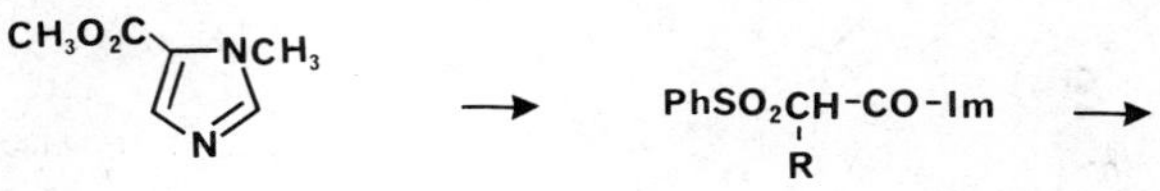

CH$_3$O$_2$C—NCH$_3$ ... imidazole ... $\longrightarrow$ ... PhSO$_2$CH-CO-Im ... $\longrightarrow$
                                                         |
                                                         R

227

228 R = H
229 R = o-Nitrobenzyl

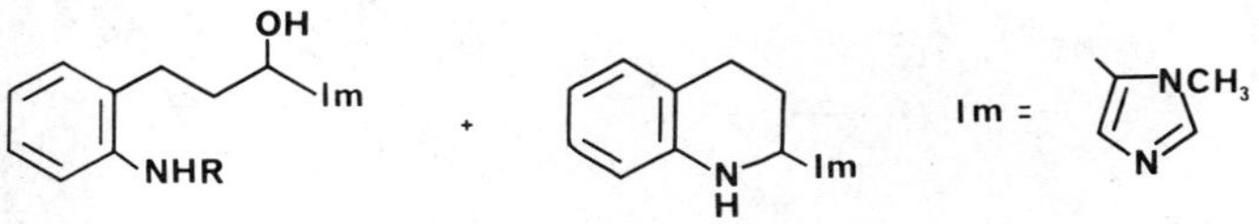

OH
... Im

NHR   +   ... N—Im   Im =   ... NCH$_3$
                                           H   N

230 R = H
232 R = COCH$_2$CH$_2$CO$_2$CH$_3$ **231**
233 R = COCH$_2$CH$_2$CH=CH$_2$

synthesis of isomacrorine (**210**). On the other hand, **230** could be selectively acylated on nitrogen to the amide **232**; chromic acid oxidation to the ketone followed by heating to form the imide gave a synthetic compound identical with dehydroisolongistrobine (**223**). This compound does indeed show three carbonyl peaks (1665, 1710, and 1780 cm^{-1}) in the IR.

Finally, intermediate **230** allowed an easy synthesis of isolongistrobine. Acylation with 4-pentenoyl chloride gave amide **233**. After oxidation to the ketone, cleavage of the double bond with NaIO$_4$—OsO$_4$ afforded the aldehyde, which exists as the carbinolamine tautomer **222**. The revised structures for the longistrobine alkaloids and their degradation products are shown.

8. MISCELLANEOUS IMIDAZOLE ALKALOIDS

8.1. Cypholophine

Cypholophine (**234**) is the major alkaloid of the New Guinea rain forest shrub *Cypholophus friesianus* (family *Urticaceae*) [216]. Extraction of the leaves with methanol gave a total of 0.08–0.11% of a mixture of **234** and its *O*-acetyl derivative **235**. Cypholophine, C$_{18}$H$_{26}$N$_2$O$_3$, is an optically inactive base, mp 126–127°; the infrared spectrum shows —OH and —NH groups but no carbonyl absorption. Acetylation affords the *O*-acetyl derivative identical with the minor alkaloid **235**. The presence of an NH group is evident from the IR (3500 cm^{-1}) and NMR (singlet, δ 7.40, exchangeable with D$_2$O) spectra of **234**. Permanganate oxidation affords 3,4-dimethoxybenzoic acid in high yield, accounting for the remaining oxygen atoms.

CH₃O—

CH₃O—

N CH₃

RO—(CH₂)₄ NH

234 R = H CYPHOLOPHINE
235 R = COCH₃

CH₃O—

CH₃O—

N CH₃

N

236

CH₃O—

CH₃O—

H₃C N

N

237

CH₃O—

CH₃O—

COR

238 R = CL
239 R = CHBR-CH₃

N⁺

240

N

241

The 100-MHz NMR spectrum is particularly informative, showing (a) a methyl group attached to a double bond or aromatic ring, (b) a 4-hydroxybutyl side-chain, (c) a four-proton singlet for a —CH₂CH₂— unit joining two aromatic rings, and (d) suggesting the presence of a 2,4,5-trisubstituted imidazole ring.

When cyclolophine is heated with *p*-toluenesulfonyl chloride in pyridine, a mixture of two isomeric products is formed, both of which lack the —OH and —NH groups. They are believed to be the cyclization products **236** and **237** resulting from displacement of tosylate ion by either nitrogen, and so place the hydroxybutyl chain on the imidazole carbon between the nitrogen atoms.

The structure deduced was consistent with the main peaks of the mass spectrum, and was confirmed by an imaginative synthesis. The bromoketone **239**, prepared by successive reaction of acid chloride **238** with diazoethane and ethereal HBr, was heated in a sealed tube with 2-iminotetrahydropyran hydrochloride and methanolic ammonia. Cypholophine was isolated from the reaction mixture in 6% overall yield.

8.2. Octahydrodipyrido [1,2-*a*: 1′,2′-*c*]imidazol-10-ium bromide

This unusual imidazolium salt [217] was isolated from two species of *Orchidacea, Dendrobium anosmum Lindl*, and *D. parishii* Rchb. f. Fresh plants were extracted with methanol; chromatography of the extracts and conversion to the bromide on an ion exchange column gave **240**, mp 164–65°, analyzing

SCHEME 5

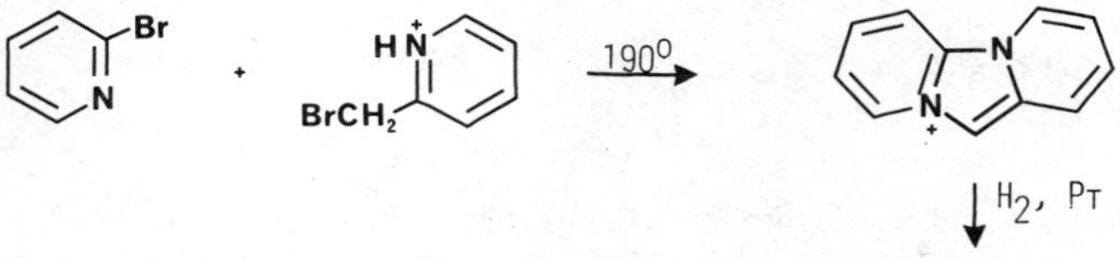

240

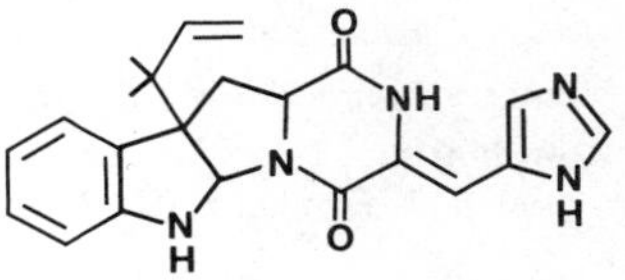

242 Roquefortine

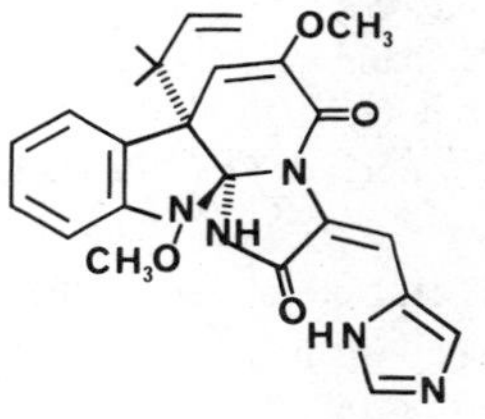

243 Oxaline

for $C_{11}H_{17}BrN_2$. The UV spectrum (λ_{max} 221 nm, log ϵ 3.78) revealed the aromatic system. Reduction with LiAlH$_4$ or Raney nickel alloy in NaOH gave the saturated (and therefore tricyclic) base **241**. The NMR spectrum of **240** showed one aromatic, eight aliphatic, and eight allylic protons, suggesting the imidazole structure. A straightforward synthesis (Scheme 5) confirmed structure **240**. Independent corroboration has been provided by an X-ray crystal structure [218].

8.3. Roquefortine and Oxaline

These two related *Penicillium* metabolites are more properly classified under indole alkaloids, but are mentioned here because of the dehydrohistidine unit, uncommon in natural products. Roquefortine (**242**) was isolated from *P. roqueforti* and the structure (with stereochemical assignments uncertain) proposed from spectroscopic data [219]. It is a potent neurotoxin. Oxaline (**243**)

was obtained by cultivating *P. oxalicum* on maize meal; its structure was proven
by X-ray analysis [220].

REFERENCES

1. A. R. Battersby and H. T. Openshaw, in *The Alkaloids*, Vol. III, R. H. F. Manske and H. L.
    Holmes, Eds., Academic Press, New York, 1953, p. 201. An update in this series appeared
    while this chapter was in press: L. Maat and H. C. Beyerman, in *The Alkaloids*, Vol. XXII,
    A. Brossi, Ed., Academic Press, New York, 1983, Chap. 5.
2. H. D. Fowler, *Nature* **193,** 582 (1962).
3. S. Jurisson, *Tartu Riikliku Ulikooli Toim.* **270,** 71 (1971); *Chem. Abstr.* **76,** 23018 (1972).
4. D. Ackermann and H. G. Menssen, *Hoppe-Seyler's Z. Physiol. Chem.* **317,** 144 (1959).
5. V. Erspamer, T. Vitali, M. Roseghini, and J. M. Cei, *Experientia* **19,** 346 (1963); ibid, *Arch.
    Biochem. Biophys.* **105,** 620 (1964); V. Erspamer, M. Roseghini, and J. M. Cei, *Biochem.
    Pharmacol.* **13,** 1083 (1964).
6. W. Appel and E. Werle, *Arzneim. Forsch.* **9,** 22 (1959).
7. D. Ackermann, F. Holtz, and H. Reinwein, *Z. Biol.* **82,** 278 (1924).
8. P. H. List, *Arch. Pharm.* **291,** 502 (1958); **290,** 517 (1957).
9. R. T. Major and F. Dürsch, *J. Org. Chem.* **23,** 1564 (1958).
10. D. Ackermann and P. H. List, *Hoppe-Seyler's Z. Physiol. Chem.* **308,** 274 (1957).
11. J. C. Madgwick, B. J. Ralph, J. S. Shannon, and J. J. Sims, *Arch Biochem. Biophys.* **141,**
    766 (1970).
12. H. H. Tallen, W. H. Stein, and S. Moore, *J. Biol. Chem.* **206,** 825 (1954).
13. W. R. D. Rangeley and R. A. Lawrie, *J. Food Technol.* **11,** 143 (1976).
14. D. H. Cocks, P. O. Dennis, and T. H. Nelson, *Nature* **202,** 184 (1964).
15. A. Noordam, L. Maat, and H. C. Beyerman, *Rec. Trav. Chim. Pays-Bas* **97,** 293 (1978).
16. F. Kutscher, *Zentr. Physiol.* **24,** 775 (1910).
17. H. Pauly, *Hoppe-Seyler's Z. Physiol. Chem.* **42,** 508 (1904).
18. R. Engeland and F. Kutscher, *Zentr. Physiol.* **26,** 569 (1912).
19. C. Reuter, *Hoppe-Seyler's Z. Physiol. Chem.* **78,** 167 (1912).
20. A. Küng, *Hoppe-Seyler's Z. Physiol. Chem.* **91,** 241 (1914).
21. D. Ackermann and P. H. List, *Hoppe-Seyler's Z. Physiol. Chem.* **313,** 30 (1958).
22. P. H. List and H. G. Menssen, *Arch. Pharm.* **292,** 260 (1959).
23. D. Ackermann and H. G. Menssen, *Hoppe-Seyler's Z. Physiol. Chem.* **318,** 212 (1960).
24. D. Ackermann, P. H. List, and H. G. Menssen, *Hoppe-Seyler's Z. Physiol. Chem.* **314,** 33
    (1959).
25. M. C. Tanret, *C. R. Acad. Sci. Paris* **149,** 222 (1909).
26. G. Barger and A. J. Ewins, *J. Chem. Soc.* **99,** 2336 (1911).
27. E. Winterstein and C. Reuter, *Hoppe-Seyler's Z. Physiol. Chem.* **86,** 234 (1913).
28. G. Barger and A. J. Ewins, *Biochem. J.* **7,** 204 (1913).
29. S. Akabori, *Ber.* **66,** 151 (1933).
30. H. Heath, A. Lawson, and C. Rimington, *J. Chem. Soc.,* 2215 (1951).
31. J. N. Ashley and C. R. Harrington, *J. Chem. Soc.* **133,** 2586 (1930).
32. D. B. Melville, D. S. Genghof, E. Inamine, and V. Kovalenko, *J. Biol. Chem.* **233,** 9 (1956).
33. P. H. List and H. Hetzel, *Planta Medica* **8,** 105 (1960).
34. S. R. Benedict, E. B. Newton, and J. A. Behre, *J. Biol. Chem.* **67,** 267 (1926).

35. B. A. Eagles and T. B. Johnson, *J. Am. Chem. Soc.* **49**, 575 (1927).

36. E. B. Newton, S. R. Benedict, and H. D. Dakin, *J. Biol. Chem.* **72**, 367 (1927).

37. D. B. Melville, W. H. Horner, and R. Lubschez, *J. Biol. Chem.* **206**, 221 (1954).

38. H. Heath and J. Wildy, *Biochem. J.* **64**, 612 (1956); ibid **63**, 1P (1956).

39. J. Wildy and H. Heath, *Biochem. J.* **65**, 220 (1957).

40. D. B. Melville, S. Eich, and M. L. Ludwig, *J. Biol. Chem.* **224**, 871 (1957).

41. H. Heath and J. Wildy, *Biochem. J.* **68**, 407 (1958).

42. D. B. Melville, M. L. Ludwig, E. Inamine, and J. R. Rachele, *J. Biol. Chem.* **234**, 1195 (1959).

43. F. Kutscher and D. Ackermann, *Hoppe-Seyler's Z. Phyisol. Chem.* **221**, 38 (1933).

44. D. Ackermann, *Hoppe-Seyler's Z. Physiol. Chem.* **295**, 1 (1953).

45. J. H. Welsh and P. B. Prock, *Biol. Bull.* **115**, 551 (1958).

46. D. Ackermann and H. G. Menssen, *Hoppe-Seyler's Z. Physiol. Chem.* **322**, 198 (1960).

47. D. Ackermann, *Hoppe-Seyler's Z. Physiol. Chem.* **296**, 286 (1954).

48. D. Ackermann and R. Janka, *Hoppe-Seyler's Z. Physiol. Chem.* **294**, 93 (1953).

49. D. Ackermann and P. H. List, *Hoppe-Seyler's Z. Physiol. Chem.* **318**, 281 (1960).

50. A. J. Weinheimer, E. K. Metzner, and M. L. Mole, Jr., *Tetrahedron* **29**, 3135 (1973).

51. D. Ackermann, *Hoppe-Seyler's Z. Physiol. Chem.* **65**, 504 (1910).

52. P. H. List and H. Reith, *Hoppe-Seyler's Z. Physiol. Chem.* **319**, 17 (1960).

53. L. Minale, G. Cimino, S. DeStefano, and G. Sodano, *Prog. Chem. Org. Nat. Prod.* **33**, 1 (1976).

54. J. H. Welsh, *Ann. Rev. Pharmacol.* **4**, 293 (1964).

55. V. Erspamer and O. Benati, *Science* **117**, 161 (1953); ibid, *Biochem. Z.* **324**, 64 (1953).

56. C. Pasini, A. Vercellone, and V. Erspamer, *Justus Liebigs Ann. Chem.* **578**, 6 (1952).

57. J. T. Baker and V. Murphy, *Compounds from Marine Organisms*, Vols. 1 and 2, CRC Press, Cleveland, 1976, 1981.

58. M. J. Keyl, I. A. Michaelson, and V. P. Whittaker, *J. Physiol.* **139**, 434 (1957).

59. M. Roseghini, V. Erspamer, L. Ramorino M., and J. E. Gutierrez, *Eur. J. Biochem.* **12**, 468 (1970).

60. M. Roseghini and M. Fichman, *Comp. Gen. Pharmacol.* **4**, 251 (1973).

61. J. A. Bender, K. DeRiemer, T. E. Roberts, R. Rushton, P. Boothe, H. S. Mosher, and F. A. Fuhrman, *Comp. Gen. Pharmacol.* **5**, 191 (1974).

62. V. Erspamer, *Arch. Exp. Path. Pharmak.* **218**, 142 (1953).

63. V. P. Whittaker, *Ann. N. Y. Acad. Sci.* **90**, 695 (1960).

64. M. Roseghini, *Experientia* **27**, 1008 (1971).

65. J. T. Baker and C. C. Duke, *Tetrahedron Lett.*, 1233 (1976).

66. C. C. Duke, J. V. Eicholzer, and J. K. Macleod, *Tetrahedron Lett.*, 5047 (1978).

67. T. Takita, T. Yoshioka, Y. Muraoka, K. Maeda, and H. Umezawa, *J. Antibiotics* **24**, 795 (1971).

68. T. Takita, Y. Muraoka, T. Yoshioka, A. Fujii, K. Maeda, and H. Umezawa, *J. Antibiotics* **25**, 755 (1972).

69. G. Koyama, H. Nakamura, Y. Muraoka, T. Takita, K. Maeda, H. Umezawa, and Y. Iitaka, *J. Antibiotics* **26**, 109 (1973).

70. B. Mehta and M. M. Bokadia, *Chem. Ind. (London)*, 98 (1981).

71. D. Ackermann and M. Mohr, *Z. Biol.* **98**, 37 (1937).

72. D. Ackermann, *Hoppe-Seyler's Z. Physiol. Chem.* **328**, 275 (1962).

73. D. Ackermann and E. Müller, *Hoppe-Seyler's Z. Physiol. Chem.* **268**, 277 (1941).

74. D. Ackermann, *Hoppe-Seyler's Z. Physiol. Chem.* **276**, 268 (1942).

75. D. Ackermann and S. Skraup, *Hoppe-Seyler's Z. Physiol. Chem.* **284**, 129 (1949).

76. G. D. Andreetti, L. Cavalca, and P. Sgarabotto, *Gazz. Chim. Ital.* **101**, 625 (1971).

77. G. D. Andreetti, P. Domiano, A. Gaetani, and A. Musatti, *Ric. Sci.* **38**, 1102 (1968).

78. V. Erspamer and G. F. Erspamer, *Riv. Biol.* **58**, 259 (1965); *Chem. Abstr.* **65**, 4139 (1966).

79. T. Vitali and G. Bertaccini, *Gazz. Chim. Ital.* **94**, 296 (1964).

80. G. G. Habermehl and W. Ecsy, *Heterocycles* **5**, 127 (1976).

81. H.-J. Preusser, G. Habermehl, M. Sablofski, and D. Schmall-Haury, *Toxicon* **13**, 285 (1975).

82. J. S. Fitzgerald, *Austral. J. Chem.* **17**, 375 (1964).

83. C. Poupat and T. Sevenet, *Phytochemistry* **14**, 1881 (1975).

84. S. R. Johns and J. A. Lamberton, *Austral. J. Chem.* **20**, 555 (1967).

85. S. R. Johns, J. A. Lamberton, J. W. Loder, A. H. Redcliffe, and A. A. Sioumis, *Austral. J. Chem.* **22**, 1309 (1969)

86. H. Rosenberg and A. G. Paul, *Tetrahedron Lett.*, 1039 (1969); ibid, *Phytochemistry* **9**, 655 (1970).

87. Y. Nagao, K. Seno, T. Miyasaka, K. Kawabata, S. Takao, and E. Fujita, *Tennen Yuki Kagobutsu Toronkai Koen Yoshishu, 22nd*, 554 (1979); *Chem. Abstr.* **93**, 204909 (1980).

88. H. Rosenberg and A. G. Paul, *Lloydia* **34**, 372 (1971).

89. H. Horan and D. G. O'Donovan, *J. Chem. Soc. (C)*, 2083 (1971).

90. H. Rosenberg and A. G. Paul, *J. Pharm. Sci.* **62**, 403 (1973).

91. H. Rosenberg, S. J. Stohs, and A. G. Paul, *Phytochemistry* **13**, 823 (1974).

92. H. Rosenberg and S. J. Stohs, *Lloydia* **37**, 313 (1974).

93. H. Rosenberg and S. J. Stohs, *Phytochemistry* **15**, 501 (1976).

94. S. R. Johns and J. A. Lamberton, *Chem. Comm.* 312 (1966); ibid, *Austral. J. Chem.* **20**, 555 (1967).

95. A. Mondon, *Chem. Ber.* **92**, 1461 (1959); cf. B. Belleau, *Can. J. Chem.* **35**, 651 (1957).

96. F. B. Power and T. Callan, *J. Chem. Soc.* **99**, 1993 (1911).

97. A. Aebi, *Helv. Chim. Acta* **39**, 1495 (1956).

98. C. Djerassi, J. Herrán, H. N. Khastgir, B. Riniker, and J. Romo, *J. Org. Chem.* **21**, 1510 (1956).

99. F. A. Kincl, J. Romo, G. Rosencranz, and F. Sondheimer, *J. Chem. Soc.* 4163 (1956).

100. C. Djerassi, C. Bankiewicz, A. L. Kapoor, and B. Riniker, *Tetrahedron* **2**, 168a (1958).

101. S. Raman, J. Reddy, W. N. Lipscomb, A. L. Kapoor, and C. Djerassi, *Tetrahedron Lett.*, 357 (1962).

102. S. Raman, J. Reddy, and W. N. Lipscomb, *Acta. Cryst.* **16**, 364 (1963).

103. R. P. Panzica and L. B. Townsend, *J. Am. Chem. Soc.* **95**, 8737 (1973).

104. R. Mechoulam, F. Sondheimer, A. Melera, and F. A. Kincl, *J. Am. Chem. Soc.* **83**, 2022 (1961).

105. D. Ackermann, O. Timpe, and K. Poller, *Hoppe-Seyler's Z. Physiol. Chem.* **183**, 1 (1929).

106. W. Linneweh, A. W. Keil, and F. A. Hoppe-Seyler, *Hoppe-Seyler's Z. Physiol. Chem.* **183**, 11 (1929).

107. W. Keil, *Hoppe-Seyler's Z. Physiol. Chem.* **187**, 1 (1930).

108. G. Barger and F. Tutin, *Biochem. J.* **12**, 402 (1918).

109. P. Hepp, *Justus Liebigs Ann. Chem.* **215**, 344 (1882).

110. F. L. Pyman, *J. Chem. Soc.* 183 (1930).

111. H. R. Matthews and H. Rapoport, *J. Am. Chem. Soc.* **95**, 2297 (1973).

112. I. R. McManus, *J. Biol. Chem.* **225**, 325 (1957).

113. Y. Seki, T. Nakamura, and Y. Okami, *J. Biochem.* **67**, 389 (1970).

114. G. Cimino, S. DeStefano, and L. Minale, *Comp. Biochem. Physiol.* **47B**, 895 (1974).

115. L. E. Fellows, E. A. Bell, and G. S. King, *Phytochemistry* **16**, 1399 (1977).

116. L. E. Fellows, R. C. Hider, and E. A. Bell, *Phytochemistry*, **16**, 1957 (1977).

117. S. Horii and Y. Kameda, *J. Antibiotics* **21**, 655 (1968).

118. K. Iinuma, S. Kondo, K. Maeda, and H. Umezawa, *J. Antibiotics* **28**, 613 (1975).

119. S. Tsuji, S. Kusomoto, and T. Shiba, *Chem. Lett.*, 1281 (1975).

120. K. Wiesner, Z. Valenta, B. S. Hurlbert, F. Bickelhaupt, and L. R. Fowler, *J. Am. Chem. Soc.* **80**, 1521 (1958); L. R. Fowler, Z. Valenta, and K. Wiesner, *Chemistry and Industry*, 95 (1962).

121. N. K. Hart, S. R. Johns, and J. A. Lamberton, *Chem. Comm.*, 1484 (1969); N. K. Hart, S. R. Johns, J. A. Lamberton, and R. I. Willing, *Austral. J. Chem.* **23**, 1679 (1970).

122. F. Khuong-Huu, J. P. LeForestier, and R. Goutarel, *Tetrahedron* **28**, 5207 (1972); F. Khuong-Huu, J.-P. LeForestier, G. Maillard, and R. Goutarel, *C. R. Acad. Sci. Paris* **270C**, 2070 (1970).

123. M. Cesario and J. Guilheim, *C. R. Acad. Sc. Paris* **271C**, 1552 (1970).

124. S. Forenza, L. Minale, R. Riccio, and E. Fattorusso, *Chem. Comm.*, 1129 (1971).

125. G. Cimeno, S. DeStefano, L. Minale, and G. Sodano, *Comp. Biochem. Physiol.* **50B**, 279 (1975).

126. E. E. Garcia, L. E. Benjamin, and R. I. Fryer, *J.C.S. Chem. Comm.*, 78 (1973).

127. D. J. Faulkner and R. J. Andersen, in *The Sea*, Vol. 5, E. D. Goldberg, Ed., Wiley-Interscience, New York, 1974, p. 679.

128. R. P. Walker, D. J. Faulkner, D. Van Engen, and J. Clardy, *J. Am. Chem. Soc.* **103**, 6772 (1981).

129. G. M. Sharma and P. R. Burkholder, *Chem. Comm.*, 151 (1971); G. Sharma and B. Magdoff-Fairchild, *J. Org. Chem.* **25**, 4118 (1977).

130. G. M. Sharma, B. Vig, and P. R. Burkholder, *Proc. Conf. on Food and Drugs from the Sea*, New York, 1969, p. 307.

131. L. H. Foley and G. Büchi, *J. Am. Chem. Soc.* **104**, 1776 (1982).

132. L. Cariello, S. Crescenzi, G. Prota, S. Capasso, F. Giordano, and L. Mazzarella, *Tetrahedron* **30**, 3281 (1974); L. Cariello, S. Crescenzi, G. Prota, F. Giordano, and L. Mazzarella, *J. C. S. Chem Comm.*, 99 (1973).

133. L. Cariello, S. Crescenzi, G. Prota, and L. Zanetti, *Experientia* **30**, 849 (1974).

134. L. Cariello, S. Crescenzi, G. Prota, and L. Zanetti, *Tetrahedron* **30**, 3611 (1974).

135. L. Cariello, S. Crescenzi, G. Prota, and L. Zanetti, *Tetrahedron* **30**, 4191 (1974).

136. Y. Komoda, S. Kaneko, M. Yamamato, M. Ishikawa, A. Itai, and Y. Iitaka, *Chem. Pharm. Bull.* **23**, 2464 (1975).

137. M. Braun, G. Büchi, and D. F. Bushey, *J. Am. Chem. Soc.*, **100**, 4208 (1978); M. Braun and G. Buchi, ibid **98**, 3049 (1976).

138. L. C. Dunn, Y.-M. Chang, and K. N. Houk, *J. Am. Chem. Soc.* **98**, 7095 (1976).

139. E. Hardy, *Bull. Soc. Chim. France* [2] **24**, 497 (1875).

140. A. W. Gerrard, *Pharm. J.* [3] **5**, 865, 965 (1875); **6**, 227 (1876); **7**, 225 (1877).

141. H. A. D. Jowett, *J. Chem. Soc.* **77**, 473 (1900).

142. A. Petit and M. Polonovski, *Bull. Soc. Chim. France* [3] **17**, 553, 554, 702 (1897); ibid, *J. Pharm* **5**, 370, 430, 475; **6**, 8 (1897).

143. K. Baeschlin, J. C. Etter, and H. Moll, *Pharm. Acta Helv.* **44**, 301 (1969).

144. P.-H. Chung, T.-F. Chin, and J. L. Lach, *J. Pharm. Sci.* **59**, 1300 (1970).

145. R. T. Koda, F. J. Dea, K. Fung, C. Elison, and J. A. Biles, *J. Pharm. Sci.* **62**, 2021 (1973).

146. H. A. D. Jowett, *J. Chem. Soc.* **77**, 851 (1900); **79**, 1331 (1901).

147. A. Pinner and E. Kohlhammer, *Ber.* **33**, 1424, 2357 (1900); **34**, 727 (1901).

148. H. A. D. Jowett, *J. Chem. Soc.* **79**, 1346 (1901).

149. A. Pinner and R. Schwarz, *Ber.* **35**, 2441 (1902).

150. H. A. D. Jowett, *J. Chem. Soc.* **83**, 438 (1903).

151. F. L. Pyman, *J. Chem. Soc.* **121**, 2616 (1922).

152. S. Akabori and S. Numano, *Ber.* **66**, 159 (1933).

153. H. A. D. Jowett, *J. Chem. Soc.* **87**, 794 (1905).

154. W. Langenbeck, *Ber.* **57**, 2072 (1924).

155. W. Döpke and G. d'Heureuse, *Tetrahedron Lett.*, 1807 (1968).

156. S. I. Zavialov, *Dokl. Akad. Nauk, S.S.S.R.* **82**, 257 (1952).

157. R. K. Hill and S. Barcza, *Tetrahedron* **22**, 2889 (1966).

158. P. W. Codding and M. N. G. James, Abstract 02, Amer. Cryst. Assoc. Meeting, June 1973, Ser. 2, Vol. 1.

159. K. Nagarajan, C. Weissmann, H. Schmid, and P. Karrer, *Helv. Chim. Acta* **46**, 1212 (1963).

160. T. D. Inch and G. J. Lewis, *Carbohydrate Res.* **22**, 91 (1972).

161. E. Merck, *E. Merck's Jahresber.*, 11 (1896); 5 (1897).

162. R. Burtles, F. L. Pyman, and J. Roylance, *J. Chem. Soc.* **127**, 581 (1925); E. Späth and E. Kunz, *Ber.* **58**, 513 (1925).

163. A. E. Tschitschibabin and N. A. Preobrashenski, *Ber.* **63**, 460 (1930).

164. K. N. Welch, *J. Chem. Soc.* **134**, 1370 (1931); J. K. Mehrotra and A. N. Dey, *J. Indian Chem. Soc.* **38**, 888 (1961).

165. A. M. Poljakowa and N. A. Preobrashenski, *J. Gen. Chem. USSR* **12**, 266 (1942); *Chem. Abstr.* **37**, 3055 (1943).

166. N. A. Preobrashenski and W. A. Preobrashenski, *Ber.* **68**, 847 (1935).

167. A. M. Poljakowa and W. A. Preobrashenski, *Ber.* **67**, 710 (1934); N. A. Preobrashenski, A. M. Poljakowa, and W. A. Preobrashenski, *Ber.* **68**, 844 (1935).

168. N. A. Preobrashenski, A. M. Poljakowa, and W. A. Preobrashenski, *Ber.* **68**, 850 (1935); **69**, 1835 (1936).

169. A. N. Dey, *J. Chem. Soc.*, 1057 (1937).

170. A. V. Chumachenko, M. E. Maurit, A. D. Treboganov, G. V. Smirnova, R. B. Teplinskaya, L. V. Vorkova, E. N. Zvonkova, and N. A. Preobrashenski, *Dokl. Chem., Akad. Nauk. SSSR* **178**, 182 (1968); A. V. Chumachenko, E. N. Zvonkova, and N. A. Preobrashenski, *J. Org. Chem. USSR* **5**, 571 (1969).

171. J. I. DeGraw, *Tetrahedron* **28**, 967 (1972).

172. J. I. DeGraw, J. S. Engstrom, and E. Willis, *J. Pharm. Sci.* **64**, 1700 (1973).

173. N. A. Preobrashenski, A. F. Wompe, and W. A. Preobrashenski, *Ber.* **66**, 1187 (1933); N. A. Preobrashenski, A. F. Wompe, W. A. Preobrashenski, and M. N. Schtschukina, *Ber.* **66**, 1536 (1933).

174. N. A. Preobrashenski, W. A. Preobrashenski, and A. M. Poljakowa, *Bull. Acad. Sci. URSS*, 983 (1936); M. M. Katsnel'son, A. M. Poljakowa, N. A. Preobrashenski, and W. A. Preobrashenski, Russ. Patent 47,693; *Chem. Abstr.* **33**, 3400 (1939).

175. N. A. Preobrashenski, M. E. Maurit, and G. V. Smirnova, *Dokl. Akad. Nauk SSSR* **81**, 613 (1951); *Chem. Abstr.* **47**, 4345 (1953).

176. A. V. Chumachenko, E. N. Zvonkova, and R. P. Evstigneeva, *J. Org. Chem. USSR* **8**, 1112 (1972).

177. W. Döpke and U. Mücke, *Z. Chem.* **13**, 104 (1973).

178. W. Döpke and U. Mücke, *Z. Chem.* **13**, 177, 297 (1973).

179. A. Noordam, L. Maat, and H. C. Beyerman, *Rec. Trav. Chim. Pays-Bas* **98**, 425 (1979).

180. J. M. van Rossum, M. J. W. J. Cornelissen, C. T. P. DeGroot, and J. A. T. M. Hurkmans, *Experientia* **16**, 373 (1960).

181. L. L. Simpson and D. R. Curtis, *Neuropoisons*, Vol. 2, Plenum Press, New York, 1974, p. 234.

182. J. Jarzebruski, *Farm. Pol.* **29**, 1107 (1973); *Chem. Abstr.* **81**, 41376 (1974).

183. H. Möhrle, J. Tenczer, H. Kessler, and G. Zimmermann, *Arch. Pharm.* **308**, 11 (1975).

184. H. A. Staab and A. Mannshreck, *Tetrahedron Lett.*, 913 (1962).

185. A. S. C. Wan, *J. Chromatogr.* **60**, 371 (1971).

186. S. Ebel, W. D. Mikulla, and K. H. Wersel, *Deutsch. Apoth. Ztg.* **111**, 931 (1971); *Chem. Abstr.* **75**, 80317 (1971); V. Massa, F. Gal, P. Susplugas, and G. Maestre, *Trav. Soc. Pharm. Montpellier* **30**, 267 (1970); *Chem. Abstr.* **75**, 25455 (1971).

187. J. Kalman, K. Toth, and D. Kuttel, *Acta Pharm. Hung.* **41**, 267 (1971); *Chem. Abstr.* **76**, 50003 (1972).

188. J. Jarzebinski, Z. Zakrzewski, and B. Nowiszewska-Jasinska, *Acta Pol. Pharm.* **30**, 485 (1973); *Chem. Abstr.* **81**, 6291 (1974).

189. T. Bican-Fister, B. Pavelic, and J. Merkas, *Acta Pharm. Jugoslav.* **23**, 17 (1973); *Chem. Abstr.* **79**, 9946 (1973).

190. A. Ben-Bassat and D. Lavie, *Israel J. Chem.* **10**, 385 (1972).

191. E. Brochmann-Hanssen, P. Schmid, and J. D. Benmaman, *J. Pharm. Sci.* **54**, 783 (1965).

192. E. Kunec-Vajic and K. Weber, *Croat. Chem. Acta* **37**, 211 (1965); *Chem. Abstr.* **64**, 638 (1966).

193. F. L. Pyman, *J. Chem. Soc.* **101**, 2260 (1912).

194. E. Leger and F. Roques, *C. R. Acad. Sci. Paris* **155**, 1088 (1912).

195. H. W. Voigtländer and W. Rosenberg, *Arch. Pharm.* **292**, 579 (1959).

196. W. Löwe and K.-H. Pook, *Justus Liebigs Ann. Chem.*, 1476 (1973).

197. E. Tedeschi, J. Kamionsky, D. Zeider, S. Fackler, S. Sarel, V. Usieli, and J. Deutsch, *J. Org. Chem.* **39**, 1864 (1974).

198. P. Pfeffer, S. African Patent 67 03,807; *Chem. Abstr.* **70**, 31682 (1969).

199. A. M. Poljakowa, W. A. Preobrashenski, and N. A. Preobrashenski, *J. Gen. Chem. USSR* **9**, 1402 (1939); *Chem. Abstr.* **34**, 1658 (1940); N. A. Dryamova, S. I. Zav'yalov, and N. A. Preobrashenski, *Zhur. Obshchei Khim.* **18**, 1773 (1948); *Chem. Abstr.* **43**, 2625 (1949).

200. J. K. Mehrotra and A. N. Dey, *J. Indian Chem. Soc.* **38**, 971 (1961).

201. H. Link and K. Bernauer, *Helv. Chim. Acta* **55**, 1053 (1972).

202. W. E. Oberhänsli, *Cryst. Struct. Comm.* **1**, 203 (1972).

203. S. Sarel, V. Usieli, and E. Tedeschi, *Tetrahedron Lett.*, 97 (1975).

204. H. Link, K. Bernauer, and W. E. Oberhänsli, *Helv. Chim. Acta* **57**, 2199 (1974).

205. H.-W. Voigtländer, G. Balsam, M. Engelhardt, and L. Pohl, *Arch. Pharm.* **311**, 927 (1978).

206. R. Robinson, *The Structural Relations of Natural Products*, Clarendon Press, Oxford, 1955, p. 96.

207. B. N. Ames, H. K. Mitchell, and M. B. Mitchell, *J. Am. Chem. Soc.* **75**, 1015 (1953).

208. E. Brochmann-Hanssen, M. A. Nunes, and C. K. Olah, *Planta Medica* **28**, 1 (1975).

209. F. Khuong-Huu, X. Monseur, G. Ratle, G. Lukacs, and R. Goutarel, *Tetrahedron Lett.*, 1757 (1973).

210. L. Tchissambou, M. Bénéchie, and F. Khuong-Huu, *Tetrahedron Lett.*, 1801 (1978).

211. R. R. Arndt, A. Jordaan, and V. P. Joynt, *J. Chem. Soc.*, *Suppl. II*, 5969 (1964).

212. A. Jordaan, V. P. Joynt, and R. R. Arndt, *J. Chem. Soc.*, 3001 (1965).

213. V. P. Joynt, R. R. Arndt, A Jordaan, K. Biemann, and J. L. Occolowitz, *J. Chem. Soc.* (B), 980 (1966).

214. R. R. Arndt, S. H. Eggers, and A. Jordaan, *Tetrahedron* **25**, 2767 (1969).

215. M. A. Wuonola and R. B. Woodward, *J. Am. Chem. Soc.* **95**, 284, 5098 (1973); ibid, *Tetrahedron* **32**, 1085 (1976).

216. N. K. Hart, S. R. Johns, J. A. Lamberton, J. W. Loder, and R. H. Nearn, *Chem. Commun.*, 441 (1970); ibid, *Austral. J. Chem.* **24**, 857 (1971).

217. K. Leander and B. Lüning, *Tetrahedron Lett.*, 905 (1968).

218. E. Söderberg and P. Kierkegaard, *Acta Chem. Scand.* **24**, 397 (1970).

219. P. M. Scott, M. A. Merrien, and J. Polonsky, *Experientia* **32**, 140 (1976); T. N. Medvedeva, M. V. Arinbasarov, and T. F. Solov'eva, *Sint. Issled. Biol. Soedin. Tezisy Dokl. Konf. Molodykh Uch.*, *6th*, 68 (1978); *Chem. Abstr.* **92**, 181447 (1980).

220. D. W. Nagel, K. G. R. Pachler, P. S. Steyn, P. L. Wessels, G. Gafner, and G. J. Kruger, *J. Chem. Soc. Chem. Comm.*, 1021 (1974).

Chapter Three

Quinolizidine Alkaloids of the Leguminosae: Structural Types, Analysis, Chemotaxonomy, and Biological Activities

A. Douglas Kinghorn
Department of Medicinal Chemistry and Pharmacognosy
College of Pharmacy
University of Illinois at Chicago
Chicago, Illinois

Manuel F. Balandrin
Plant Resources Institute
University Research Park
Salt Lake City, Utah

CONTENTS

1. INTRODUCTION

The Leguminosae (Fabaceae, pea family) is the third largest family of flowering plants after the Compositae and Orchidaceae, and consists of about 650 genera and 18,000 species [1]. Plants in this family are indigenous to all climatic conditions ranging from equatorial rain forests to the verges of dry and cold deserts [1]. Legumes are of extreme importance to mankind, because they are used not only as human foodstuffs and livestock forage and fodder, but also to conserve soil and fix nitrogen, and as sources of timber, fuel, oils, dyes, gums, and resins [1–9]. Plants of the Leguminosae rank second in economic importance only to those of the Gramineae, and the demand for legumes is likely to escalate as larger human populations begin to utilize more marginal agricultural lands [1, 2, 4, 5, 10, 11].

The Leguminosae is divided into three subfamilies, namely, the Caesalpinioideae, the Mimosoideae, and the Papilionoideae (Faboideae), with the latter two groups considered to be derived from the former [1, 12, 13]. The taxonomic borderlines between these subfamilies are not clearly delineated, because of the absence of marked extinction gaps, and the fact that some modern genera exhibit transitional morphological characteristics [1, 12, 13]. The largest legume subfamily is the Papilionoideae, which embraces approximately 440 genera and 12,000 species in 32 tribes, as recently reclassified by Polhill [14].

Over 450 alkaloids have been reported to occur in plants of the Leguminosae, with the majority of such compounds occurring in papilionaceous species [15–17]. About 5, 15, and 15%, respectively, of the species in the Caesalpinioideae, Mimosoideae, and Papilionoideae are known to biosynthesize alkaloids [16, 17]. Quinolizidine (lupine) alkaloids are the largest single group of legume alkaloids. They appear to be restricted in distribution to species in the more primitive tribes of the Papilionoideae, and have not so far been isolated from any plant in the Caesalpinioideae or Mimosoideae [15–18]. Lupine alkaloids have thus been shown to be of some use in establishing phylogenetic relationships at the generic and tribal levels in the Papilionoideae [14, 15, 17–20]. These compounds are also important because of their toxicity for humans and live-stock as constituents of poisonous plants [21–27], and because some of them exhibit potentially useful pharmacological activities [28–31].

In this chapter, the known legume quinolizidine alkaloids will be listed, and mention made of recent biosynthetic studies on these compounds that have contributed to an overall perspective of their biogenetic relationships. In addition, uses of modern analytical methods for identification of legume quinolizidine alkaloids will be described, with particular emphasis placed on recent studies utilizing gas chromatography/mass spectrometry (GC/MS). Next, the chemotaxonomic implications of these compounds in the tribes and genera of the Leguminosae will be considered, with mention made of the biological activities of quinolizidine alkaloids observed in higher animals. Finally, a short discussion of the hypothesized role of quinolizidine alkaloids in the plant will be presented. Many of the topics covered in this chapter have not been extensively reviewed by others. In contrast, certain other aspects of lupine alkaloids, such as their chemistry [32–46], methods of synthesis [32–40, 42–46], and detailed botanical occurrence [15, 17, 32–40, 42–46], as well as their spectroscopic characteristics, comprising optical rotary dispersion and circular dichroism spectra [38, 47–51], ultraviolet [40, 45], infrared [34, 38, 40, 43, 45, 46, 52–55], and mass spectra [34, 35, 39, 40, 42, 43, 46, 56–68], and proton [34, 38, 42–44, 46, 69–74], carbon-13 [40, 45, 75–77], and nitrogen-15 [78] nuclear magnetic resonance spectra, have been dealt with at length by others, especially in the excellent annual reviews published by the Royal Society of Chemistry [32–40, 42–46].

2. STRUCTURAL TYPES OF LEGUME QUINOLIZIDINE ALKALOIDS

There have been only two previous attempts to list comprehensively the structures of quinolizidine alkaloids found in the Leguminosae, with both being part of larger compilations of all the alkaloids found in this family [15, 17]. In the following paragraphs the structural variation of 161 legume quinolizidine alkaloids will be considered, with compounds grouped according to their postulated order of biogenetic advancement and/or structural complexity as follows: (1) alkaloids of the lupinine type (quinolizidines with simple substituents); (2) leontidine-type alkaloids; (3) sparteine-/lupanine-type quinolizidine alkaloids; (4) esters of sparteine-/lupanine-type alkaloids; (5) tricyclic degradation products of the sparteine-/lupanine-type alkaloids; (6) pyridone quinolizidine bases; (7) matrine-type quinolizidine alkaloids; (8) *Ormosia*-type quinolizidine alkaloids; (9) dimeric quinolizidine alkaloids; and (10) quinolizidine alkaloids whose structures do not fit into classifications 1 through 9. Obviously, until the biosynthetic origin of each of these compound classes is known, any attempt at their classification will be somewhat arbitrary. However, we hope that these lists of structures, which embrace the known legume quinolizidine alkaloid constituents up to mid-1982, will be of use to researchers in this area.

The quinolizidine alkaloid groups mentioned above are tabulated, respectively, in Figs. 1 through 10. For all compounds cited, a definitive structural reference and/or a reference for the initial isolation from a legume source are included in the figures. Wherever possible, compound rotatory directions are provided and are related to appropriate configurational notations in the struc-

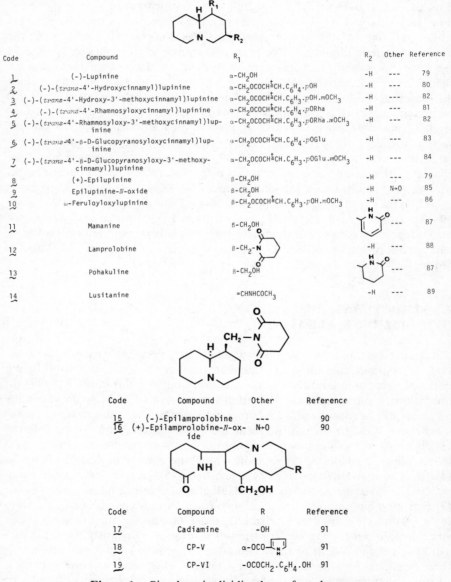

Code	Compound	R₁	R₂	Other	Reference
1	(-)-Lupinine	α-CH₂OH	-H	---	79
2	(-)-(trans-4'-Hydroxycinnamyl)lupinine	α-CH₂OCOCH≐CH.C₆H₄.pOH	-H	---	80
3	(-)-(trans-4'-Hydroxy-3'-methoxycinnamyl)lupinine	α-CH₂OCOCH≐CH.C₆H₃.pOH.mOCH₃	-H	---	82
4	(-)-(trans-4'-Rhamnosyloxycinnamyl)lupinine	α-CH₂OCOCH≐CH.C₆H₄.pORha	-H	---	81
5	(-)-(trans-4'-Rhamnosyloxy-3'-methoxycinnamyl)lupinine	α-CH₂OCOCH≐CH.C₆H₃.pORha.mOCH₃	-H	---	82
6	(-)-(trans-4'-β-D-Glucopyranosyloxycinnamyl)lupinine	α-CH₂OCOCH≐CH.C₆H₄.pOGlu	-H	---	83
7	(-)-(trans-4'-β-D-Glucopyranosyloxy-3'-methoxycinnamyl)lupinine	α-CH₂OCOCH≐CH.C₆H₃.pOGlu.mOCH₃	-H	---	84
8	(+)-Epilupinine	β-CH₂OH	-H	---	79
9	Epilupinine-N-oxide	β-CH₂OH	-H	N→O	85
10	ω-Feruloyloxylupinine	β-CH₂OCOCH≐CH.C₆H₃.pOH.mOCH₃	-H	---	86
11	Mamanine	β-CH₂OH		---	87
12	Lamprolobine	β-CH₂-N	-H	---	88
13	Pohakuline	β-CH₂OH		---	87
14	Lusitanine	=CHNHCOCH₃	-H	---	89

Code	Compound	Other	Reference
15	(-)-Epilamprolobine	---	90
16	(+)-Epilamprolobine-N-oxide	N→O	90

Code	Compound	R	Reference
17	Cadiamine	-OH	91
18	CP-V	α-OCO-	91
19	CP-VI	-OCOCH₂.C₆H₄.OH	91

Figure 1.   Simple quinolizidine bases from legumes.

tures. However, such information is not always known, especially when new structures are elucidated without compound isolation, for example, by GC/MS. For the ease of presentation of the *Ormosia* alkaloid structures (Fig. 8), only one antipodal series is provided for each compound, which may not correspond to the naturally occurring form. The compounds briefly discussed in Sections 2.1 through 2.10 in the text will primarily be those reported as novel legume constituents since 1978, since existing reviews [15, 17] have already described new legume quinolizidine alkaloids reported prior to this date.

2.1. Quinolizidines with Simple Substituents

Compounds related in structure to lupinine (**1**) are shown in Fig. 1. Additional substituted esters of (-)-lupinine (**3, 5–7**) were isolated from young seedlings of *Lupinus luteus* by Murakoshi et al. [82, 84]. This same Japanese group has also obtained two further alkaloids structurally related to lupinine, namely, epilamprolobine and its *N*-oxide (**15, 16**), from *Sophora tomentosa* leaves, where they were found to occur with a number of known alkaloids [90]. The 10-carbon compounds (-)-lupinine (**1**) and (+)-epilupinine (**8**) were the first legume quinolizidine alkaloids whose absolute configurations were determined [33, 79].

2.2. Leontidine-Type Quinolizidine Alkaloids

Derivatives of leontidine (**26**), a quinolizidine/indolizidine alkaloid, have only been known to occur in legumes (Fig. 2) since the discovery of three such compounds in *Camoensia maxima* roots in 1975 [93]. The structural scope of this

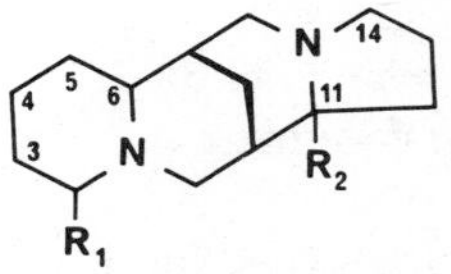

Code	Compound	R_1	R_2	Others	Reference
20	11-Epileontidane	-H	α-H	6β-H	92
21	Camoensidine	=O	α-H	6β-H	93
22	12-Hydroxycamoensidine	=O	α-H	6β-H, 12-OH	94
23	Camoensine	=O	α-H	$\Delta^{3,4}$, $\Delta^{5,6}$	93
24	12α-Hydroxycamoensine	=O	α-H	$\Delta^{3,4}$ $\Delta^{5,6}$, 12α-OH	94
25	Tetrahydroleontidine	=O	β-H	6β-H	92
26	Leontidine	=O	β-H	$\Delta^{3,4}$, $\Delta^{5,6}$	93
27	12-Hydroxy-16-methoxy-11:12, 13:14-tetrade-hydrocamoensine	=O	---	$\Delta^{3,4}$, $\Delta^{5,6}$, $\Delta^{11,12}$, $\Delta^{13,14}$, 12-OH, 16-OCH$_3$	94

Figure 2. Leontidine-type quinolizidine alkaloids from legumes.

alkaloid class in the family has been extended by the isolation of further representatives (**22, 24, 27**) from a second species in the genus, *C. brevicalyx* [94]. An additional unprecedented compound in the legumes, tetrahydroleontidine (**25**), was detected by GC/MS with its C(11) epimer, camoensidine (**21**), as well as the new compound, 11-epileontidane (**20**), in a stem-leaf extract of *Maackia amurensis* [92].

2.3. Sparteine-/Lupanine-Type Quinolizidine Alkaloids

Further representatives of this, the most structurally diverse group of legume quinolizidine alkaloids (Fig. 3), have been identified recently by our group by GC/MS, including 4α-hydroxysparteine (**29**) from *Acosmium panamense* [95], 10,17-dioxo-β-isoparteine (**47**) from *Lupinus sericeus* [108], and 5,6-dehydro-α-isolupanine (**60**) from *Lupinus bicolor* ssp. *microphyllus* [117]. In addition, chamaetine (**49**), the 4α-epimer of nuttalline (**50**), was found to be characteristic of six *Chamaecytisus* species by Daily and coworkers [109]. Calpurmenine (12β,13α-dihydroxylupanine) (**53**) was isolated from the South African species, *Calpurnia aurea* ssp. *sylvatica*, and structurally confirmed using X-ray crystallography [112, 113].

2.4. Esters of Sparteine-/Lupanine-Type Alkaloids

The esters of the sparteine-/lupanine-type quinolizidine alkaloids that have been isolated to date from legumes are listed in Fig. 4. In recent years only one such isolate has been discovered, namely, the 13α-pyrrolylcarboxylic acid ester of calpurmenine (**84**), which was found accompanying its parent alkaloid in *Calpurnia aurea* ssp. *sylvatica* [112, 113].

2.5. Tricyclic Degradation Products of
Sparteine-/Lupanine-Type Alkaloids

The structures of a group of tricyclic (nonpyridone) alkaloids that are postulated degradation products of the tetracyclic lupine alkaloids are shown in Fig. 5. Virgiboidine (**90**) was found as a novel constituent of two species in the African genus, *Virgilia* [125]. Further studies on a lupine alkaloid formerly known as "dehydroalbine" using various spectral and X-ray methods have enabled this compound to be structurally revised as albine (**91**), a compound that has no double bond between N(12) and C(11), and with the allyl group attached to C(13) rather than C(11) [126]. A tricyclic compound, LC-2 (**93**), related biogenetically to multiflorine (**64**), was obtained from the Australian lupine, *Lupinus cosentinii* [127].

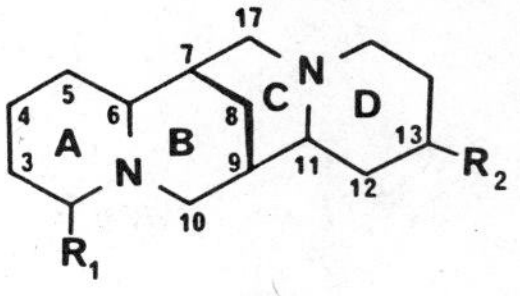

Code	Compound	R_1	R_2	Others	Reference
28	(-)-Sparteine	-H	-H	6β-H, 11α-H	79
29	(-)-4α-Hydroxysparteine	-H	-H	6β-H, 11α-H, 4α-OH	95
30	7-Hydroxysparteine	-H	-H	6β-H, 11α-H, 7β-OH	96,97
31	(+)-Retamine	-H	-H	6β-H, 11α-H, 12β-OH	34
32	13-Hydroxysparteine (Thermopsamine)	-H	α-OH	6β-H, 11α-H	98,99
33	(-)-Virgiline	-H	α-OH	6β-H, 11α-H, 10=O	100
34	(-)-Lindenianine	-H	β-OH	6β-H, 11α-H, 10=O	101
35	(-)-Aphylline	-H	-H	6β-H, 11α-H, 10=O	101
36	(+)-17-Oxosparteine	-H	-H	6β-H, 11α-H, 17=O	102
37	10,17-Dioxosparteine	-H	-H	6β-H, 11α-H, 10=O, 17=O	103
38	11,12-Dehydrosparteine	-H	-H	6β-H, Δ^{11,12}	92
39	(+)-Monspessulanine	-H	-H	6β-H, Δ^{11,12}, 17=O	104
40	(-)-Aphyllidine	-H	-H	Δ^{5,6}, 11α-H, 10=O	105
41	(-)-Oxoaphyllidine	-H	-H	Δ^{5,6}, 11α-H, 10=O, 17=O	34
42	(-)-α-Isosparteine	-H	-H	6β-H, 11β-H	79
43	Epiaphylline	-H	-H	6β-H, 11β-H, 10=O	103
44	(+)-β-Isosparteine	-H	-H	6α-H, 11α-H	79
45	(-)-7-Hydroxy-β-isosparteine	-H	-H	6α-H, 11α-H, 7α-OH	106
46	10-Oxo-β-isosparteine	-H	-H	6α-H, 11α-H, 10=O	15,107
47	10,17-Dioxo-β-isosparteine	-H	-H	6α-H, 11α-H, 10=O, 17=O	108
48	(+)-Lupanine	=O	-H	6β-H, 11α-H	79
49	(+)-Chamaetine (4α-hydroxylupanine)	=O	-H	6β-H, 11α-H, 4α-OH	109
50	(+)-Nuttalline (4β-hydroxylupanine)	=O	-H	6β-H, 11α-H, 4β-OH	110
51	(+)-13-Hydroxylupanine	=O	α-OH	6β-H, 11α-H	34
52	13-Ethoxylupanine	=O	α-OC$_2$H$_5$	6β-H, 11α-H	111
53	Calpurmenine	=O	α-OH	6β-H, 11α-H, 12β-OH	112,113
54	Jamaidine (13-epihydroxylupanine)	=O	β-OH	6β-H, 11α-H	114
55	4,13-Dihydroxylupanine	=O	-OH	6β-H, 11α-H, 4-OH	115
56	CP-II (10,13-dihydroxylupanine)	=O	-OH	6β-H, 11α-H, 10-OH	91
57	(+)-13-Epimethoxylupanine	=O	β-OCH$_3$	6β-H, 11α-H	34
58	(+)-17-Oxolupanine	=O	-H	6β-H, 11α-H, 17=O	34
59	5,6-Dehydrolupanine	=O	-H	Δ^{5,6}, 11α-H	116
60	(+)-α-Isolupanine	=O	-H	6β-H, 11β-H	79
61	5,6-Dehydro-α-isolupanine	=O	-H	Δ^{5,6}, 11β-H	117
62	Argyrolobine	-OH	-H	Δ^{5,6}, 11α-H, 10=O	105
63	Lupanoline	β-OH	-H	6α-H, 11α-H, 17=O	118

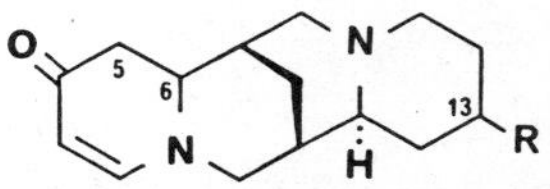

Code	Compound	R	Other	Reference
64	(-)-Multiflorine	-H	6β-H	110
65	13-Hydroxymultiflorine	α-OH	6β-H	34
66	5,6-Dehydro-13-hydroxy-multiflorine	α-OH	Δ^{5,6}	34

Figure 3. Legume quinolizidine alkaloids of the sparteine-/lupanine-type.

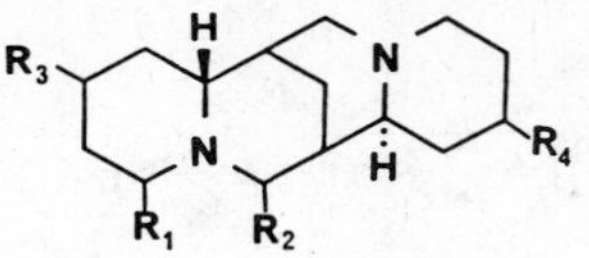

Code	Compound	R_1	R_2	R_3	R_4	Reference
67	(-)-O-(2-Pyrrolylcarbonyl)-virigiline	-H	=O	-H	α-OCO-[pyrrole]	100
68	(+)-13-Benzoyloxylupanine	=O	-H	-H	α-OCO.C$_6$H$_5$	34
69	(+)-13-*trans*-Cinnamoyloxylupanine	=O	-H	-H	α-OCOCH≝CH.C$_6$H$_5$	34
70	(+)-13-*cis*-Cinnamoyloxylupanine	=O	-H	-H	α-OCOCH≝CH.C$_6$H$_5$	34
71	Cadiaine	=O	-H	-H	α-OCOCH$_2$.C$_6$H$_4$.pOH	119
72	(+)-Calpurnine (Oroboidine)	=O	-H	-H	α-OCO-[pyrrole]	34
73	Cineverine	=O	-H	-H	α-OCO.C$_6$H$_3$.mOCH$_3$.pOCH$_3$	120
74	Cinegalline	=O	-H	-H	α-OCO-⟨OH,OCH$_3$,OH⟩	115
75	Cinegalleine	=O	-H	-H	α-OCO-⟨OCH$_3$,OCH$_3$,OH⟩	115
76	Cinevanine	=O	-H	-H	α-OCO.C$_6$H$_3$.mOCH$_3$.pOH	115
77	Isocinevanine	=O	-H	-H	α-OCO.C$_6$H$_3$.mOH.pOCH$_3$	121
78	O-Formylcinegalleine	=O	-H	-H	α-OCO-⟨OCH$_3$,OCH$_3$,OCHO⟩	115
79	Sarodesmine	=O	-H	-H	α-OCO-⟨OCH$_3$,OCH$_3$,OCH$_3$⟩	115
80	(+)-13α-Tigloyloxylupanine	=O	-H	-H	α-OCOC(CH$_3$)=CHCH$_3$	34
81	CP-III	=O	-OH	-H	α-OCO-[pyrrole]	91
82	Catalauverine	=O	-H	-OH	α-OCO.C$_6$H$_3$.mOCH$_3$.pOCH$_3$	115
83	Catalaudesmine	=O	-H	-OH	α-OCO-⟨OCH$_3$,OCH$_3$,OCH$_3$⟩	115

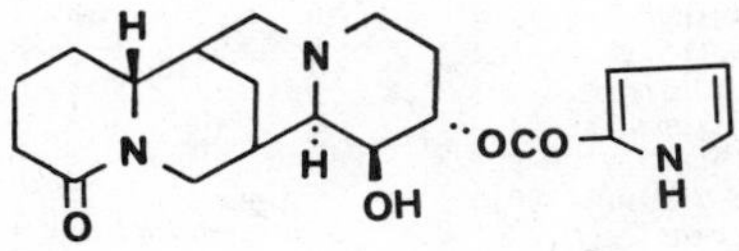

84   13α-(2'-Pyrrolylcarbonyl)calpurmenine   (Ref. 112)

Figure 4. Esters of sparteine-/lupanine-type quinolizidine alkaloids found in legumes.

2.6. Pyridone Quinolizidine Bases

In Fig. 6 the structures of the pyridone bases so far obtained as natural products from legumes are listed. The absolute configurations of a number of compounds of this type were established in 1961 [79, 161]. In recent isolation and structural studies, Murakoshi et al. have obtained *N*-ethylcytisine (**97**) and 12-cytisine-acetic acid (**102**) from extracts of *Echinosophora koreensis* [130] and *Euchresta japonica* [134], respectively. The methyl ester of **102** was considered by its discoverers to be an artifact [134]. Other new compounds of this type that have

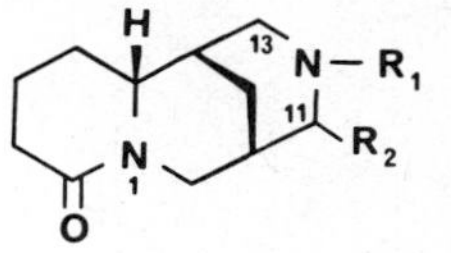

Code	Compound	R_1	R_2	Other	Reference
85	Tetrahydrorhombifoline	-(CH$_2$)$_2$CH:CH$_2$	-H	---	34
86	N-Methyltetrahydrocytisine	-CH$_3$	-H	---	122,123
87	Angustifoline (Jamaicensine)	-H	β-CH$_2$CH:CH$_2$	---	114
88	12,13-Dehydroangustifoline	---	β-CH$_2$CH:CH$_2$	Δ^{12,13}	124
89	11-Oxotetrahydrorhombifoline	-(CH$_2$)$_2$CH:CH$_2$	=O	---	58

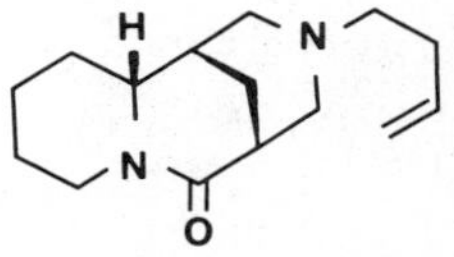

90 Virgiboidine (Ref. 125)

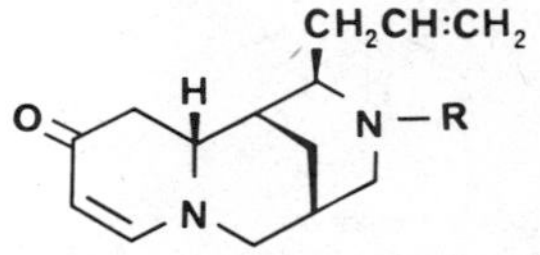

	R	Reference
91 Albine	-H	126
92 N-Methylalbine	-CH$_3$	34

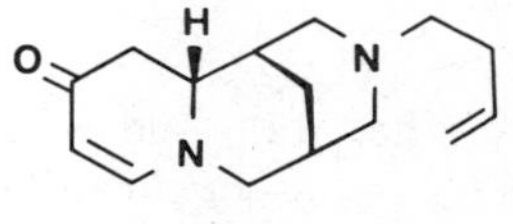

93 LC-2 (Ref. 127)

Figure 5. Tricyclic degradation products of sparteine-/lupanine-type quinolizidine alkaloids found in legumes.

been found as minor constituents of legume alkaloid fractions are 11-allylcytisine (**103**) [135] and 13-acetoxyanagyrine (**107**). The latter compound, while first suggested as a constituent of *Baptisia australis* leaves by GC/MS detection [138], was later confirmed as a constituent of two other legume extracts by chromatographic and spectroscopic comparison with the product obtained on acetylation of baptifoline (**106**) [139].

2.7. Matrine-Type Quinolizidine Alkaloids

A large number of tetracylic bases (Fig. 7), structurally related to matrine (**111**), have now been found as legume constituents. Recently, two such compounds,

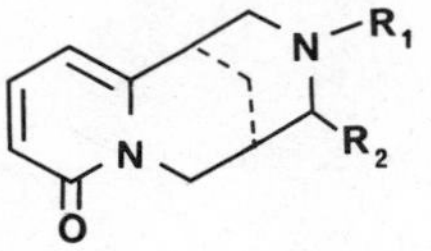

Code	Compound	R_1	R_2	Other	Reference
94	(-)-Cytisine	-H	-H	---	128
95	Cytisine-N-oxide	-H	-H	N→O	129
96	(-)-N-Methylcytisine	-CH$_3$	-H	---	128
97	(-)-N-Ethylcytisine	-CH$_2$CH$_3$	-H	---	130
98	(-)-Rhombifoline	-(CH$_2$)$_2$CH:CH$_2$	-H	---	128
99	(-)-N-Formylcytisine	-CHO	-H	---	131
100	(-)-N-Acetylcytisine	-COCH$_3$	-H	---	132
101	(-)-N-(3-Oxobutyl)cytisine	-(CH$_2$)$_2$COCH$_3$	-H	---	133
102	(-)-12-Cytisineacetic acid	-CH$_2$COOH	-H	---	134
103	11-Allylcytisine	-H	α-CH$_2$CH:CH$_2$	---	135
104	(-)-Tinctorine (Alteramine)	-CH$_3$	α-CH$_2$CH:CH$_2$	---	123,136

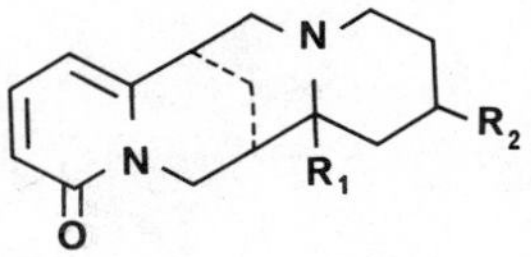

Code	Compound	R_1	R_2	Reference
105	(-)-Anagyrine	β-H	-H	79
106	(-)-Baptifoline	β-H	β-OH	137
107	13-Acetoxyanagyrine	β-H	β-OCOCH$_3$	138,139
108	(-)-Epibaptifoline	β-H	α-OH	34
109	(-)-Thermopsine	α-H	-H	79
110	(-)-Argentamine	α-H	β-OH	140

Figure 6.   Legume pyridone quinolizidine bases.

$\Delta^{5,17}$-dehydromatrine-N-oxide, (**128**) and $5\alpha,9\alpha$-dihydroxymatrine (**132**), were obtained from species in the genus *Euchresta* [145, 147], which thus represents only the third genus in the Papilionoideae where the matrine-type quinolizidine alkaloids have been reported to date. Also, the pyridone base, 7,8-dehydro-sophoramine (**127**), as well as 13, 14-dehydrosophoridine (**122**) were obtained as constituents of the same species, *Sophora alopecuroides*, by different groups [141, 144]. An unnamed stereoisomer (**120**) of sophocarpine (**118**) was isolated and characterized as a constituent of the epigeal parts of *Sophora flavescens* [142].

2.8.   *Ormosia*-Type Quinolizidine Alkaloids

The structures of all the *Ormosia*-type quinolizidine alkaloids found to date in the Leguminosae are shown in Fig. 8, with compounds **134–136** being examples of depiperidyl-*Ormosia* alkaloids. Recently, we showed that sweetinine, a partially characterized compound first isolated nearly 20 years earlier from the bark of *Acosmium panamense* (Benth.) Yakovlev (= *Sweetia panamensis*

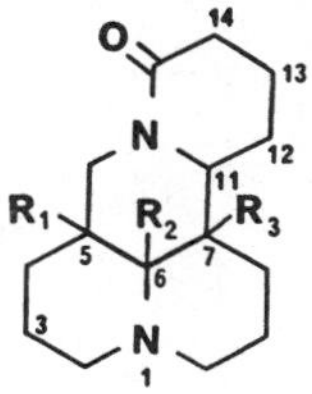

Code	Compound	R_1	R_2	R_3	Others	Reference
111	(+)-Matrine	α-H	α-H	α-H	11β-H	141
112	(+)-Matrine-N-oxide	α-H	α-H	α-H	11β-H, N_1→0	141
113	(+)-Allomatrine	α-H	β-H	α-H	11β-H	141
114	(-)-Sophoridine	β-H	α-H	α-H	11β-H	142
115	(-)-Sophoridine-N-oxide	β-H	α-H	α-H	11β-H, N_1→0	143
116	(+)-Isosophoridine	β-H	β-H	α-H	11β-H	142
117	(+)-Isomatrine	β-H	β-H	β-H	11β-H	141
118	(-)-Sophocarpine	α-H	α-H	α-H	11β-H, $\Delta^{13,14}$	141
119	(+)-Sophocarpine-N-oxide (Sophocarpidine)	α-H	α-H	α-H	11β-H, $\Delta^{13,14}$, N_1→0	141
120	Unnamed (stereoisomer of sophocarpine)	β-H	α-H	α-H	11β-H, $\Delta^{13,14}$	142
121	(-)-Sophoramine (Sophochrysine)	α-H	α-H	α-H	$\Delta^{11,12}$ $\Delta^{13,14}$	141
122	(-)-13,14-Dehydrosophoridine	β-H	α-H	α-H	11β-H, $\Delta^{13,14}$	141,144
123	(+)-13,14-Dehydrosophoridine-N-oxide	β-H	α-H	α-H	11β-H, $\Delta^{13,14}$, N_1→0	142,143
124	(+)-Isosophoramine	α-H	β-H	α-H	$\Delta^{11,12}$ $\Delta^{13,14}$	141
125	(-)-Neosophoramine	β-H	α-H	α-H	$\Delta^{11,12}$ $\Delta^{13,14}$	141
126	(+)-Lehmannine	β-H	β-H	β-H	11α-H, $\Delta^{12,13}$	141
127	(-)-7,8-Dehydrosophoramine	α-H	α-H	---	$\Delta^{7,8}$, $\Delta^{11,12}$, $\Delta^{13,14}$	141
128	(+)-5,17-Dehydromatrine-N-oxide	---	α-H	α-H	11β-H, $\Delta^{5,17}$, N_1→0	145
129	3α-Hydroxysophoridine	β-H	α-H	α-H	11β-H, 3α-OH	146
130	(+)-Sophoranol	α-OH	α-H	α-H	11β-H	141
131	(+)-Sophoranol-N-oxide	α-OH	α-H	α-H	11β-H, N_1→0	141
132	(+)-5α,9α-Dihydroxymatrine	α-OH	α-H	α-H	11β-H, 9α-OH	147
133	Sophorbenzamine	-H	-H	-H	$\Delta^{11,12}$, $\Delta^{13,14}$, $14-CH_2 \cdot C_6H_5$	148

Figure 7. Matrine-type quinolizidine alkaloids from legumes.

Benth.) is the racemate of the known *Ormosia*-alkaloid, 6-epipodopetaline (138) [155, 156]. Other isolates of this compound have been designated "amazonine" and "Alkaloid R-6" [152, 153]. Also, both the racemate and levoenantiomer of 6-epipodopetaline were isolated from the bark of *Podopetalum ormondii* by McLean and associates [154]. In the last-mentioned study, a number of constant melting point quasiracemate mixtures of the major alkaloidal constituents were obtained, since *Ormosia* alkaloids have a strong tendency to cocrystallize during standard isolation procedures [154]. Moreover, they may also be isolated in partially racemic forms [154]. The aminal derivative of sweetinine, (±)-homo-6-epipodopetaline (147), was also found as a constituent of various plant parts of *A. panamense* and *A. subelegans;* it has not been previously found to occur as a natural product [161]. The structure of homodasycarpine (149), a saturated alkaloid with the same carbon skeleton as the last-named compound, and a constituent of *Ormosia costulata*, was established by X-ray crystallography [163].

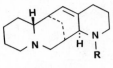

Code	Compound	R	Reference
134	Aloperine	-H	149
135	N-Methylaloperine	-CH₃	150
136	N-Allylaloperine	-CH₂CH:CH₂	149

Code	Compound	R_1	R_2	R_3	Reference
137	18-Epiamazonine	α-H	α-H	α-H	151
138	6-Epipodopetaline (= Amazo- nine = Alkaloid R-6 = Sweetinine)	α-H	α-H	β-H	152-156
139	Dihydroormojanine	α-H	β-H	α-H	151,157
140	(-)-Podopetaline	β-H	α-H	β-H	158

Code	Compound	R_1	R_2	R_3	R_4	Reference
141	(-)-Piptanthine	α-H	α-H	β-H	β-H	159
142	(-)-Templetine	α-H	β-H	β-H	β-H	159
143	16-Epiormosanine	β-H	α-H	α-H	β-H	160
144	18-Epiormosanine	β-H	α-H	β-H	α-H	122
145	(+)-Ormosanine (Piptamine)	β-H	α-H	β-H	β-H	159
146	Dasycarpine	β-H	β-H	β-H	α-H	157

Code	Compound	R_1	R_2	R_3	R_4	Other	Reference
147	Homo-6-epipodopetaline	α-H	--	β-H	α-H	$\Delta^{16,17}$	161
148	Jamine (Homo-ormosanine)	α-H	β-H	β-H	β-H	----	162
149	Homodasycarpine	β-H	β-H	α-H	β-H	----	163

Figure 8. *Ormosia*-type quinolizidine alkaloids from legumes.

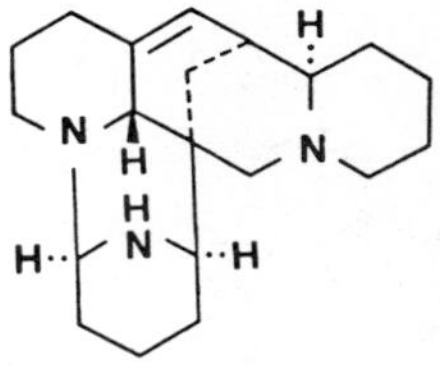

150 Ormojanine (Ref. 157)

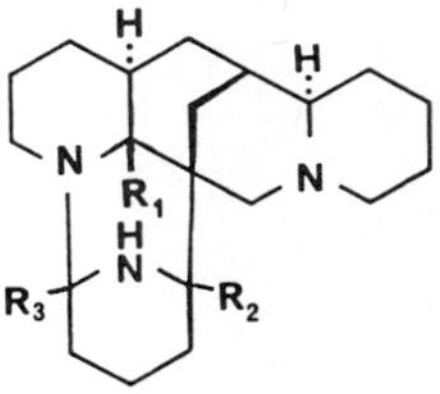

Code	Compound	R_1	R_2	R_3	Reference
151	Ormosajine	α-H	β-H	β-H	164
152	Panamine	β-H	α-H	α-H	159

Figure 8. (*Continued*)

2.9. Quinolizidine Alkaloid Dimers

The structures of the quinolizidine alkaloid dimers encountered during isolation studies on legume extracts are shown in Fig. 9. A collaborative carbon-13 and two-dimensional proton NMR study, involving several institutions [169], has shown that the known *Ormosia* alkaloid dimer, ormosinine (158), consists of one unit of panamine (152) and one unit of ormosanine (145), and not two units of panamine as formerly thought [15]. Thus, both ormojine (157) and ormosinine have the same carbon skeleton, but are stereoisomeric at three of the asymmetric carbon atoms present in each monomeric unit [168, 169].

2.10. Quinolizidine Alkaloids of Miscellaneous Structure

A compound originally termed tsukushinamine, isolated from *Sophora franchetiana*, was assigned a structure on the basis of spectral data that represents a new class of tetracyclic quinolizidine alkaloids with 15 carbon atoms [170] (Fig. 10). Later work using X-ray crystallography allowed the assignment of an unambiguous structure for this compound, now known as tsukushinamine A (159) [171]. Two further compounds of this type have also been isolated from this species, namely tsukushinamine B (160) and tsukushinamine C (161) [172].

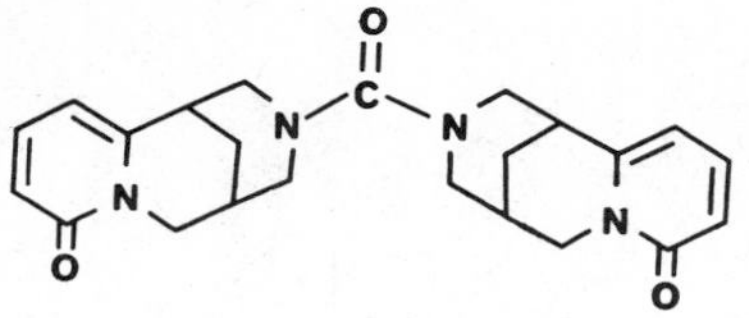

153   Argentine (Ref. 38) 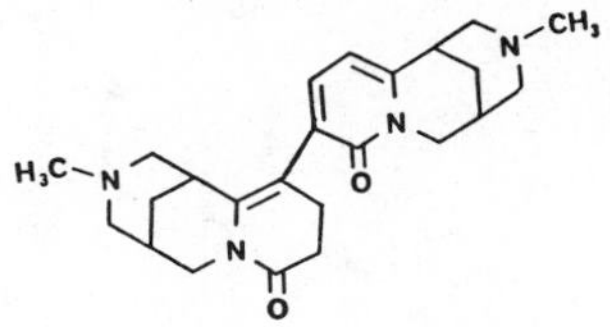   154   Dimethamine (Ref. 165)

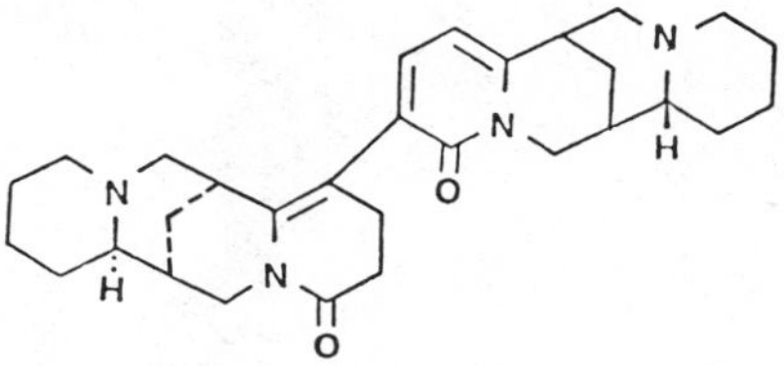

155   Dithermamine (Ref. 166) 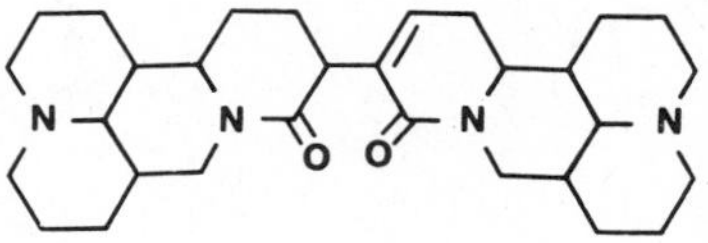   156   Goebeline (Ref. 167)

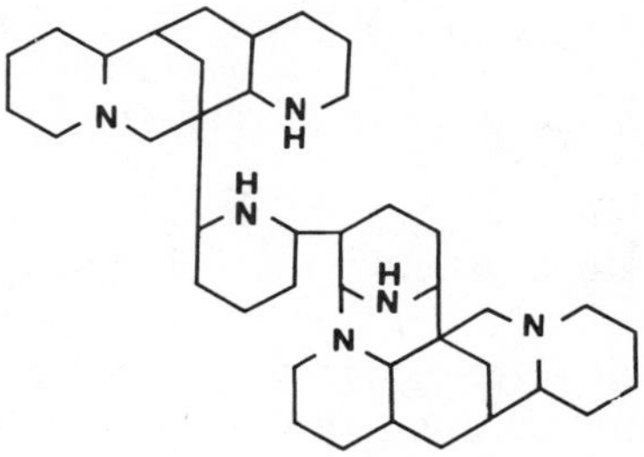

157   Ormojine   (Dimer of ormosajine) (Ref. 168)
158   Ormosinine (Ref. 169)

Figure 9.   Dimeric quinolizidine alkaloids from legumes.

3.   BIOGENETIC RELATIONSHIPS OF LEGUME QUINOLIZIDINE ALKALOIDS

It has been known for some time, as a result of radiotracer feeding experiments, that legume quinolizidine alkaloids are biosynthesized from the basic amino acid lysine, via cadaverine, its decarboxylation product [19, 23, 173–183]. Recent studies by Hartmann, Wink, and coworkers using cell suspension cultures of *Lupinus polyphyllus*, *Baptisia australis*, and *Sarothamnus scoparius*, as well as differentiated plants, have considerably expanded our knowledge of the bio-

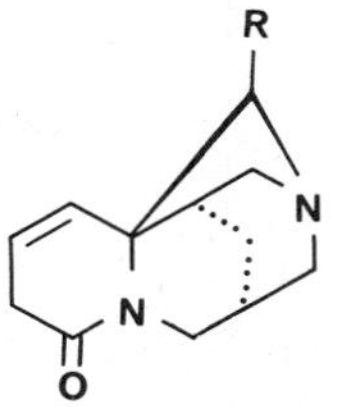

159 (-)-Tsukushinamine-A (R = β-CH$_2$CH:CH$_2$) (Ref. 170-172)

160 (-)-Tsukushinamine-B (R = α-CH$_2$CH:CH$_2$) (Ref. 172)

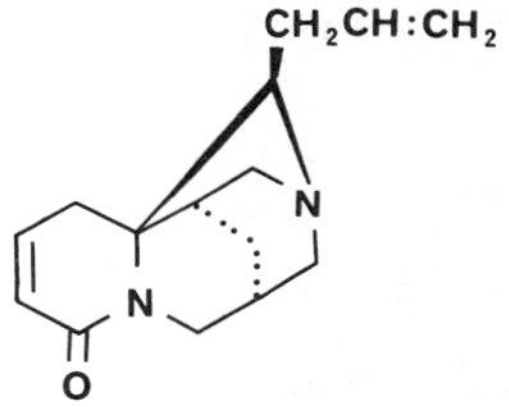

161 (-)-Tsukushinamine-C (Ref. 172)

Figure 10. Legume quinolizidine alkaloids of miscellaneous structure.

synthetic pathways leading to the sparteine-/lupanine-type of tetracyclic quino-lizidine alkaloids [138, 184–198]. These authors have been able to conclusively demonstrate that the biosynthesis of these compounds is controlled by two key enzymes which are localized in leaf chloroplasts, that is, lysine decarboxylase and 17-oxosparteine synthase [184–186]. Lysine decarboxylase catalyzes the conversion of lysine to cadaverine and is found in all parts of lupine plants [193]. 17-Oxosparteine synthase is a membrane-associated, pyruvate-dependent, cadaverine–pyruvate transaminating enzyme system that is localized exclu-sively in chloroplasts [184–186]. This enzyme catalyzes the conversion of three cadaverine units directly into 17-oxosparteine (**36**), in a coordinated, "chan-nelled" process which does not release free intermediates [185]. The participa-tion of diamine oxidase [175] in this pathway has been conclusively ruled out, and Δ^1-piperideine, tripiperidine [183, 199, 200], and tetrahydroanabasine have been shown *not* to be free intermediates in tetracyclic lupine alkaloid biosyn-thesis [184, 185]. Although Δ^1-piperideine has been observed to be incorporated into both sparteine-/lupanine- [199] and matrine-type [200] tetracyclic quino-lizidine alkaloids, one can argue that these observations may actually result from the incorporation of 5-aminopentanal, the open form of Δ^1-piperideine [185]. These new experimental findings are thus inconsistent with the previous biogenetic schemes proposed by Schütte [176], Nowacki and Waller [19, 179–182], and Spenser et al. [199, 200].

The key product and the first tetracyclic alkaloid to be synthesized is thus 17-oxosparteine (**36**), which is the immediate precursor of lupanine (**48**) [138, 185, 186, 189, 191, 194]. In whole plants, leaf chloroplasts convert cadaverine directly into lupanine, without releasing free intermediates, in a channeled process involving a sequence of membrane-associated enzymes [186]. Lupanine (**48**) is, in turn, the precursor of sparteine (**28**), 13-hydroxylupanine (**51**), and its esters, and the saturated tricyclic alkaloids [angustifoline (**87**) and tetrahydrorhombifoline (**85**)], [201], as well as the α-pyridone alkaloids and their respective tricyclic analogs [e.g., rhombifoline (**98**) and tinctorine (**104**), via 5,6-dehydrolupanine (**59**)] [138, 189, 191, 194]. This work thus confirms the "independent" biosynthesis of sparteine (**28**) and lupanine (**48**), and establishes that sparteine is not the precursor of lupanine as had previously been believed, but vice versa [138, 186, 189, 191, 194]. The independent biosynthesis of sparteine and lupanine had been suggested by previous findings [202]. 5,6-Dehydrolupanine (**59**) has previously been found to be the intermediate between lupanine and anagyrine (**105**) [116, 203], and it has been suggested that 5,6-dehydro-α-isolupanine (**61**) is probably the corresponding intermediate between α-isolupanine (**60**) and thermopsine (**109**) [117, 204]. A summary of Hartmann and Wink's findings on the biosynthesis of tetracyclic quinolizidine alkaloids is shown in Figs. 11*a* and 11*b*.

In addition, Hartmann and Wink have suggested that bicyclic lupinine-type alkaloids could be biosynthesized by a mechanism similar to that proposed for

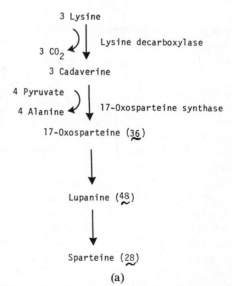

(a)

Figure 11. Summary of recent studies by Hartmann and Wink on tetracyclic lupine alkaloid biosynthesis. (*a*) Overview. (*b*) Stages in the transformation of cadaverine to 17-oxosparteine by 17-oxosparteine synthase [185].

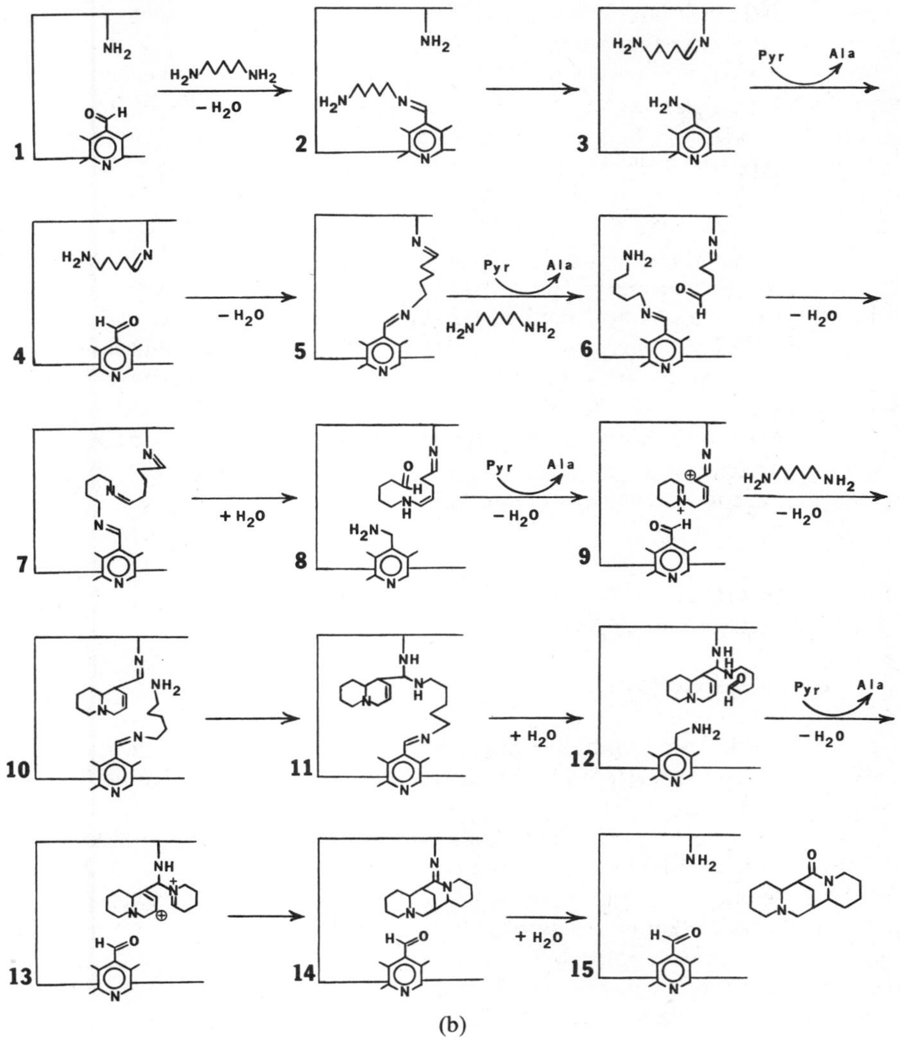

(b)

Figure 11. (*Continued*)

17-oxosparteine, but in which the bicyclic intermediate formed from two cadaverine units is "prematurely" released [185]. It also seems reasonable to assume that the tetracyclic matrine- (**111–133**) and aloperine- (**134–136**) type alkaloids may be biosynthesized in a "channeled" manner analogous to that of 17-oxosparteine. This process would exclude the participation of Δ^1-piperideine or the tripiperidine proposed by Leeper et al. for the biogenesis of matrine (**111**) [200]. Similarly, the C_{20}-pentacyclic *Ormosia* alkaloids could likely be biosynthesized "directly" from four cadaverine units, ruling out the intermediacy of a piperideine tetramer that has recently been proposed [205].

Since the enzymes involved in lysine biosynthesis and both lysine decarboxy-lase and 17-oxosparteine synthase are localized in leaf chloroplasts, lysine and quinolizidine alkaloid biosynthesis occur in the same subcellular compartment [186, 188, 190, 191, 193]. Studies with cell suspension cultures have established that functional chloroplasts are required for quinolizidine alkaloid production, and have demonstrated that alkaloid biosynthesis undergoes a diurnal rhythm [192, 195, 197]. Maximum alkaloid biosynthesis has been observed to occur during the day (or periods of illumination) and to drop off considerably at night [192, 195, 197]. This finding confirms previous observations [182] that suggest that lupine alkaloid biosynthesis undergoes light-dependent regulation [192, 195, 197]. In whole plants, lupine alkaloids are biosynthesized in the leaves and are then translocated to other plant parts such as the seeds, pods, and roots [192, 193, 197]. Alkaloid concentration in mature leaves has been observed to drop by about 90% during leaf senescence, indicating that the alkaloids are not being stored, but are in a dynamic state of turnover and/or translocation [193, 197].

4. ANALYTICAL METHODS FOR LEGUME QUINOLIZIDINE ALKALOIDS

4.1. Combined GC/MS

Combined gas chromatography/mass spectrometry (GC/MS) has been increas-ingly used over the last decade for the convenient analysis of lupine [92, 95, 103, 108, 117, 138, 139, 161, 187, 189, 194, 196, 201, 204, 206–209] and *Ormosia* [92, 153, 155, 156, 161, 210] quinolizidine alkaloids. Such compounds are ther-mostable and volatile under gas chromatographic conditions [211, 212]. This sensitive technique is applicable to the qualitative analysis of individual com-ponents of crude alkaloid fractions and is normally able to resolve lupine alka-loid diastereoisomeric pairs, with *cis, cis* isomers having a shorter column resi-dence time than *cis, trans* isomers [92, 103, 117, 204]. For work of a chemotaxo-nomic nature, GC/MS is particularly suitable, since in such studies it is desir-able to identify all the alkaloids that may have accumulated at a specific time and site of a collection in a specific part of a species, rather than only the most abundant compounds present [139]. Also, the use of GC/MS may enable the experimenter to rule out the presence of a particular alkaloid group in the plant material being examined [139]. Impressive separations of lupine alkaloids have been obtained using the high-column efficiencies achieved in capillary GC/MS [138, 187, 189, 194, 196].

In recent work carried out in our laboratory, we have found OV-17 (50% phenyl–50% methylsilicones) to be a satisfactory stationary phase for legume alkaloid separation during GC/MS, because of its high resolving power, as well as its high thermostability and consequent limited tendency for column bleeding [92, 95, 108, 139, 156, 161]. In Table 1, the GC/MS and TLC profiles of 50 naturally occurring and semisynthetic quinolizidine alkaloids are presented, together with data for the dipiperidyl alkaloid, ammodendrine, which is commonly found in legume alkaloid fractions. These data were obtained by injection into the apparatus of 1–5% w/v solutions in methylene chloride of the compounds tested. Retention times are expressed relative to lupanine (**48**) ($RR_t = 1.00$), the most common lupine alkaloid we have encountered. In most cases, fragment ion relative abundance data computed after GC/MS is similar to that obtained by direct probe analysis of these compounds. The need for the use of a combination of GC/MS and TLC analytical data for accurate compound identification may be vividly shown by referring to data we have obtained for tetrahydrorhombifoline (**85**) and its structural isomer, *N*-methylangustifoline, a compound obtained by treatment of angustifoline (**87**) with

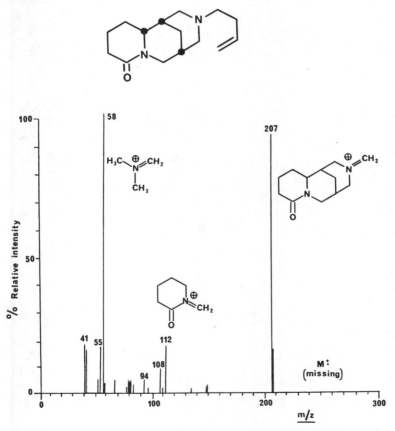

Figure 12. Electron-impact low-resolution mass spectrum of tetrahydrorhombifoline.

Table 1. Analytical Data (GC/MS and TLC) of Some Papilionaceous Quinolidizine Alkaloids and Their Derivatives

Compound	$RR_i{}^a$	Prominent Mass Spectral Fragment Peaks by GC/MS[b] (m/z, % relative abundance)	$R_f{}^c$		
			S_1	S_2	S_3
13-Acetoxyanagyrine (**107**)	1.81	302 (M^+, 19), 243 (21), 160 (9), 146 (27), 121 (10), 112 (9), 97 (18), 96 (100), 43 (22), 41 (23)	0.37	0.68	0.20
4α-Acetoxysparteine[d]	0.83	292 (M^+, 40), 251 (37), 233 (27), 195 (81), 136 (93), 135 (100), 98 (94), 97 (60), 96 (98), 41 (64)	—	—	—
4β-Acetoxysparteine[d]	0.80	M^+, not observed, 233 ($M^+ - 59$, 49), 136 (16), 135 (22), 134 (18), 110 (11), 98 (100), 97 (15), 96 (31), 43 (48), 41 (55)	—	—	—
N-Acetylcytisine (**100**)	1.45	232 (M^+, 36), 190 (15), 160 (19), 147 (84), 146 (100), 134 (24), 108 (22), 82 (25), 44 (100), 43 (68)	0.41	0.48	—
Ammodendrine[e]	0.58	208 (M^+, 48), 179 (43), 165 (80), 137 (42), 136 (55), 123 (65), 110 (79), 109 (68), 94 (40), 43 (100)	0.17	0.23	0.37
Anagyrine (**105**)	1.36	244 (M^+, 22), 243 (5), 229 (3), 160 (7), 146 (13), 136 (9), 122 (7), 98 (100), 97 (10), 41 (22)	0.49	0.66	0.28
Angustifoline (**87**)	0.96	M^+, not observed, 193 ($M^{+} - 41$, 100), 150 (20), 112 (96), 94 (13), 84 (11), 80 (18), 68 (10), 55 (26), 44 (21), 41 (25)	0.41	0.63	—
Aphyllidine (**40**)	0.91	246 (M^+, 42), 245 (19), 137 (14), 136 (18), 135 (15), 134 (16), 110 (16), 98 (100), 97 (55), 42 (22)	0.61	0.81	0.61
Aphylline (**35**)	1.07	248 (M^+, 35), 247 (33), 220 (43), 191 (19), 137 (46), 136 (100), 98 (31), 97 (40), 96 (35), 84 (32), 41 (47)	0.60	0.80	0.52
Baptifoline (**106**)	1.90	260 (M^+, 27), 241 (4), 160 (13), 146 (27), 145 (11), 134 (9), 114 (100), 96 (31), 70 (37), 41 (33)	0.40	0.33	0.05
Camoensidine (**21**)	0.93	234 (M^+, 52), 233 (44), 136 (59), 135 (56), 134 (22), 122 (100), 110 (21), 96 (39), 84 (52), 41 (45)	0.16	0.31	0.39
Cytisine (**94**)	0.96	190 (M^+, 57), 160 (23), 148 (35), 147 (82), 146 (100), 134 (32), 109 (22), 83 (25), 44 (93), 41 (57)	0.32	0.42	0.07
5,6-Dehydrolupanine (**59**)	0.95	246 (M^+, 18), 163 (4), 148 (4), 136 (5), 134 (7), 98 (100), 97 (34), 96 (8), 84 (11), 41 (25)	0.48	0.71	0.51

124

Compound		MS data: m/z (rel. int.)			
11,12-Dehydrosparteine (38)	0.41	232 (M^+, 26), 175 (12), 148 (22), 137 (35), 134 (75), 98 (100), 97 (88), 96 (34), 55 (30), 41 (68)	—	0.02	—
10,17-Dioxo-β-isosparteine (47)	1.42	262 (M^+, 26), 234 (6), 206 (6), 151 (22), 150 (100), 124 (27), 111 (30), 110 (27), 84 (66), 41 (38)	—	—	—
Epiaphylline (43)	0.88	248 (M^+, 35), 247 (33), 220 (43), 191 (19), 137 (46), 136 (100), 98 (31), 97 (40), 96 (35), 84 (32), 41 (47)	0.65	0.85	0.57
13-Epihydroxylupanine (54)	1.51	264 (M^+, 55), 247 (23), 166 (33), 165 (61), 152 (100), 150 (21), 114 (24), 113 (31), 112 (38), 94 (20)	0.34	0.43	0.16
11-Epileontidane (20)	0.23	220 (M^+, 25), 179 (42), 150 (11), 137 (91), 122 (35), 110 (23), 98 (96), 96 (52), 84 (100), 83 (75)	0.05	0.07	0.67
Epilupinine (8)	0.09	169 (M^+, 32), 168 (26), 152 (61), 138 (50), 124 (27), 111 (34), 110 (37), 97 (68), 83 (100), 55 (58)	0.24	0.10	0.49
13-Epimethoxylupanine (57)	1.29	278 (M^+, 35), 263 (58), 247 (72), 179 (42), 166 (81), 148 (43), 134 (43), 112 (57), 55 (100), 41 (95)	0.35	0.68	0.51
6-Epipodopetaline (138)	1.14	315 (M^+, 26), 257 (8), 231 (30), 217 (8), 149 (26), 98 (48), 97 (21), 96 (16), 84 (100), 56 (18)	0.17	0.40	0.60
N-Formylcytisine (99)	1.52	218 (M^+, 52), 160 (18), 148 (23), 147 (58), 146 (100), 134 (17), 117 (13), 109 (17), 82 (15), 44 (73)	0.40	0.44	—
Homo-6-epipodopetaline (147)	1.34	327 (M^+, 96), 312 (9), 284 (9), 243 (29), 229 (51), 122 (39), 98 (100), 84 (39), 55 (46), 41 (51)	0.24	0.58	0.60
13-Hydroxylupanine (51)	1.51	264 (M^+, 24), 246 (48), 165 (37), 152 (100), 149 (23), 137 (32), 134 (57), 112 (30), 55 (56), 41 (46)	0.28	0.24	0.16
4α-Hydroxysparteine (29)	0.68	250 (M^+, 27), 233 (5), 209 (35), 153 (100), 136 (41), 114 (71), 98 (56), 97 (34), 55 (24), 41 (39)	0.04	0.05	0.50
4β-Hydroxysparteine[d]	0.70	250 (M^+, 23), 233 (11), 209 (37), 153 (100), 136 (50), 110 (34), 98 (65), 97 (45), 84 (26), 55 (57)	0.04	0.05	0.50
α-Isolupanine (60)	0.94	248 (M^+, 42), 247 (30), 150 (32), 149 (50), 136 (100), 110 (22), 98 (37), 97 (22), 84 (22), 41 (54)	0.51	0.69	0.50
α-Isosparteine (42)	0.28	234 (M^+, 20), 193 (22), 150 (15), 137 (57), 110 (28), 98 (100), 97 (33), 84 (29), 55 (49), 41 (68)	0.08	0.07	0.80

Table 1. (*Continued*)

Compound	$RR_t{}^a$	Prominent Mass Spectral Fragment Peaks by GC/MS[b] (m/z, % relative abundance)	$R_f{}^c$		
			S_1	S_2	S_3
β-Isosparteine (**44**)	0.41	234 ($M^{+\cdot}$, 20), 193 (16), 150 (13), 137 (100), 136 (30), 98 (82), 97 (41), 84 (29), 55 (49), 41 (23)	0.11	0.15	0.79
Jamine (**148**)	1.44	329 ($M^{+\cdot}$, 89), 328 (46), 246 (90), 245 (36), 231 (89), 150 (35), 98 (80), 96 (44), 84 (100), 41 (77)	0.47	0.71	0.80
Lamprolobine (**12**)	1.00	264 ($M^{+\cdot}$, 21), 222 (11), 152 (43), 138 (100), 136 (25), 111 (39), 110 (53), 97 (61), 83 (74), 41 (38)	0.36	0.59	0.43
Lupanine (**48**)	1.00	248 ($M^{+\cdot}$, 39), 247 (21), 150 (37), 149 (49), 136 (100), 110 (24), 98 (30), 97 (28), 84 (19), 41 (40)	0.30	0.65	0.51
Lupinine (**1**)	0.09	169 ($M^{+\cdot}$, 32), 168 (28), 152 (54), 138 (52), 97 (66), 96 (51), 83 (100), 82 (43), 55 (57), 41 (57)	0.25	0.36	0.52
Matrine (**111**)	1.20	248 ($M^{+\cdot}$, 73), 247 (100), 219 (8), 205 (29), 177 (23), 150 (63), 136 (26), 98 (26), 96 (67), 41 (54)	0.26	0.52	0.35
N-Methylangustifoline[d]	0.83	$M^{+\cdot}$, not observed, 208 (14), 207 ($M^{+\cdot} - 41$, 100), 112 (45), 108 (8), 94 (7), 58 (26), 55 (14), 42 (12), 41 (18)	0.48	0.82	—
N-Methylcytisine (**96**)	0.80	204 ($M^{+\cdot}$, 7), 160 (2), 146 (3), 96 (3), 82 (3), 59 (4), 58 (100), 57 (3), 42 (8), 41 (7)	0.53	0.61	0.23
Multiflorine (**64**)	1.28	246 ($M^{+\cdot}$, 52), 189 (11), 164 (10), 148 (25), 134 (100), 110 (32), 98 (32), 84 (33), 55 (47), 41 (62)	0.27	0.35	0.26
Nuttalline (**50**)	1.09	264 ($M^{+\cdot}$, 44), 263 (24), 150 (45), 149 (32), 148 (35), 136 (100), 134 (38), 110 (26), 98 (48), 41 (62)	0.26	0.56	0.44
Ormosanine (**145**)	1.19	317 ($M^{+\cdot}$, 22), 234 (18), 233 (14), 221 (9), 219 (62), 151 (20), 110 (7), 98 (40), 96 (16), 84 (100)	0.06	0.08	0.76
Ormosinine (**158**)	1.32	$M^{+\cdot}$, not observed, 315 ($M^{+\cdot} - 315$, 27), 233 (13), 231 (11), 218 (19), 217 (100), 189 (13), 98 (64), 96 (21), 84 (17), 41 (13)	0.07	0.09	0.81
10-Oxo-β-isosparteine (**46**)	1.04	248 ($M^{+\cdot}$, 42), 220 (21), 150 (21), 136 (45), 123 (18), 110 (66), 98 (84), 97 (100), 84 (32), 41 (39)	—	—	—

			S_1	S_2	S_3
17-Oxolupanine (**58**)	1.35	262 ($M^{+\cdot}$, 36), 151 (23), 150 (100), 136 (11), 112 (29), 110 (41), 97 (33), 84 (31), 55 (29), 41 (22)	0.60	0.80	0.54
17-Oxosparteine (**36**)	0.86	248 ($M^{+\cdot}$, 32), 220 (26), 150 (19), 136 (47), 123 (30), 110 (76), 98 (91), 97 (100), 84 (24), 41 (75)	0.61	0.80	0.55
11-Oxotetrahydrorhombifoline (**89**)	1.53	262 ($M^{+\cdot}$, 14), 222 (14), 221 (100), 162 (5), 150 (19), 112 (15), 96 (6), 55 (29), 44 (21), 41 (22)	—	—	—
Panamine (**152**)	1.39	315 ($M^{+\cdot}$, 20), 233 (12), 231 (11), 218 (16), 217 (100), 189 (11), 137 (5), 98 (29), 96 (9), 84 (7)	0.21	0.27	0.46
Retamine (**31**)	0.62	$M^{+\cdot}$, not observed, 232 ($M^{+\cdot} - 18$, 10), 207 (15), 175 (10), 135 (18), 134 (24), 98 (100), 97 (28), 96 (21), 84 (17), 41 (25)	0.25	0.18	0.65
Rhombifoline (**98**)	1.03	$M^{+\cdot}$, not observed, 204 (8), 203 ($M^{+\cdot} - 41$, 55), 160 (12), 146 (14), 136 (8), 122 (8), 98 (15), 58 (100), 55 (20), 41 (34)	—	—	—
Sparteine (**28**)	0.32	234 ($M^{+\cdot}$, 17), 193 (24), 150 (12), 137 (87), 110 (23), 98 (100), 97 (41), 84 (22), 55 (23), 41 (39)	0.08	0.07	0.80
Tetrahydroleontidine (**25**)	0.78	234 ($M^{+\cdot}$, 27), 233 (19), 136 (44), 135 (29), 122 (54), 96 (53), 84 (72), 83 (52), 55 (67), 41 (100)	—	—	—
Tetrahydrorhombifoline (**85**)	0.83	$M^{+\cdot}$, not observed, 208 (9), 207 ($M^{+\cdot} - 41$, 58), 112 (9), 108 (10), 94 (4), 58 (100), 55 (9), 42 (14), 41 (16)	0.57	0.82	0.49
Thermopsine (**109**)	1.29	244($M^{+\cdot}$, 26), 243 (5), 229 (4), 160 (8), 146 (13), 136 (11), 122 (10), 98 (100), 97 (12), 41 (19)	0.63	0.77	0.35

[a] Retention data relative to that of lupanine (**50**); obtained by GC/MS on 3% OV-17 on Gas Chrom Q (100–120), temperature programmed 180–310°C, 4°C/min, He flow rate, 18 mL/min.
[b] Mass spectra obtained at 70 eV on a Varian MAT-112S instrument, equipped with a Varian 166 data system, linked to a Varian Aerograph Model 1440 gas chromatograph.
[c] Thin-layer chromatographic data obtained on silica gel G plates; S_1, methanol–28% ammonia (65:1), S_2, chloroform–methanol–28% ammonia (85:15:1), S_3, cyclohexane–diethylamine (7:3).
[d] Not known to date as a natural product.
[e] A dipiperidyl alkaloid commonly found in alkaloid extracts of legumes.

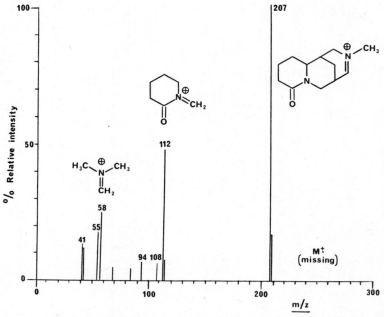

Figure 13.   Electron-impact low-resolution mass spectrum of *N*-methylangustifoline.

methyl iodide in acetone [201]. Both compounds were found to exhibit the same retention time on OV-17, and to produce identical mass spectral fragment peaks that varied only in intensity (Figs. 12 and 13). However, these compounds are separable by analytical TLC (Table 1) [201].

We have experienced only partial success in separating the C_{20}-*Ormosia* alkaloids by GC/MS. This is unfortunate, since these compounds have a tendency to cocrystallize, and thus fractional crystallization can not always be used successfully for their purification [154, 161, 210]. However, greater resolution of these compounds may be obtained by reaction with formaldehyde or phosgene to produce more rigid C_{21}-hexacyclic analogs [92, 156, 161]. The GC/MS methodology we have utilized to date requires considerable patience to complete the identification of esters of the sparteine-/lupanine-type alkaloids, since these compounds may require over an hour for elution from the chromato-

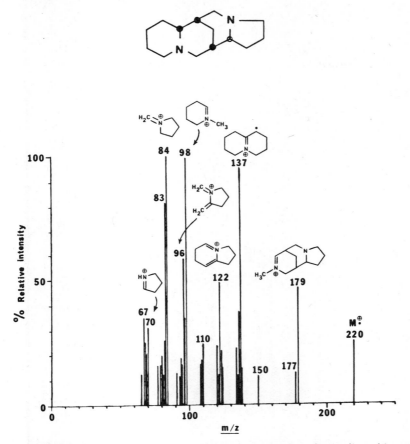

Figure 14.   Electron-impact low-resolution mass spectrum of 11-epileontidane.

graphic column, under the conditions cited in Table 1. In addition, these compounds may be lost during normal drying procedures of plant material containing lupine alkaloids, due to enzymatic hydrolysis [196].

The GC/MS technique is applicable not only to the identification of known legume alkaloids, but may also be used to establish the structures of new compounds, providing chemical interconversions involving compounds of known structure are conducted. The structures of the new natural products 11-epileontidane (**20**), 4α-hydroxysparteine (**29**), 10,17-dioxo-β-isosparteine (**47**), and 11,12-dehydrosparteine (= 5,6-dehydro-α-isoparteine) (**38**) were recently elucidated in this manner [92, 95, 108, 161]. Postulated mass spectral fragmentation patterns are represented in Figs. 14 and 15 for 11-epileontidane (**20**) and camoensidine (**23**), which are the quinolizidine/indolizidine analogs, respectively, of the common lupine alkaloids sparteine (**28**) and lupanine (**48**).

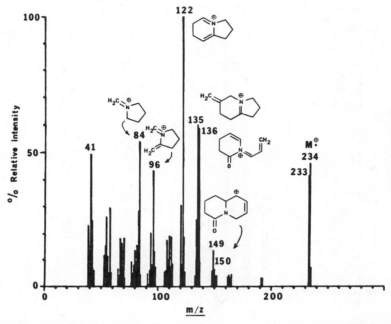

Figure 15. Electron-impact low-resolution mass spectrum of camoensidine.

4.2. Miscellaneous Methods of Qualitative and Quantitative Analysis of Lupine Alkaloids

An excellent general reagent for the detection of quinolizidine alkaloids on thin-layer plates is Munier's modification of Dragendorff's reagent [213]. However, certain dilactam representatives of this group of compounds (e.g., alkaloids **37, 47,** and **58**) produce only faint color reactions with this reagent, since they are neutral compounds [181, 182].

Because of the widespread interest in legumes containing quinolizidine alkaloids, many investigators have devised analytical methods for their identification and/or quantitation, using techniques such as precipitation reactions, titration, thin-layer chromatography, gas–liquid chromatography, and high-pressure liquid chromatography, either individually or in combination [92, 95, 98, 103, 108, 116, 138, 139, 161, 187, 189, 193, 201, 203, 204, 206–209, 211–227].

5. CHEMOTAXONOMIC IMPLICATIONS OF LEGUME QUINOLIZIDINE ALKALOIDS

Quinolizidine alkaloids have been reported to occur in only 10 of the most primitive tribes of the Papilionoideae (Fig. 16). The known distribution of these compounds in legumes is summarized in Table 2. Clearly from this table several types of quinolizidine alkaloid are useful for establishing phylogenetic relationships at the tribal and generic levels in the Papilionoideae, because of their restricted distribution within the subfamily [14, 15, 17–20, 92, 139, 204, 228]. Notable in this regard are the simple quinolizidine bases shown in Fig. 1, especially compounds related to pohakuline (**13, 17–19**), the leontidine-type quinolizidine alkaloids (**20–27**), the matrine series (**111–133, 156**), and the *Ormosia*-type alkaloids (**134–152, 157, 158**). In certain cases, the absence or cooccurrence of specific alkaloid structural groups also have important chemotaxonomic significance, as will be discussed in the following paragraphs.

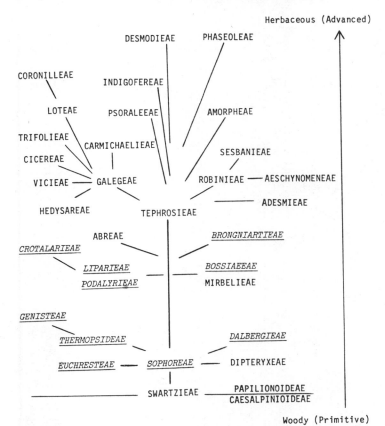

Figure 16. Tribal relationships in the Papilionoideae according to Polhill and the distribution of quinolizidine-alkaloid bearing genera (indicated in italics and underlined).

Table 2. Distribution of Quinolizidine Alkaloids in Genera of the Legume Subfamily Papilionoideae

Tribe	Genus	Simple	Leontidane Type	Sparteine/Lupanine Type[a]	Ester Type[b]	Pyridone Base	Matrine Type	Ormosia Type	Dimer
Sophoreae	*Acosmium*			+					
	Cadia			+	+			+	
	Camoensia		+						
	Ormosia			+		+		+	+
	Pericopsis			+		+			
	Clathrotropis			+		+			
	Diplotropis			+		+			
	Bolusanthus			+		+			
	Calpurnia			+	+	+			
	Maackia		+	+		+			
	Cladrastis	+							
	Sophora	+		+		+	+	+[c]	+
	Ammodendron			+		+			+
Euchresteae	*Euchresta*	+					+		
Dalbergieae	*Dalbergia*			+					
Thermopsideae	*Anagyris*			+		+			
	Piptanthus			+		+			
	Ammopiptanthus			+				+	
	Thermopsis	+		+		+			
	Baptisia			+		+			+

Tribe	Genus						
Genisteae	*Lupinus*			+	+	+	+
	Argyrolobium			+		+	
	Adenocarpus					+	
	Laburnum			+		+	
	Petteria			+		+	
	Chamaecytisus			+	+	+	
	Cytisus			+		+	
	Spartium		+	+		+	
	Retama			+		+	
	Genista			+	+	+	
	Teline			+		+	
	Ulex			+		+	
Podalyrieae	*Virgilia*				+	+	+
	Podalyria					+	
Liparieae	*Liparia*					+	
Crotalarieae	*Lebeckia*					+	
Bossiaeeae	*Lamprolobium*						+
	Templetonia	+		+		+	+
	Hovea	+		+		+	
Brongniartieae	*Harpalyce*						+

[a] Including tricyclic postulated decomposition products of sparteine/lupanine-type quinolizidine alkaloids.
[b] Comprising esters of sparteine/lupanine-type quinolizidine alkaloids.
[c] Depiperidyl–*Ormosia* alkaloids only detected in this genus.

Our current state of knowledge regarding the phylogenetic relationships among quinolizidine alkaloid-bearing tribes and genera is summarized in Fig. 17. The most primitive papilionates known to biosynthesize quinolizidine alkaloids belong to the *Cadia* group of genera (also referred to as the subtribe Cadieae or Acosmieae), the basal group of the tribe Sophoreae [14, 229]. In turn, the Sophoreae constitute the base group of the Papilionoideae as a whole, and are considered to be ancestral to all the rest of the quinolizidine alkaloid-bearing tribes of the subfamily [1, 12, 14, 229] (Fig. 17).

The South American genus *Acosmium* is the most primitive taxon of the Leguminosae known to biosynthesize quinolizidine alkaloids, and lies close to the borderline between the Caesalpinioideae and the Papilionoideae [95, 155, 161, 229]. Fossilized *Ascomium* species have been identified in tertiary strata in

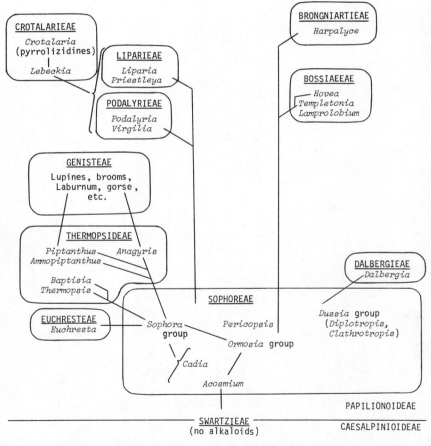

Figure 17. Phylogenetic relationships of quinolizidine-alkaloid bearing tribes and genera in the Papilionoideae.

South America, a finding which suggests that the biosynthetic pathways leading to the quinolizidine alkaloids are at least 50–65 million years old, extending back to the time of the divergence of the three legume subfamilies at the beginning of the Tertiary period [1, 12, 14, 161].

Our studies on *Acosmium* species have revealed that these primitive rain forest trees accumulate tetracyclic quinolizidine alkaloids, such as sparteine (**28**), α-isosparteine (**42**), 4α-hydroxysparteine (**29**), multiflorine (**64**), lupanine (**48**), and aphylline (**35**), as well as the tricyclic degraded alkaloids angustifoline (**87**) and tetrahydrorhombifoline (**85**), and *Ormosia*-type alkaloids related in structure to aloperine (**134**), 6-epipodopetaline (**138**), and ormosanine (**145**) [95, 156, 161]. 4α-Hydroxysparteine and multiflorine are undoubtedly biogenetically related in *Acosmium*, since one of these compounds is always found to the exclusion of the other [161]. Notably absent from all *Acosmium* extracts sampled so far are dipiperidyl alkaloids such as ammodendrine, simple quinolizidine alkaloids of the lupinine-type, "pre-pyridone" alkaloids such as 5,6-dehydrolupanine (**59**), and α-pyridone bases such as anagyrine (**105**) and cytisine (**94**) [161]. Pyridone bases are generally regarded as the most biogenetically advanced members of the lupine class of alkaloids, and their absence from *Acosmium* strongly suggests that the mutations that gave rise to the biosynthetic pathways leading to these compounds had not yet appeared when this taxon evolved [161]. In light of the recent findings of Wink and Hartmann [185], the first biosynthetic pathways to evolve in the Leguminosae were probably those leading directly to the tetracyclic and pentacyclic quinolizidines, with concomitant routes of degradation from tetracyclics to tricyclics. The development of pathways toward the dipiperidyl and lupinine-type alkaloids, as well as the oxidative pathways producing the pyridones must therefore have occurred subsequent to the evolution of this primitive genus [161].

In the slightly more advanced African genus, *Cadia*, *Ormosia*-type quinolizidine alkaloids appear to be absent, but various tetracyclic alkaloids and their esters, pohakuline-type compounds (**17–19**), and lupinine-type compounds, as exemplified by lusitanine (**14**), make their appearance [17, 119]. *Cadia* also appears to be devoid of pyridone alkaloids. Thus, it appears that plants in the papilionaceous subtribe Cadieae (Acosmieae) lack the enzyme systems necessary for oxidative reactions that would produce pyridone alkaloids.

In the more advanced *Ormosia* group of genera, the dipiperidyl alkaloid ammodendrine first appears, cooccurring with saturated tricyclic and tetracyclic lupine alkaloids, as well as *Ormosia* alkaloids [92, 161]. However, pyridones still do not appear to accumulate in this genus, with isolated reports of N-methylcytisine (**96**) possibly attributable to cases of taxonomic misidentification [15, 17, 122, 230].

The *Ormosia*-type quinolizidine alkaloid theme is perpetuated in *Sophora* (in the form of the depiperidyl representatives, **134–136**, in *S. alopecuroides*), *Ammopiptanthus*, *Hovea*, and *Templetonia* [17]. Recent revisions of the genus *Piptanthus* have now clarified the taxonomic status of the closely related genera *Piptanthus* and *Ammopiptanthus* [231, 232], and *Piptanthus* typically accumulates pyridone alkaloids, with *Ormosia*-type alkaloids apparently absent, whereas the opposite seems to occur in the case of *Ammopiptanthus* [17].

The pohakuline-type compounds (**13, 17–19**) that occur in *Cadia* and *Sophora* [17] may be biogenetically related to the pentacyclic *Ormosia* alkaloids. Geissman and Crout [177] have proposed a pohakuline-type intermediate in a biogenetic scheme for pentacyclic *Ormosia* alkaloids.

The quinolizidine/indolizidine alkaloids **20–27** have so far been detected only in the genera *Camoensia* [93, 94] and *Maackia* [92], both of the tribe Sophoreae. These alkaloids may be biogenetically related to C_{14} compounds such as angustifoline (**87**) or 11-allylcytisine (**103**).

Alkaloids of the tsukushinamine type (**159–161**), which have only been found in *Sophora franchetiana* to date, are presumed to be derived from anagyrine-type precursors, such as baptifoline (**106**), with which they cooccur [170–172]. Baptifoline-type compounds are also thought to be the biogenetic precursors for *N*-(3-oxobutyl)cytisine (**101**) and rhombifoline (**98**) [172]. Similarly, compounds such as calpurmenine (**53**) could be the biogenetic precursors for alkaloids such as tetrahydrorhombifoline (**85**).

The matrine-type alkaloids seem to be primarily distributed within the legumes in the *Sophora* group of genera (*Sophora, Ammothamnus, Echinosophora, Goebelia, Keyserlingia*) [17]. The recent discovery of this class of compounds in *Euchresta* [145, 147] therefore establishes a phylogenetic link between this genus and the *Sophora* group. No representatives of the matrine series of alkaloids were detected by GC/MS in a recent study on a species in the genus *Retama* [139]. Previous identifications of matrines in *Retama* species using TLC data alone [15] may thus be considered spurious, and the occurrence of these compounds in this genus may be regarded as unlikely.

Dimeric quinolizidine alkaloids such as dimethamine (**154**), dithermamine (**155**), and goebeline (**156**) appear to be highly restricted in distribution [17], and may be of chemotaxonomic value even at the species level. Hesse has suggested that such dimeric compounds may be formed via a radical coupling mechanism [233]. On the other hand, a urea such as argentine (**153**) could be an artifact formed by the reaction of cytisine (**94**) and traces of phosgene present in the chloroform used during isolation work.

Recent studies have suggested that tetracyclic and tricyclic quinolizidine alkaloids may be further metabolized in legumes by ring-B cleavage. (+)-Kuraramine, a dipiperidyl derivative, has been suggested as a metabolite of *N*-methylcytisine (**96**) in *Sophora flavescens* [234]. Similarly, the dipiperidyl alkaloidal acid, virgidivarine, appears to be a metabolite of virgiboidine (**90**) [125, 235]. These proposed metabolites bear the same relationship to their parent alkaloids that mamanine (**11**) and pohakuline (**13**) bear to thermopsine (**109**) and α-isolupanine (**60**), respectively. The relatively common legume dipiperidyl alkaloid, ammodendrine, may thus be a metabolite of a lupine alkaloid such as lupanine (**48**).

The occurrence of quinolizidine alkaloids in the genus *Lebeckia* of the tribe Crotalarieae [236] supports the view that the homologous pyrrolizidine alkaloids arose in the genus *Crotalaria* from a mutation of the quinolizidine alkaloid pathway such that ornithine became the alkaloid precursor instead of lysine [182]. The recent discovery of the toxic indolizidine alkaloid swainsonine and its *N*-oxide in the leguminous genera *Astragalus, Oxytropis,* and *Swainsona* [237] suggests a biosynthetic pathway intermediate between the quinolizidine and pyrrolizidine routes of biosynthesis.

6. BIOLOGICAL ACTIVITIES OF LEGUME QUINOLIZIDINE ALKALOIDS IN HIGHER ANIMALS

The toxicity of legume quinolizidine alkaloids to range animals is well known [22, 25, 27, 108, 204, 206, 218, 221, 222, 238–240]. Sheep appear to be especially susceptible to intoxication, with symptoms that include nervousness, incoordination, dyspnea, ataxia, convulsions, coma, and even death through respiratory paralysis [22, 25]. Quinolizidine alkaloid intoxication does not appear to be cumulative, and animals may eat large amounts of alkaloid-bearing legumes, provided that they do not consume a lethal dose on any one occasion [22, 25]. The highest incidence of livestock losses to quinolizidine alkaloid poisoning occur in the autumn, during the seeding stage [22, 25]. This is logical, since seeds are known to accumulate the highest quantities of alkaloids of all plant parts, probably as a means of defense against predation [22, 182, 193].

The pyridone alkaloids such as cytisine (**94**) and anagyrine (**105**) are more acutely toxic than the corresponding saturated alkaloids such as sparteine (**28**) and lupanine (**48**). Both cytisine and anagyrine have been implicated as teratogens in higher animals, and anagyrine has been shown to be the cause of "crooked calf disease" of cattle [206, 218, 221, 222, 238–240]. Although relatively rare, human poisoning cases by legume quinolizidine alkaloids, sometimes with fatal results, are occasionally reported [22, 24, 26].

Cytisine (**94**) has been shown to possess nicotine-like activity, while *N*-methylcytisine (**96**) is much less toxic [227, 241, 242]. Alkaloids such as sparteine (**28**) have long been known to possess antiarrhythmic ("cardiotonic") and uterotonic properties in both experimental animals and in humans [28, 30, 243–248]. Experimental evidence indicates that the pharmacological actions of sparteine, especially on the uterus, are mediated via prostaglandins [247]. In view of the uterotonic and teratogenic potential of certain lupine quinolizidine alkaloids, the fact that the alkaloids from the seeds of *Lupinus termis* were shown to have a detrimental effect on spermatogenesis in rats is of interest [249]. These alkaloids were also shown to have a growth-inhibiting effect when fed to the same test animals [250]. However, the alkaloids of *L. termis* have been shown to exert a mild hypoglycemic effect on diabetic rats [251].

The quinolizidine alkaloids and amino acids from *Sophora secundiflora* seeds have been shown to have a marked synergistic toxic effect when coadministered to experimental animals [242]. The seeds of this species (mescalbeans) have been reported to possess hallucinogenic properties in humans [223,

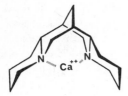

Figure 18.   Mode of binding of calcium by α-isosparteine.

252, 253]. *Genista canariensis*, a species known to accumulate pyridone alkaloids such as cytisine (**94**), has also been claimed to be a mild hallucinogen[254]. Studies with quinolizidine alkaloids such as sparteine (**28**), multiflorine (**64**), and lupinine (**1**) have established that they have a depressant effect on the central nervous systems of experimental animals [255–257]. In addition, human subjects who cannot rapidly metabolize sparteine show signs of CNS activity, such as diplopia, blurred vision, dizziness, and headache [258].

Ormosia-alkaloids have been claimed to possess hypnotic and analgesic ("morphine-like") activities in humans[259], but this activity has not been confirmed in recent studies [161]. Piptanthine (**141**) is claimed to possess pharmacological properties resembling those of sparteine (**28**), but with a lower potency [260]. Oxypanamine, an incompletely structurally characterized air-oxidation product of the *Ormosia*-alkaloid panamine (**152**), was found to have hypotensive, vasodepressor, and histamine-like activities in test animals [261–266]. A study of the acute toxicity of *Ormosia* seeds in experimental animals demonstrated that the alkaloid fraction induced symptoms of toxicity which included lethargy, flaccidity, pareses, and partial loss of the righting reflex [267]. Histopathological findings in those animals that succumbed to the *Ormosia* treatment were restricted to the lungs [267].

The pharmacological activity of the sparteines has been postulated to be related to their ability to bind or chelate divalent cations such as calcium [268–273]. Diamines such as sparteine and its stereoisomers can act as ligands for divalent cations such as magnesium and calcium (Fig. 18) [268–273]. In addition, sparteine has been implicated in membrane processes involving potassium conductance [274–276].

The pharmacological properties of a number of legume quinolizidine alkaloids are summarized in Table 3. Pharmacological evaluations of several legume alkaloids revealed no useful analgesic, antiinflammatory [279], or neuroleptic activities [280].

7.   FUNCTIONS OF QUINOLIZIDINE ALKALOIDS IN LEGUMES

Although at one time alkaloids were thought to represent metabolic waste products or forms of nitrogen storage, these speculations are no longer accepted and alkaloids are now generally regarded as toxic defense compounds which plants biosynthesize in order to ward off predators [182, 281]. The quinolizidine

Table 3. **Examples of Biological Activity of Legume Quinolizide Alkaloids**

Alkaloid	Structure	Type of Activity	Reference
Sparteine	28	Oxytocic; uterotonic	30, 243, 247, 248
		Antiarrhythmic	28, 245
		Diuretic	23, 174
		Respiratory depressant/stimulant	29, 30
		Hypoglycemic	251
Retamine	31	Uterotonic	34
		Hypotensive	34
		Diuretic	34
Lupanine	48	Antiarrhythmic	245
		Hypotensive	277
		Hypoglycemic	251
13-Hydroxylupanine	51	Antiarrhythmic	245
		Hypotensive	277
Calpurnine	72	Ichthyotoxic	278
		Antiarrhythmic	31
		Hypotensive	31
Multiflorine	64	CNS depressant	256, 257
Anagyrine	105	Teratogenic (calves)	206, 218, 221, 222 238
Cytisine	94	Teratogenic (chicks, rabbits)	218, 238
		Respiratory stimulant (nicotine-like activity)	24, 25, 29, 241
		Hallucinogenic	223, 252
		Oxytocic, uterotonic	30

alkaloids lupinine (1), sparteine (28), lupanine (48), 13-hydroxylupanine (52), and 17-hydroxylupanine are toxic to, and inhibit the growth of, the hemipterous insect *Acyrthosiphon pisum*, an aphid that normally feeds on quinolizidine alkaloid-free peas and beans [282, 283]. In addition, lupinine (1) has been shown to be a toxic feeding deterrent and growth inhibitor for the two-striped grasshopper, *Melanoplus bivittatus* (Orthoptera) [282, 284]. Recent studies by Wink and Hartmann have confirmed that, in general, aphids avoid plants with high quinolizidine alkaloid concentrations in favor of "sweet" lupines and other legumes [193]. These authors have also found that phloem-feeding aphids contain quinolizidine alkaloids [197]. Thrips (*Frankliniella* spp.) preferentially feed on sweet lupines, and this behavior has been recommended as an extremely rapid and accurate biological screening technique for selecting low-alkaloid containing individuals [285, 286]. A study of Colorado populations of herbaceous lupines revealed different alkaloid profiles which could be correlated to the degree of susceptibility of these populations to attack by the larvae of the lycaenid butterfly, *Glaucopsyche lygdamus* [219]. The distinct alkaloid patterns observed were considered to represent alternative chemical defense strategies. Thus, individual variability in alkaloid profiles would constitute an antispecialist chemical defense mechanism which impedes or prevents selection of resistant strains of phytophagous insect pests. This seems to confirm the hypothesis that quinolizidine alkaloids act as natural insecticides which afford plants protection from generalist ("browsing") herbivores because of their toxicity or unpalatability [219]. These findings concur with those of other workers, suggesting that quinolizidine alkaloids are general antipredator feeding deterrents which serve as defense compounds [182, 193]. On the other hand, sparteine (**28**) in *Sarothamnus scoparius* acts as a feeding stimulant for the specialized aphid, *Acyrthosiphon spartii* (Hemiptera) [282, 287, 288]. Larvae of the blue lycaenid butterfly, *Plebejus icarioides*, are obligate feeders on several alkaloid-rich *Lupinus* species [182, 289]. In this case the butterfly larvae readily tolerate the lupine alkaloids present in these plants, although they are not essential to their diet for survival [182]. These findings suggest that certain specialized insects have coevolved with and adapted to "bitter" legumes and are now able to detoxify and/or sequester the quinolizidine alkaloids from their diet, perhaps to some ecological advantage. This parallels the findings with some other toxic plant constituents, such as the well-documented case of Monarch butterflies which accumulate and metabolize cardenolides from the *Asclepias* (milkweed) species on which they normally feed [290].

Quinolizidine alkaloids are also general feeding deterrents for herbivorous mammals. European hares tend to avoid eating bitter lupines, but readily consume "sweet" varieties with a lower alkaloid content [182], and range animals avoid feeding on quinolizidine–alkaloid bearing legumes if other forages are available [22]. Therefore, the role of these compounds in legume defense seems to be established beyond doubt [182, 291].

REFERENCES

1. R. M. Polhill, P. H. Raven, and C. H. Stirton, in R. M. Polhill and P. H. Raven, Eds., *Advances in Legume Systematics*, Royal Botanic Gardens, Kew, U.K., 1981, p. 1.

2. R. J. Summerfield and A. H. Bunting, Eds., *Advances in Legume Science*, Royal Botanic Gardens, Kew, U.K., 1980.

3. J. A. Duke, *Handbook of Legumes of World Economic Importance*, Plenum Press, New York, 1980.

4. O. N. Allen and E. K. Allen, *The Leguminosae, a Source Book of Characteristics, Uses, and Nodulation*, University of Wisconsin Press, Madison, Wisconsin, 1981.

5. D. Isely, *Econ. Bot.* **36**, 46 (1982).

6. L. Kaplan, *Econ. Bot.* **35**, 240 (1981).

7. T. Hymowitz and C. A. Newell, *Econ. Bot.* **35**, 272 (1981).

8. P. Felker, *Econ. Bot.* **35**, 174 (1981).

9. E. C. Saxon, *Econ. Bot.* **35**, 163 (1981).

10. *Underexploited Tropical Plants with Promising Economic Value*, National Academy of Sciences, Washington, D.C., 1975.

11. *Tropical Legumes: Resources for the Future*, National Academy of Sciences, Washington, D.C., 1979.

12. P. H. Raven and R. M. Polhill, in R. M. Polhill and P. H. Raven, Eds., *Advances in Legume Systematics*, Royal Botanic Gardens, Kew, U.K., 1981, p. 27.

13. D. Isely and R. M. Polhill, *Taxon* **29**, 105 (1980).

14. R. M. Polhill, in R. M. Polhill and P. H. Raven, Eds., *Advances in Legume Systematics*, Royal Botanic Gardens, Kew, U.K., 1981, p. 191.

15. J. A. Mears and T. J. Mabry, in J. B. Harborne, D. Boulter, and B. L. Turner, Eds., *Chemotaxonomy of the Leguminosae*, Academic Press, New York, 1971, p. 73.

16. S. J. Smolenski and A. D. Kinghorn, in R. M. Polhill and P. H. Raven, Eds., *Advances in Legume Systematics*, Royal Botanic Gardens, Kew, U.K., 1981, p. 579.

17. A. D. Kinghorn and S. J. Smolenski, in R. M. Polhill and P. H. Raven, Eds., *Advances in Legume Systematics*, Royal Botanic Gardens, Kew, U.K., 1981, p. 585.

18. C. M. R. Gomes, O. R. Gottlieb, R. C. Gottlieb, and A. Salatino, in R. M. Polhill and P. H. Raven, Eds., *Advances in Legume Systematics*, Royal Botanic Gardens, Kew, U.K., 1981, p. 465.

19. E. K. Nowacki and G. R. Waller, *Rev. Latinoam. Quím.* **8**, 49 (1977).

20. A. Salatino and O. R. Gottlieb, *Biochem. Syst. Ecol.* **8**, 133 (1980).

21. F. Galinovsky, *Fortschr. Chem. Org. Naturst.* **8**, 245 (1951).

22. J. M. Kingsbury, *Poisonous Plants of the United States and Canada*, Prentice-Hall, Englewood Cliffs, N.J., 1964.

23. G. A. Swan, *An Introduction to the Alkaloids*, Wiley, New York, 1967.

24. H. G. H. Richards and A. Stephens, *Med. Sci. Law* **10**, 260 (1970).

25. E. C. G. Clarke, in R. H. F. Manske, Ed., *The Alkaloids, Chemistry and Physiology*, Vol. 12, Academic Press, New York, 1970, p. 514.

26. J. Schmidlin-Mészáros, *Mitt. Geb. Lebensmittelunters. Hyg.* **64**, 194 (1973).

27. S. J. Smolenski, A. D. Kinghorn, and M. F. Balandrin, *Econ. Bot.* **35**, 321 (1981).

28. E. L. McCawley, in R. H. F. Manske, Ed., *The Alkaloids, Chemistry and Physiology*, Vol. 5, *Pharmacology*, Academic Press, New York, 1955, p. 79.

29. M. J. Dallemagne and C. Heymans, in R. H. F. Manske, Ed., *The Alkaloids, Chemistry and Physiology*, Vol. 5, *Pharmacology*, Academic Press, New York, 1955, p. 109.

30. A. K. Reynolds, in R. H. F. Manske, Ed., *The Alkaloids, Chemistry and Physiology*, Vol. 5, *Pharmacology*, Academic Press, New York, 1955, p. 163.

31. E. Lindner, J. Kaiser, and U. Schacht, *Arzneim.-Forsch.* **26**, 1651 (1976).

32. N. J. Leonard, in R. H. F. Manske and H. L. Holmes, Eds., *The Alkaloids, Chemistry and Physiology*, Vol. 3, Academic Press, New York, 1953, p. 119.

33. N. J. Leonard, in R. H. F. Manske, Ed., *The Alkaloids, Chemistry and Physiology*, Vol. 7, Academic Press, New York, 1960, p. 253.

34. F. Bohlmann and D. Schumann, in R. H. F. Manske, Ed., *The Alkaloids, Chemistry and Physiology*, Vol. 9, Academic Press, New York, 1967, p. 175.

35. J. E. Saxton, *Alkaloids* (*London*) **1**, 86 (1971).

36. J. E. Saxton, *Alkaloids* (*London*) **2**, 79 (1972).

37. J. E. Saxton, *Alkaloids* (*London*) **3**, 95 (1973).

38. J. E. Saxton, *Alkaloids* (*London*) **4**, 104 (1974).

39. J. E. Saxton, *Alkaloids* (*London*) **5**, 93 (1975).

40. M. F. Grundon, *Alkaloids* (*London*) **6**, 90 (1976).

41. Z. Valenta and H. J. Liu, in K. Wiesner and D. H. Hey, Eds., *International Review of Science, Organic Chemistry, Series Two, volume 9, Alkaloids*, Butterworths, Boston, 1976, p. 1.

42. M. F. Grundon, *Alkaloids* (*London*) **7**, 69 (1977).

43. M. F. Grundon, *Alkaloids* (*London*) **8**, 66 (1978).

44. M. F. Grundon, *Alkaloids* (*London*) **9**, 69 (1979).

45. M. F. Grundon, *Alkaloids* (*London*) **10**, 66 (1981).

46. M. F. Grundon, *Alkaloids* (*London*) **11**, 63 (1981).

47. S. Iskandarov and S. Y. Yunusov, *Khim. Prir. Soedin.* **6**, 494 (1970).

48. S. Iskandarov, R. A. Shaimardanov, and S. Y. Yunusov, *Khim. Prir. Soedin.* **7**, 636 (1981).

49. A. I. Ishbaev, K. A. Aslanov, A. S. Sadykov, and M. A. Ramazanova, *Khim. Prir. Soedin.* **8**, 328 (1972).

50. W. Klyne, P. M. Scopes, R. N. Thomas, J. Skolik, J. Gawroński, and M. Wiewiórowski, *J. Chem. Soc., Perkin Trans. I* **1974**, 2565.

51. W. Wysocka and J. Gawroński, *J. Chem. Soc., Perkin Trans. I* **1979**, 1948.

52. J. Skolik, P. J. Krueger, and M. Wiewiórowski, *Tetrahedron* **24**, 5439 (1968).

53. J. Skolik, M. Wiewiórowski, and P. J. Krueger, *J. Mol. Struct.* **5**, 461 (1970).

54. Z. U. Petrochenko, A. I. Begisheva, K. A. Aslanov, and A. S. Sadykov, *Khim. Prir. Soedin.* **8**, 798 (1972).

55. T. K. Yunusov, A. P. Matveeva, V. B. Leont'ev, F. G. Kamaev, K. A. Aslanov, and A. S. Sadykov, *Khim. Prir. Soedin.* **8**, 200 (1972).

56. N. Neuner-Jehle, H. Nesvadba, and G. Spiteller, *Monatsh. Chem.* **95**, 687 (1964).

57. N. Neuner-Jehle, D. Schumann, and G. Spiteller, *Monatsh. Chem.* **98**, 836 (1967).

58. S. McLean, A. G. Harrison, and D. G. Murray, *Can. J. Chem.* **45**, 751 (1967).

59. D. Schumann, N. Neuner-Jehle, and G. Spiteller, *Monatsh. Chem.* **99**, 390 (1968).

60. H. M. Fales, H. A. Lloyd, and G. W. A. Milne, *J. Am. Chem. Soc.* **92**, 1590 (1970).

61. S. Iskandarov, Y. V. Rashkes, D. D. Kamalitdinov, and S. Y. Yunusov, *Khim. Prir. Soedin.* **5**, 331 (1969).

62. Y. D. Cho and R. O. Martin, *Arch. Mass Spectral Data* **2**, 328 (1971).

63. Y. D. Cho and R. O. Martin, *Arch. Mass Spectral Data* **2**, 732 (1971).

64. N. S. Wulfson, Z. S. Ziyavidinova, and V. G. Zaikin, *Org. Mass Spectrom.* **7**, 1313(1973).

65. V. G. Zaikin, Z. S. Ziyavidinova, and N. S. Vul'fson, *Izv. Akad. Nauk. SSSR, Ser. Chim.* **23**, 1734 (1974).

66. N. S. Vul'fson, Z. S. Ziyavidinova, and V. G. Zaikin, *Khim. Geterotsikl. Soedin.* **10**, 251 (1974).

67. N. S. Vul'fson and V. G. Zaikin, *Usp. Khim.* **45**, 1870 (1976).

68. Y. K. Kushmuredov, F. S. Éshbaev, A. K. Kasymov, and S. Kuchkarov, *Khim. Prir. Soedin.* **15**, 353 (1979).

69. H. P. Hamlow, S. Okuda, and N. Nakagawa, *Tetrahedron Lett.* **1964**, 2553.

70. F. Bohlmann, D. Schumann, and H. Schulz, *Tetrahedron Lett.* **1965**, 173.

71. F. Bohlmann, D. Schumann, and C. Arndt, *Tetrahedron Lett.* **1965**, 2705.

72. F. G. Kamaev, V. B. Leont'ev, K. A. Aslanov, Y. A. Ustynyuk, and A. S. Sadykov, *Khim. Prir. Soedin.* **10**, 744 (1974).

73. R. Cahill and T. A. Crabb, *Org. Magn. Reson.* **4**, 283 (1972).

74. A. S. Sadykov, F. G. Kamayev, V. A. Korenevsky, V. B. Leont'ev, and Y. A. Ustynyuk, *Org. Magn. Reson.* **4**, 837 (1972).

75. F. Bohlmann and R. Zeisberg, *Chem. Ber.* **108**, 1043 (1975).

76. D. Tourwé and G. Van Binst, *Heterocycles* **9**, 507 (1978).

77. T. A. Crabb, in G. A. Webb, Ed., *Annual Reports on NMR Spectroscopy*, Vol. 8, Academic Press, New York, 1978, p. 71.

78. S. N. Y. Fanso-Free, G. T. Furst, P. R. Srinivasan, R. L. Lichter, R. B. Nelson, J. A. Panetta, and G. W. Gribble, *J. Am. Chem. Soc.* **101**, 1549 (1979).

79. S. Okuda, K. Tsuda, and H. Kataoka, *Chem. Ind. (London)* **1961**, 1115.

80. I. Murakoshi, K. Sugimoto, J. Haginiwa, S. Ohmiya, and H. Otomasu, *Phytochemistry* **14**, 2714 (1975).

81. I. Murakoshi, F. Kakegawa, K. Toriizuka, J. Haginiwa, S. Ohmiya, and H. Otomasu, *Phytochemistry* **16**, 2046 (1977).

82. I. Murakoshi, K. Toriizuka, J. Haginiwa, S. Ohmiya, and H. Otomasu, *Phytochemistry* **17**, 1817 (1978).

83. I. Murakoshi, K. Toriizuka, J. Haginiwa, S. Ohmiya, and H. Otomasu, *Chem. Pharm. Bull.* **27**, 144 (1979).

84. I. Murakoshi, K. Toriizuka, J. Haginiwa, S. Ohmiya, and H. Otomasu, *Phytochemistry* **18**, 699 (1979).

85. J. Peterson, *Aust. J. Exp. Biol. Med. Sci.* **41**, 123 (1963).

86. H. Podkowinska and M. Wiewióroski, *Bull. Acad. Polon. Sci., Ser. Sci. Biol.* **13**, 623 (1965). [Through *Chem. Abstr.* **64**, 18025 (1966).]

87. M. M. Kadooka, M. Y. Chang, H. Fukami, P. J. Scheuer, J. Clardy, B. A. Solheim, and J. P. Springer, *Tetrahedron* **32**, 919 (1976).

88. N. K. Hart, S. R. Johns, and J. A. Lamberton, *Chem. Commun.* **1968**, 302.

89. K. Wicky and E. Steinegger, *Pharm. Acta Helv.* **40**, 658 (1965).

90. I. Murakoshi, E. Kidoguchi, M. Nakamura, J. Haginiwa, S. Ohmiya, K. Higashiyama, and H. Otomasu, *Phytochemistry* **20**, 1725 (1981).

91. J. L. van Eijk and M. H. Radema, *Tetrahedron Lett.* **1976**, 2053.

92. A. D. Kinghorn, M. F. Balandrin, and L.-J. Lin, *Phytochemistry* **21**, 2269 (1982).

93. J. Santamaria and F. Khoung-Huu, *Phytochemistry* **14**, 2501 (1975).

94. P. G. Waterman and D. F. Faulkner, *Phytochemistry* **21**, 215 (1982).

95. M. F. Balandrin and A. D. Kinghorn, *Heterocycles* **19**, 1931 (1982).

96. F. Fraga, I. M. Gavilán, A. Durán, E. Seoane, and I. Ribas, *Tetrahedron* **11**, 78 (1960).

97. J. M. H. Pinkerton and L. K. Steinrauf, *J. Org. Chem.* **32**, 1828 (1967).

98. M. F. Cranmer and T. J. Mabry, *Phytochemistry* **5**, 1133 (1966).

99. V. I. Vinogradova, S. Iskandarov, and S. Y. Yunusov, *Khim. Prir. Soedin.* **7**, 463 (1971).

100. G. C. Gerrans and J. Harley-Mason, *Chem. Ind. (London)* **1963**, 1433.

101. T. Nakano, A. Castaldi Spinella, and A. Morales-Méndez, *J. Org. Chem.* **39**, 3584 (1974).

102. L. Jusiak, E. Soczewinski, and A. Waksmunsdzki, *Acta Polon. Pharm.* **24**, 619 (1967).

103. J. N. Anderson and R. O. Martin, *J. Org. Chem.* **41**, 3441 (1976).

104. E. P. White, *J. Chem. Soc.* **1964**, 4613.

105. Y. Tsuda and L. Marion, *Can. J. Chem.* **42**, 764 (1964).

106. M. Carmack, S. I. Goldberg, and E. W. Martin, *J. Org. Chem.* **32**, 3045 (1967).

107. S. I. Goldberg, V. Balthis, and B. Kabadi, Abstract presented at the Chemical Society Autumn Meeting, 1969, Southampton, U.K.

108. I.-C. Kim, M. F. Balandrin, and A. D. Kinghorn, *J. Agric. Food Chem.* **30**, 796 (1982).

109. A. Daily, H. Dutschewska, N. Mollov, S. Spassov, and D. Schumann, *Tetrahedron Lett.* **1978**, 1453.

110. S. I. Goldberg and A. H. Lipkin, *J. Org. Chem.* **37**, 1823 (1972).

111. M. H. Radema, *Planta Med.* **28**, 143 (1975).

112. M. H. Radema, J. L. van Eijk, W. Vermin, A. J. de Kok, and C. Romers, *Phytochemistry* **18**, 2063 (1979).

113. W. J. Vermin, A. J. de Kok, C. Romers, M. H. Radema, and J. L. van Eijk, *Acta Crystallogr.*, *Sect. B* **B35**, 1839 (1979).

114. H. A. Lloyd and E. C. Horning, *J. Org. Chem.* **25**, 1959 (1960).

115. G. Faugeras and R. R. Paris, *Plant. Méd. Phytothér.* **5**, 134 (1971).

116. Y. D. Cho and R. O. Martin, *Can. J. Chem.* **49**, 265 (1971).

117. A. D. Kinghorn and S. J. Smolenski, *Planta Med.* **38**, 280 (1980).

118. B. P. Moore and L. Marion, *Can. J. Chem.* **31**, 187 (1953).

119. J. L. van Eijk and M. H. Radema, *Pharm. Weekbl.* **111**, 1285 (1976).

120. G. Faugeras and M. Paris, *Ann. Pharm. Franc.* **26**, 265 (1968).

121. G. Faugeras, R. R. Paris, and E. Valdes-Bermejo, *C. R. Acad. Sci. Paris, Sér. C.* **273**, 1372 (1971).

122. S. McLean, P. K. Lau, S. K. Cheng, and D. G. Murray, *Can. J. Chem.* **49**, 1976 (1971).

123. D. Knöfel and H. R. Schütte, *J. Prakt. Chem.* **312**, 887 (1970).

124. M. D. Bratek and M. Wiewiórowski, *Roczn. Khim.* **33**, 1187 (1959). [Through *Chem. Abstr.* **54**, 9979 (1960).]

125. J. L. van Eijk and M. H. Radema, *Planta Med.* **44**, 224 (1982).

126. J. Wolinska-Mocydlarz and M. Wiewiórowski, *Bull. Acad. Polon. Sci., Ser. Sci. Chim.* **24, 613 (1976).**

127. A. B. Beck, B. H. Goldspink, and J. R. Knox, *J. Nat. Prod.* **42**, 385 (1979).

128. S. Okuda, K. Tsuda, and H. Kataoka, *Chem. Ind.* (*London*) **1961**, 1751.

129. M. Seoane and I. Ribas, *An. R. Soc. Españ. Fis. Quim. B* **47**, 625 (1951). [Through *Chem. Abstr.* **46**, 5790 (1952).]

130. I. Murakoshi, M. Watanabe, J. Haginiwa, S. Ohmiya, and H. Otomasu, *Phytochemistry* **21**, 1470 (1982).

131. S. Ohmiya, H. Otomasu, I. Murakoshi, and J. Haginiwa, *Phytochemistry* **13**, 643 (1974).

132. S. Ohmiya, H. Otomasu, I. Murakoshi, and J. Haginiwa, *Phytochemistry* **13**, 1016 (1974).

133. I. Murakoshi, K. Fukuchi, J. Haginiwa, S. Ohmiya, and H. Otomasu, *Phytochemistry* **16**, 1460 (1977).

134. S. Ohmiya, H. Otomasu, J. Haginiwa, and I. Murakoshi, *Phytochemistry* **18**, 649 (1979).

135. W. J. Keller and G. M. Hatfield, *Phytochemistry* **18**, 2068 (1979).

136. R. A. Shaimardanov, S. Iskandarov, and S. Y. Yunusov, *Khim. Prir. Soedin.* **6**, 276 (1970).

137. S. Okuda, I. Murakoshi, H. Kamata, Y. Kashida, J. Haginiwa, and K. Tsuda, *Chem. Pharm. Bull.* **13**, 482 (1965).

138. M. Wink, T. Hartmann, L. Witte, and H. M. Schiebel, *J. Nat. Prod.* **44**, 14 (1981).

139. M. F. Balandrin, E. F. Robbins, and A. D. Kinghorn, *Biochem. Syst. Ecol.* **10**, 307 (1982).

140. P. H. Ngok, Y. K. Kushmuradov, K. A. Aslanov, A. S. Sadykov, Z. S. Ziyavitdinova, V. G. Zaikin, and N. S. Vul'fson, *Khim. Prir. Soedin.* **6**, 111 (1970).

141. A. Ueno, K. Morinaga, S. Fukushima, and S. Okuda, *Chem. Pharm. Bull.* **26**, 1832 (1976).

142. K. Morinaga, A. Ueno, S. Fukushima, M. Namikoshi, Y. Iitaka, and S. Okuda, *Chem. Pharm. Bull.* **26**, 2483 (1978).

143. S. Kuchkarov, Y. K. Kushmuradov, K. A. Aslanov, and S. A. Sadykov, *Khim. Prir. Soedin.* **13**, 581 (1977).

144. Y. K. Kushmuradov, S. Kushchkarov, and K. A. Aslanov, *Khim. Prir. Soedin.* **14**, 231 (1978).

145. S. Ohmiya, H. Otomasu, J. Haginiwa, and I. Murakoshi, *Phytochemistry* **17**, 2021 (1978).

146. T. E. Monakhova, N. F. Proskurnina, O. N. Tolkachev, V. S. Kabanov, and M. E. Perel'son, *Khim. Prir. Soedin.* **9**, 59 (1973).

147. S. Ohmiya, K. Higashiyama, H. Otomasu, I. Murakoshi, and J. Haginiwa, *Phytochemistry* **18**, 645 (1979).

148. B. A. Abdusalamov, O. A. Khoroshkova, and K. A. Aslanov, *Khim. Prir. Soedin.* **12**, 71 (1976).

149. O. N. Tolkachev, T. E. Monakhova, V. I. Sheichenko, V. S. Kabanov, O. G. Fesenko, and N. F. Proskurnina, *Khim. Prir. Soedin.* **11**, 30 (1975).

150. S. Kuchkarov, Y. K. Kushmuradov, A. I. Begisheva, and K. A. Aslanova, *Sb. Nauch. Tr. Tashkent Un-t* **1976**, 108. [Through *Chem. Abstr.* **87**, 2564g (1977).]

151. D. T.-W. Chu, Ph.D. Dissertation, University of New Brunswick, Fredericton, N.B. Canada, 1971.

152. P.-T. Cheng, S. McLean, R. Misra, and S. C. Nyburg, *Tetrahedron Lett.* **1976**, 4245.

153. M. F. Mackay and B. J. Poppleton, *Cryst. Struct. Commun.* **9**, 805 (1980). [Through *Chem. Abstr.* **93**, 177642g (1980).]

154. S. McLean, R. Misra, V. Kumar, and J. A. Lamberton, *Can. J. Chem.* **59**, 34 (1981).

155. T. J. Fitzgerald, J. B. LaPidus, and J. L. Beal, *Lloydia* **27**, 107 (1964).

156. M. F. Balandrin and A. D. Kinghorn, *J. Nat. Prod.* **44**, 619 (1981).

157. P. Deslongchamps, Z. Valenta, and J. S. Wilson, *Can. J. Chem.* **44**, 2539 (1966).

158. M. F. Mackay, L. Satzke, and A. M. Mathieson, *Tetrahedron* **31**, 1295 (1975).

159. J. R. Cannon, J. R. Williams, J. F. Blount, and A. Brossi, *Tetrahedron Lett.* **1974**, 1683.

160. M. F. Mackay, M. J. McCall, and B. J. Poppleton, *J. Cryst. Mol. Struct.* **6**, 125 (1976).

161. M. F. Balandrin, Ph.D. Dissertation, University of Illinois at the Medical Center, Chicago, 1982.

162. J. K. Frank, E. N. Deusler, N. N. Thayer, R. Heckendorn, K. L. Rinehart, Jr., I. C. Paul, and R. Misra, *Acta Crystallogr.*, *Sect. B* **B34**, 2316 (1978).

163. A. H.-J. Wang, E. N. Deusler, N. N. Thayer, R. Heckendorn, K. L. Rinehart, Jr., and I. C. Paul, *Acta Crystallogr.*, *Sect. B* **B34**, 2319 (1978).

164. C. H. Hassall and E. M. Wilson, *J. Chem. Soc.* **1964**, 2657.

165. S. Iskandarov, V. I. Vinogradova, R. A. Shaimardanov, and S. Y. Yunusov, *Khim. Prir. Soedin.* **8**, 218 (1972).

166. V. I. Vinogradova, S. Iskandarov, and S. Y. Yunusov, *Khim. Prir. Soedin.* **8**, 87 (1972).

167. S. Iskandarov, B. Sadykov, Y. V. Rashkes, and S. Y. Yunusov, *Khim. Prir. Soedin.* **8**, 347 (1972).

168. A. P. Davies and C. H. Hassall, *Tetrahedron Lett.* **1966**, 6291.

169. N. S. Bhacca, M. F. Balandrin, A. D. Kinghorn, T. A. Frenkiel, R. Freeman, and G. A. Morris, *J. Am. Chem. Soc.*, **105**, 2538 (1983).

170. S. Ohimya, K. Higashiyama, H. Otomasu, J. Haginiwa, and I. Murakoshi, *Chem. Pharm. Bull.* **27**, 1055 (1979).

171. J. Bordner, S. Ohmiya, H. Otomasu, J. Haginiwa, and I. Murakoshi, *Chem. Pharm. Bull.* **28**, 1965 (1980).

172. S. Ohmiya, K. Higashiyama, H. Otomasu, J. Haginiwa, and I. Murakoshi, *Phytochemistry* **20**, 1997 (1981).

173. K. W. Bentley, *The Alkaloids* (*The Chemistry of Natural Products, Vol. 1*), Wiley-Interscience, New York, 1957.

174. T. Robinson, *The Biochemistry of Alkaloids*, Springer-Verlag, New York, 1968.

175. I. D. Spenser, *Compr. Biochem.* **20**, 231 (1968).

176. H. R. Schütte, in K. Mothes and H. R. Schütte, Eds., *Biosynthese der Alkaloide*, VEB Deutscher Verlag der Wissenschaften, Berlin, 1969, p. 324.

177. T. A. Geissman and D. H. G. Crout, *Organic Chemistry of Secondary Plant Metabolism*, Freeman, Cooper, San Francisco, 1969.

178. I. D. Spenser, in S. W. Pelletier, Ed., *Chemistry of the Alkaloids*, Van Nostrand Reinhold, New York, 1970, p. 669.

179. E. K. Nowacki and G. R. Waller, *Phytochemistry* **14**, 155 (1975).

180. E. K. Nowacki and G. R. Waller, *Phytochemistry* **14**, 161 (1975).

181. E. K. Nowacki and G. R. Waller, *Phytochemistry* **14**, 165 (1975).

182. G. R. Waller and E. K. Nowacki, *Alkaloid Biology and Metabolism in Plants*, Plenum Press, New York, 1978.

183. E. Leete, in E. A. Bell and B. V. Charlwood, Eds., *Secondary Plant Products* (*Encyclopedia of Plant Physiology, New Series, Vol. 8*), Springer-Verlag, New York, 1980, p. 72.

184. M. Wink and T. Hartmann, *FEBS Letters* **101**, 343 (1979).

185. M. Wink, T. Hartmann, and H.-M. Schiebel, *Z. Naturforsch.* **34C**, 704 (1979).

186. M. Wink, T. Hartmann, and L. Witte, *Z. Naturforsch.* **35C**, 93 (1980).

187. M. Wink, L. Witte, H.-M. Schiebel, and T. Hartmann, *Planta Med.* **38**, 238 (1980).

188. T. Hartmann, M. Wink, G. Schoofs, and S. Teichmann, *Planta Med.* **39**, 282 (1980).

189. M. Wink, T. Hartmann, and L. Witte, *Planta Med.* **40**, 31 (1980).

190. T. Hartmann, G. Schoofs, and M. Wink, *FEBS Letters* **115**, 35 (1980).

191. M. Wink and T. Hartmann, *Planta Med.* **40**, 149 (1980).

192. M. Wink and T. Hartmann, *Planta Med.* **42**, 116 (1981).

193. M. Wink and T. Hartmann, *Z. Pflanzenphysiol.* **102**, 337 (1981).

194. M. Wink, L. Witte, and T. Hartmann, *Planta Med.* **43**, 342 (1981).

195. M. Wink and T. Hartmann, *Plant Cell Rep.* **1**, 6 (1981).

196. M. Wink, H. M. Schiebel, L. Witte, and T. Hartmann, *Planta Med.* **44**, 15 (1982).

197. M. Wink and T. Hartmann, *Z. Naturforsch.* **37C**, 369 (1982).

198. R. B. Herbert, *Alkaloids* (*London*) **11**, 1 (1981).

199. W. M. Golebiewski and I. D. Spenser, *J. Am. Chem. Soc.* **98**, 6726 (1976).

200. F. J. Leeper, G. Grue-Sørensen, and I. D. Spenser, *Can. J. Chem.* **59**, 106 (1981).

201. M. F. Balandrin and A. D. Kinghorn, *J. Nat. Prod.* **44**, 495 (1981).

202. Y. D. Cho, R. O. Martin, and J. N. Anderson, *J. Am. Chem. Soc.* **93**, 2087 (1971).

203. Y. D. Cho and R. O. Martin, *Can. J. Biochem.* **49**, 971 (1971).

204. A. D. Kinghorn, M. A. Selim, and S. J. Smolenski, *Phytochemistry* **19**, 1705 (1980).

205. G. A. Cordell, *Introduction to Alkaloids, a Biogenetic Approach*, Wiley, New York, 1981.

206. R. F. Keeler, *Teratology* **7**, 31 (1973).

207. W. J. Keller and S. G. Zelenski, *J. Pharm. Sci.* **67**, 430 (1978).

208. W. J. Keller, *Phytochemistry* **19**, 2233 (1980).

209. W. J. Keller, *J. Nat. Prod.* **44**, 357 (1981).

210. R. Misra, W. Wong-Ng, P.-T. Cheng, S. McLean, and S. C. Nyburg, *J. Chem. Soc. Chem. Commun.* **1980**, 659.

211. Y. D. Cho and R. O. Martin, *Anal. Biochem.* **44**, 49 (1971).

212. H. A. Lloyd, H. M. Fales, P. F. Highet, W. J. A. VandenHeuvel, and W. C. Wildman, *J. Am. Chem. Soc.* **82**, 3791 (1960).

213. R. Munier, *Bull. Soc. Chim. Biol.* **35**, 1225 (1953).

214. W. Plarre and B. Scheidereiter, *Z. Pflanzenzüchtg.* **74**, 89 (1975).

215. L. P. Ruiz, Jr., *N.Z.J. Agric. Res.* **20**, 51 (1977).

216. E.-M. Karlsson and H.-W. Petter, *J. Chromatogr.* **155**, 218 (1978).

217. W. Wysocka, *J. Chromatogr.* **116**, 235 (1976).

218. R. F. Keeler, *Teratology* **7**, 23 (1973).

219. P. M. Dolinger, P. R. Ehrlich, W. L. Fitch, and D. E. Breedlove, *Oecologia (Berlin)* **13**, 191 (1973).

220. E. Steinegger, U. P. Schunegger, and H. Fallert, *Pharm. Acta Helv.* **51**, 170 (1976).

221. R. F. Keeler, *J. Toxicol. Environ. Health* **1**, 887 (1976).

222. R. F. Keeler, E. H. Cronin, and J. L. Shupe, *J. Toxicol. Environ. Health* **1**, 899 (1976).

223. G. M. Hatfield, L. J. J. Valdes, W. J. Keller, W. L. Merrill, and V. H. Jones, *Lloydia* **40**, 374 (1977).

224. J. L. Ruiz, Jr., *N.Z.J. Agric. Res.* **21**, 241 (1978).

225. N. Ota and Y. Mino, *Shoyakugaku Zasshi* **33**, 140 (1979). [Through *Chem. Abstr.* **92**, 169114a (1980).]

226. D. von Baer and W. Feldheim, *Bol. Soc. Chil. Quím.* **25**, 85 (1980).

227. G. M. Hatfield, W. J. Keller, and J. M. Rankin, *J. Nat. Prod.* **43**, 164 (1980).

228. A. Salatino and O. R. Gottlieb, *Biochem. Syst. Ecol.* **9**, 267 (1981).

229. R. M. Polhill, in R. M. Polhill and P. H. Raven, Eds., *Advances in Legume Systematics*, Royal Botanic Gardens, Kew, U.K., 1981, p. 213.

230. V. E. Rudd, *Contrib. U.S. Natl. Herb.*, *Smithson. Inst.*, *U.S. Natl. Mus.* **32**, 279 (1965).

231. B. L. Turner, *Brittonia* **32**, 281 (1980).

232. B. L. Turner, in R. M. Polhill and P. H. Raven, Eds., *Advances in Legume Systematics*, Royal Botanic Gardens, Kew, U.K., 1981, p. 403.

233. M. Hesse, *Alkaloid Chemistry*, Wiley, New York, 1981.

234. I. Murakoshi, E. Kidoguchi, J. Haginiwa, S. Ohmiya, K. Higashiyama, and H. Otomasu, *Phytochemistry* **20**, 1407 (1981).

235. J. L. van Eijk, A. J. de Kok, C. Romers, and D. Seykens, *Planta Med.* **44**, 221 (1982).

236. G. C. Gerrans, A. S. Howard, and M. J. Nattrass, *Phytochemistry* **15**, 816 (1976).

237. R. J. Molyneux and L. F. James, *Science* **216**, 190 (1982).

238. R. F. Keeler, *Lloydia* **38**, 56 (1975).

239. R. F. Keeler, in R. F. Keeler, K. R. Van Kampen, and L. F. James, Eds., *Effects of Poisonous Plants on Livestock*, Academic Press, New York, 1978, p. 397.

240. A. M. Davis, *J. Range Manage.* **35**, 81 (1982).

241. R. B. Barlow and L. J. McLeod, *Br. J. Pharmacol.* **35**, 161 (1969).

242. M. Izaddoost, B. G. Harris, and R. W. Gracy, *J. Pharm. Sci.* **65**, 352 (1976).

243. K. N. Garg, S. Sharma, and S. Bala, *Jap. J. Pharmacol.* **23**, 195 (1973).

244. F. Binning, *Arzneim.-Forsch.* **24**, 752 (1974).

245. M. Raschack, *Arzneim.-Forsch.* **24**, 753 (1974).

246. K. Engelmann, W. Raake, and A. Petter, *Arzneim.-Forsch.* **24**, 759 (1974).

247. F. S. Abtahi, F. J. Auletta, D. Sadeghi, B. Djahanguire, and A. Scommegna, *Prostaglandins* **16**, 473 (1978).

248. J. R. Crout, *Fed. Regist.* **43**, 58634 (1978).

249. R. I. Tannous and S. N. Nayfeh, *Aust. J. Biol. Sci.* **22**, 1071 (1969).

250. R. I. Tannous, S. Shadarevian, and J. W. Cowan, *J. Nutr.* **94**, 161 (1968).

251. J. Shani, A. Goldschmied, B. Joseph, Z. Ahronson, and F. G. Sulman, *Arch. Int. Pharmacodyn. Ther.* **210**, 27 (1974).

252. W. J. Keller, *Phytochemistry* **14**, 2305 (1975).

253. J. L. Díaz, *Annu. Rev. Pharmacol. Toxicol.* **17**, 647 (1977).

254. J. Fadiman, *Econ. Bot.* **19**, 383 (1965).

255. S. I. Goldberg, *Psychopharmacol. Bull.* **5**, 23 (1969).

256. S. I. Goldberg and M. S. Sahli, *J. Med. Chem.* **10**, 124 (1966).

257. S. I. Goldberg and R. F. Moates, *Phytochemistry* **6**, 137 (1967).

258. M. Eichelbaum, N. Spannbrucker, and H. J. Dengler, *Eur. J. Clin. Pharmacol.* **16**, 189 (1979).

259. K. Hess and F. Merck, *Ber. Dtsch. Chem. Ges.* **52B**, 1976 (1919).

260. E. A. Trutneva and V. V. Berezhinskaya, *Farmakol. Toksikol.* **23**, 445 (1960), [Through *Chem. Abstr.* **55**, 11673d (1961).]

261. H. A. Lloyd and E. C. Horning, *J. Am. Chem. Soc.* **80**, 1506 (1958).

262. N. C. Moran, G. P. Quinn, and W. M. Butler, Jr., *Fed. Proc.* **15**, 462 (1956).

263. G. P. Quinn, W. M. Butler, Jr., and N. C. Moran, *Fed. Proc.* **15**, 470 (1956).

264. N. C. Moran, G. P. Quinn, and W. M. Butler, Jr., *Fed. Proc.* **16**, 324 (1957).

265. N. C. Moran, G. P. Quinn, and W. M. Butler, Jr., *J. Pharmacol. Exp. Ther.* **125**, 73 (1959).

266. N. C. Moran, G. P. Quinn, and W. M. Butler, Jr., *J. Pharmacol. Exp. Ther.* **125**, 85 (1959).

267. K. Genest, A. Lavalle, and E. Nera, *Arzneim.-Forsch.* **21**, 888 (1971).

268. G. Fraenkel, C. Cottrell, J. Ray, and J. Russell, *J. Chem. Soc. D., Chem. Commun.* **1971**, 273.

269. S. F. Mason and R. D. Peacock, *J. Chem. Soc., Dalton Trans.* **1973**, 226.

270. E. Boschmann, L. M. Weinstock, and M. Carmack, *Inorg. Chem.* **13**, 1297 (1974).

271. G. Fraenkel, B. Appleman, and J. G. Ray, *J. Am. Chem. Soc.* **96**, 5113 (1974).

272. J. T. Wrobleski and G. J. Long, *Inorg. Chem.* **16**, 704 (1977).

273. J. Skolik, U. Majchrzak-Kuczyńska, and M. Wiewiórowski, *Bull. Acad. Polon. Sci., Sér. Sci. Chim.* **26**, 741 (1978).

274. M. Ohta and T. Narahashi, *J. Pharmacol. Exp. Ther.* **187**, 47 (1973).

275. C. L. Schauf, C. A. Colton, J. S. Colton, and F. A. Davis, *J. Pharmacol. Exp. Ther.* **197**, 414 (1976).

276. A. L. Kleinhaus, *J. Physiol. (London)* **299**, 309 (1980).

277. M. Mazur, P. Polakowski and A. Szadowska, *Acta Physiol. Polon.* **17**, 299 (1966). [Through *Chem. Abstr.* **65**, 11163f (1966f)].

278. G. Faugeras, R. Paris, M. Debray, J. Bourgeois, and C. Delabos, *Plant. Méd. Phytothér.* **9**, 37 (1975).

279. H. R. Kaplan, R. E. Wolke, and M. H. Malone, *J. Pharm. Sci.* **56**, 1385 (1967).

280. T. L. Nucifora and M. H. Malone, *Arch. Int. Pharmacodyn.* **191**, 345 (1971).

281. D. A. Levin and B. M. York, Jr., *Biochem. Syst. Ecol.* **6**, 61 (1978).

282. L. M. Schoonhoven, *Rec. Adv. Phytochem.* **5**, 197 (1972).

283. J. Krzymańska, *Biul. Inst. Ochr. Rosl.* **36**, 237 (1967).

284. K. L. S. Harley and A. J. Thorsteinson, *Can. J. Zool.* **45**, 305 (1967).

285. L. Forbes and E. W. Beck, *Agron. J.* **46**, 528 (1954).

286. A. Gustafsson and I. Gadd, *Hereditas* **53**, 15 (1965).

287. B. D. Smith, Ph.D. Dissertation, University of London, London, 1957.

288. B. D. Smith, *Nature (London)* **212**, 213 (1966).

289. J. C. Downey and D. B. Dunn, *Ecology* **45**, 172 (1964).

290. C. N. Roeske, J. N. Seiber, L. P. Brower, and C. M. Moffitt, *Rec. Adv. Phytochem.* **10**, 93 (1976).

291. D. H. Janzen, in R. M. Polhill and P. H. Raven, Eds., *Advances in Legume Systematics*, Royal Botanic Gardens, Kew, U.K., 1981, p. 951.

Chapter Four

Chemistry and Pharmacology of Maytansinoid Alkaloids

Cecil R. Smith, Jr., and Richard G. Powell

Northern Regional Research Center
Agricultural Research Service
U.S. Department of Agriculture
Peoria, Illinois 61604

CONTENTS

1. INTRODUCTION

The maytansinoids are a class of complex alkaloids also belonging to the ansamacrolide group of natural products [1]. The ansamacrolides are diverse in nature, and many are antibiotics. In general, these compounds have 19- to 25-membered rings which are unsymmetrical. Typically, they contain a variety of linkages: carbon–carbon single and double bonds, lactone or lactam linkages, and carbonyl groups. In many cases, the ring joins two nonadjacent positions of an aromatic moiety; alkyl or other substituents usually occur at different locations on the ring. Most ansamacrolides are bacterial metabolites, and maytansinoids were the first of this group to be found in higher plants. Leboeuf [2] pointed out the constitutional similarity between maytansinoids and the ansamacrolide antibiotic geldanamycin (**1**).

The maytansinoids have been reviewed previously by Komoda and Kishi [3] as well as by Leboeuf [2].

2. NATURALLY OCCURRING MAYTANSINOID ALKALOIDS

2.1. Maytansinoids from Higher Plants

Maytansine (**2**) was the first maytansinoid to be isolated and characterized. Kupchan and coworkers [4, 5] isolated maytansine from ethanolic extracts of *Maytenus buchananii* and *M. serrata* (Celastraceae) as well as *Putterlickia verrucosa* (Celastraceae). Initially, they began intensive investigation of the crude ethanolic extract of *M. serrata* because of its high level of activity in experimental tumor systems. Fractionation procedures consisted of partitioning between various solvent pairs, open column chromatography, thin-layer chromatography (TLC), and high-pressure liquid chromatography (HPLC). This fractionation was monitored by KB cell culture and P388 lymphocytic leukemia systems; had maytansine not been highly active in these bioassays, isolation of this compound occurring at so low a concentration (0.2 mg/kg of plant material) probably would not have been possible.

Because of the small amount of **2** on hand and its structural complexity, X-

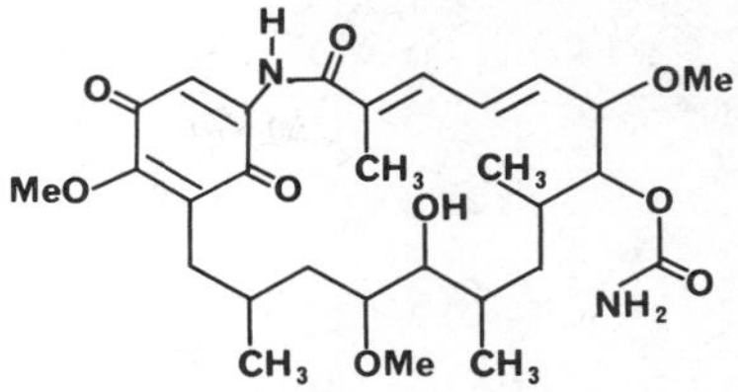

1 Geldanamycin

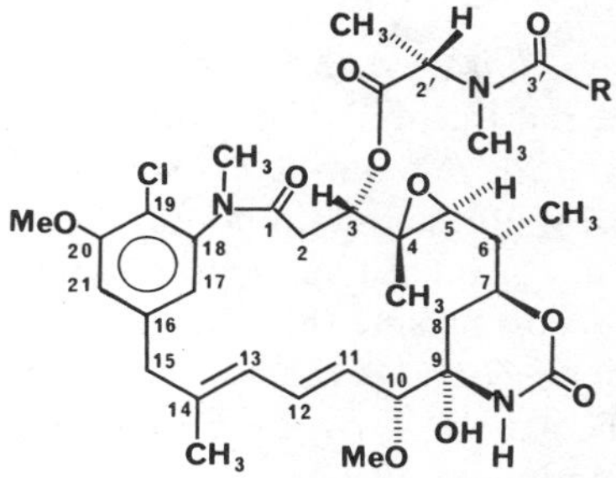

2 Maytansine, R = CH$_3$

3 Maytanprine, R = CH$_2$CH$_3$

4 Maytanbutine, R = CH(CH$_3$)$_2$

5 Maytanvaline, R = CH$_2$CH(CH$_3$)$_2$

ray crystallographic analysis was necessary [4, 5]. After it was discovered that 2 could be converted to methyl or ethyl ethers, 2 was converted to a crystalline 2-bromopropyl ether that was used for structure determination by X-ray crystallography [4, 5]. The structure and stereochemistry of maytansine are summarized in formula 2. Maytansine is the archetype of a new series of ansamacrolides distinguished by a central 19-membered lactam ring onto which two six-membered rings are fused—one in the northwest corner, a benzenoid ring with chloro and methoxyl substituents. The other, in the southeast corner, is a cyclic urethane. In addition, there are epoxide and conjugated diene functions and various methyl groups attached. Through the C(3) hydroxyl, the macrocycle is esterified to N-acetyl-N-methyl-alanine.

Several structural variants of maytansine have been discovered, all with the same central macrocycle but with different ester side-chains; other maytansinoids occur without an ester side-chain. There are also differences as to whether the nitrogen attached at C(18) has a methyl group or hydrogen attached. Following the discovery of maytansine, Kupchan and coworkers [6] isolated and identified maytanprine (3), maytanbutine (4), maytanvaline (5), and maytanacine (6) in the course of their investigations of extracts from

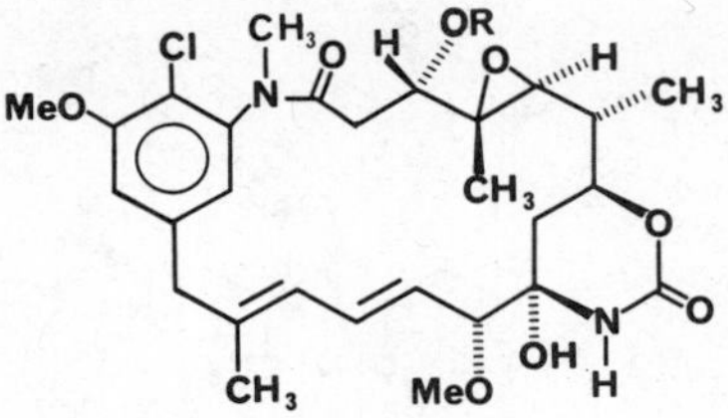

6 Maytanacine, R = Ac

10 Maytansinol, R = H

Maytenus sp. and *Putterlickia verrucosa.* Compounds **3, 4,** and **5** differ from maytansine only in the nature of the terminal acyl group of the nitrogen-containing ester side-chain. Maytanacine (**6**) was the first naturally occurring compound of this series found to contain no nitrogen in its ester side-chain. In addition, Kupchan et al. [6] identified three maytansinoids lacking an ester sidechain—maysine (**7**), normaysine (**8**), and maysenine (**9**), as well as maytansinol (**10**). In compounds **7** and **8**, the hydroxyl group at C(3) is replaced by a double bond. In **9**, the 4,5-epoxy group also is replaced by a double bond, thus forming a conjugated diene system.

Under the terminology recommended by Kupchan and coworkers [7], all ansamacrolides related to maytansine are "maytansinoids," and those maytansinoids lacking the ester side-chain are "maytansides." Apparently the term "maytanside" is to be applied both to compounds with an unesterified C(3)

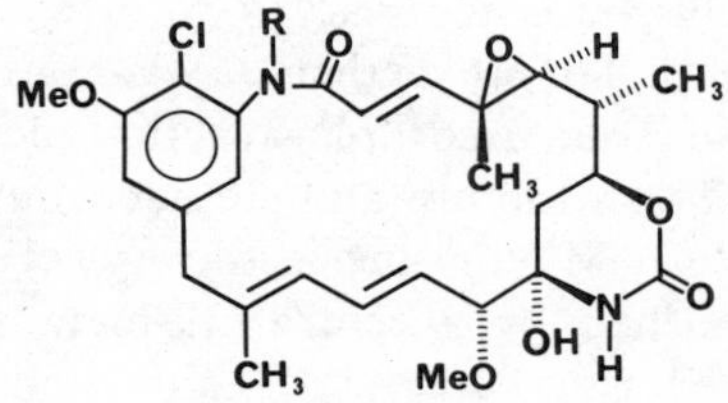

7 Maysine, R = CH₃

8 Normaysine, R = H

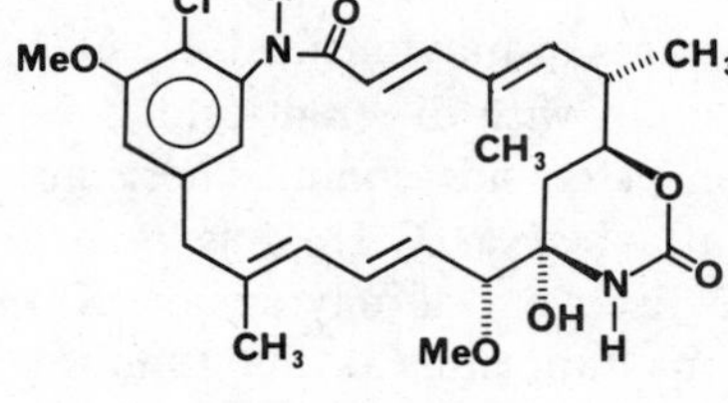

9 Maysenine

hydroxyl (e.g., **10**) and the corresponding elimination products such as **7, 8,** and **9.**

Some time later, two new maytansinoids were discovered which lacked an *N*-methyl group on the C(18) nitrogen. From extracts of *Maytenus buchananii,* Sneden and Beemsterboer [8] characterized a minor maytansinoid as normaytansine (**11**). The mass spectrum (MS) of normaytansine had similarities to that of normaysine (**8**) suggesting the absence of an *N*-methyl group in the macrocyclic moiety; this inference was confirmed by NMR spectroscopy. Still later, Sneden et al. [9a] reported another minor but more exotic normaytansinoid—normaytancyprine (**12**)—isolated from *Putterlickia verrucosa.*

11 Normaytansine, R = CH₃

12 Normaytancyprine, R =

Normaytancyprine has an ester side-chain terminated by a dimethylcyclopropyl group; it is the first and only maytansinoid thus far found to contain a cyclopropane ring. On the basis of their spectral data, Sneden et al. were unable to determine the configurations of methyl groups on this three-membered ring.

Higher plants also have yielded maytansinoids with an oxygen function at C(15). The first of these, colubrinol (**13**) and colubrinol acetate (**14**), were isolated from *Colubrina texensis* (Rhamnaceae) by Wani et al. [10] along with maytanbutine (**4**), previously isolated from *Maytenus* sp. [6]. Compounds **13** and **14** were shown to be interconvertible by straightforward acetylation or deacetylation reactions. Later, Kupchan and coworkers [6] characterized a related maytansinoid, maytanbutacine (**15**), from *Maytenus serrata.* Maytanbutacine differs from colubrinol acetate (**14**) in the nature of the ester side-chain, and was the second maytanside ester from a higher plant that lacks nitrogen in the ester side-chain.

From seeds of *Trewia nudiflora* (Euphorbiaceae), Powell and coworkers obtained an ethanolic extract with pronounced biological activity. From this extract, they isolated two series of new maytansinoids. One series, exemplified by trewiasine (**16**), dehydrotrewiasine (**17**), and demethyltrewiasine (**18**), has a

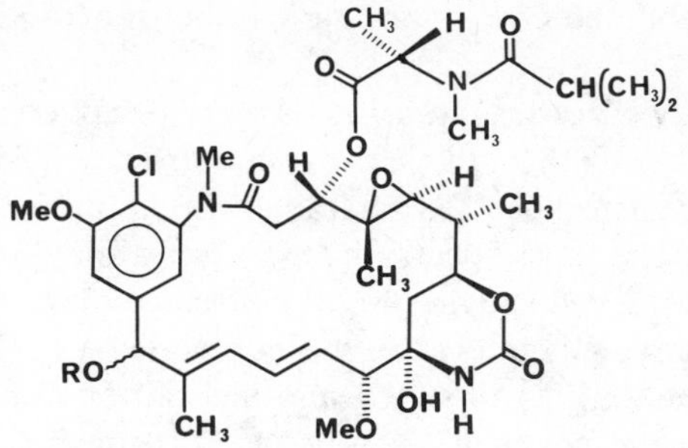

13 Colubrinol, R = H

14 Colubrinol acetate, R = Ac

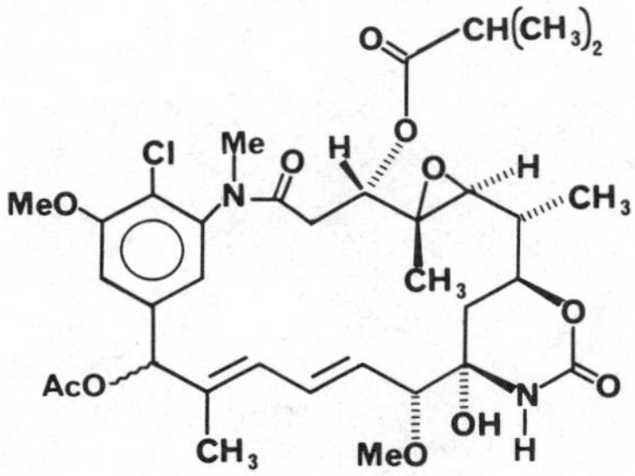

15 Maytanbutacine

methoxyl substituent at C(15) and a variety of ester side-chains [11]. Trewiasine, by far the most abundant of the *Trewia* maytansinoids, has the same ester side-chain as maytanbutine (**4**). Demethyltrewiasine lacks an *N*-methyl group in the ester side-chain, as evidenced by NMR and mass spectroscopy.

Members of the second series of *Trewia*-derived maytansinoids character-

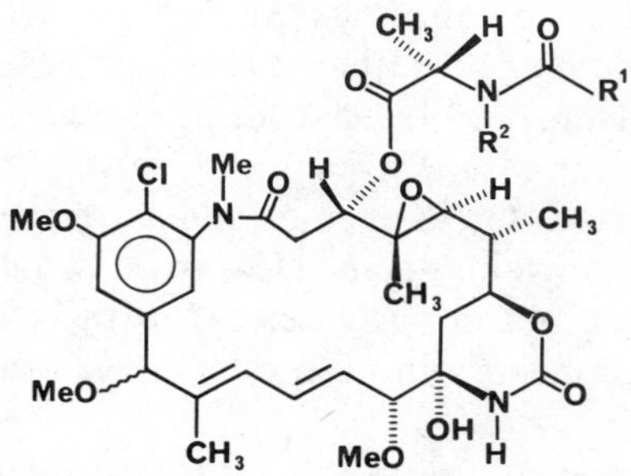

16 Trewiasine, $R^1 = CH(CH_3)_2$; $R^2 = CH_3$

17 Dehydrotrewiasine, $R^1 = C(CH_3)=CH_2$; $R^2 = CH_3$

18 Demethyltrewiasine, $R^1 = CH(CH_3)_2$; $R^2 = H$

19 Treflorine, R = H
20 Trenudine, R = OH

ized by Powell et al. [12] contain two fused macrocyclic rings; in addition to the 19-membered ring characteristic of all previously known maytansinoids, they have a 12-membered ring joining C(3) and the amide nitrogen at C(18), as exemplified by treflorine (19), trenudine (20), and N-methyltrenudone (21). Compounds 19, 20, and 21 can be envisioned as being formally derived from trewiasine (16) by abstraction of two hydrogens to establish a covalent bond between the "missing" C(4′) and C(18) N-methyl groups of 16. The absence of these two methyl groups was evident in the NMR spectra of 19, 20, and 21. These compounds with two macrocycles contain more oxygens than 16—one additional in 19, and two in the case of 20. These extra oxygens are represented by hydroxyl groups at C(4′) and C(5′). On the other hand, N-methyltrenudone (21) also has more oxygens than 16, a hydroxyl at C(4′) and a carbonyl function at C(5′). Under conditions of mild alkaline hydrolysis, 19, 20, and 21 undergo a hydrolysis-elimination reaction at the ester linkage, but the other terminus of the 12-membered ring remains intact so that a product such as compound 22 is formed.

21 N-Methyltrenudone

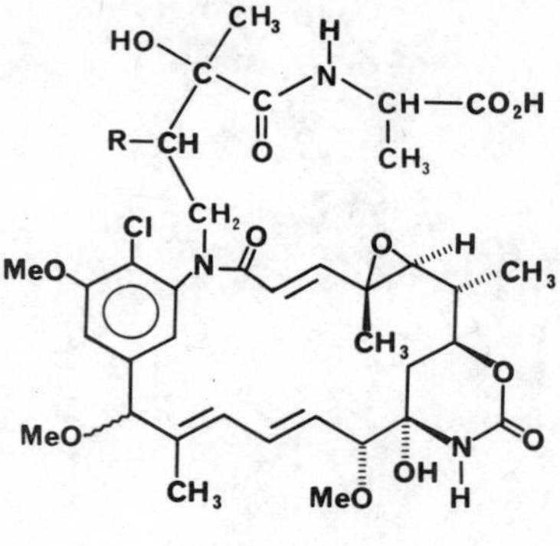

22

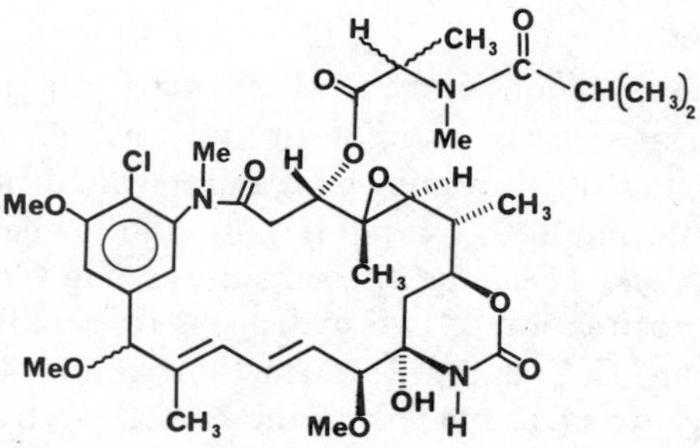

23 10-Epitrewiasine

Powell and coworkers [13] identified two other compounds as minor maytansinoid components of *Trewia nudiflora*—10-epitrewiasine (**23**) and nortrewiasine (**24**). 10-Epitrewiasine was found to have the same molecular formula as trewiasine (**16**), and the NMR spectra of the two were similar in most respects. However, the ^{1}H NMR spectra differed significantly in the coupling constants for protons at C(10) and C(11), signifying a different configuration at C(10). There were also small but significant differences in chemical shifts of these protons. ^{13}C NMR spectra of **16** and **23** supported this conclusion, since shifts for C(8), C(9), C(10), C(11), and C(12) differed noticeably, as would be expected if there were a configurational difference at C(10). In this respect, 10-epitrewiasine differs from all other naturally occurring maytansinoids. Nortrewiasine (**24**) was identified by its NMR spectra; its ^{1}H NMR spectrum lacked a signal characteristic of an *N*-methyl group at C(18). MS indicated that **24** contained one less CH_2 than **16**, consistent with the structure shown.

From a chemotaxonomic standpoint, it is noteworthy that the genera of higher plants thus far known to yield maytansinoids are members of three closely related orders of the plant kingdom—Celastrales, Rhamnales, and Euphorbiales [14].

24 Nortrewiasine

2.2. Maytansinoids from Microorganisms

Several years after the discovery of maytansinoids in higher plants, a group of Japanese workers encountered another group of maytansinoids in the fermentation broth of a *Nocardia* species, most of which were new compounds (see Table 1) [15–20]. Most of these compounds were esterified at the C(3) hydroxyl with simple acyl groups containing two to five carbons, but no nitrogen. One series, called the ansamitocins (6, 25a–d, Table 1), are all esters of maytansinol (10); one of these, referred to as "ansamitocin P-0," is identical with maytansinol, as "ansamitocin P-1" is with maytanacine (6). However, in later investigations of these *Nocardia*-derived maytansinoids, some structural features of the maytansinol moiety were modified. In compounds 26c–g, also in 27g, 28, and 29, the *N*-methyl group is absent, replaced by NH, while in compounds 27a–d, the C(14) methyl group is oxygenated to give a hydroxy-methyl function. Compounds 27e–g have hydroxyl substituents in the ester side-chain, and 28–30 lack the aromatic chlorine, the *N*-methyl group, or both. Two new compounds (PHO-3 and epi-PHO-3) have a hydroxyl at C(15), a site also oxygenated in colubrinol (13) and in *Trewia* maytansinoids (16–24). Structural correlations with previously known maytansinoids were achieved by correlation of physical properties, especially MS and NMR spectra.

2.3. Stereochemistry of Maytansinoids

As noted previously, the stereochemistry and absolute configuration of maytansine (2) were determined by X-ray crystallographic analysis of its 3-bromopropyl ether [5]. Maytansinoids are notoriously prone to be microcrys-talline and their complexity, together with their wide range of reactive functional groups, make it difficult to derivatize these compounds. As a result, none from higher plants besides 2 has been examined crystallographically, either in the free state or as a derivative. Accordingly, knowledge of the stereochemistry of various maytansinoids depends heavily on interrelating

Table 1. Structures of Ansamitocins and Analogs

Compound No.	Name	R_1	R_2	R_3	R_4
10	Ansamitocin P-0 (maytansinol)	H	H	H	CH_3
6	Ansamitocin P-1 (maytanacine)	$COCH_3$	H	H	CH_3
25a	Ansamitocin P-2	$COCH_2CH_3$	H	H	CH_3
25b	Ansamitocin P-3	$COCH(CH_3)_2$	H	H	CH_3
25c	Ansamitocin P-3'	$COCH_2CH_2CH_3$	H	H	CH_3
25d	Ansamitocin P-4	$COCH_2CH(CH_3)_2$	H	H	CH_3
26a	PHO-3	$COCH(CH_3)_2$	H	OH	CH_3
26b	epi-PHO-3	$COCH(CH_3)_2$	H	OH	CH_3
26c	PND-0	H	H	H	H
26d	PND-1	$COCH_3$	H	H	H
26e	PND-2	$COCH_2CH_3$	H	H	H
26f	PND-3	$COCH(CH_3)_2$	H	H	H
26g	PND-4	$COCH_2CH(CH_3)_2$	H	H	H
27a	PHM-1	$COCH_3$	OH	H	CH_3

	Name				
27b	PHM-2	COCH$_2$CH$_3$	OH	H	CH$_3$
27c	PHM-3	COCH(CH$_3$)$_2$	OH	H	CH$_3$
27d	PHM-4	COCH$_2$CH(CH$_3$)$_2$	OH	H	CH$_3$
27e	P4-βHY	COCH$_2$C(OH)(CH$_3$)$_2$	H	H	CH$_3$
27f	P4-γHY	COCH$_2$CH(CH$_2$OH)CH$_3$	H	H	CH$_3$
27g	PND-4-βHY	COCH$_2$C(OH)(CH$_3$)$_2$	H	H	H

Compound No.	Name	X	Y
28	deClQND-0	H	H
29	QND-0	Cl	H
30	deClQ-0	H	CH$_3$

them with **2**. ^{1}H and ^{13}C NMR have been most useful as stereochemical probes and have demonstrated that the central 19-membered macrocyle has the same relative configuration at all chiral centers in these compounds, with the sole exception of 10-epitrewiasine (**23**). Since natural maytansinoids are strongly levorotatory in all cases examined, it seems fairly certain that the macrocyclic moiety in all these compounds shares the absolute configuration established by Bryan et al. [5]. With the possible exception of 10-epitrewiasine (**23**), the work of Bryan et al. also defines the configuration of C(2′) of the ester side-chain in those that are *N*-acyl-L-alanine derivatives. In those maytansinoids with two fused macrocycles (**19, 20,** and **21**), the stereochemistry of chiral centers in the 12-membered ring cannot be specified at present. Izawa and coworkers [20] (see Table 1) isolated and characterized two microbial transformation products—PHO-3 (**26a**) and epi-PHO-3 (**26b**)—from ansamitocin P-3, and ascertained that PHO-3 is identical with deacetylmaytanbutacine (derived from **15**); thus the C(15) oxygen function in **15** has the *R* (or *β*) configuration as in **26a**.

3.   TECHNIQUES FOR SEPARATION AND CHARACTERIZATION OF MAYTANSINOID ALKALOIDS

Maytansinoids are essentially neutral compounds with marked lipophilic character and their isolation normally involves extraction into an organic solvent followed by several solvent partitioning steps and extensive chromatography. Isolations are guided by an appropriate bioassay such as the KB *in vitro* system or the PS leukemia *in vivo* system [21]. Fractionation of *Maytenus serrata* and *M. buchananii* extracts were described in detail by Kupchan et al. [6]; Nettleton et al. [22] outlined a similar scheme for fractionation of an extract from *M. rothiana*. A considerably less complex scheme was used by Powell et al. [12] in their studies of *Trewia nudiflora* constituents as illustrated in Fig. 1. Izawa et al. have detailed the isolation of ansamitocins P-1, P-2, P-3, P-3′, P-4, and 15 related metabolites from the culture broth of a mutant *Nocardia* species [19].

3.1.   Chromatographic Techniques

High-performance liquid chromatography (HPLC) has been widely used to analyze various maytansinoid mixtures and methods are available for isolating gram quantities of these compounds. A quantitative procedure for determining maytansine (**2**), maytanprine (**3**), and maytanbutine (**4**) in extracts of *Maytenus ilicifolia* has been described [23]. Concentrates from plant samples as small as 9 g were chromatographed on μ Porasil columns using a solvent system of methylene chloride–isopropanol–water (96:4:0.5); the minimum

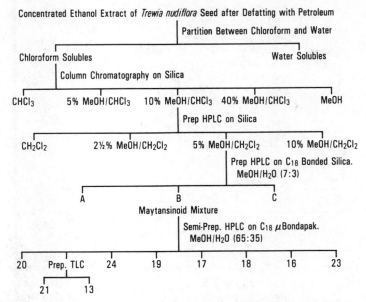

Figure 1. General procedure for isolation of maytansinoids from *Trewia nudiflora* seed.

detectable concentration of any single component was reported to be 5 ng. Nettleton et al. used a similar system in their large-scale isolation of maytansine from *M. rothiana* [22]. A concentrate derived from 400 kg of seed yielded 7.6 g of maytansinoids, after chromatography using methylene chloride–isopropanol–water (95:5:0.5), and further chromatography on silica using ethyl acetate–isopropanol (95:5) yielded 5.2 g of highly purified maytansine. The elution order in these systems is maytanbutine, maytanprine, maytansine.

Several HPLC procedures were used in investigating constituents of *Trewia nudiflora* [11, 12]. Concentration of a maytansinoid-rich fraction was first achieved using silica columns and a stepwise gradient of increasing methanol in dichloromethane. Fractions of interest eluted with methanol–dichloromethane (5:95). Further separations were carried out on reversed-phase (C_{18}) columns either with a stepwise gradient of increasing methanol in water or isocratically with methanol–water (7:3). Purified compounds were then isolated by HPLC on a C_{18} μ Bondapak column using methanol–water (65:35). Figure 2 illustrates the types of separations realized by reversed-phase HPLC of *T. nudiflora* constituents.

Analytical and preparative procedures for separating ansamitocins from *Nocardia* species have been developed by Izawa et al. [19, 24]. Excellent separations of up to 5-g samples of P-2 (**25a**), P-3 (**25b**), and P-4 (**25d**) were

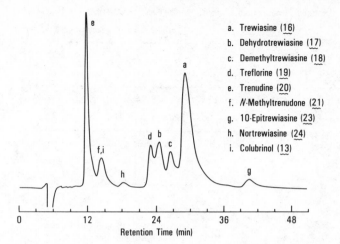

a. Trewiasine (16)
b. Dehydrotrewiasine (17)
c. Demethyltrewiasine (18)
d. Treflorine (19)
e. Trenudine (20)
f. N-Methyltrenudone (21)
g. 10-Epitrewiasine (23)
h. Nortrewiasine (24)
i. Colubrinol (13)

Figure 2. HPLC of *Trewia* maytansinoids on C_{18} μ-Bondapak using MeOH/H_2O (65:35). Mixture was prepared by recombination of *Trewia* maytansinoids in approximately the same proportions as found in *Trewia* seed except that the proportion of trewiasine (16) was reduced by two-thirds. Reprinted with permission from reference 13.

achieved on Kieselgel 60 columns eluted with hexane–ethyl acetate (1:7) saturated with water, using peak "shaving" and recycling techniques. Quantitative analyses were carried out by reversed-phase HPLC on C_{18} μ Bondapak columns using methanol–water (69:31).

Preparative TLC has also been extensively used to purify many of the maytansinoids, ansamitocins, and their reaction products. Kupchan et al. [6] found several stationary phases and solvent systems useful in their investigations of *Maytenus* and *Putterlickia* species. Preparative TLC of enriched fractions on alumina developed with ethyl acetate–methanol (93:7) gave concentration of activity in a band near R_f 0.5 and further PTLC on silica gel developed with 3% methanol in ethyl acetate gave a band (R_f 0.2) highly enriched in maytansine (2). Bands of slightly higher R_f contained maytanprine (3) and maytanbutine (4). Further purification of maytansine was achieved on Chrom AR plates developed with 5% methanol in ethyl acetate and isolation of maytanvaline (5) was accomplished on Chrom AR developed with 5% methanol in chloroform. Wani et al. [10] isolated colubrinol (13) and colubrinol acetate (14) by repeated PTLC of a concentrate from *Colubrina texensis* on alumina and silica gel. Investigations of the maytansinoids from *Trewia nudiflora* included final purification of these compounds by preparative TLC on silica plates developed with 5% methanol in dichloromethane [12]. Isolation of normaytancyprine (12) [9a] was achieved by successive TLC in the following systems (TLC plates, developing solvent): Chrom AR (20% benzene in ethyl acetate); silica gel (20% benzene in ethyl acetate); alumina (6% methanol in benzene); and silica gel (4% methanol in chloroform).

In their studies of the ansamitocins, Asai et al. reported separations of various components by preparative TLC on silica plates developed in ethyl acetate saturated with water [16]. Satisfactory analyses of individual components were obtained on silica plates developed with chloroform–methanol (9:1) and with ethyl acetate–methanol (19:1). In general, the R_f increased with increasing lipophilic character of the C(3) ester moiety in this series (P-1 < P-2 < P-3 < P-3' < P-4). Izawa et al. have reported TLC separations of these and 16 metabolites from a mutant *Nocardia* species and compared their R_f values in three systems: silica developed with chloroform–methanol (9:1), silica developed with ethyl acetate saturated with water, and RP-18 (reversed phase) developed with methanol–water (4:1) [19].

3.2. Typical Reactions

Early attempts to prepare derivatives of maytansine (**2**) in methanol or ethanol solutions gave corresponding methyl or ethyl ether derivatives [4]. Further investigations revealed that etherification of the C(9) hydroxyl (carbinolamide) group was a typical reaction of maytansides and that these reactions were reversible in aqueous 2 N hydrochloric acid. General procedures for preparation of maytansinoid ethers involve solution of the maytansinoid in an appropriate dry solvent along with the desired alcohol or thiol and a small amount of trifluoroacetic or *p*-toluenesulfonic acid for 3 days at room temperature. The desired derivative is normally isolated from the reaction mixture by TLC on silica gel [25]. Maytansinoid ethers reported to date include the following: maytansine 9-methyl ether (**31**), maytansine 9-ethyl ether (**32**), maytansine 9-propyl thioether (**33**), maytansine 9-bromopropyl ether (**34**), maytanbutine 9-methyl ether (**35**), maytanbutine 9-ethyl ether (**36**), maytanbutine 9-propyl thioether (**37**), and maysine 9-methyl ether (**38**).

31 Maytansine 9-methyl ether, R^1 = COCHMeNMeCOMe; R^2 = OMe

32 Maytansine 9-ethyl ether, R^1 = COCHMeNMeCOMe; R^2 = OCH$_2$Me

33 Maytansine 9-propyl thioether, R^1 = COCHMeNMeCOMe; R^2 = SCH$_2$CH$_2$Me

34 Maytansine 9-bromopropyl ether, R^1 = COCHMeNMeCOMe; R^2 = OCH$_2$CH$_2$CH$_2$Br

35 Maytanbutine 9-methyl ether, R^1 = COCHMeNMeCOCH(Me)$_2$; R^2 = Me

36 Maytanbutine 9-ethyl ether, R^1 = COCHMeNMeCOCH(Me)$_2$; R^2 = OCH$_2$Me

37 Maytanbutine 9-propyl thioether, R^1 = COCHMeNMeCOCH(Me)$_2$; R^2 = SCH$_2$CH$_2$Me

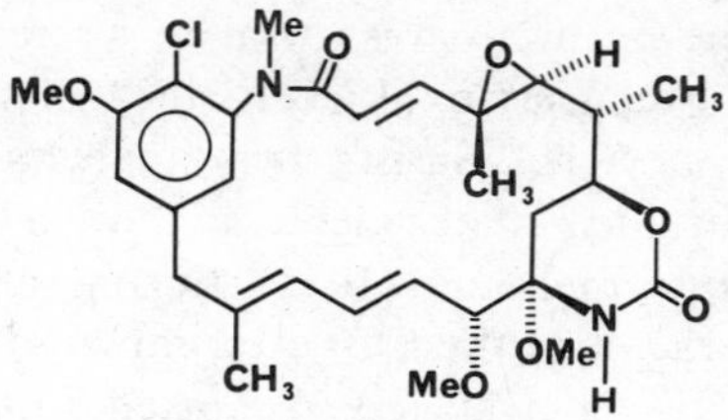

38 Maysene 9-methyl ether

Kupchan et al. [6] observed that hydrolysis of maytanvaline (**5**) with sodium carbonate in 50% aqueous methanol for several hours at room temperature gave maysine (**7**) and *N*-isovaleryl-*N*-methylalanine (**39**). Similarly, normaysine (**8**) was obtained from hydrolysis of both normaytansine (**11**) [8] and

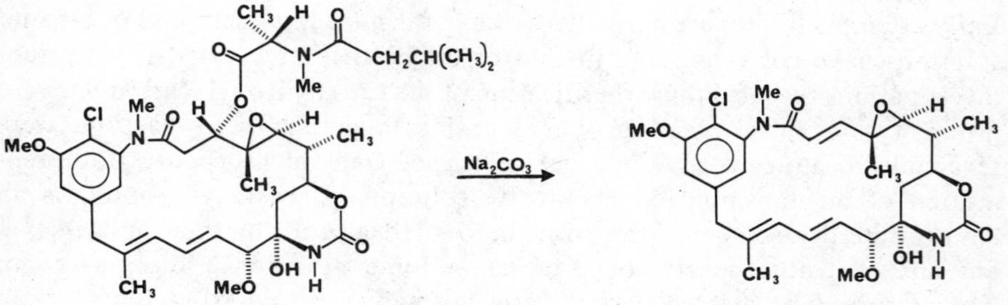

5 Maytanvaline

7 Maysine

normaytancyprine (**12**) [9], and trewiasine (**16**) has yielded trewsine (**40**) [11]. By contrast, hydrolysis of both treflorine (**19**) and trenudine (**20**), compounds in which the C(3) ester groups are hindered by incorporation into additional macrocyclic rings, proceeds at a much slower rate. Reaction periods of 3–4

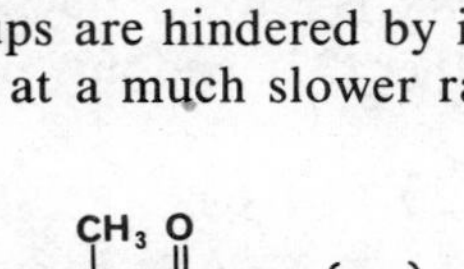

39 *N*-isovaleryl-*N*-methylalanine

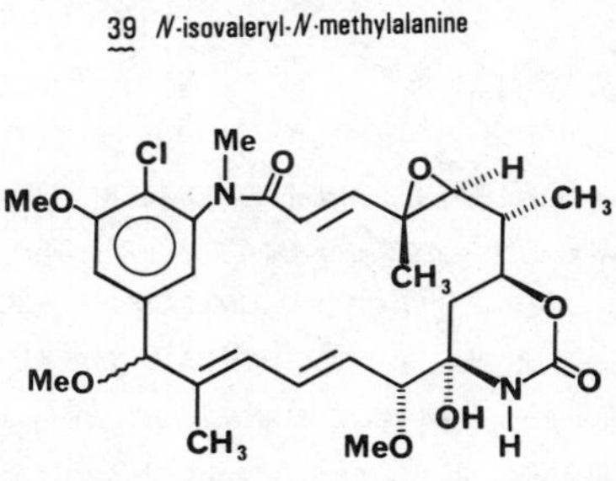

40 Trewsine

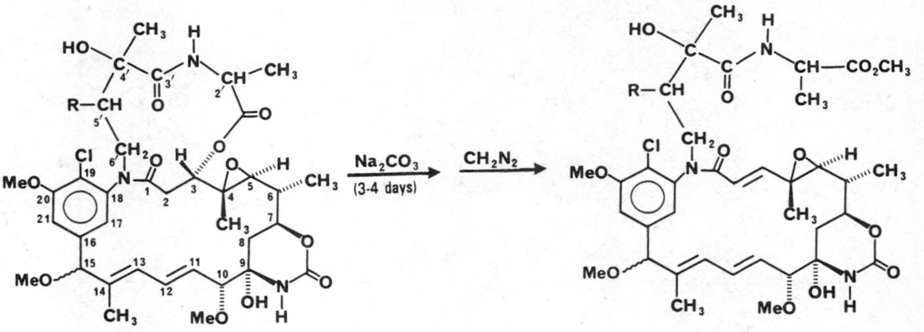

19 Treflorine, R = H

20 Trenudine, R = OH

41 R = H

42 R = OH

days are required for these compounds to yield reasonable amounts of their hydrolysis products which were identified as methyl esters (**41** and **42**) [12] after treatment with diazomethane.

Maytansinoids having an ester group at C(15) may be selectively hydrolyzed, leaving the C(3) ester group intact. For example, Wani et al. [10] reported conversion of colubrinol acetate (**14**) to colubrinol (**13**), and Kupchan et al. [6] were able to prepare deacetylmaytanbutacine (**43**) from maytanbutacine (**15**) similarly. The latter conversion was accomplished using a solution of maytanbutacine and sodium bicarbonate in methanol–water (1:1) for 43 h at room temperature.

Several oxidation and reduction reactions of maytansinoids have been reported. Oxidations include conversion of colubrinol (**13**) to the corresponding C(15) ketone (**44**) using manganese dioxide [10] and oxidation of deacetylmaytanbutacine (**43**) to the C(15) ketone (**45**) with Jones reagent [6]. Conversion of normaysine (**8**) to maysenine (**9**), that is, replacement of the 4,5-epoxide with a 4,5-double bond, was accomplished by means of chromous chloride in acetic acid [6].

Although not available by conventional hydrolysis procedures, alcohols corresponding to maytanside esters can be prepared by lithium aluminum hy-

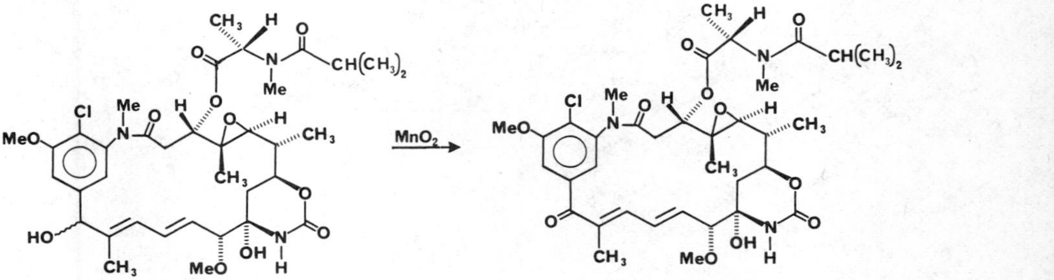

13 Colubrinol

44

43 Deacetylmaytanbutacine   45

dride reduction in tetrahydrofuran at low temperatures. This procedure has given reasonable yields of maytansinol (**10**) from maytanbutine (**4**) [6] and from several of the ansamitocins [16]; trewiasinol (**46**) has also been obtained from trewiasine (**16**) [26].

Maytansinol (**10**) was an early target for esterification in studies to assess the effects of variation in structure of the C(3) ester moiety on biological activity [27]. In general, simple C(3) esters are prepared by treating maytansinol with the appropriate anhydride in pyridine. Kupchan et al. prepared maytansinol 3-acetate (maytanacine or ansamitocin P-1) (**6**), maytansinol 3-propionate (ansamitocin P-2) (**25a**), maytansinol 3-bromoacetate (**47a**), maytansinol 3-trifluoroacetate (**47b**), and maytansinol 3-crotonate (**47c**) [25]. Similarly, Wani and coworkers were able to convert colubrinol (**13**) to the corresponding C(15)

8 Normaysine   9 Maysenine

4 Maytanbutine, R = H   10 Maytansinol, R = H
16 Trewiasine, R = OCH₃   46 Trewiasinol, R = OCH₃

10 Maytansinol   6 Maytansinol acetate

acetate (**14**). Japanese workers have patented procedures utilizing maytansinol to prepare maytansine and a variety of maytansinoid esters [28, 29].

47a, Maytansinol 3-bromoacetate, R = COCH₂Br

b, Maytansinol 3-triflouroacetate, R = COCF₃

c, Maytansinol 3-crotonate, R = COCH=CHMe

3.3. X-ray Crystallography

The reversible interrelation of maytansine (**2**) and its 3-bromopropyl ether derivative (**34**) made the latter an attractive target for X-ray crystallographic analysis [5]. The structure was solved by the heavy atom method and the unit cell was found to contain four molecules of $C_{37}H_{51}BrClN_3O_{10}$. Absolute stereochemistry of maytansine was established as 3*S*, 4*S*, 5*S*, 6*R*, 7*S*, 9*S*, 10*R*, and 2'*S*.

Chemical correlations, spectral studies, and optical rotation data indicate that most of the known maytansinoids have the same absolute configuration at all chiral centers. One exception is 10-epitrewiasine (**23**) which has the 10*S* rather than 10*R* configuration [13]. Maytansinoids possess a roughly rectangular 19-membered ring in which the two longer sides are nearly parallel, separated by about 5.4 Å so that there is a hole in the center of the ring. The face opposite the ester residue is predominantly hydrophobic, while the opposite face is more hydrophilic; this may be important to the way in which these compounds bind to cell constituents. The symmetrical nature, chemically enforced rigidity, and absence of serious strain in these compounds suggest that his conformation may be maintained in solution.

Stereochemistry of PHO-3 (**26a**) and epi-PHO-3 (**26b**) at C(15) originally deduced from NMR experiments, was confirmed by an X-ray analysis of epi-PHO-3. Thus, epi-PHO-3 has the 15S configuration and PHO-3 the 15R configuration [20]. Conformation of epi-PHO-3 in the crystalline state was similar to that of maytansine. The physicochemical properties of PHO-3 (**26b**) demonstrated that it is identical with deacetylmaytanbutacine (**43**), and conclusions based on NMR studies indicate that the other known naturally occurring maytansinoids bearing an oxygen function at C(15) all belong to the R series.

3.4. Nuclear Magnetic Resonance

3.4.1. ^{1}H NMR Spectra.   Maytansinoids have been studied extensively by ^{1}H NMR spectroscopy, and unequivocal assignments have been made for all signals in most of the known compounds. Spectra obtained at high field, 300–470 MHz, were particularly useful in structural determinations of several compounds isolated from *Trewia nudiflora* [12].

Proton assignments for maytansine (**2**) and several representative maytansinoids are given in Table 2. Variations in substitution of the maytansinoid ring system are readily apparent. For example, methoxyl substitution at C(15), in compounds such as **16, 19, 20, 21,** and **40,** shifts the remaining C(15) proton signal downfield to near δ 4.8 and the additional methoxyl singlet appears near δ 3.4. This substitution causes only minor changes for protons in other regions of these molecules. If there is a hydroxyl substituent rather than methoxyl at this C(15) position, the accompanying C(15) proton signal occurs at δ 5.49 in colubrinol (**13**), at 5.37 in ansamitocin PHO-3 (**29**), at 5.11 in ansamitocin epi-PHO-3 (**30**), and at 5.37 in deacetylmaytanbutacine (**43**). Esterification of the hydroxyl, as in maytanbutacine (**15**), shifts the H(15) signal downfield to δ 6.21. Inversion of configuration at C(10), for example, in trewiasine (**16**) as compared with epitrewiasine (**23**), shifts the C(10) proton signal downfield from δ 3.51 to δ 3.72 while H(11) shifts from δ 5.72 to 5.90. Marked differences in coupling betweeen protons on carbons 10 and 11 are also noted; $J_{10,11} = 9.1$ Hz in trewiasine (**16**), and $J_{10,\ 11} = 3.5$ Hz in epitrewiasine (**23**). Maytansine 9-methyl ether (**31**) differs from maytansine (**2**) in that the C(9) hydroxyl signal in the ^{1}H spectrum of maytansine (**2**) is replaced in **31** by an O-methyl singlet at δ 3.41.

Maytansinoids esterified at C(3) normally exhibit a multiplet in the δ 4.5 to 4.8 region assigned to H(3). This resonance appears upfield, near δ 3.5, in maytansinol (**10**) and trewiasinol (**44**) since they have a free OH at C(3). Maysine (**7**), trewsine (**40**), and related maytansides having 2,3-double bonds lack the normal signals ascribed to protons on these two carbons but exhibit new vinyl proton doublets near δ 5.6 and 6.4. Compounds having an aromatic N-methyl group (18-NCH$_3$) exhibit a methyl singlet near δ 3.16; this characteristic signal is absent in maytansinoids lacking this N-methyl group,

Table 2. ^{1}H NMR of Some Representative Maytansinoids[a]

Proton Assignments	Maytansine (2)	Colubrinol (13)	Trewiasine (16)	10-Epi-trewiasine (23)	Trewsine (40)	Treflorine (19)	Trenudine (20)	N-Methyl-trenudone (21)	(41)	(42)
2_A	2.16 dd, J = 14.4, 3.1	2.18 dd, J = 14.3, 3.0	2.18 dd, J = 14.3, 3.0	2.17 dd, J = 14.4, 3.0	5.67 d, J = 15.5	2.21 dd, J = 14.9, 3.9	2.22 dd, J = 13.6, 3.7	2.09 dd, J = 14.7, 3.6	5.56 d, J = 15.4	5.62 d, J = 15.6
2_B	2.60 dd, J = 14.4, 12.1	2.56 dd, J = 14.3, 12.0	2.55 dd, J = 14.3, 12.2	2.58 dd, J = 14.4, 12.2	—	2.50 dd, J = 14.9, 12.0	2.53 dd, J = 13.6, 12.0	2.62 dd, J = 14.7, 12.0	—	—
3	4.75 d, J = 12.1, 3.1	4.75 dd, J = 12.0, 3.0	4.75 dd, J = 12.2, 3.0	4.83 dd, J = 12.2, 3.0	6.42 d, J = 15.5	4.51 d, J = 12.0, 3.9	4.63 dd, J = 12.0, 3.7	4.51 dd, J = 12.0, 3.6	6.29 d, J = 15.4	6.42 d, J = 15.6
4 CH$_3$	0.78 s	0.77 s	0.76 s	0.81 s	1.04 s	0.72 s	0.80 s	0.82 s	1.03 s	1.05 s
5	3.00 d, J = 9.6	3.02 d, J = 9.6	3.01 d, J = 9.6	2.86 d, J = 9.6	2.66 d, J = 9.8	3.07 d, J = 9.8	2.98 d, J = 9.8	3.06 d, J = 9.8	2.62 d, J = 9.8	2.63 d, J = 9.9
6 CH$_3$	1.27 d, J = 6.4	1.27 d, J = 6.1	1.26 d, J = 6.2	1.26 d, J = 6.3	1.31 d, J = 6.4	1.29 d, J = 6.4	1.30 d, J = 6.4	1.29 d, J = 5.3	1.35 d, J = 6.4	1.34 d, J = 6.6
7	4.26 m	4.27 m	4.28 m	4.23 m	4.30 m	4.22 m	4.21 m	4.18 m	4.35 m	4.34 m
10	3.47 d, J = 9.1	3.51 d, J = 9.0	3.51 d, J = 9.1	3.72 d, J = 3.5	3.46 d, J = 9.4	3.52 d, J = 9.0	3.55 d, J = 8.8	3.54 d, J = 8.8	3.46 d, J = 9.3	3.48 d, J = 9.0
11	5.65 dd, J = 15.3, 9.1	5.71 dd, J = 15.3, 9.0	5.72 dd, J = 15.3, 9.1	5.90 dd, J = 15.3, 3.5	5.58 dd, J = 15.3, 9.4	5.59 dd, J = 15.2, 9.0	5.68 dd, J = 15.4, 8.8	5.68 dd, J = 15.4, 8.8	5.69 dd, J = 15.0, 9.3	5.63 dd, J = 15.4, 10.0
12	6.41 dd, J = 15.3, 11.1	6.45 dd, J = 15.3, 11.0	6.46 dd, J = 15.3, 11.1	6.49 dd, J = 15.3, 11.1	6.44 dd, J = 15.3, 10.9	6.48 dd, J = 15.2, 10.9	6.52 dd, J = 15.4, 10.8	6.52 dd, J = 15.4, 10.7	6.46 dd, J = 15.0, 11.0	6.44 dd, J = 15.4, 10.0

Table 2. (*Continued*)

Proton Assignments	Maytansine (2)	Colubrinol (13)	Trewiasine (16)	10-Epi-trewiasine (23)	Trewsine (40)	Treflorine (19)	Trenudine (20)	N-Methyl-trenudone (21)	(41)	(42)
13	6.65 d $J = 11.1$	6.95 d $J = 11.0$	6.98 d $J = 11.1$	6.69 d $J = 11.1$	6.24 d $J = 10.9$	6.60 d $J = 10.9$	6.72 d $J = 10.8$	6.67 d $J = 10.7$	6.38 d $J = {<}12$	6.27 d $J = 10.0$
14 CH$_3$	1.62 s	1.59 s	1.52 s	1.51 s	1.56 s	1.57 s	1.55 s	1.53 s	1.58 s	1.58 s
15$_A$	3.10 d $J = 12.4$	5.47 s	4.86 s	4.78 s	4.67 s	4.95 s	4.88 s	4.87 s	4.82 s	4.71 s
15$_B$	3.63 d $J = 12.4$	—	—	—	—	—	—	—	—	—
17	6.72 d $J = 1.7$	6.55 d $J = 1.5$	6.54 d $J = 1.5$	6.55 d $J = 1.5$	6.60 d $J = 1.6$	7.26 d $J = 1.6$	7.26 d $J = 1.6$	7.25 d $J = 1.5$	6.97 d $J = 1.5$	6.73 d $J = 1.5$
21	6.81 d $J = 1.7$	7.34 d $J = 1.5$	7.22 d $J = 1.5$	7.23 d $J = 1.5$	7.20 d $J = 1.6$	7.47 d $J = 1.6$	7.27 d $J = 1.6$	7.54 d $J = 1.5$	7.25 d $J = 1.5$	7.22 d $J = 1.5$
10 OCH$_3$[b]	3.33 s	3.34 s	3.35 s	3.36 s	3.34 s	3.39 s	3.42 s	3.40 s	3.37 s	3.37 s
15 OCH$_3$[b]	—	—	3.37 s	3.36 s	3.37 s	3.41 s	3.43 s	3.42 s	3.40 s	3.38 s
20 OCH$_3$	3.96 s	4.00 s	3.99 s	4.00 s	4.00 s	4.02 s	4.00 s	3.99 s	4.01 s	4.01 s
18 NCH$_3$	3.18 s	3.16 s	3.16 s	3.16 s	3.25 s	—	—	—	—	—
2'	5.32 q	5.36 m	5.37 m	4.80 m	—	4.79 m	4.95 m	5.57 q	4.28 m	4.53 m
2'CH$_3$	1.29 d $J = 6.9$	1.26 d $J = 6.6$	1.28 d $J = 6.8$	1.34 d $J = 6.8$	—	1.33 d $J = 6.9$	1.34 d $J = 7.0$	1.29 d $J = 7.0$	1.43 d $J = 7.1$	1.47 d $J = 7.2$
2'NCH$_3$	2.84 s	2.87 s	2.88 s	2.97 s	—	—	—	2.75 s	—	—
4'	2.09 s	2.76 m	2.76 m	2.75 m	—	—	—	—	—	—
4'CH$_3$	—	1.04 d $J = 6.5$	1.06 d $J = 6.6$	1.05 d $J = 6.6$	—	1.40 s	1.48 s	1.53 s	1.40 s	1.53 s

4'CH$_3$	—	1.1 d $J = 6.8$	1.12 d $J = 6.8$	1.09 d $J = 6.8$	—	—	—	—
5'$_A$	—	—	—	1.45 m $J = 14.2,$ 3.0	3.93 m $J = 2.5,$ 2.4	—	1.77 dt $J = 14.8,$ 4.8, 4.8	4.06 $J = 7.0,$ 2.3
5'$_B$	—	—	—	2.78 m $J = 14.2,$ 3.0	—	—	2.46 m $J = 14.8,$ 9.5, 4.8	—
6'$_A$	—	—	—	3.03 m $J = 14.2,$ 3.0	3.55 dd $J = 15.0$	4.17 d $J = 14.6$	3.28 dt $J = 14.2,$ 4.8, 4.8	3.43 dd $J = 14.4,$ 7.0
6'$_B$	—	—	—	4.46 m $J = 14.2,$ 3.0	4.49 dd $J = 15.0$	4.51 d $J = 14.6$	4.25 $J = 14.2,$ 9.5, 4.8	4.21
2'NH	—	—	—	7.06 d $J = 10.7$	7.67 d $J = 9.9$	—	7.51 d $J = 7.4$	7.58 d $J = 8.0$
CO$_2$CH$_3$	—	—	—	—	—	—	3.80 s	3.77 s

[a]Chemical shifts (δ) are expressed in ppm from internal tetramethylsilane, and coupling constants (J) are expressed in Hz. Spectra were recorded in deuteriochloroform solution. In all of the above, the C(6) and C(8) proton signals occur at approximately δ 1.3 and are obscured by other signals in this region.
[b]These assignments may be reversed.

171

including normaytansine (11), nortrewiasine (24), normaysine (8), treflorine (19), trenudine (20), N-methyltrenudone (21), and normaytancyprine (12).

Maytanacine (ansamitocin P-1) (6) is distinguished in the NMR by an acetate methyl signal at δ 2.18, and most other ansamitocins give proton spectra typical of simple propionate (25a), butyrate (25c), isobutyrate (25b), and isovalerate (25d) esters of maytansinol. Maytansine and other maytanside esters containing nitrogen in the C(3) ester side-chain are N-substituted deriv-atives of L-alanine; thus, maytansine is the N-methyl-N-acetylalanine ester of maytansinol. The 2' proton of maytansine is apparent at δ 5.32, the 2' methyl group resonates at δ 1.29, and the 2'NCH₃ singlet appears at δ 2.84. This latter signal is replaced by an NH doublet in the δ 7.0–7.6 region in compounds lacking a 2'NCH₃ group; in these, the NH doublet is coupled to H(2') (7–10 Hz).

Maytanbutine (4), colubrinol (13), trewiasine (16), and several other compounds from *Trewia* all possess the same N-acyl group [COCH(CH₃)₂] in the ester side-chain. This group is characterized in the ¹H spectrum by a signal near δ 2.76 (H-4') and by two methyl doublets near δ 1.06 and 1.12. In contrast, dehydrotrewiasine, with a C=C in the side-chain, gives a vinyl methyl singlet at δ 1.92 (4'-CH₃) and two vinyl proton singlets at δ 5.02 and 5.22 (4'-CH₂). Treflorine (19) has a C(4') hydroxyl group that results in a downfield shift of the 4' methyl signal to δ 1.40. High-field ¹H NMR spectra were invaluable in determining the structures of trenudine (20) and N-methyltrenudone (21) with their additional oxygen functions in the 12-membered ring; portions of the 470-MHz spectra of (19) and (20) are shown in Figure 3 with appropriate assignments. Trenudine (20) exhibits an additional oxygenated methine signal at δ 3.93 (H-5') and slight downfield shifts of the H-6' signals. In N-methyltrenudone (21), the isolated methylene [C(6')] appears as two doublets at δ 4.14 and 4.51 (J = 14.6 Hz) [12].

3.4.2. ¹³C NMR Spectra.   ¹³C NMR measurements on maytansinoids also provide valuable structural information. Spectra of maytansine (2) recorded by Powell et al. [11] and by Sneden et al. [9a] are identical, within experimental error, and most signals have been unequivocally assigned. Some representative ¹³C spectra of maytansinoids are given in Table 3 and a number of others have been reported by Wallace and Sneden [9b]. Maytansine (2) exhibits three carbonyl signals in the δ 168–171 region (ester and amide) and the carbi-nolamide carbonyl signal is evident at δ 152.3. Compounds lacking a C(3) ester side-chain, such as trewsine (40), show only two carbonyl signals (δ 164.2 and 152.2). Typical maytansinoids, such as 2, have 10 olefinic and aromatic carbons; assignments may be calculated from their multiplicities and (in the case of aromatic protons) from published substituent constants. The chlorine-bearing carbon, C(19), has a long relaxation time and is readily apparent as a sharp signal of reduced amplitude. The maytansinoid ring system also has one methine signal near δ 38.9 [C(6)], and three methylene signals characteristic of C(2), C(8), and C(15) (δ 37.4, 36.2, and 46.6). Epoxy-bearing carbons [C(4) and

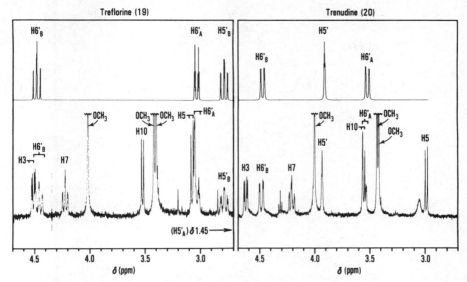

Figure 3. Portions of the 470-MHz ^{1}H NMR spectra of treflorine (**19**) and trenudine (**20**) recorded in CDCl$_3$. Upper traces are simulated spectra using parameters of Table 1. Reprinted with permission from reference 12. Copyright 1982 American Chemical Society.

C(5)] and other oxygenated carbons [C(3), C(7), C(9), and C(10)] have been assigned on the basis of their shifts and multiplicities. The two *O*-methyl signals of **2** are evident at δ 56.6, and three *C*-methyl signals [those attached to C(4), C(6), and C(14)] were also assigned from their shifts and multiplicities. The spectrum of **2** also exhibits an aromatic *N*-methyl signal at δ 35.4 and the side-chain *N*-CH$_3$ (2′) is evident at δ 31.8.

Comparison of the ^{13}C spectrum of trewiasine (**16**) with that of maytansine (**2**) confirmed that both compounds possessed a 19-membered maytansinoid ring system, but it was apparent that trewiasine (**16**) had an additional methoxyl substituent located at C(15). This fact was evidenced by a marked downfield shift of the C(15) signal from δ 46.6 in **2** to δ 86.7 in **16,** and by an additional *O*-methyl signal at δ 56.3 in **16**. Trewiasine (**16**) also exhibits additional signals due to carbons in the C(3) ester moiety. In compounds bearing a hydroxyl at C(15), such as ansamitocins PHO-3 and epi-PHO-3, the C(15) signal appears at δ 79.2 or 77.6, depending on stereochemistry [20]. Evidence that (**23**) is the C(10) epimer of trewiasine (**16**) was also apparent from ^{13}C spectra. Most obvious differences between the two spectra were shifts in signals for C(10) and those α and β to C(10), along with a downfield shift of one methoxyl signal to δ 58.4 in the spectrum of **23**.

Reduction of C(3) esters to corresponding carbinols, such as conversion of trewiasine (**16**) to trewiasinol (**44**), gives significant upfield shifts of signals assigned to C(3) and adjacent carbons. In compounds resulting from

Table 3. ^{13}C NMR of Some Representative Maytansinoids[a]

Carbon Assignment	Maytansine (2), [11]	Maytanprine 3, [9b]	Trewiasine (16), [11]	Maytanacine 6, [9b]	PHO-3 26a, [20]	10-Epitrewiasine (23), [13]	Teflorine (19), [12]	Trenudine (20), [12]	N-Methyl Trenudone (21), [12]	Maysine (7), [9b]
2	32.4 t	32.5 t	32.4 t	32.8 t	33.5 t	32.6^b	32.5 t	33.1 t	32.4 t	121.9 d
3	78.1 d	78.1 d	78.2 d	77.0 d	77.4 d	78.2	78.5 d	78.5 d	78.8 d	147.5 d
4	60.1 s	60.1 s	60.0 s	60.3 s	61.4 s	60.2	59.4 s	59.5 s	59.5 s	59.7 s
5	67.2 d	67.2 d	67.7 d	66.4 d	67.3 d	67.3	67.1 d	66.9 d	66.7 d	66.9 d
6	38.9 d	39.0 d	38.9 d	38.5 d	39.3 d	38.8	37.8 d	37.9 d	38.1 d	38.7 d
7	74.1 d	74.2 d	74.1 d	74.3 d	75.4 d	74.3	74.0 d	74.0 d	71.4 d	75.0 d
8	36.2 t	36.4 t	36.3 t	36.0 t	37.2 t	31.8^b	36.1 t	36.2 t	36.2 t	35.5 t
9	80.6 s	81.0 s	80.7 s	81.1 s	81.2 s	82.7	80.8 s	80.8 s	81.0 s	81.2 s
10	86.6 d	88.8 d	85.5 d	88.3 d	89.4 d	83.2	88.7 d	88.6 d	88.6 d	88.6 d
11	127.8 d	127.9 d	129.9 d	128.3 d	130.7 d	128.8	128.9 d	129.3 d	129.3 d	127.2 d
12	133.2 d	133.3 d	132.5 d	132.2 d	133.1 d	130.2	132.6 d	132.4 d	132.6 d	133.0 d
13	125.3 d	125.6 d	128.0 d	124.5 d	125.7 d	126.2	126.5 d	127.2 d	127.3 d	124.6 d
14	142.1 s	139.0 s^b	142.1 s	139.9 s^b	142.1 s^b	142.5	141.9 s	141.9 s	141.7 s	140.3 s^b
15	46.6 t	46.5 t	86.7 d	47.2 t	79.2 d	87.1	87.0 d	86.6 d	86.9 d	46.7 t
16	141.2 s	142.3 s^b	141.3 s	142.7 s^b	145.2 s^b	141.4	140.7 s	141.5 s	141.4 s	142.0 s^b
17	122.2 d	122.4 d	120.3 d	122.2 d	120.0 d	120.2	121.9 d	120.7 d	120.7 d	122.2 d
18	139.2 s	141.2 s^b	139.0 s	140.1 s^b	142.8 s^b	136.9	139.1 s	140.3 s	140.4 s	140.5 s^b
19	118.7 s	119.0 s	118.9 s	119.0 s	119.8 s	119.2	118.2 s	118.2 s	118.7 s	119.3 s

	1	2	3	4	5	6	7	8	9	10
20	155.9 s	156.1 s	156.3 s	156.2 s	157.2 s	156.4	156.2 s	156.0 s	156.3 s	156.4 s
21	113.1 d	113.4 d	109.0 d	113.5 d	109.7 d	109.1	109.0 d	109.0 d	108.6 d	112.7 d
C=O	170.9 s	173.3 s	176.7 s	—	170.5 s	177.1	176.1 s	175.0 s	171.4 s	—
C=O	170.3 s	171.0 s	170.9 s	169.1 s	176.8 s	170.9	171.8 s	174.9 s	171.1 s	—
C=O	168.8 s	168.7 s	168.8 s	168.7 s	—	169.0	170.9 s	170.8 s	169.5 s	164.3 s
C=O	152.3 s	152.2 s	152.4 s	152.2 s	154.4 s	154.8	152.5 s	152.6 s	152.5 s	152.1 s
OCH$_3$	56.6 2q	56.6 2q	56.3–7 3q	56.7 2q	56.8 2q	56.3–7, 58.4	56.3–8 3q	56.4–9 3q	56.7–9 3q	56.6 2q
CH$_3$	15.5 q	15.4 q	14.6 q	15.8 q	14.7 q	14.7	15.7 q	16.2 q	14.4 q	16.0 q
CH$_3$	14.6 q	14.5 q	13.1 q	14.5 q	—	13.2	14.3 q	14.2 q	12.9 q	14.2 q
CH$_3$	13.3 q	13.3 q	12.0 q	—	10.3 q	12.3	12.1 q	12.0 q	12.3 q	14.8 q
CH$_3$	12.1 q	12.2 q	10.0 q	12.1 q	12.6 q	10.0	9.8 q	9.9 q	10.1 q	—
18 NCH$_3$	35.4 q	35.3 q	35.2 q	35.6 q	35.9 q	35.4	—	—	—	36.0 q
2'NCH$_3$	31.8 q	30.6 q	30.4 q	—	—	30.5	—	—	30.2 q	—
2'	52.2 d	52.3 q	52.4 d	20.9 q	34.6 d	53.8	46.3 d	46.7 d	52.6 d	—
4'	21.9 q	26.7 q	30.4 d	—	—	30.5	73.4 s	72.2 s	78.5 s	—
4'CH$_3$	—	9.1 q	19.4 q	—	18.3 q[c]	19.2	28.9 q	28.5 q	23.6 q	—
4'CH$_3$	—	—	18.8 q	—	20.4 q[c]	18.8	—	—	—	—
5'	—	—	—	—	—	—	35.6 t	79.1 d	204.0 s	—
6'	—	—	—	—	—	—	43.8 t	52.4 t	57.8 t	—

[a] Chemical shifts (δ) are expressed in ppm from tetramethylsilane. Proton decoupled and off-resonance decoupled spectra were recorded in deuteriochloroform solution.

[b] Assignment may be reversed.

elimination of the C(3) ester moiety, as in trewsine (**40**), signals for C(2) and C(3) shift markedly downfield in accord with an additional double bond. A signal near δ 35.4 is indicative of an *N*-methyl group (aromatic amide) and the upfield carbonyl signal, usually near δ 154, has been attributed to the carbinolamide group [20, 26].

Various C(3) esters of maytansinoids have been studied by ^{13}C NMR. Assignments are relatively straightforward in the ansamitocins as most of these compounds are simple acyl esters of maytansinol. A signal near δ 52.3 is attributed to C(2′) in maytansine (**2**), trewiasine (**16**), and similar compounds. This C(2′) signal shifts upfield to near δ 46.5 in the exceptional cases of treflorine (**19**) and trenudine (**20**), probably due to conformational effects. Presence of an *N*-methyl group (2′*N*-CH₃) is indicated by a signal near δ 31.0 and a peak near δ 13.4 has been assigned to the methyl group attached to C(2′) [9b]. The C(4′) methyl signal of maytansine (δ 21.9) is replaced by signals typical of an isopropyl group (δ 30.4, 19.4, 18.8) in the ^{13}C spectrum of trewiasine (**16**) and by five signals of a 1,2-dimethyl-substituted cyclopropyl group in normaytancyprine (**12**) [9a]. Several additional features are present in spectra of treflorine (**19**) and other compounds in which the C(3) ester is bound in a second ring. In treflorine (**19**), C(4′) is oxygenated and appears at δ 73.4; additional methylene signals are present at δ 35.6 [C(5′)] and 43.8 [C(6′)]. Hydroxylation of C(5′), as in trenudine (**20**), shifts the 5′ signal to δ 79.1, and conversion of C(5′) to a carbonyl, as in *N*-methyltrenudone (**21**), shifts this signal far downfield to δ 204.0. Appropriate shifts are also noted for signals of carbons adjacent to C(5) as one progresses from treflorine (**19**) to trenudine (**20**), and to *N*-methyltrenudone (**21**) [12].

3.5. Mass Spectra

Electron impact mass spectra (EIMS) of maytansinoids yield considerable structural information although molecular ions usually are not observed. The most significant high-mass ion usually occurs at $M^+ - 61$ [$M^+ - (H_2O + HNCO)$, abbreviated as $M^+ - a$]. This loss is attributed to the carbinolamide portion of the molecule [6]. Subsequent elimination of the C(3) substituent (ROH) gives another characteristic ion, $M^+ - (a + b)$. Other typical ions normally appear at $M^+ - (a + CH_3)$, $M^+ - (a + Cl)$, $M^+ - (a + b + CH_3)$, and $M^+ - (a + b + Cl)$ [6, 16]. Compounds bearing a methoxyl substituent at C(15) exhibit an additional ion corresponding to $M^+ - (a + b + OCH_3)$ and those having an additional hydroxyl function at any one of several positions yield an ion at $M^+ - (a + H_2O)$ [4, 5]. Some typical fragmentations of maytansine (**2**) are summarized in Figure 4.

Mass spectra of maytansinoids also contain information concerning the nature of the C(3) substituents. Prominent ions normally occur at $b - (OH)$ and $b - (CO_2H)$. Compounds with an *N*-methyl group in the ester side-chain

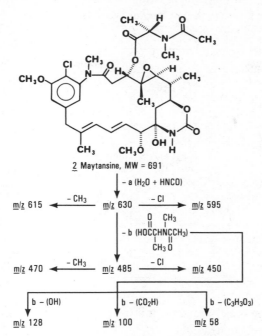

Figure 4. Characteristic fragments observed in the electron impact mass spectrum of maytansine.

yield a major ion at m/z 58 $[CH_3CH\!=\!NHCH_3]^+$ and those lacking the N-methyl group give a corresponding ion at m/z 44 $[CH_3CH\!=\!NH_2]^+$. These latter two ions are essentially absent in mass spectra of maytansinoids lacking the amide group in the C(3) ester. Treflorine (**19**), trenudine (**20**), and N-methyltrenudone (**21**) have C(3) substituents bound to the usual maytansinoid ring at a second position, via the aromatic amide nitrogen, and fail to exhibit ions at $M^+ - (a + b)$, $b - (OH)$, and $b - (CO_2H)$. However, these compounds yield characteristic ions at $M^+ - (a + OCH_3 + Cl + CO_2H)$. Electron impact mass spectra of the products obtained from base hydrolysis of treflorine and trenudine (**41** and **42**) were useful in confirming their structures [12] because a side-chain derived from the 12-membered ring remains attached even after the hydrolysis-elimination reaction. Ester **41** gave an intense ion at m/z 202 $(C_9H_{16}NO_4)$, representing the entire ester side-chain severed from the aromatic amide nitrogen and two other prominent ions presumed to arise from cleavage between C(3′) and C(4′) (m/z 572) and between C(4′) and C(5′) (m/z 528). Fragment ions were present in the spectrum of **42** at m/z 588 [cleavage between C(3′) and C(4′)], m/z 570 (588 − H$_2$O), and m/z 544 [cleavage between C(4′) and C(5′)].

Negative ion chemical ionization mass spectra of maytansinoids yield additional information; most importantly, substantial molecular anions (M^-)

Table 4.   Negative Ion Chemical Ionization Mass Spectra of Maytansinoids

Compound		M^-	$(M - a)^-$	$[M - (a + b)]^-$	$(b - H)^-$
				m/z (relative intensity)	
(2)	Maytansine	691 (2)	630 (100)	485 (35)	144 (52)
(3)	Maytanprine	705 (2)	644 (100)	485 (31)	158 (53)
(4)	Maytanbutine	719 (3)	658 (100)	485 (31)	172 (70)
(5)	Maytanvaline	733 (6)	672 (100)	485 (33)	186 (51)
(8)	Normaysine	532 (1)	471 (100)	—	—
(13)	Colubrinol	735 (4)	674 (100)	501 (79)	172 (47)
(16)	Trewiasine	749 (41)	688 (85)	515 (31)	172 (100)
(17)	Dehydrotrewiasine	747 (3)	686 (100)	515 (18)	170 (17)
(18)	Demethyltrewiasine	735 (1)	674 (100)	515 (14)	158 (25)
(19)	Treflorine	749 (2)	688 (100)	—	—
(20)	Trenudine	765 (1)	704 (100)	—	—
(21)	N-Methyltrenudone	777 (1)	716 (100)	—	—
(23)	10-Epitrewiasine	749 (4)	688 (100)	515 (3)	172 (10)
(24)	Nortrewiasine	735 (9)	674 (59)	501 (100)	172 (23)

are observed. The most abundant ion in these spectra is often $(M - a)^-$ and, with exceptions for compounds **19, 20,** and **21,** $[M - (a + b)]$ is also prominent. In contrast to eims results, fragments eliminated from C(3) are directly observed as major peaks at $(b - H)^-$. Negative ion spectra of some representative maytansinoids are summarized in Table 4. Relative intensities of the ions vary with heating profile, and reported spectra were observed at the point of highest M^- signal [13].

4.   SYNTHESIS OF MAYTANSINOID ALKALOIDS

4.1.   Total Syntheses

To date, total syntheses of naturally occurring maytansinoids or their derivatives have been reported by three different groups—Corey and his associates at Harvard University, Meyers and his research group at Colorado State University, and also Isobe and his associates at Nagoya University, Japan.

4.1.1.   Corey Synthesis.   Corey and coworkers undertook the synthesis of N-methylmaysenine (**48**), the N-methyl derivative of **9**, by a series of reactions aimed at optimum stereospecificity [30]. Their sequence began with tri-O-acetyl-D-glucal (**49**), whose acetyl groups were removed with sodium methoxide. The resulting methanolic solution of unsaturated triol was treated with mercuric acetate to provide an acetoxy methoxymercuration product which,

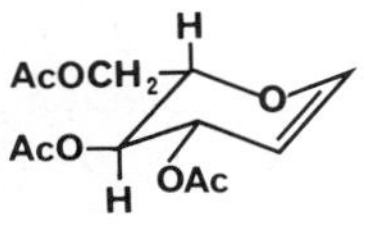

49

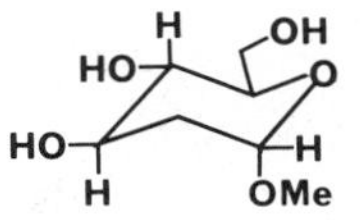

50

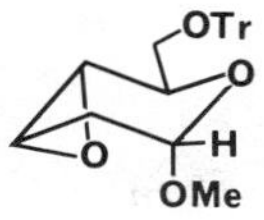

51

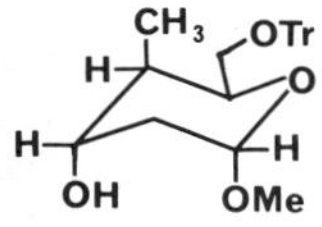

52

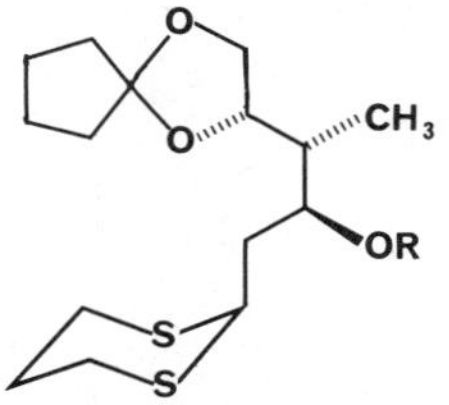

53 R = H
54 R = MEM

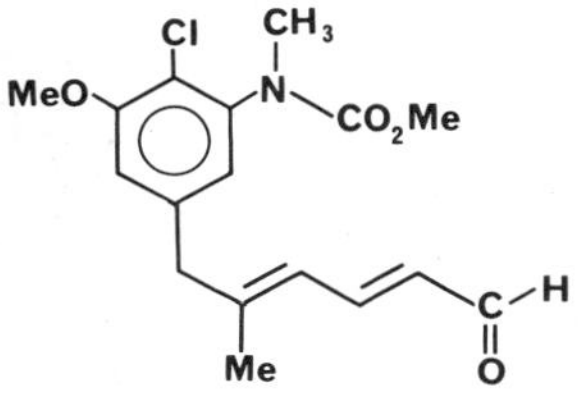

55

in turn, was converted to the corresponding chloromercuri derivative; this was reduced with sodium borohydride to give **50**. Treatment of **50** with trityl chloride gave a 6-*O*-trityl derivative which was converted to epoxide **51** by successive treatment with sodium hydride and triisopropylbenzenesulfonylimidazole. Epoxide **51** was converted to **52** by treatment with methyllithium in the presence of cuprous iodide. By reaction with propane-1,3-dithiol in hydrochloric acid–chloroform, the pyranose ring of **52** was opened to give a 1,3-dithiane; the trityl ether linkage also was cleaved in the process. In the resulting product, the hydroxyl groups at C(5) and C(6) were ketalized with a cyclopentanone equivalent (1-ethoxycyclopentene) in a reaction catalyzed by boron trifluoride. The product (**53**) from this reaction sequence was treated with β-methoxyethoxymethyl (MEM) chloride to give **54** with the C(3) hydroxyl blocked. Intermediate **54** represented the eastern portion of the maytansinoid macrocycle and was designed to be condensed with synthon **55** embodying the eastern and southern zones of the desired macrocycle.

Corey's strategy for synthesis of *N*-methylmaysenine (**48**) called for condensation of dithiaketal **54** with another intermediate, dienal **55**; the latter was to form the western and southern portions of the macrocycle [30].

For preparation of the aromatic moiety, the northwest corner of the macrocycle, Corey and associates [31] began their sequence with gallic acid (**56**).

CO_2H

HO   OH
   OH

56

OC_2H_5

$CO_2C_2H_5$

57

X

$NHCH_2C_6H_5$

$CO_2C_2H_5$

58a, X = H

b, X = Cl

Cl CH_3
RO NCH_2Ph

$CO_2C_2H_5$

59a, R = H

b, R = Me

Cl
MeO   NHMe

$CO_2C_2H_5$

60

Cl
MeO NHR_2

$O{=}C{-}OR_1$

60 $R^1 = CH_2CH_3$; $R^2 = CH_3$

61 $R^1 = CH_3$; $R^2 = CHO$

Cl
MeO   NHMe

CH_2OH

62

Cl CH_3
MeO $N{-}CO_2CH_3$

CH_2X

63a, X = OH

b, X = I

This starting material was converted to enone ester **57** by a Birch reduction followed by reaction with acidic ethanol. Treatment of **57** with *N*-methylbenzylamine produced an enamino ketone, **58a**. The requisite chlorine was introduced into **58a** by reaction with *t*-butylhypochlorite to give **58b**. Aromatization of **58b** was accomplished by reaction with benzeneselenenyl bromide in the presence of lithium diethylamide to provide **59a**. Ester **59a** was methylated by methyl iodide in the presence of potassium carbonate to give methoxy ester **59b**, which underwent hydrogenolysis in the presence of palladium to afford the debenzylatyed end product, **60**.

Synthesis of intermediate **55** proceeded by further elaboration of aromatic moiety **60** [32]. Reduction of either amide **61** or amino ester **60** with lithium aluminum hydride afforded alcohol **62**. After treatment of **62** with methyl chloroformate followed by potassium carbonate and then by ethanolic sodium hydroxide, an alcohol (**63a**) was obtained; **63a** was transformed into benzylic iodide **63b** by mesylation followed by treatment with sodium iodide.

Another intermediate (**66b**) was required for coupling with **63b** en route to *N*-methylmaysenine and maytansine (**2**). Synthesis of **66b** was accomplished by

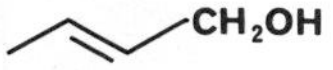

64

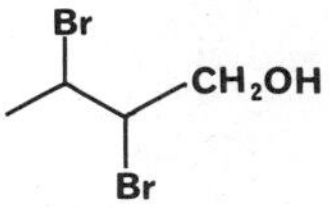

65

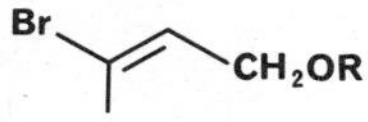

66a, R = H
b, R = THP

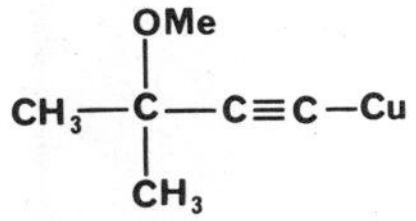

67

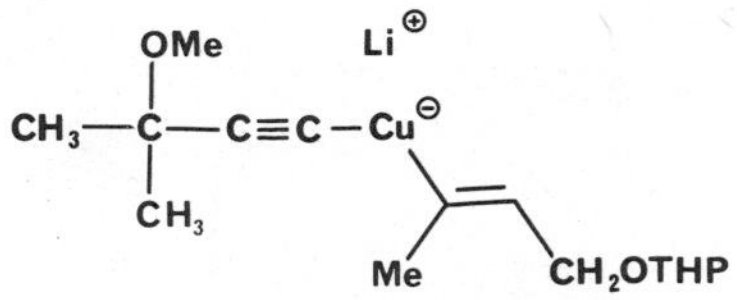

68

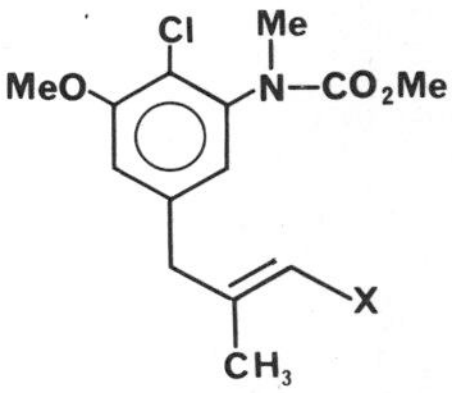

69a, X = CH₂OTHP
b, X = CH₂OH
c, X = CHO

brominating *E*-crotyl alcohol (**64**) to provide *erythro*-2,3-dibromobutane-1-ol, **65** [32]. Dehydrobromination of **65** was effected with lithium diisopropylamide and hexamethylphosphoric amide in THF to afford *E*-3-bromo-2-buten-1-ol (**66a**); the hydroxyl group was blocked by reaction with dihydropyran to give **66b**. Cuprous acetylide **67** was prepared by treating 3-methoxy-3-methyl-1-butyne with butyllithium followed by cuprous iodide. The acetylide was coupled with **66b** to give mixed Gilman reagent **68**, and **68** was condensed with benzylic iodide **63b** to provide **69a**. Compound **69a** was deprotected with methanolic toluensulfonic acid to give **69b** which, in turn, was oxidized with manganese dioxide in methylene chloride to aldehyde **69c**.

The elaboration of **69c** to give dienal **55** was accomplished by a special technique in which **69c** reacted with the α-trimethylsilyl derivative of acetaldehyde *N*-*t*-butylimine (generated from the corresponding α-lithio derivative) [33]. Hydrolysis of the *t*-butylimine under appropriate conditions provided the

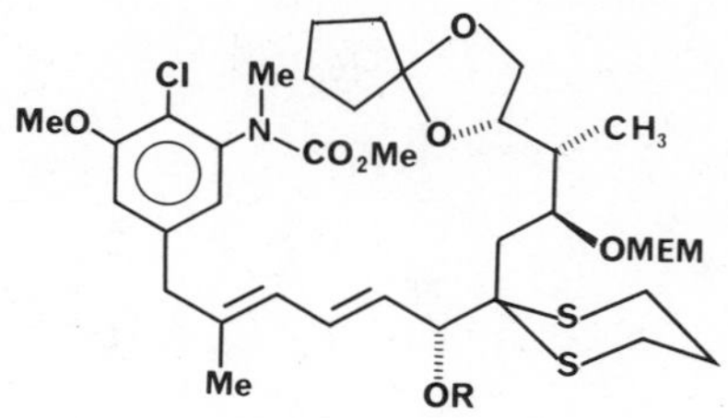

70a, R = H

b, R = Me

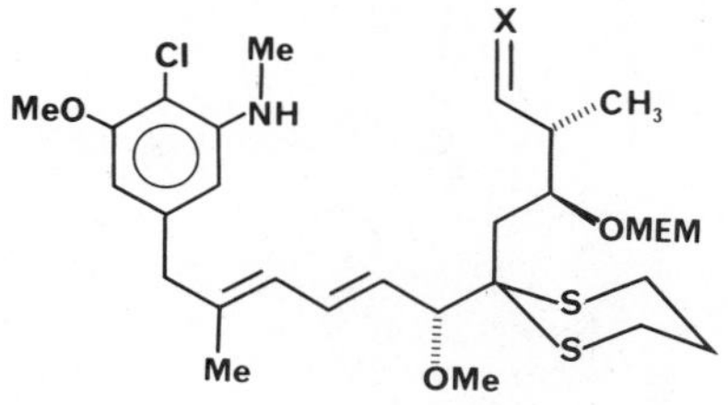

71a, X = O

b, X = C$<$CHO / Me

desired dienol (55). Intermediate 55 had in place the essential features of the western and southwestern zones of the maytansinoid macrocycle. Accordingly, the stage was set for condensing 54 with 55 to be followed by the final steps en route to *N*-methylmasenine (48) [30].

Fully protected dithiane 54 was lithiated with butyllithium and tetramethyl-ethylenediamine in THF, and this derivative was treated briefly with dienal 55. This process yielded a mixture of 70a and its C(10) epimer in nearly equal amounts; the two isomers were separated chromatographically. The correct epimer was methylated with methyl iodide in the presence of sodium hydride to give 70b. After removal of the *N*-carbomethoxy function with lithium thiomethoxide, the ketal group masking the 1,2-diol function in the northeast corner was removed with dilute perchloric acid. The resulting diol grouping was cleaved with lead tetraacetate to provide 71a, an unstable aldehyde purified by conventional techniques, but applied at low temperatures.

In elaborating the northern facade of *N*-methylmaysenine, Corey and his group [30] again resorted to the special technique [33] used for preparing α,β-unsaturated aldehyde 69c via an *N*-butylimine intermediate. The product, enal 71b, was further elongated to 72a through a Wittig-type reaction with the lithio derivative of dimethyl methoxycarbonylmethanephosphonate. The carbome-thoxy function of 72a was hydrolyzed with tetra-*n*-butylammonium hydroxide

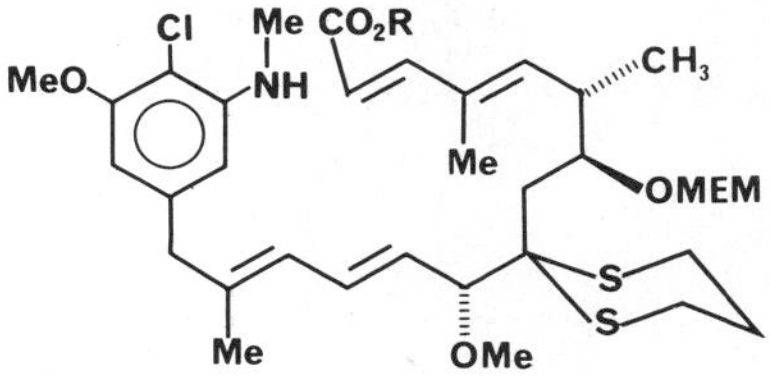

72a, R = Me

b, R = H

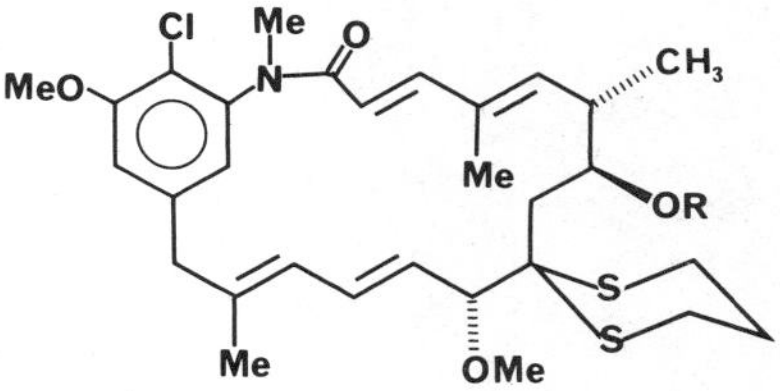

73a, R = MEM

b, R = H

to give **72b** with free carboxyl group. The culminating step in formation of the macrocycle—through an amide linkage—was accomplished by using a mixed anhydride with mesitylenesulfonic acid to activate the carboxyl function and thus generate **73a**. The protecting MEM group was removed with aqueous sulfuric acid to give **73b**. The racemic form of **73b** had been synthesized earlier by Corey et al. [34] using a similar route.

At this point, Corey's synthesis of (−)-*N*-methylmaysenine (**74**) was complete except for elaboration of the carbinolamide moiety in the southeast corner [30]. Compound **73b** was acylated with *p*-nitrophenylchloroformate and the resulting product was treated with ammonia to give a carbamate. The carbonyl-masking dithiane moiety was eliminated with mercuric chloride–calcium chloride, and a subsequent ring closure provided (−)-**74.**

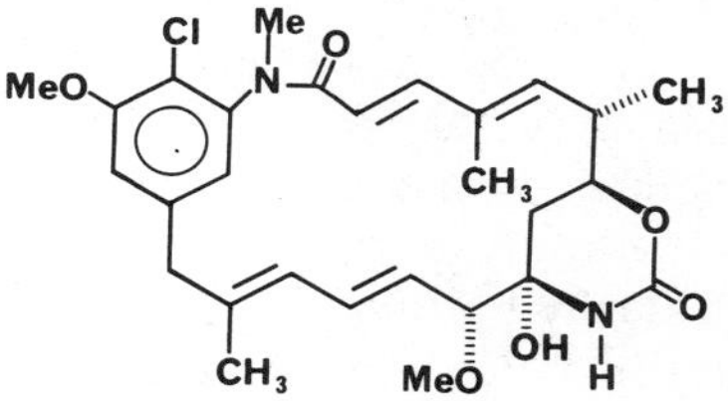

74

$$CH_3 - \langle\bigcirc\rangle - \overset{O}{\underset{\parallel}{S}}CH_2\overset{O}{\underset{\parallel}{C}} - O - \langle\bigcirc\rangle$$

75

76a, $R^1 = C_6H_5$; $R^2 = H$

   b, $R^1 = C_6H_5$; $R^2 = SiMe_2tBu$

   c, $R^1 = H$; $R^2 = SiMe_2tBu$

Although N-methylmaysenine (**74**) possesses all the requisite features of the maytansinoid macrocycle, it lacks an acyloxy function at C(1), shown to be essential for biological activity [25]. Accordingly, it was deemed important to prepare another maytansinoid with the required functionality at C(1). (R)-$(+)$-p-Tolyl phenoxycarbonylmethyl sulfoxide (**75**) was converted to its magnesium derivative by treatment with t-butyl magnesium chloride, and this was condensed with dienal **71b**, a key intermediate in Corey's earlier sythesis of **74** [35]. The α-sulfinyl group was removed by reductive cleavage with aluminum amalgam; this reaction was highly stereoselective and afforded the 3S-hydroxy ester **76a** in high yield. Intermediate **76a** was silylated to give **76b** which was, in turn, hydrolyzed with lithium hydroxide to give **76c**. Cyclization of **76c** was effected by the same mixed anhydride approach used by the Corey group for **73a** and provided the desired lactam [30].

The C(7) oxygen function of **76c** was deprotected by a novel procedure through an intermediate thioether which was converted to the desired 7-hydroxy compound. Elaboration of the carbinolamide moiety then proceeded essentially as in the conversion of **73b** to **74**. Desilylation of the 3-hydroxyl function of the product (**77a**) was accomplished with hydrogen fluoride to give **77b**; this derivative was converted to the corresponding 9-O-methyl ether by reaction with methanol catalyzed by p-toluenesulfonic acid. The product, 4,5-deoxymaytansinol (**77c**), then awaited introduction of the 4,5-epoxy function, a common feature of naturally occurring maytansinoids. This step was accomplished stereospecifically by treating **77c** with t-butyl hydroperoxide in the presence of oxyvanadium (IV) bis(acetylacetonate), and represented a formal total synthesis of maytansine (**2**) [35].

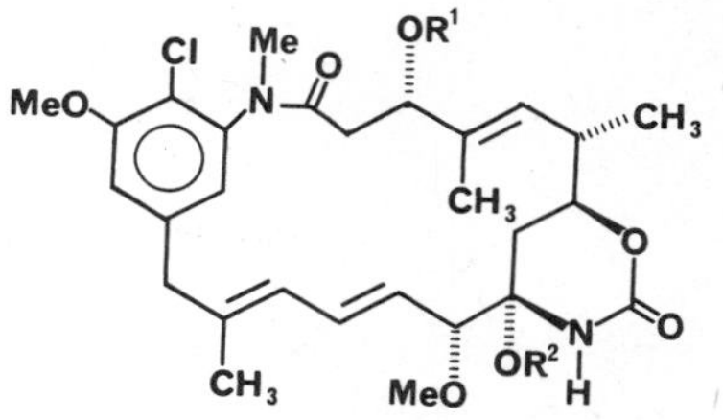

77a, R^1 = SiMe$_2$tBu; R^2 = H

b, R^1 = R^2 = H

c, R^1 = H; R^2 = Me

4.1.2. Meyers Synthesis.

Meyers and his associates carried out an enantiose-lective convergent total synthesis of (−)-maysine (7) based on the retrosynthetic strategy of dissecting 7 into three zones to be synthesized separately; these intermediates were then condensed to form the macrocycle nearly completed [36].

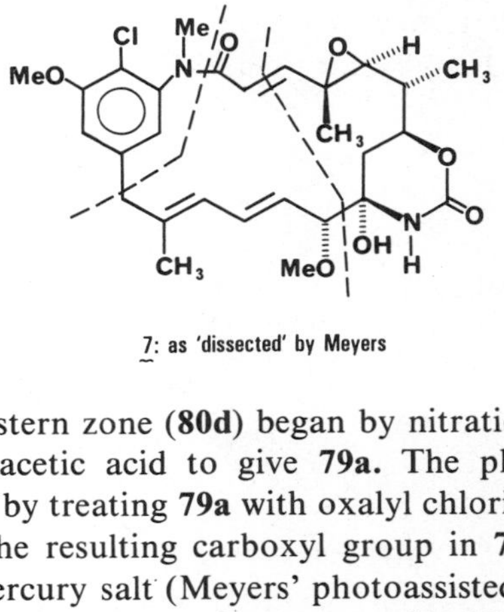

7: as 'dissected' by Meyers

Synthesis of the western zone (80d) began by nitration of methyl vanillate (78) with nitric acid–acetic acid to give 79a. The phenolic hydroxyl was replaced with chlorine by treating 79a with oxalyl chloride. The product (79b) was saponified, and the resulting carboxyl group in 79c was replaced with bromine through a mercury salt (Meyers' photoassisted modification [37] of the Cristol–Firth–Hunsdiecker reaction) to provide 79d. The nitro group of 79d was reduced with iron powder in ethanolic hydrochloric acid to give 80a, a substituted aniline. At this stage, the Meyers group introduced an amino-protecting function expected to withstand a large number of steps to follow, yet removable under mild conditions resulting in only gaseous products. The first step in this process was reaction of 80a with phenyl chloroformate, yielding urethane 80b as an intermediate. An exchange of urethane groups was then effected with trimethylsilylethanol and potassium t-butoxide, thus providing 80c. Substitution of the amino group was then completed with methyl iodide and potassium t-butoxide to give 80d.

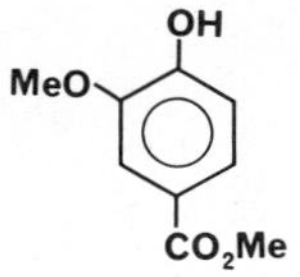

78

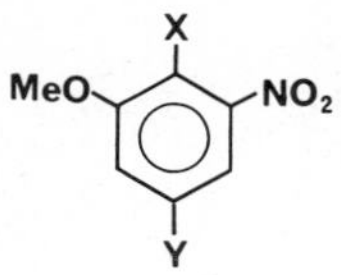

79a, X = OH; Y = CO₂Me   80a, R¹ = R² = H

b, X = Cl; Y = CO₂Me   b, R¹ = H; R² = CO₂Ph

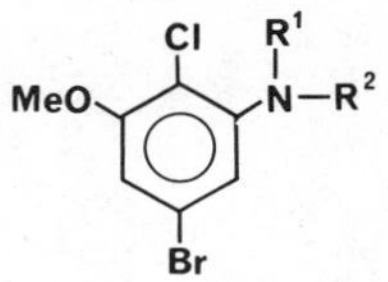

c, X = Cl; Y = CO₂H   c, R¹ = H; R² = CO₂(CH₂)₂SiMe₃

d, X = Cl; Y = Br   d, R¹ = Me; R² = CO₂(CH₂)₂SiMe₃

The Meyers group next turned their attention to the southern zone of the maytansinoid macrocycle, for which β-bromoacrolein (**87c**) was desired as a synthon. Tribromoacetate **82** was secured by free radical addition of tribromomethane to vinyl acetate (**81**), and **82** was converted to β-bromoacrolein diethyl acetal (**84**) under the influence of *t*-butoxide. Because of its instability, **85** was not isolated, but was generated in ether solution (acetal cleavage with oxalic acid), and this solution was added directly to potassium triethyl 2-phosphonopropionate (**86**). The resulting ester (**87a**) was reduced with diisobutylaluminum hydride to furnish alcohol **87b,** which was heated with *N*-bromosuccinimide and dimethyl sulfide to yield the target dibromo compound, **87c.**

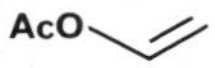

81 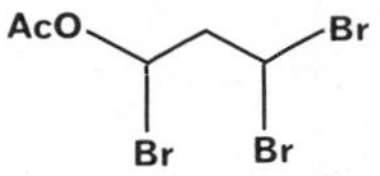   82 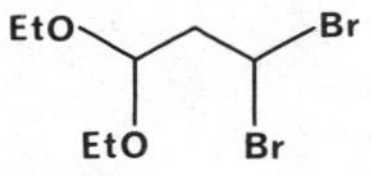   83

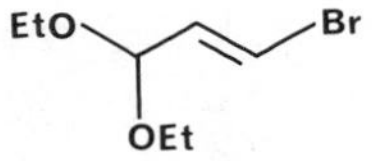

84 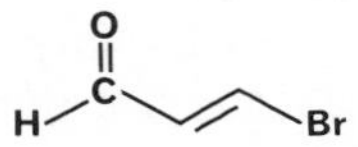   85 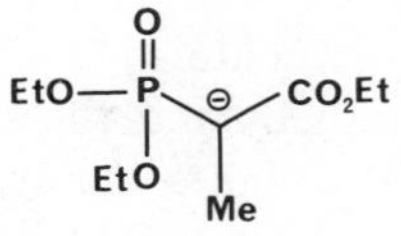   86

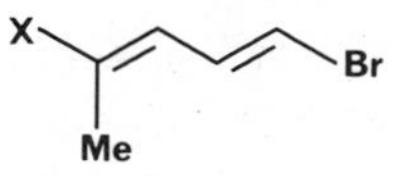

87a, X = CO₂Et

b, X = CH₂OH

c, X = CH₂Br

MeO — (aromatic ring) — Cl, Me, N–CO₂–(chain)–SiMe₃; lower substituent CH₂–C(Me)=CH–CH=CH–Br

88

HO / CO₂H, Me — **89**

EEO / CO₂EE, Me — **90**

EEO / OR, Me — **91a, R = H**
b, R = CH₂Ph

Meyers and associates [36] were then ready to complete a major fragment representing the southern and western zones of maytansinoids. Aromatic moiety **80d** was treated with butyllithium and, by means of a mixed cuprate intermediate, then was coupled to dibromodiene **87c** to form **88**.

As a starting point for the northern zone of the maytansinoid macrocycle, Meyers et al. [36] chose S(+)-3-hydroxy-2-methylpropionic acid (**89**). They expected that consecutive chiral centers could be added to this substrate stereospecifically, and that C(2) of acid **89** ultimately would become C(6) in the completed macrocycle. Treatment of **89** with ethyl vinyl ether resulted in simultaneous esterification and blocking of the C(3) hydroxyl group to give **90** (EE = ethoxyethyl). Ester **90** was reduced with lithium aluminum hydride to give alcohol **91a** whose newly generated hydroxyl then was protected by reaction with benzyl bromide to yield **91b.** Subsequently, the other hydroxyl group of **91b** was selectively deprotected by acid hydrolysis to provide **92a,** which was oxidized to aldehyde **92b** by Swern's "activated" dimethyl sulfoxide reagent [38]. Aldehyde **92b** was treated with the lithio propylimine of cyclohexylamine to provide a new α,β-unsaturated aldehyde, **93;** this compound was reduced with sodium borohydride to give alcohol **94.** At this stage, the essential epoxide substituent was introduced enantioselectively by the Sharpless method [39], in which **94** was treated with t-butyl hydroperoxide, titanium tetraisopropoxide, and diethyl tartarate to yield **95a.** The free hydroxyl of **95a** was protected with t-butyldimethyl silyl reagent to give **95b,** whose benzyl group was removed with sodium–liquid ammonia. The product, **95c,** was shown by HPLC to be >99% pure enantiomerically.

Elaboration of the eastern zone began with oxidation of **95c** by Collins' chromium trioxide–pyridine reagent to provide aldehyde **96.** The lithio derivative of ethyl dithioacetate was condensed with **96** to give **97,** and the

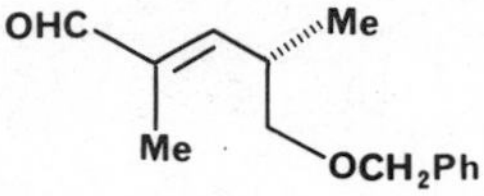

93

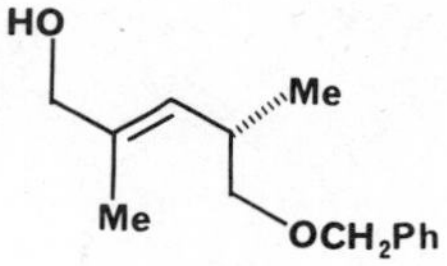

94

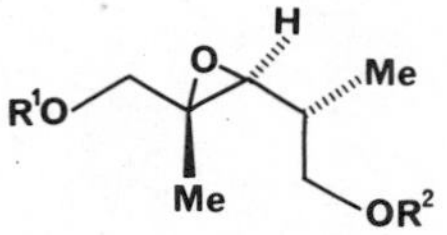

95a, R¹ = H; R² = CH₂Ph

   b, R¹ = SiMe₂tBu; R² = CH₂Ph

   c, R¹ = SiMe₂tBu; R² = H

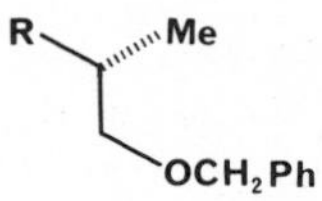

92a, R = CH₂OH

  b, R = CHO

newly formed hydroxyl was blocked with ethyl vinyl ether. Dithio ester **97** was treated with ethyl magnesium iodide; this Grignard reagent underwent thiophilic addition to generate **98**, a carbanion equivalent which was formylated with 2-(*N*-formyl-*N*-methyl)-aminopyridine to afford aldehyde **99**.

The stage was now set for coupling two major fragments—**88** (southwest portion) with **99** (northeast portion). Bromodiene **88** was converted to its lithio derivative by treatment with *t*-butyllithium; this derivative was condensed with **99** to yield a 1:1 mixture of C(10α) and C(10β) epimers, which were separated

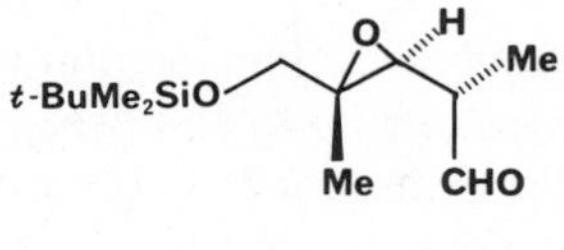

96

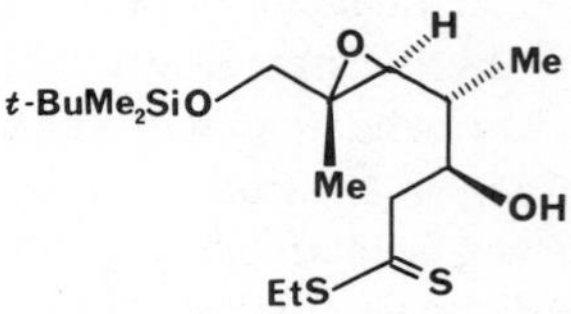

97

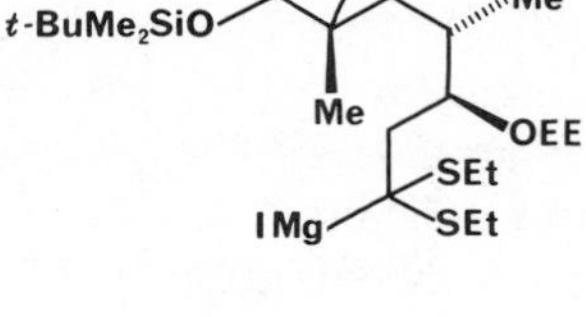

98

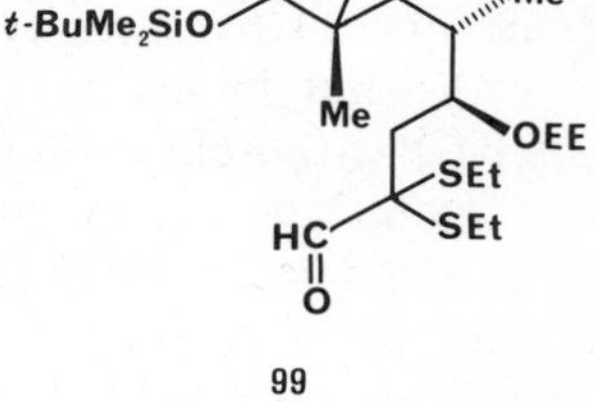

99

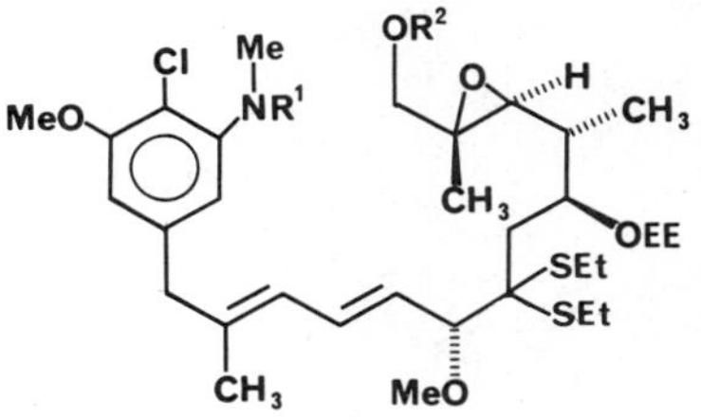

100a, $R^1 = COCH_2CH_2SiMe_3$; $R^2 = tBuMe_2Si$

b, $R^1 = R^2 = H$

chromatographically. The Meyers group applied the Horeau method to determine which epimer had the required 10α configuration. Since the C(10) hydroxy compounds were somewhat unstable, they were converted to the corresponding methyl ethers by treatment with sodium hydride and methyl iodide; like their hydroxylic precursors, these ethers were readily separable by chromatography. The C(10α) methoxy epimer, represented by **100a**, was

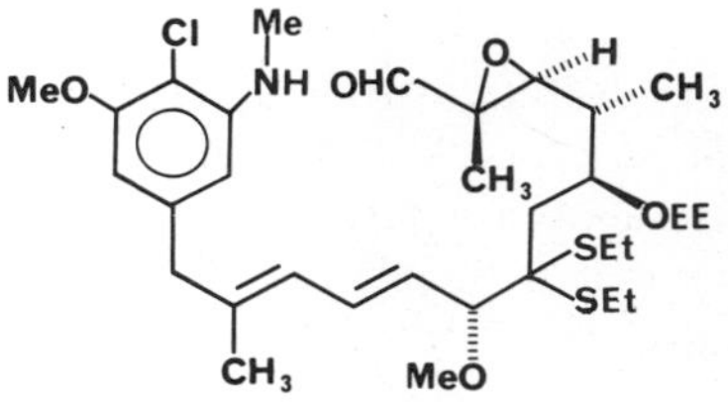

101a, R = H

b, R = $COCH_2P(OEt)_2$

deprotected by reaction with tetrabutylammonium fluoride to give **100b**. Oxidation of **100b** to **101a** was accomplished by a special technique using an alkoxy Grignard reagent and 1,1'-azodicarbonyl piperazine [40]. Aldehyde **101a** was then acylated at the amino group with 1-(diethylphosphono)acetyl chloride to provide **101b**. Ring closure was effected by treating **101b** with potassium *t*-butoxide to give **102a**. The carbonyl group at C(9), masked as a

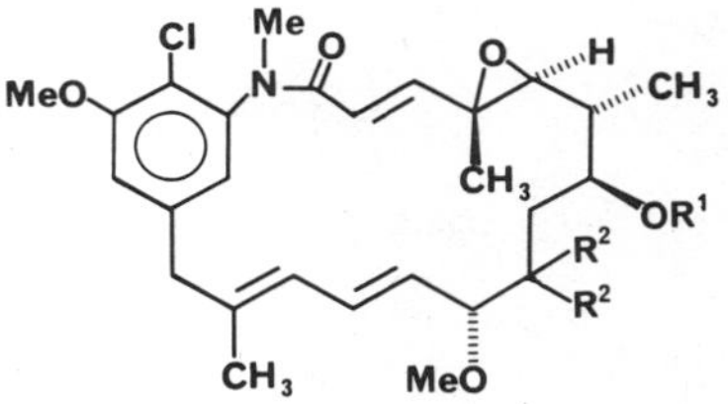

102a, $R^1 = EE$; $R^2 = SEt$

b, $R^1 = H$; $R^2 = =O$

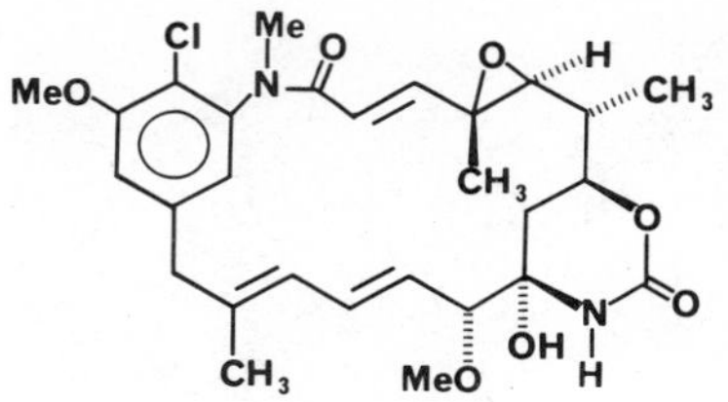

7

diethyl dithioacetal, was regenerated by treatment with calcium carbonate–mercuric chloride followed by hydrochloric acid. The final step to provide maysine (**7**) was accomplished by treating **102b** with phosgene followed by ammonia; in this manner, the cyclic urethane in the southeast corner was set in place.

4.1.3. Isobe Synthesis. A third total synthesis of maytansinoids has been carried out by Isobe and his associates at Nagoya University, Japan. This group gave its attention first to elaborating the southeast corner of the maytansinoid architecture, and chose acrolein dimer (**103a**) as their starting material [41]. This dimer was treated with lithium bis(trimethylsilyl)phenyl thioether; after dehydration of the initial aldol condensation product, olefinic intermediate **103b** was isolated (a separable mixture of Z and E isomers). Each of these isomers was treated with phenylselenyl chloride in the presence of dichloromethane and ethylene glycol monomethyl ether and pyridine to give

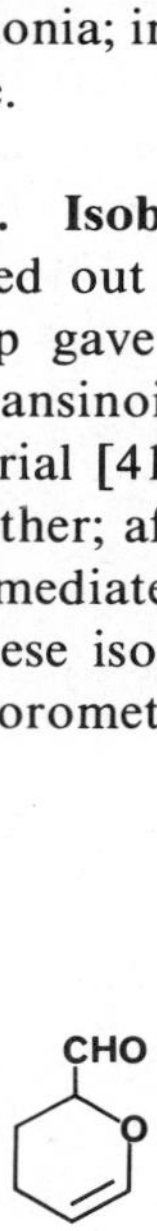

103a

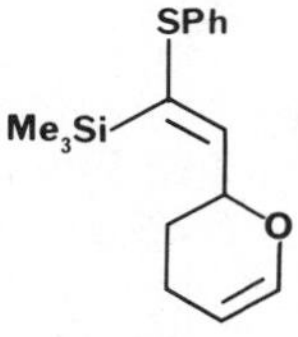

103b

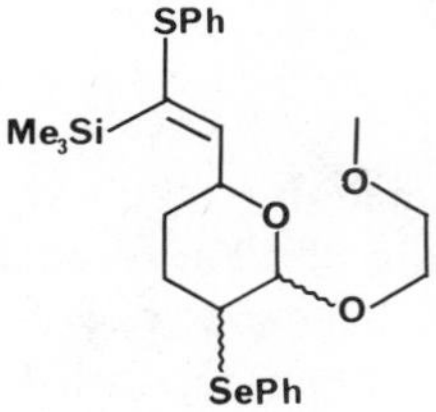

103c

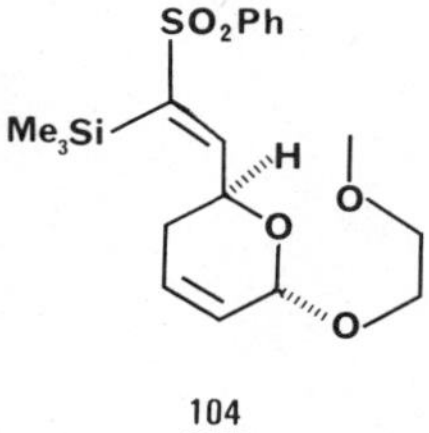

104

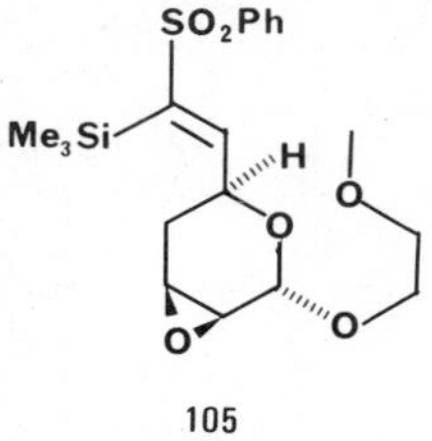

105

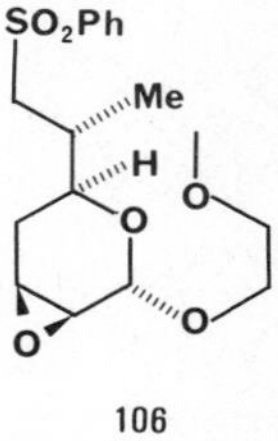

106

107

108

Z- and E-103c, which were oxidized with m-chloroperbenzoic acid to sulfone
104. Further oxidation with m-chloroperbenzoic acid produced epoxide 105.
Heteroconjugated addition of methyllithium to 105 was carried out and
yielded 106 [42]. The side-chain which is to form the northern facade of the
macrocycle was further elaborated by condensing 106 with the carbanion of 4-
bromo-2-pentene. When the product, 107, underwent reaction with sodium p-
anisyl oxide followed by a large excess of methyl iodide, 108 was produced by a
selective opening of the oxirane ring. Compound 108 was treated successively
with 2-chloroethanol, camphorsulfonic acid, and trimethyl orthoformate to
give 109; the latter was oxidized with chromium trioxide–pyridine to a ketone
which was ketalized with trimethyl orthoformate, yielding 110a. The 2-
chloroethyl protecting group was then transformed into 2-phenylsulfonylethyl
by reaction with sodium thiophenolate followed by m-chloroperbenzoic acid.
The resulting glycoside (110b) was so exceptionally alkali-labile that when
treated with sodium borohydride, ring cleavage as well as reduction occurred
to provide open-chain product 111a. One hydroxyl was protected by
acetylation and the other by silylation to afford 111b. In the next step of the
sequence, aldehyde 112 was generated from 111b by ozonolysis. Aldehyde 112
was converted to another aldehyde, 113, by successive reactions with (a)
pyridinium tosylate, trimethyl orthoformate, and methanol to acetalize the
carbonyl function; (b) sodium methoxide to hydrolyze the acetate function;
and (c) chromium trioxide–pyridine to oxidize the resulting hydroxyl. At this

109

110a, R = CH₂CH₂Cl

b, R = CH₂CH₂SO₂Ph

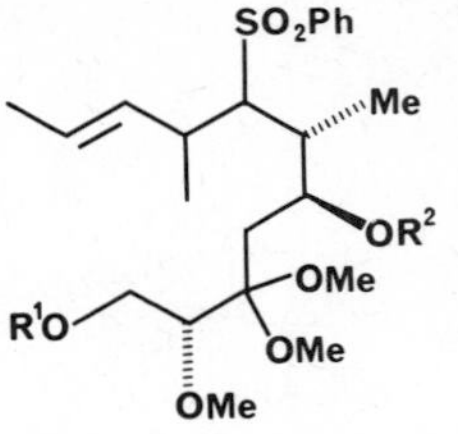

111a, R¹ = R² = H

b, R¹ = Ac; R² = SiMe₂tBu

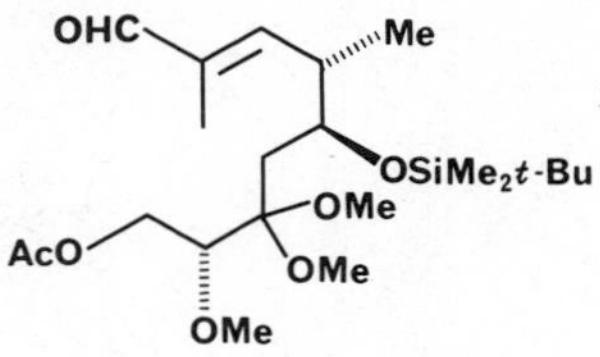

112

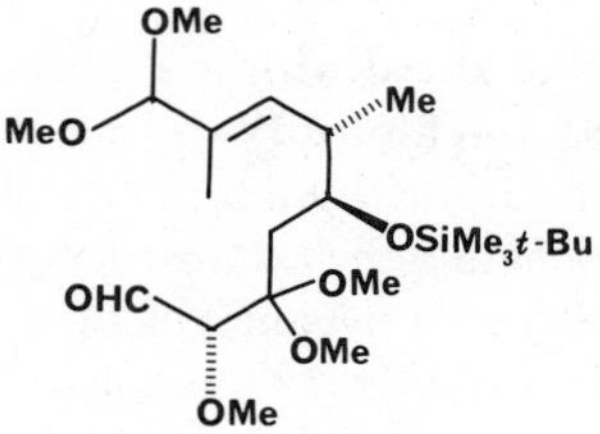

113

point, the fragment that was to become the southeastern zone of the maytansinoid macrocycle was ready for condensation with an ylide (**115e**) representing the aromatic northwestern zone.

Isobe and coworkers [42] approached the aromatic moiety for the northwest corner through benzyl iodide derivative **114a;** this iodide was reacted with 4-lithio-4-(phenylsulfonyl)-1-pentene to provide **114b.** Ozonolysis of **114b** yielded an unsaturated aldehyde, **115a** (predominantly the *E* isomer), which was reduced with sodium borohydride to give alcohol **115b;** this alcohol was, in turn, converted to bromide **115c** by treatment with phosphorus tribromide, lithium bromide, and collidine. The bromine was then displaced by triphenyl

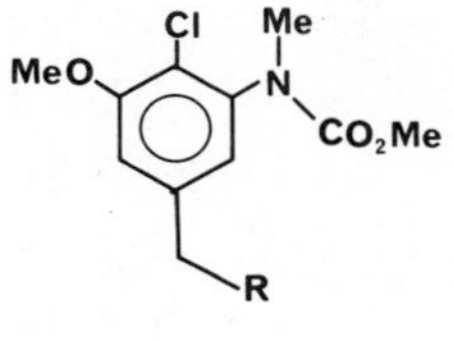

114a, R = I

b, R = CMe—CH₂CH=CH₂
            |
          SO₂Ph

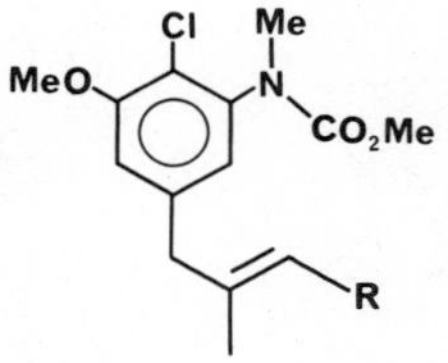

115a, R = CHO

b, R = CH₂OH

c, R = CH₂Br

d, R = CH₂PPh₃

e, R = CH=PPh₃

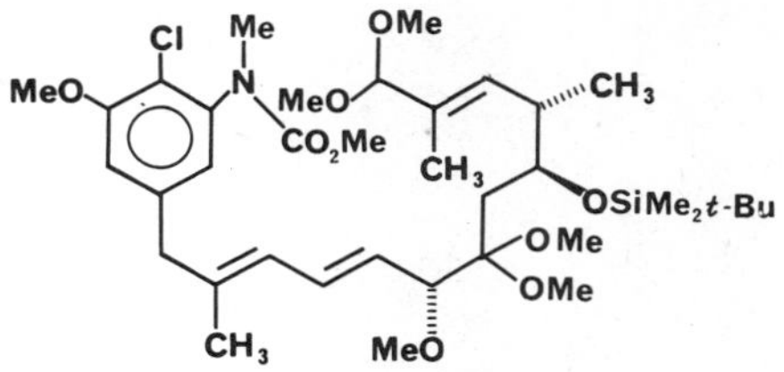

116

phosphine to give **115d,** which was treated with *n*-butyllithium to furnish ylide **115e.**

The stage now was set for condensing ylide **115e** with aldehyde **113** by a Wittig reaction. The dimethylacetal grouping was removed from **116** by mild treatment with aqueous acetic acid to give unsaturated aldehyde **117a.** This aldehyde was treated first with sodium borohydride to reduce the carbonyl function, then with potassium hydroxide to eliminate the *N*-carbomethoxy group and yield **117b.** Then **117b** was epoxidized regio- and stereoselectively with titanium tetraisopropoxide and *t*-butyl hydroperoxide to a single epoxide,

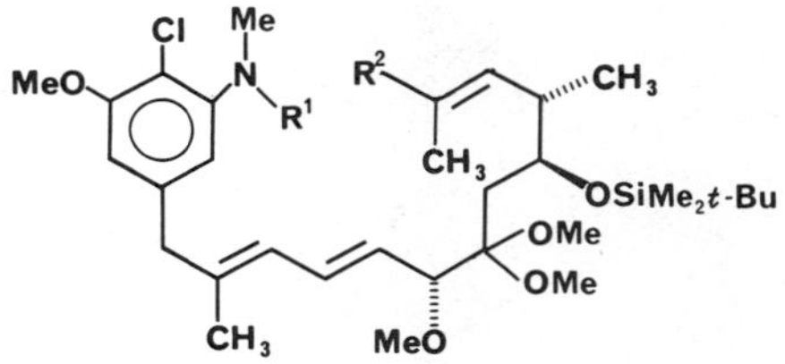

117a, R^1 = CO$_2$Me; R^2 = CHO

b, R^1 = H; R^2 = CH$_2$OH

118a. Epoxy alcohol **118a** then was converted to aldehyde **118b** by an aldol reaction [sic], and this product was treated with lithium ethyl acetate to provide **119a.** The sequence gave the correct stereochemistry (α) for the newly formed hydroxyl at C(3), and this hydroxyl was protected by dimethyl *t*-

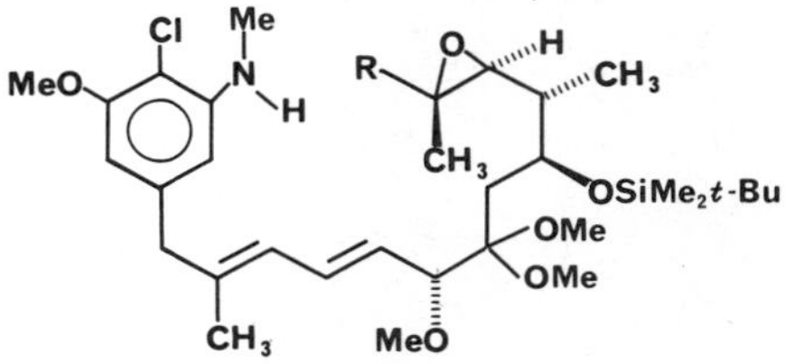

118a, R = CH$_2$OH

b, R = CHO

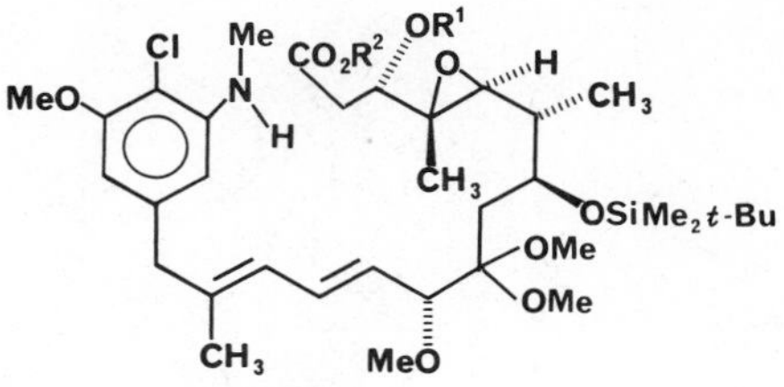

119a, R¹ = H; R² = Et

   b, R¹ = SiMe₂tBu; R² = Et

   c, R¹ = SiMe₂tBu; R² = H

butylsilyl chloride (producing **119b**); then the carbethoxy group was subjected to alkaline hydrolysis, yielding **119c**. Ring closure to **120a** was effected with mesitylene sulfonyl chloride, and desilylation was accomplished with tetrabutylammonium fluoride to provide **120b**. The dimethylacetal grouping then was hydrolyzed under mild acidic conditions to give ketone **121**.

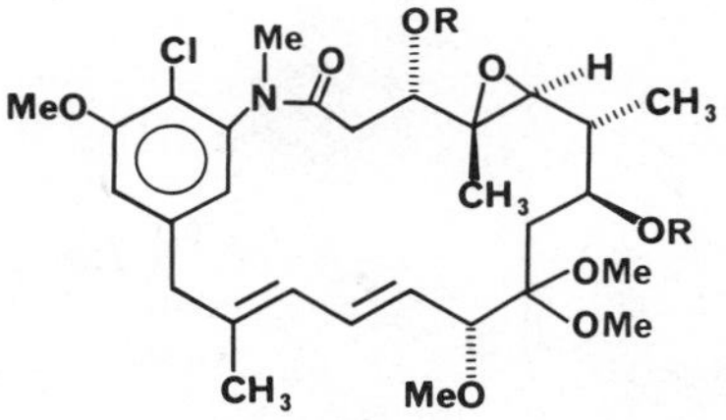

120a, R = SiMe₂tBu

   b, R = H

At this stage, only the carbinolamide ring in the southeast corner was required to complete the structure. This moiety was constructed by treating **121** with *p*-nitrophenylchloroformate, then with ammonia, to yield maytansinol (**10**).

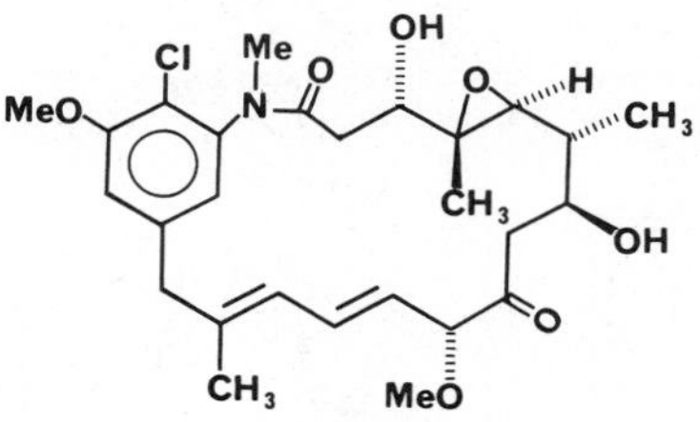

121

4.1.4. Total Synthesis—An Overview of Three Routes. In comparing the routes employed by Corey, Meyers, Isobe, and their respective associates, it is, of course, apparent that all three routes differ in details. However, all three do exhibit certain similar features. In their construction of the maytansinoid macrocycle, all three groups attached two major fragments stepwise, the first point of attachment being in the southern zone. All three groups added the carbinolamide ring in the southeast corner last. All three kept the C(7)-hydroxyl blocked until just before this ring was added, and kept C(9) in the form of an acetalized carbonyl.

4.1.5. Other Approaches to Synthesis of Maytansinoids. In addition to the three completed total syntheses discussed in this chapter, several other groups have initiated synthetic work directed to the same end. Two Chinese groups, both at the Shanghai Institute of Materia Medica, have synthesized major fragments, or precursors of fragments, of the maytansinoid macrocycle. Pan et al. [43] constructed a synthon representing the northeastern zone, while Zhou et al. [44] synthesized the southwestern moiety. A report of these two fragments being combined has not come to the authors' attention as of this writing. Other approaches have been contributed by Fried and coworkers [45, 46], Edwards and Ho [47, 48], Barton and coworkers [49, 50], Götschi and coworkers [51], Ganem and coworkers [52, 53], as well as Samson et al. [54] and Sirat et al. [55]. Some of these synthetic approaches were summarized by Komoda and Kishi in their review [3].

4.2. Microbial and Chemical Transformations of Maytansinoids

After their initial discovery that certain strains of *Nocardia* synthesize a variety of maytansinoids, the scientific staff of Takeda Chemical Industries (Osaka, Japan) undertook an extensive program of research on maytansinoids which included both microbial transformations and *in vitro* chemical reactions. The Takeda staff has taken out a plethora of patents on these transformation products, and some of their new derivatives are described only in patents as of this writing. The Takeda group has ascertained that, with the exception of deacylation, these microbial reactions are mediated by enzymes within the cells and not by cell-free extracts.

22-O-Demethyl Derivatives. Several microorganisms, especially *Bacillus megaterium,* convert the aromatic methoxyl group at C(20) to a hydroxyl group [56]. There are definite differences in yield among various maytansinoids with different bacteria. While *B. megaterium* gives the highest conversion of ansamitocin P-4 (**25d**), *Streptomyces platensis* gives the highest with maytansinol (**10**) [56].

N-Demethylation Products. *Streptomyces minutiscleroticus* enzymes effect the demethylation of the amide nitrogen and its replacement with hydrogen [56]. Yields were variable among substrates tested, ranging from 40% with ansamitocin P-3 (25b) to zero with maytansinol (10).

15-Hydroxylation Products. When fermented with a strain of *Streptomyces sclerotialus,* ansamitocin P-3 was hydroxylated nonstereospecifically at C(15) so that both epimers (PHO-3 and epi-PHO-3) were isolated [20]. By ^{13}C NMR and X-ray crystallographic studies, the configuration of these two epimers at C(15) were established as *R* for PHO-3 and *S* for epi-PHO-3.

9-Thiomaytansinoids. Kupchan and coworkers had demonstrated several years earlier that the tertiary C(9) OH undergoes facile replacement by OR and SR groups to form ethers [25]. The Japanese workers showed that this tertiary hydroxyl can be replaced chemically with SH by treating the substrate compound with hydrogen sulfide or phosphorus pentasulfide (P_2S_5) [57].

Dechloromaytansinoids. The Takeda research group achieved chemical dechlorination of maytansinoid esters by reduction with excess lithium aluminum hydride in THF at fairly low temperatures [58]. The ester linkage is, of course, cleaved concurrently.

4,5-Deoxymaytansinoids. The 4,5-epoxide function of certain maytanside esters was eliminated chemically, leaving an *E*-double bond, by the reagent titanium trichloride [59].

The Takeda workers have prepared various other maytansinoid derivatives, including various 22-alkoxy ethers [60], various acyl derivatives of 15-hydroxy compounds [61], and a wide range of C(3) esters [62–64].

5. ANTINEOPLASTIC PROPERTIES OF MAYTANSINOID ALKALOIDS

5.1. Results in Experimental Tumor Systems

Maytansine (2), trewiasine (16), and other C(3) esters of maytansinol (10) and trewiasinol (46) exhibit potent cytotoxicity against cells derived from human carcinoma of the nasopharynx (KB). Average ED_{50} values for these materials are normally in the range of 10^{-4} to 10^{-6} μg/mL [13, 65]. Maytansinoid compounds lacking a C(3) ester grouping, that is, maysine (7), normaysine (8), maysenine (9), and maytansinol (10), give ED_{50} values in the 10^{-1}–10^{-2} μg/mL range and compounds having the C(9) hydroxyl bound in an ether linkage (31–38) are an intermediate group having ED_{50} values of 10^{-2}–10^{-4} μg/mL [25].

Maytansine (2) is highly active in a variety of murine tumor systems including B16 melanoma, colon 26, L1210 leukemia, Lewis lung carcinoma, and P388 leukemia. The compound is also effective in increasing the life span of mice bearing the mast cell tumor P815 and the plasma cell tumor YPC-1, and also against Walker 256 carcinosarcoma in rats. In the P388 system (daily i.p. injections, days 1-9), maytansine is active over a wide range of doses (100-0.4 μg/kg/inj) with the optimal dose of 25 μg/kg/inj giving an increase in life span (ILS) of 120% [65]. Comparable activity has been noted for other natural (3, 4, 5, 6, 25a-d) and semisynthetic (47a, 47c) C(3) esters of maytansinol (10). Ansamitocin P-3 (25b) also has demonstrated activity against sarcoma 180, Ehrlich carcinoma, and P815 mastocytoma [3]. Trewiasine (16) yields ILS values of 26-68% in the 1-31 μg/kg dosage range, against PS leukemia and other maytansinoids having methoxyl (17, 18, 19, 20, 21, 23, 24) or hydroxyl (13) substitution at C(15) give comparable results [13]. Trewiasine (16) and trenudine (20) are also markedly active against B1 melanoma; 16 gives ILS values of 65-107% in the 4-32 μg/kg range, and 20 gives ILS values of 37-112% in the 1-16 μg/kg range [11, 12]. Of the maytansinoids tested to date, none has shown any marked superiority over maytansine.

5.2. Clinical Studies

Phase 1 clinical trials of maytansine began in the United States in late 1975. These initial studies to evaluate therapeutic potential were conducted in patients with far advanced disease which was refractory to conventional therapy or for which there was no available form of beneficial therapy known. At the University of Texas System Cancer Center, 60 patients were administered doses ranging from 0.01 to 0.9 mg/m^2 i.v. for 3 days. Antitumor activity was detected in one patient each with melanoma, breast carcinoma, and head and neck clear cell carcinoma. Principal toxic effects encountered were nausea, vomiting, diarrhea, and occasionally, stomatitis and alopecia [66].

Chabner et al. [67] reported the dose-limiting toxicity for maytansine to be 2 mg/m^2 using a single i.v. infusion given every 3 weeks. Toxic side effects manifested were profound weakness, diarrhea, nausea, and vomiting. Out of 28 patients treated in his studies, responses were observed in one patient with non-Hodgkin's lymphoma, one with ovarian cancer, and two with acute lymphocytic leukemia.

Of 40 adult patients evaluated at the Mayo Clinic, no tumor responses were seen [68]. However, most treated tumors, colorectal and lung types, were relatively drug resistant to single-agent therapy. In addition to previously noted toxicities, mild hematologic, neuro-, and possibly cardiac toxicity was noted.

Blum and Kahlert treated 38 adult solid-tumor patients with five daily bolus injections of maytansine repeated at 21-day intervals. Gastrointestinal toxicity was dose related and dose limiting at doses of $\geq$0.5 mg/m^2. Dose-related neurotoxicity was observed but no drug-related myelosuppression or change in serum creatinine level was seen. Of 16 patients evaluable for response, two with breast cancer had therapeutic benefit [69].

Chahinian et al. reported neurologic toxicity as prominent in a study of 71 patients with leukemia, lymphoma, or carcinoma [70]. Neurologic toxicity occurred with unacceptable frequency and severity when a cumulative dose of 6 mg/m^2 was exceeded at weekly doses ranging from 0.4 to 1.3 mg/m^2 by i.v. bolus. Gastrointestinal toxicity was frequently observed at dose levels ≥ 1.1 mg/m^2/week and a few patients experienced severe hepatotoxic effects and myelosuppression. Objective response was seen in three patients including one each with thyroma, large-cell anaplastic lung carcinoma, and diffuse histiocytic lymphoma. Minor response was seen in one patient with carcinoma of the colon and in one with lymphosarcoma cell leukemia; all responses were of short duration, and occurred in three patients who previously had received vincristine.

The response rate observed in phase II studies of maytansine has proved disappointingly low although interpretation of results must be tempered by the knowledge that patient populations studied are heavily pretreated, often preterminal, and unlikely to respond to any therapy. For example, O'Connell et al. [71] were unable to demonstrate any antitumor effect of maytansine in a study of 31 patients with advanced colorectal carcinoma. In another study [72], no complete or partial responses were observed in 33 evaluable patients with melanoma and only one partial response was seen in 21 evaluable patients with breast cancer. Toxicities were similar to those observed in phase I studies.

In a study conducted by the Southwest Onocology Group, 41 patients with advanced drug-resistant breast cancer were treated with a 5-day intermittent i.v. infusion of maytansine repeated every 21 days. One patient had a partial regression of pulmonary disease and seven had transient responses of <50% reduction in tumor or stable disease [73]. The conclusion was that maytansine has minimal, if any, efficacy as a single agent for the treatment of advanced breast cancer at a dose and schedule causing considerable toxicity.

In a phase II study with advanced lymphomas, three partial responders were noted among 31 patients entered in the trial, and toxicity was considered to be acceptable [74]. The 10% response rate offers little encouragement that maytansine will be of significant value in the treatment of Hodgkin's disease or non-Hodgkin's lymphoma. Maytansine was administered to a total of 107 patients with various tumors in a phase I-II study in Michigan [75]. Gastrointestinal and neurologic toxic reactions were dose related, and myelosuppression occurred only in previously treated patients. A patient with squamous cell carcinoma of the lung responded for 5 weeks and another patient with adenocarcinoma of the lung responded for 4 weeks in the phase I portion of this study. One partial remission lasting 14 weeks was seen in the phase II trial with a patient having malignant melanoma. Because of the relative absence of myelosuppression, it was felt that incorporation of maytansine in combination chemotherapeutic regimens was warranted, provided that significant activity could be demonstrated in single-agent trials.

5.3. Structure-Activity Relationships

A number of maytansine analogs were studied by Kupchan and coworkers for antileukemic, cytotoxic, antitubulin, and antimitotic activity [25]. From their data, it is apparent that a C(3) ester is necessary for activity, although considerable variation in the nature of the ester group is possible with little effect on overall activity. The ansamitocins (25a, 25b, 25d) and other simple C(3) esters of maytansinol (6, 47a, 47c) show antileukemic activity comparable to maytansine, and compounds such as 19, 20, and 21, in which the C(3) ester linkage forms part of an additional macrocyclic ring, also show similar activity. Presence of an oxygenated function at C(15) apparently gives some reduction of cytotoxicity but has little effect on P388 or B1 activity [13]. Microbial conversion of ansamitocin P-3 (25b) to the corresponding 20-O-demethyl derivative enhances activity against P388 leukemia [56]. Maytansinoids which lack a C(3) ester moiety, for example, 7–10, are devoid of PS activity even when tested at much higher dose levels than the active compounds discussed above.

Maytansinoids inhibit cilia regeneration of partially deciliated *Tetrahymena pyriformis* W; and activity is dependent on the nature of the acyl group at the C(3) position [76]. Marked reduction of activity was observed upon introduction of a hydroxyl into the C(3) acyl moiety of P-4 (25d) [18]. Introduction of hydroxyl at C(15), or conversion of the C(14) methyl to hydroxymethyl, also gives reduced activity. Since deClQND-O (28) inhibited cilia regeneration at 50 μg/mL and its derivative acetylated at C(3) showed complete inhibition at 8 μg/mL, an epoxide group at the 4,5-position may not be essential for activity in the maytansinoids.

Etherification of the C(9) hydroxyl in maytansine or maytanbutine (31, 35) results in a marked decrease in antileukemic activity. These ethers begin to show PS activity only at doses ca. 400-fold greater than the minimum dose required for the parent maytansinoid. Ready conversion of the C(9) ethers to the parent maytansinoids *in vitro* suggests that the activity of these compounds may be due to hydrolysis *in vivo*. The C(9) carbinolamide moiety appears to function as an alkylating agent [7], and some synthetic carbinolamides (which themselves show significant antineoplastic activity) are known to alkylate nucleic acids rapidly under slightly acidic conditions [77]. The proposed mechanism involves loss of water from a carbinolamide such as 122, forming a reactive azomethine lactone (123) that rapidly reacts with a biological nucleophile. This mechanism is supported by the observed facile acid–catalyzed exchange of the OH group of carbinolamides for alkoxy and thioalkyl groups.

122   123

6. PHARMACOLOGY OF MAYTANSINOID ALKALOIDS

Several groups have studied the mechanism by which maytansinoids exert their inhibitory effects. Maytansine has been found to produce a mitotic arrest similar to that of agents such as vincristine or colchicine [78]. The drug interferes with the formation of microtubules and also causes microtubule depolymerization. Maytansine was ca. 100-fold more effective than vincristine in inhibiting mitosis in sea urchin eggs [79]. Binding of maytansine to crude rat brain tubulin is dependent on temperature and ionic strength and is reversible. Maytansine and vincristine appear to share a common binding site, although an additional site specific for maytansine seems to be present also. An *in vitro* study of the vagus nerve of cats has shown that maytansine induces alterations of the neurofibrillar elements concomitant with a partial blockage of fast axoplasmic transport [80]. Effects of maytansine and vinblastine on the alkylation of tubulin sulfhydryls are consistent with the view that the two drugs are competing for the same overlapping sites, but suggests that the nature of the binding is different [81]. The ansamitocins apparently inhibit tumor growth by interrupting the assembly of microtubular structures of cells in a manner similar to that of maytansine [82]. Antitubulin activities of a number of maytansinoids have been studied using inhibition of polymerization of bovine brain tubulin, depolymerization of the once polymerized tubulin, and immunofluorescent staining of cytoplasmic microtubules in A31 cells [83]. These studies indicated that a C(3) ester moiety is not important for the manifestation of antitubulin activity *in vitro* but that it is essential for activity at cellular levels. An ester attached at C(3) apparently plays an important role in increasing permeation of these compounds into living cells. Data obtained on the competition of various maytansinoids in the binding of tritium-labeled vincristine and rat brain tubulin also emphasizes the importance of an ester group at the C(3) position [84].

7. PEST CONTROL PROPERTIES OF MAYTANSINOID ALKALOIDS

Freedman et al. [85] reported that ethanolic extracts of *Trewia nudiflora* seed act as antifeedants for the spotted cucumber beetle (*Diabrotica undecimpunctata howardi* Barber) and the European corn borer [*Ostrina nubilalis* (Hübner)]. Also indicated were morphogenic effects on the codling moth [*Laspeyresia pomonella* (L.)], disruption of the normal life cycle of the redbanded leafroller [*Argyrotaenia velutinana* (Walker)], and reduction in progeny of the plum curculio [*Conotrachelus nenuphar* (Herbst)]. In addition, *Trewia* extracts were toxic to the striped cucumber beetle [*Acalymma vittatum* (*F.*)] and gave 100% control of the chicken body louse [*Menacanthus stramineus* (Nitzsch)] from 5 to 28 days. Maytansinoids (16–21) were responsible for the activity against these insects and trewiasine (16) exhibited an LD_{50} of 7.4 ppm when incorporated into the diet of *O. nubilalis*. When trewiasine was applied topically to codling moth larvae at doses of 1.0, 0.1, and 0.01%, it caused mortalities of 90, 70, and 10%, respectively. Mortality occurred mainly in the larval stage and development time was lengthened in all three treatments [86]. The fact that trewiasine also acts as an antifeedant or repellent complicates toxicological analysis and the exact mode of action of trewiasine against these insects remains unknown.

8. BIOSYNTHESIS

The ansamycin antibiotics of microbial origin and the maytansinoids from higher plants are structurally and biogenetically related [2, 87]. For several microbial ansamycins, including geldanamycin (1) [88], rifamycin S [89], and streptovaricin D [90], the ansa chain has been shown to be polyketide in origin, as is part of the nucleus in those cases where it is naphthalenoid. White et al. [89] suggested that the remaining segment of the naphthalenoid ansamycin nuclei arises from the same $C_7 N_1$ precursor as the benzenoid nuclei of geldanamycin and the maytansinoids. After considerable effort, the key intermediate for the aminobenzenoid nucleus of the ansamycins and maytansinoids has been established as 3-amino-5-hydroxybenzoic acid [91, 92, 93] which arises via shikimate precursors. Possible biosynthetic pathways to most of the known ansamycins have been outlined in detail by Ghisalba and Nüesch [91].

REFERENCES

1. F. J. Antosz, "Ansamacrolides," in *Encyclopedia of Chemical Technology*, M. Grayson and D. Eckroth, Eds., 3rd. ed., Vol. 2, Wiley, New York, 1978, pp. 852–870.

2. M. Leboeuf, *Plant. Med. Phytother.* **12**, 53 (1978).

3. Y. Komoda and T. Kishi, "Maytansinoids," in *Anticancer Agents Based on Natural Product Models*, J. M. Cassady and J. D. Douros, Eds., Academic Press, New York, 1980, pp. 353–389.

4. S. M. Kupchan, Y. Komoda, W. A. Court, G. J. Thomas, R. M. Smith, A. Karim, C. J. Gilmore, R. C. Haltiwanger, and R. F. Bryan, *J. Am. Chem. Soc.* **94**, 1354 (1972).

5. R. F. Bryan, C. J. Gilmore, and R. C. Haltiwanger, *J. Chem. Soc., Perkin Trans. 2*, 897 (1973).

6. S. M. Kupchan, Y. Komoda, A. R. Branfman, A. T. Sneden, W. A. Court, G. J. Thomas, H. P. J. Hintz, R. M. Smith, A. Karim, G. A. Howie, A. K. Verma, Y. Nagao, R. G. Dailey, V. A. Zimmerly, and W. C. Sumner, Jr., *J. Org. Chem.* **42**, 2349 (1977).

7. S. M. Kupchan, Y. Komoda, A. R. Branfman, R. G. Dailey, and V. A. Zimmerly, *J. Am. Chem. Soc.* **96**, 3706 (1974).

8. A. T. Sneden and G. L. Beemsterboer, *J. Nat. Prod.* **43**, 541 (1980).

9. (a) A. T. Sneden, W. C. Sumner and S. M. Kupchan, *J. Nat. Prod.* **45**, 624 (1982); (b) W. A. Wallace and A. T. Sneden, *Org. Magn. Reson.* **19**, 31 (1982).

10. M. C. Wani, H. L. Taylor, and M. E. Wall, *J. Chem. Soc. Chem. Commun.*, 390 (1973).

11. R. G. Powell, D. Weisleder and C. R. Smith, *J. Org. Chem.* **46**, 4398 (1981).

12. R. G. Powell, D. Weisleder, C. R. Smith, J. Kozlowski, and W. K. Rohwedder, *J. Am. Chem. Soc.* **104**, 4929 (1982).

13. R. G. Powell, C. R. Smith, R. D. Plattner, and B. E. Jones, *J. Nat. Prod.* **46**, 660 (1983).

14. A. J. Cronquist, *Evolution and Classification of Flowering Plants,* New York Botanical Garden, Bronx, N.Y., 1968, pp. 226, 260–263.

15. E. Higashide, M. Asai, K. Ootsu, S. Tanida, Y. Kozai, T. Hasegawa, T. Kishi, Y. Sugino, and M. Yoneda, *Nature* **270**, 721 (1977).

16. M. Asai, E. Mizuta, M. Izawa, K. Haibara, and T. Kishi, *Tetrahedron* **35**, 1079 (1979).

17. S. Tanida, T. Hasegawa, K. Hatano, E. Higashide, and M. Yoneda, *J. Antibiot.* **33**, 192 (1980).

18. S. Tanida, M. Izawa, and T. Hasegawa, *J. Antibiot.* **34**, 489 (1981).

19. M. Izawa, S. Tanida, and M. Asai, *J. Antibiot.* **34**, 496 (1981).

20. M. Izawa, Y. Wada, F. Kasahara, M. Asai and T. Kishi, *J. Antibiot.* **34**, 1591 (1981).

21. R. I. Geran, N. H. Greenberg, M. M. MacDonald, A. M. Schumacher, and B. J. Abbott, *Cancer Chemother. Rep., Part 3* **3**, 1 (1972).

22. D. E. Nettleton, Jr., D. M. Balitz, M. Brown, J. E. Mosley, and R. W. Myllymaki, *J. Nat. Prod.* **44**, 340 (1981).

23. M. S. Ahmed, H. H. S. Fong, D. D. Soejarto, R. H. Dobberstein, D. P. Waller, and R. Moreno-Azorero, *J. Chromatogr.* **213**, 340 (1981).

24. M. Izawa, K. Haibara, and M. Asai, *Chem. Pharm. Bull.* **28**, 789 (1980).

25. S. M. Kupchan, A. T. Sneden, A. R. Branfman, G. A. Howie, L. I. Rebhun, W. E. McIvor, R. W. Wang, and T. C. Schnaitman, *J. Med. Chem.* **21**, 31 (1978).

26. R. G. Powell and C. R. Smith, Jr., unpublished results.

27. S. M. Kupchan, A. R. Branfman, A. T. Sneden, A. K. Verma, R. G. Dailey, Jr., Y. Komoda, and Y. Nagao, *J. Am. Chem. Soc.* **97**, 5294 (1975).

28. N. Hashimoto and T. Kishi, U.S. Patent 4,137,230 (1979).

29. N. Hashimoto and T. Kishi, U.S. Patent 4,190,580 (1980).

30. E. J. Corey, L. O. Weigel, A. R. Chamberlin, and B. Lipshutz, *J. Am. Chem. Soc.* **102**, 1439 (1980).

31. E. J. Corey, H. F. Wetter, A. P. Kozikowski, and A. V. Rama Rao, *Tetrahedron Lett.*, 777 (1977).

32. E. J. Corey, M. G. Bock, A. P. Kozikowski, A. V. Rama Rao, D. Floyd and B. Lipshutz, *Tetrahedron Lett.,* 1051 (1978).

33. E. J. Corey, D. Enders, and M. G. Bock, *Tetrahedron Lett.,* 7 (1976).

34. E. J. Corey, L. O. Weigel, D. Floyd, and M. G. Bock, *J. Am. Chem. Soc.* **100**, 2916 (1978).

35. E. J. Corey, L. O. Weigel, A. R. Chamberlin, H. Cho, and D. H. Hua, *J. Am. Chem. Soc.* **102**, 6613 (1980).

36. A. I. Meyers, A. L. Campbell, D. L. Comins, K. A. Babiak, M. P. Fleming, R. Henning, M. Heuschmann, J. P. Hudspeth, J. M. Kane, P.J. Reider, D. M. Roland, K. Shimizu, K. Tomioka, and R. D. Walkup, *J. Am. Chem. Soc.* **105**, 5015 (1983).

37. A. I. Meyers and M. P. Fleming, *J. Org. Chem.* **44**, 3405 (1979).

38. A. J. Mancuso, S. L. Huangard, and D. Swern, *J. Org. Chem.* **43**, 2480 (1980).

39. T. Katsuki and K. B. Sharpless, *J. Am. Chem. Soc.,* **102**, 5976 (1980).

40. K. Narasaka, A. Morikawa, K. Saigo and T. Mukaiyama, *Bull. Chem. Soc. Japan,* **50**, 2773 (1977).

41. M. Isobe, M. Kitamura and T. Goto, *Tetrahedron Lett.* **22**, 239 (1981).

42. M. Isobe, M. Kitamura and T. Goto, *J. Am. Chem. Soc.* **104**, 4997 (1982).

43. Pan Bai-chuan, Zhang Hong, Pan Chung-ying, Shu Yun, Wang Khoo-fu, and Gao Yi-sheng, *Hua Hsueh Hsueh Pao* **38**, 502 (1980).

44. Zhou Qi-ting, Bai Dong-lu, Sun Han-li, Yang You-chun, Du Yuan-hua, Xu Yi-bao, Chen Mei-qing, and Gao Yi-sheng, *Hua Hsueh Hsueh Pao* **38**, 507 (1980).

45. W. J. Elliott and J. Fried, *J. Org. Chem.* **41,** 2469 (1976).

46. G. Gormley, Y. Y. Chan, and J. Fried, *J. Org. Chem.* **45,** 1447 (1980).

47. O. E. Edwards and P.-T. Ho, *Can. J. Chem.* **55,** 371 (1977).

48. P.-T. Ho, *Can. J. Chem.* **58,** 861 (1980).

49. D. H. R. Barton, S. D. Géro and C. D. Maycock, *J. Chem. Soc. Commun.*, 1089 (1980).

50. D. H. R. Barton, M. Bénéchie, F. Khuong-Huu, P. Potier, and V. Reyna-Pinedo, *Tetrahedron Lett.* **23,** 651 (1982).

51. E. Götschi, F. Schneider, H. Wagner, and K. Bernauer, *Helv. Chim. Acta* **60,** 1416 (1977).

52. J. E. Foy and B. Ganem, *Tetrahedron Lett.*, 775 (1977).

53. R. Bonjouklian and B. Ganem, *Tetrahedron Lett.*, 2835 (1977).

54. M. Samson, P. DeClercq, H. deWilde, and M. Vandewalle, *Tetrahedron Lett.*, 3195 (1977).

55. H. M. Sirat, E. J. Thomas, and J. D. Wallis, *J. Chem. Soc. Perkin Trans. 1,* 2885 (1982).

56. K. Nakahama, M. Izawa, M. Asai, M. Kida, and T. Kishi, *J. Antibiotics* **34,** 1581 (1981).

57. N. Hashimoto and H. Shimadzu, European Patent Appl. 65,730 (May 20, 1981).

58. M. Asai, K. Nakahama, and M. Izawa, U.S. Patent 4,307,016 (Dec. 22, 1981).

59. H. Akimoto and A. Kawai, U.S. Patent 4,371,533 (Feb. 1, 1983).

60. O. Miyashita and H. Akimoto, U.S. Patent 4,309,428 (March 2, 1982).

61. M. Asai, K. Nakahama, and M. Izawa, U.S. Patent 4,364,866 (Dec. 21, 1982).

62. O. Miyashita and H. Akimoto, U.S. Patent 4,308,269 (Dec. 29, 1981).

63. O. Miyashita and H. Akimoto, U.S. Patent 4,317,821 (March 2, 1982).

64. N. Hashimoto and T. Kishi, U.S. Patent 4,265,814 (May 5, 1981).

65. J. Douros, M. Suffness, D. Chiuten, and R. Adamson, "Maytansine," in *Advances in Medical Oncology Research and Education* **5,** *Basis for Cancer Therapy* **1,** B. W. Fox, Ed., Pergamon Press, New York, 1979, pp. 59–73.

66. F. Cabanillas, V. Rodríguez, S. W. Hall, M. A. Burgess, G. P. Bodey, and E. J. Freireich, *Cancer Treat. Rep.* **62,** 425 (1978).

67. B. A. Chabner, A. S. Levine, B. L. Johnson and R. C. Young, *Cancer Treat. Rep.* **62,** 429 (1978).

68. R. T. Egan, J. N. Ingle, J. Rubin, S. Frytak, and C. G. Moertel, *J. Natl. Cancer Inst.* **60,** 93 (1978).

69. R. H. Blum and T. Kahlert, *Cancer Treat. Rep.* **62,** 435 (1978).

70. A. P. Chahinian, C. Nogeire, T. Ohnuma, M. L. Greenberg, M. Sivak, I. S. Jaffrey, and J. F. Holland, *Cancer Treat. Rep.* **63,** 1953 (1979).

71. M. J. O'Connell, A. Shani, J. Rubin, and C. G. Moertel, *Cancer Treat. Rep.* **62,** 1237 (1978).

72. F. Cabanillas, G. P. Bodey, M. A. Burgess, and E. J. Freireich, *Cancer Treat. Rep.* **63,** 507 (1979).

73. J. A. Neidhart, L. R. Laufman, C. Vaughn, and J. D. McCracken, *Cancer Treat. Rep.* **64,** 675 (1980).

74. S. Rosenthal, D. T. Harris, J. Horton, and J. H. Glick, *Cancer Treat. Rep.* **64,** 1115 (1980).

75. R. Franklin, M. K. Samson, R. J. Fraile, H. Abu-Zahra, R. O'Bryan, and L. H. Baker, *Cancer* **46,** 1104 (1980).

76. S. Tanida, E. Higashide, and M. Yoneda, *Antimicrobial Agents and Chemotherapy* **16,** 101 (1979).

77. J. W. Lown, K. C. Majumdar, A. I. Meyers, and A. Hecht, *Bioorganic Chem.* **6,** 453 (1977).

78. F. Mandelbaum-Shavit, M. K. Wolpert-DeFillipes, and D. G. Johns, *Biochem. Biophys. Res. Comm.* **72,** 47 (1976).

79. S. Remillard, L. I. Rebhun, G. A. Howie, and S. M. Kupchan, *Science* **189,** 1002 (1975).

80. J. A. Donoso, D. F. Watson, I. E. Heller-Bettinger, and F. E. Samson, *Cancer Res.* **38**, 1633 (1978).

81. R. F. Luduena and M. C. Roach, *Arch. Biochem. Biophys.* **210**, 498 (1981).

82. K. Ootsu, Y. Kozai, M. Takeuchi, S. Ikeyama, K. Igarashi, K. Tsukamoto, Y. Sugino, T. Tashiro, S. Tsukagoshi, and Y. Sakurai, *Cancer Res.* **40**, 1707 (1980).

83. S. Ikeyama and M. Takeuchi, *Biochem. Pharmacol.* **30**, 2421 (1981).

84. J. York, M. K. Wolpert-DeFilippes, D. G. Johns, and V. S. Sethi, *Biochem. Pharmacol.* **30**, 3239 (1981).

85. B. Freedman, D. K. Reed, R. G. Powell, R. V. Madrigal, and C. R. Smith, Jr., *J. Chem. Ecol.* **8**, 409 (1982).

86. D. K. Reed, W. F. Kwolek, and C. R. Smith, Jr., *Econ. Entomol.* **76**, 641 (1983).

87. K. L. Rinehart, Jr. and L. S. Shield, *Fortschr. Chem. Org. Naturst.* **33**, 231 (1976).

88. A. Haber, R. D. Johnson, and K. L. Rinehart, Jr., *J. Am. Chem. Soc.* **99**, 3541 (1977).

89. R. J. White, E. Martinelli, G. G. Gallo, G. Lancini, and P. Beynon, *Nature* (London) **243**, 273 (1973).

90. B. Milavetz, K. Kakinuma, K. L. Rinehart, Jr., J. P. Rolls, and W. J. Hoak, *J. Am. Chem. Soc.* **95**, 5793 (1973).

91. O. Ghisalba and J. Nüesch, *J. Antibiot.* **34**, 64 (1981).

92. J. J. Kirby, I. A. McDonald, and R. W. Rickards, *J. Chem. Soc. Chem. Commun.* 768 (1980).

93. K. Hatano, S. Akiyama, M. Asai, and R. W. Rickards, *J. Antibiot.* **35**, 1415 (1982).

Chapter Five

^{13}C and Proton NMR Shift Assignments and Physical Constants of C$_{19}$-Diterpenoid Alkaloids

S. William Pelletier, Naresh V. Mody,
Balawant S. Joshi, and Lee C. Schramm
Institute for Natural Products Research
and
Department of Chemistry
University of Georgia
Athens, Georgia

CONTENTS

1. INTRODUCTION

The C$_{19}$-diterpenoid alkaloids, isolated from plants of the *Aconitum, Atragne, Consolid, Delphinium,* and *Inula* species, have been known for their pharmacological properties, complex structures and interesting chemistry for over a century [1]. These alkaloids are divided into four categories as defined below.

A. *Aconitine Type:* These alkaloids possess the skeleton (I) of aconitine, in which the C(7) position is not substituted by any other group except hydrogen, for example, aconitine, delphinine, condelphine.

B. *Lycoctonine Type:* These alkaloids possess the skeleton (I) of lycoctonine, in which the C(7) position is always oxygenated, for example, lycoctonine, browniine, delcosine, dictyocarpine.

C. *Pyrodelphinine Type:* These alkaloids possess the skeleton (I) of the pyrolysis product of delphinine, in which a double bond is present between C(8) and C(15), for example, falaconitine and mithaconitine.

D. *Heteratisine Type:* These alkaloids possess the skeleton (II) of heteratisine, in which a lactone moiety is present in ring C, for example, heteratisine, heterophylline.

The numbering system shown below is used in this catalog.

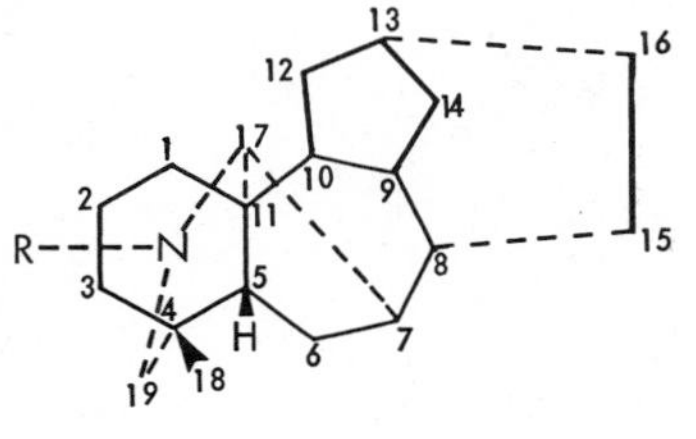

I 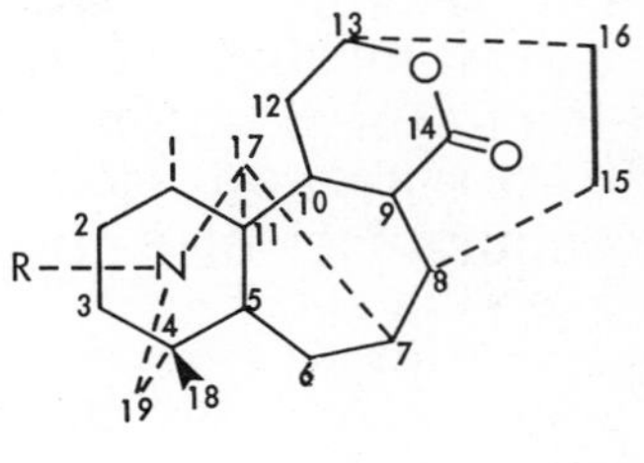   II

At the present time the structures of about 164 naturally occurring C$_{19}$-diterpenoid alkaloids have been reported. Of these, the structures of more than 60 have been elucidated within the last six years. Activity in the area of structure elucidation of these alkaloids has increased recently largely because of the availability of modern spectroscopic techniques, for example, proton NMR, ^{13}C NMR, high-resolution MS, and single-crystal X-ray analysis. Most of the recent structural work in this area has been carried out by ^{13}C NMR spectroscopy and X-ray analysis. The former technique has established the major substitution patterns of these alkaloids, a development which has helped in the unambiguous assignment of the chemical shifts of these complex alkaloids. The structures of several well-known alkaloids, for example, cammaconine [2], acomonine, and iliensine [3], have been revised recently with the aid of ^{13}C NMR spectroscopy. These cases demonstrate the significant role that ^{13}C NMR data are playing in solving the structures of the newly isolated alkaloids and in confirming the identity of known alkaloids. The systematic studies in our laboratory [4–7] on the ^{13}C NMR spectra of C$_{19}$-diterpenoid alkaloids have established a ^{13}C NMR data bank for more than 100 alkaloids and their derivatives. On the basis of these ^{13}C NMR data, computer programs capable of predicting the structures of known and new diterpenoid alkaloids from their ^{13}C NMR spectra have been developed [8].

Recently, the configuration of the C(1)-methoxyl group in lycoctonine [9] has been revised from β- to α- on the basis of the X-ray analysis of two new degradation products of lycoctonine [10], on the chemical transformation of lycoctonine to delphatine [11], and on the X-ray analysis of browniine perchlorate [11]. This configurational change in the structure of lycoctonine and the X-ray analysis of dictyocarpine have required the structures of 37 lycoctonine-type alkaloids to be revised [11].

This catalog provides proton and/or ^{13}C NMR shift assignments and physical constants for 159 naturally occurring C$_{19}$-diterpenoid alkaloids and 48 derivatives. Twenty-nine of the naturally occurring alkaloids have been identified in the literature by more than one name. Each of these duplicate names is included in alphabetical order in the *Index of Naturally Occurring C$_{19}$-diterpenoid Alkaloids and Their Derivatives*. Thus, the reader will be able to obtain information under either name. When available the following data are presented for each compound: (1) molecular formula and structure, (2) experimentally determined high-resolution mass measurement, (3) melting point, (4) specific rotation, (5) plant sources (an * at the end of the list of plant species on a structure plate indicates that additional species containing this alkaloid are listed in the *Addendum to Alkaloid-Bearing Plants* at the end of the plates), (6) proton NMR data, (7) ^{13}C NMR shift assignments, and (8) references. For certain alkaloids we were unable to secure ^{13}C data either because the data have not been determined, or have not been published. In some instances repeated requests to authors for spectral data went unanswered.

The molecular weight (exact mass) of every compound included in this catalog is calculated to five decimal points and appears in two tables compiled according to increasing CHNO and the mass values. This exact mass information will be useful for high-resolution MS analysis.

The only systematic study of proton NMR spectra of C$_{19}$-diterpenoid alkaloids is that by Tsuda and Marion [12] who established the position of the acetate and benzoate groups in aconitine, delphinine, and related alkaloids. They observed that the C(8)-acetyl group in the proton NMR spectra appears at $\delta \sim 1.3$, instead of the usual position (δ 2.0), when the benzoate group is present in the α-configuration at C(14). This unusual behavior was explained by the strong shielding effect of the aromatic nucleus on the methyl of the C(8)-acetate group. This behavior of the acetate group in the presence of the C(14)-benzoate group is useful for establishing the position of both groups. The proton shifts of the various functional groups appear in the general ranges shown in Table 1.

Jones and Benn made the initial contribution to ¹³C NMR spectroscopy of diterpenoid alkaloids [13]. On the basis of available ¹³C chemical shift data for more than 160 compounds, we have prepared a table showing the general chemical shift ranges of all carbons of the C_{19}-diterpenoid alkaloid skeleton (Table 2). Chemical shifts of quaternary carbon atoms [C(4), C(8) and C(11)] play an important role in establishing the structures of these alkaloids. The signals of C(4), C(8) and C(11) appear as singlets in the SFORD spectra in different chemical shift ranges according to their substitution patterns. A singlet between 32.0 and 35.0 ppm indicates that a methyl group is present at C(4). The —CH_2OH, —CH_2OCH_3, and —CH_2OAc groups at C(4) can be detected easily by observing a singlet between 37.0 and 41.0 ppm. The presence of various oxygenated groups at C(8) is revealed by a singlet between 72.5 and 92.5 ppm. A singlet of C(11) occurs at three different ranges depending on the neighbouring groups (see Table 2). Thus, one can establish the presence of various functional groups at C(1) and C(10) by observing the range of C(11).

Table 3, which lists the occurrence of C_{19}-diterpenoid alkaloids according to plant sources, will be useful for chemotaxonomic study on various *Aconitum* and *Delphinium* species. Only those alkaloids for which definite structures have been proposed in the literature are included in this table.

Because the role of ¹³C NMR spectroscopy in solving difficult structural problems of complex diterpenoid alkaloids has been discussed recently in several articles [1, 4–7], we are not treating this subject in detail here. We anticipate that this chapter will be useful in providing data on C_{19}-diterpenoid alkaloids (structure, physical properties, plant sources, ¹H and ¹³C NMR data, and references), which have been scattered through hundreds of papers and dozens of reviews.

1.1. Literature Cited

1. S. W. Pelletier and N. V. Mody in *The Alkaloids*, R. H. F. Manske and R. G. Rodrigo, eds., Academic Press, New York, 1979, Vol. 17, Chapter 1.

2. N. V. Mody, S. W. Pelletier, and N. M. Mollov, *Heterocycles* **14**, 1751 (1980).

3. S. W. Pelletier and N. V. Mody, *Tetrahedron* **22**, 207 (1981).

4. S. W. Pelletier and Z. Djarmati, *J. Am. Chem. Soc.* **98**, 2626 (1976).

5. S. W. Pelletier, N. V. Mody, R. S. Sawhney, and J. Bhattacharyya, *Heterocycles* **7**, 327 (1977).

6. S. W. Pelletier, N. V. Mody, and R. S. Sawhney, *Can. J. Chem.* **57**, 1652 (1979).

7. S. W. Pelletier, N. V. Mody, and O. D. Dailey, Jr., *Can. J. Chem.* **58**, 1825 (1980).

8. J. Finer-Moore, N. V. Mody, S. W. Pelletier, N. A. B. Gray, C. W. Crandell, and D. H. Smith, *J. Org. Chem.* **46**, 3399 (1981).

9. O. E. Edwards and M. Przybylska, *Can. J. Chem.* **60**, 2661 (1982).

10. M. Cygler, M. Przybylska, and O. E. Edwards, *Acta Crystallogr.* Sect. B **38**, 429, 1500 (1982).
11. S. W. Pelletier, N. V. Mody, K. I. Varughese, J. A. Maddry, and H. K. Desai, *J. Am. Chem. Soc.* **103**, 6536 (1981).
12. Y. Tsuda and L. Marion, *Can. J. Chem.* **41**, 1634 (1963).
13. A. J. Jones and M. H. Benn, *Can. J. Chem.* **51**, 486 (1973).

2. PROTON SHIFTS OF VARIOUS FUNCTIONAL GROUPS OF C₁₉-DITERPENOID ALKALOIDS

Table 1.

Functional Group	Chemical Shift Range	Multiplicity
C(4)—CH_3	0.8–1.10	Singlet
N—CH_2—CH_3	0.95–1.15	Triplet
C(8)—$OCOCH_3$[a]	1.25–1.45	Singlet
$OCOCH_3$	1.95–2.05	Singlet
$NHCOCH_3$	2.15–2.25	Singlet
N—CH_3	2.35–2.60	Singlet
C—OCH_3 (aliphatic)	3.10–3.90	Singlet
C—OCH_3 (aromatic)	3.85–3.95	Singlet
HO—C(1)—β—H	3.60–3.75	Multiplet
CH_3O—C(14)—β—H	3.60–3.85,[a] 4.1[b]	Varies
HO—C(15)—β—H	3.80–4.60	Doublet
CH_3O—C(6)—H	3.90–4.20	Varies
HO—C(14)—β—H	3.97–4.30, 4.55–4.65[b]	Varies
R—COO—C(14)—β—H	4.70–5.30	Varies
C(7)—O—CH_2—O—C(8)	4.80–5.55	Singlet or two singlets[c]
CH_3COO—C(6)—α—H	5.40–5.50	Broad singlet
Aromatic protons	6.40–8.75	Varies
NHCO—	10.10–11.10	

[a]In the presence of the C(14) aromatic ester group.
[b]When the hydroxy group is present at C(10).
[c]The separation between the singlets is 0.05–0.12 ppm.

3. ¹³C CHEMICAL SHIFTS OF VARIOUS FUNCTIONAL GROUPS OF C₁₉-DITERPENOID ALKALOIDS

Table 2.

Functional Group	Chemical Shift Range	Remarks
C(1)—β—OH	68.0–69.0 ppm	General range
C(1)—α—OH	69.5–70.0 ppm	C(10)—OH present
C(1)—α—OH	71.0 ppm	C(2)—C(3) double bond present

Table 2.   (*Continued*)

Functional Group	Chemical Shift Range	Remarks
C(1)—α—OH	72.0–73.0 ppm	General range
C(1)—β—OAc	72.0–73.0 ppm	General range
C(1)—α—OH	77.0–77.5 ppm	C(3)—C(4) epoxy group present
C(1)—α—OAc	77.0–78.0 ppm	General range
C(1)—α—OCH$_3$	77.0–80.0 ppm	C(4)—CH$_3$ present
C(1)—α—OCH$_3$	83.0–85.5 ppm	General range
C(1)—α—OCH$_3$	83.0–84.0 ppm	C(3)—α—OH present
C(1)—H	91.0 ppm	C(1)—C(19)—O— present
C(2)—H$_2$	25.5–27.0 ppm	C(1)—OCH$_3$ present C(3)—α—OH absent
C(2)—H$_2$	27.0–30.0 ppm	C(1)—α—OH present
C(2)—H$_2$	30.5 ppm	C(1)—βOH present
C(2)—H$_2$	32.0–32.5 ppm	C(1)—α—OCH$_3$ and C(3)—OAc present
C(2)—H$_2$	33.5–38.0 ppm	C(1)—α—OCH$_3$ and C(3)—OH present
C(2)—H	125.0–131.0 ppm	C(2)—C(3) double bond present
C(3)—H$_2$	29.5–30.5 ppm	C(1)—α—OH and C(4)—CH$_2$OCH$_3$ present
C(3)—H$_2$	31.0–32.5 ppm	Lycoctonine-type with C(1)—OCH$_3$ and C(4)—CH$_2$OCH$_3$ present
C(3)—H$_2$	34.5–35.5 ppm	Aconitine type with C(1)—α—OCH$_3$ and C(4)—CH$_2$OCH$_3$ present
C(3)—H$_2$	36.0–39.5 ppm	Lycoctonine type with C(1)—OCH$_3$ and C(4)—CH$_3$ present
C(3)—H	57.5–58.5 ppm	C(3)—C(4) epoxy present
C(3)—α—OH	70.5–72.0 ppm	C(1)—α—OCH$_3$ present
C(3)—H	134.0–138.0 ppm	C(2)—C(3) double bond present
C(4)—CH$_3$	32.0–35.0 ppm	C(3)—H$_2$ present
C(4)—H	35.0–36.0 ppm	C(3)—H$_2$, C(6)—H$_2$ present
C(4)—CH$_2$OR′	37.0–41.0 ppm	C(3)—H$_2$ present
C(4)—CH$_2$OCH$_3$	43.0–44.0 ppm	C(3)—α—OH present
C(4)	58.5–59.5 ppm	C(3)—C(4) epoxy present and C(19)—H$_2$
C(4)—OH	70.0–71.0 ppm	C(19)—H$_2$ present
C(5)—H	37.5–38.5 ppm	C(1)—C(19)—O— present
C(5)—H	39.0–39.5 ppm	C(1)—β—OH and C(6)—α—OCH$_3$ present

Table 2. (*Continued*)

Functional Group	Chemical Shift Range	Remarks
C(5)—H	41.0–42.0 ppm	C(1)—OH, C(3)—H$_2$, C(6)—H$_2$ present
C(5)—H	42.5–46.5 ppm	C(6)—β—OCH$_3$ present
C(5)—H	44.0–45.0 ppm	C(1)—α—OH and C(6)—α—OCH$_3$ present
C(5)—H	45.5–49.5 ppm	C(1)—OCH$_3$, C(3)—H$_2$, C(4)—H, or CH$_2$OR, CH$_3$, C(6)—H$_2$ present
C(5)—H	46.5–49.0 ppm	C(3)—α—OH and C(6)—α—OCH$_3$ present
C(5)—H	48.0–51.0 ppm	C(1)—OH, C(3)—H$_2$, C(4)—OH, C(6)—H$_2$ present
C(5)—H	48.5–50.0 ppm	C(6)—α—OCH$_3$ present
C(5)—H	50.5–52.5 ppm	C(1)—OCH$_3$ and C(6)—β—OH or OAc present
C(5)—H	52.5–53.5 ppm	C(1) ketone and C(6)—α—OCH$_3$ present
C(5)—H	56.5–57.5 ppm	C(6) ketone present
C(6)—H$_2$	23.5–27.5 ppm	C(7)—H present
C(6)—H$_2$	28.5–29.0 ppm	C(4)—H, C(7)—H present
C(6)—H$_2$	32.5–34.0 ppm	C(7)—OH present
C(6)—β—OH or OAc	72.0–73.0 ppm	C(7)—H present
C(6)—β—OH or OAc	77.0–79.5 ppm	C(7)—OCH$_2$O—C(8) present
C(6)—α—OCH$_3$	79.5–85.0 ppm	C(7)—H present
C(6)—β—OCH$_3$	89.5–92.0 ppm	Lycoctonine type
C(6)—ketone	215.5–217.5 ppm	Lycoctonine type
C(7)—H	44.0–53.0 ppm	General range
C(7)—β—OH	87.5–89.0 ppm	General range
C(7)—OCH$_2$O	90.5–93.5 ppm	C(7)—OCH$_2$O—C(8) present
C(8)—OH	72.5–74.5 ppm	C(7), C(9), and C(15) unsubstituted
C(8)—OH	75.5–76.5 ppm	C(15)—α—OH present
C(8)—OH	76.0–78.5 ppm	C(7)—OH present
C(8)—OCH$_3$	78.5–81.0 ppm	General range
C(8)—OCH$_2$O	81.5–84.0 ppm	C(14) ketone absent
C(8)—OCH$_2$O	86.0–87.5 ppm	C(14) ketone present
C(8)—OAc	85.5–86.0 ppm	C(15) unsubstituted
C(8)—OAc	91.5–92.5 ppm	C(15)—α—OH present
C(8)	146.5–148.5 ppm	C(8)—C(15) double bond present

Table 2. (*Continued*)

Functional Group	Chemical Shift Range	Remarks
C(9)—H	43.5–46.5 ppm	Aconitine type with C(14) ester
C(9)—H	45.0–46.0 ppm	Lycoctonine type with C(1)—α—OH
C(9)—H	47.0–55.0 ppm	Lycoctonine type with C(14) ketone, C(10)—OH, and C(7)—OCH$_2$O—C(8) absent
C(9)—H	58.0–59.5 ppm	C(7)—OCH$_2$O—C(8), C(10)—OH, and C(14) ketone present
C(9)—OH	77.5–78.5 ppm	General range
C(10)—H	36.5–37.5 ppm	C(9)—OH present
C(10)—H	37.5–43.0 ppm	General range, C(9)—OH absent
C(10)—H	46.0–47.5 ppm	C(8)—C(15) double bond present
C(10)—OH	79.5–82.5 ppm	General range
C(11)	47.0–51.5 ppm	General range
C(11)	55.0–56.6 ppm	C(10)—OH and C(7)—OCH$_2$O—C(8) present
C(11)	61.0 ppm	C(1) ketone present
C(12)—H$_2$	23.0–26.5 ppm	C(9)—OH present
C(12)—H$_2$	26.3–32.0 ppm	C(10) and C(13) not substituted
C(12)—H$_2$	33.5–38.0 ppm	C(13)—OH or OAc present
C(12)—H$_2$	33.5–34.5 ppm	C(1)—ketone present
C(12)—H$_2$	36.0–37.5 ppm	C(10)—OH present
C(13)—H	36.5–39.0 ppm	C(10)—OH and C(7)—OCH$_2$O—C(8) present
C(13)—H	43.0–46.5 ppm	General range
C(13)—H	48.5–50.0 ppm	C(9)—OH present
C(13)—OH	74.0–77.0 ppm	C(14)—OH or OR present
C(13)—OH	77.0–78.0 ppm	C(8)—C(15) double bond present
C(14)—OR^2	72.5–74.5 ppm	C(10)—OH present
C(14)—OR^2	74.5–77.0 ppm	C(9), C(10), and C(13) unsubstituted
C(14)—OR^2	77.5–80.5 ppm	C(13)—OH present
C(14)—OCH$_3$	80.5–82.5 ppm	C(7)—OCH$_2$O—C(8) and C(10)—O— present
C(14)—OCH$_3$	83.5–85.0 ppm	General range
C(14)—ketone	213.5–216.5 ppm	General range
C(15)—H$_2$	32.5–36.0 ppm	Lycoctonine type

Table 2. (*Continued*)

Functional Group	Chemical Shift Range	Remarks
C(15)—H$_2$	37.5–42.0 ppm	Aconitine type
C(15)—α—OH	78.5–79.0 ppm	Aconitine type
C(16)—OCH$_3$	79.5–84.5 ppm	General range
C(16)—OCH$_3$	82.5–84.5 ppm	C(13)—OH present
C(16)—OCH$_3$	83.5–84.0 ppm	C(7)- -OCH$_2$O—C(8) and C(14) ketone present
C(16)—OCH$_3$	89.5–92.0 ppm	C(15)—α—OH present
C(17)—H	60.5–63.5 ppm	Aconitine type
C(17)—H	63.5–66.5 ppm	Lycoctonine type
C(18)—H$_3$	24.5–26.0 ppm	Lycoctonine type
C(18)—H$_3$	26.0–27.5 ppm	Aconitine type
C(18)—H$_2$OH	66.5–68.5	General range
C(18)—H$_2$OCOR^3	68.5–70.5 ppm	Lycoctonine type
C(18)—H$_2$OCH$_3$	75.5–77.5 ppm	C(3)—α—OH present
C(18)—H$_2$OCH$_3$	77.5–79.5 ppm	Lycoctonine type
C(18)—H$_2$OCH$_3$	79.0–81.0 ppm	Aconitine type
C(19)—H$_2$	48.5–50.0 ppm	C(3)—α—OH present
C(19)—H$_2$	52.0–52.5 ppm	C(4)—CH$_2$OCOR present
C(19)—H$_2$	52.5–57.5 ppm	General range
C(19)—H$_2$	56.0–56.5 ppm	N—CH$_3$ present, C(13)—α—OH absent
C(19)—H$_2$	56.5–57.5 ppm	C(4)—CH$_3$ present, C(1)—α—OH absent
C(19)—H$_2$	56.5–57.0 ppm	C(1)—α—OH present
C(19)—H$_2$	60.5–62.0 ppm	C(1)—α—OH and C(4)—CH$_3$ present
C(19)—H	68.0–69.0 ppm	C(1)—C(19)—O— present
N—CH$_3$	42.5–43.0 ppm	General range
N—CH$_2$—CH$_3$	46.0–48.0 ppm	C(3)—α—OH present
N—CH$_2$—CH$_3$	48.0–51.5 ppm	General range
N—CH$_2$—CH$_3$	13.0–14.5 ppm	General range
C(1)—OCH$_3$	55.5–57.0 ppm	General range
C(6)—OCH$_3$	56.5–58.0 ppm	General range
C(14)—OCH$_3$	57.5–58.5 ppm	General range
C(16)—OCH$_3$	55.5–58.5 ppm	General range
C(16)—OCH$_3$	60.5–62.5 ppm	C(15)—α—OH present
C(18)—OCH$_3$	58.5–59.5 ppm	General range
C(7)—OCH$_2$O—C(8)	92.5–94.5 ppm	C(6) ketone absent
C(7)—OCH$_2$O—C(8)	95.5–96.5 ppm	C(6) ketone present
C(8)—OCOCH$_3$	169.0–170.0 ppm	General range
OCOCH$_3$	170.0–172.5 ppm	General range
OCOCH$_3$	21.0–22.5 ppm	General range
OCO—C$_6$H$_5$	165.5–166.5 ppm	General range
OCOC$_6$H$_5$.	128.0–133.0 ppm	General range

R^1 = H or CH$_3$ or Ac.   R^2 = H or Ac or Bz.   R^3 = CH$_3$ or C$_6$H$_5$ or C$_6$H$_4$NR'R''.

4. OCCURRENCE OF C[19]-DITERPENOID ALKALOIDS IN PLANT SPECIES

Table 3.

Aconitum alatiacum Steinb.
  Mesaconitine[1]
Aconitum angustifolium
  Aconitine[2]
  Hypaconitine[2]
  Mesaconitine[2]
Aconitum anthoroideum
  Condelphine[3]
Aconitum arcuatum
  Aconosine[4,5]
  Talatizamine[4,5]
Aconitum balfourii Stapf.
  Pseudaconitine[6]
Aconitum barbatum Pers var. *puberulum* Ledeb.
  N-Deacetylranaconitine[7,256]
  Lappaconitine[7,256]
  Lycaconitine[7,256]
  Puberaconitidine[256]
  Puberaconitine[256]
  Puberanidine[256]
  Puberanine[256]
  Ranaconitine[7, 256]
  Septentriodine[7, 256]
  Septentrionine[7, 256]
Aconitum bullatifolium Lé veillé var. *homotrichum*
  Aconitine[8]
  Bullatine B[8]
  Bullatine C[8]
  Hypaconitine[8]
Aconitum callianthum Koidz.
  Aconitine[9,10]
  Hypaconitine[9,11]
  Mesaconitine[9,12]
Aconitum cammarum Stapf. (*A. stoerckianum* Reichb.)
  Aconitine[13]
  Neoline[13]
Aconitum carmichaeli Debaux
  14-Acetyltalatizamine[14,15]
  Aconitine[14-19]
  14-Benzoylmesaconine[20]
  Fuziline[15,18,19]
  Hokbusine A[21]
  Hokbusine B[21]
  15-α Hydroxyneoline[20]

Table 3.   (*Continued*)

Aconitum carmichaeli Debaux (*Continued*)
  Hypaconitine[14-20,22]
  Isodelphinine[20]
  Isotalatizidine[14,17]
  Karakoline[14]
  Mesaconitine[16-19]
  Neoline[14,18-20]
  Senbusine A[15]
  Senbusine B[15]
  Talatizamine[14-17]
Aconitum chasmanthum Stapf.
  Chasmaconitine[23]
  Chasmanine[24-26]
  Chasmanthinine[23]
  Homochasmanine[27]
  Indaconitine[28]
Aconitum chinense Sieb.
  Aconitine[29]
  14-Benzoylaconine[29]
  14-Benzoylmesaconine[29]
  Mesaconine[29]
Aconitum confertiflorum DC
  14-Acetyltalatizamine[30]
Acontium crassicaule W. T. Wang
  Chasmanine[32,33]
  Crassicaulidine[34]
  Crassicauline A[32,33]
  Crassicaulisine[34]
  Yunaconitine[32,33]
Aconitum deinorrhizum Stapf.
  Pseudaconitine[6]
Aconitum delavyi Franch
  Delavaconitine[17,35,36]
  Yunaconitine[17]
Aconitum delphinifolium DC.
  Delphinifoline[37]
Aconitum episcopale Levl.
  Episcopalisine[38]
  Episcopalisinine[38]
  Episcopalitine[38]
  Scopaline[39]
Aconitum excelsum Reichb. (*A. leucostomum* Worosch)
  Excelsine[40]
  Lappaconitine[41]
  Mesaconitine[42]
Aconitum falconeri Stapf
  Falaconitine[43,44]

Table 3. (*Continued*)

Aconitum falconeri (*Continued*)
   Indaconitine[45]
   Mithaconitine[43,44]
   Pseudaconitine[45]
   Veratroylpseudaconine[45]
Aconitum fauriei Lé veillé and Vaniot
   Aconitine[9,46]
   Mesaconitine[9,46]
Aconitum ferox Wall
   Bikhaconitine[47,48]
   Chasmaconitine[47]
   Diacetylpseudaconitine[48]
   Indaconitine[47]
   Pseudaconitine[49,50]
   Veratroylpseudaconine[48]
Aconitum finetianum Hand-Mazz.
   Avadharidine[51,52]
   N-Deacetylfinaconitine[52]
   N-Deacetyllappaconitine[52]
   N-Deacetylranaconitine[52]
   Delcosine[53,54]
   Finaconitine[53,54]
   Lappaconitine[53,54]
   Lycoctonine[51,52]
   Ranaconitine[53,54]
Aconitum firmum Reichb. (*A. napellus* L.)
   Aconitine[55]
Aconitum fischeri Reichb. (*A. kamtschaticum* Pall.)
   Aconosine[4,5]
   Aconitine[16]
   Jesaconitine[56]
   Talatizamine[4,5]
Aconitum flavum Hand-Mazz.
   3-Acetylaconitine[57,58]
   Aconitine[57,58]
   Flavaconitine[59]
Aconitum forrestii Diels
   Aconosine[60]
   Cammaconine[60]
   Chasmaconitine[60]
   Crassicauline A[60]
   Vilmorrianine C[60]
   Yunaconitine[60]
Aconitum forrestii Stapf
   Chasmanine[61]
   Foresticine[61]
   Forestine[61]
   Liwaconitine[62]

Table 3.   (*Continued*)

Aconitum forrestii (*Continued*)
  Talatizamine[61]
  Yunaconitine[61]
Aconitum forrestii Stapf var. *albo-villosum* (Chen et Liu) W. T. Wang
  Foresaconitine[63]
Aconitum franchetti Fin. et Gagn
  Chasmaconitine[64]
  Chasmanine[64]
  Indaconitine[64]
  Ludaconitine[64]
  Talatizamine[64]
Aconitum fukutomei
  Aconitine[65]
Aconitum geniculatum Flet. et Laue, var *unguiculatum* W. T. Wang
  Pseudaconitine[66]
  Yunaconitine[66]
Aconitum gigas Lé veillé and Vaniot (*Lycoctonum gigas* Nakai)
  Gigactonine[67]
  Lycaconitine[67]
  Lycoctonine[67]
  Septentriodine[67]
  N-(Succinyl)-anthranoyllycoctonine[67]
Aconitum grossedentataum Nakai
  Aconitine[9,10]
  Hypaconitine[9,11]
  Mesaconitine[9,12]
Aconitum hakusanense Nakai
  Aconitine[9,10]
  Hypaconitine[9,11]
  Mesaconitine[9,12]
Aconitum hemsleyanum Pritz var. *circinatum* W. T. Wang
  Karakoline[66]
  Pseudaconitine[66]
  Yunaconitine[66]
Aconitum heterophyllum Wall
  6-Benzoylheteratisine[68]
  Heteratisine[68,69]
  Heterophyllidine[69]
  Heterophylline[69]
  Heterophyllisine[69,70]
Aconitum ibukiense Nakai
  Aconitine[9,10]
  Hypaconitine[9,11]
  Mesaconitine[9,12]
Aconitum japonicum Thunb.
  14-Acetyltalatizamine[71,72]
  Aconitine[71,72]

Table 3. (*Continued*)

Aconitum japonicum (*Continued*)
   Condelphine[73]
   14-Dehydrodelcosine[74-76]
   Delcosine[74-76]
   3-Deoxyaconitine[71,72]
   15-α-Hydroxyneoline[71,72]
   Hypaconitine[71,72]
   Isotalatizidine[71,72]
   Mesaconitine[71,72,75]
   Neoline[71,72]
   Takaonine[71]
   Takaosamine[71,72]
Aconitum jinyangense W. T. Wang
   14-Acetylneoline[77]
Aconitum kamtschaticum Pall. (*A. fischeri* Reichb.)
   Hypaconitine[11]
   Mesaconitine[11,46]
Aconitum karakolicum Rapcs.
   Aconifine[78-80]
   Aconitine[78-81]
   1-Benzoylkarasamine[253]
   Delsoline[253]
   Karakolidine[78,82]
   Karakoline[82-84]
   Karasamine[253]
   Monticoline[253]
   Neoline[253]
Aconitum kongboense
   Vilmorrianine A[85]
Aconitum koreanum (*A. coreanum* Levl.) Rapcs.
   Hypaconitine[17,31,86,87]
Aconitum kusnezoffii Reichb.
   Aconitine[88]
   Beiwutine[88]
   3-Deoxyaconitine[88]
   Hypaconitine[88]
   Mesaconitine[88]
Aconitum leucostomum Worosch (*A. excelsum* Reichb.)
   Hydroxylappaconine[89]
   Lapaconidine[89-95]
   Lappaconine[89]
   Lappaconitine[92-95]
Aconitum lucidusculum Nakai
   Lucaconine[96]
Aconitum lycoctonum L.
   Lycaconitine[97]
   Lycoctonine[97]

Table 3. (*Continued*)

Aconitum majimai Nakai (*A. yezoense* Nakai)
    Aconitine[46]
    Mesaconitine[12]
Aconitum manshuricum Nakai
    Mesaconitine[12]
Aconitum mitakense Nakai
    Aconitine[98]
    Chasmanine[98]
    Hypaconitine[98]
    Jesaconitine[98,99]
    Mesaconitine[98,99]
    Neoline[98]
Aconitum miyabei Nakai
    Isodelphinine[100,101]
    Sachaconitine[100,102]
Aconitum mokchangense Nakai
    Aconitine[46]
    Mesaconitine[12]
Aconitum monticola Steinb.
    Delsoline[103,105]
    1-Deoxydelsoline[106]
    Dihydromonticamine[106]
    Monticamine[107]
    Monticoline[107]
Aconitum nagarum var. *heterotrichum* f. *dielsianum* W. T. Wang
    Aconitine[108]
    3-Deoxyaconitine[108]
    Nagarine[108]
Aconitum nagarum Stapf. var. *lasiandrum* W. T. Wang
    Aconifine (10-β Hydroxyaconitine; Wang's and Zhu's Nagarine)[17,78,80,109-111]
    Aconitine[112]
    Bullatine C[112]
    3-Deoxyaconitine[112]
    Neoline[8,112,113]
Aconitum napellus L. (*A. spicatum* Donn)
    Aconitine[114,115,116]
    Benzoylaconine (Isaconitine, Pikraconitine)[117]
    3-Deoxyaconitine[118]
    Hypaconitine[10]
    Mesaconitine[2,12]
    Neoline[119]
    Neopelline[120]
Aconitum napellus ssp. *fissurae* and *superbum*
    Aconitine[2]
    Mesaconitine[2]
Aconitum nasutum Fisch and Reichb.
    Aconosine[121]

Table 3. (*Continued*)

Aconitum nemorum Popov
   14-Acetyltalatizamine[122]
   Talatizamine[122]
Aconitum orientale Mill.
   Aconorine[123]
   Avadharidine[124]
   Lappaconitine[124]
Aconitum pendulum
   3-Acetylaconitine[17]
   Aconitine[125]
   3-Deoxyaconitine[17]
   Hypaconitine[125]
   Penduline[17]
Aconitum ranuncualaefolium
   N-Deacetyllappaconitine[126]
   Lappaconitine[126]
   Ranaconitine[126]
Aconitum sachalinese F. Schmidt
   Aconitine[46]
   Jesaconitine[46,127,128]
Aconitum sachaliense var. *compactum*
   Chasmanine[129]
   Neoline[129]
Aconitum sanyoense Nakai var. *sanyoense*
   Aconitine[130]
   Hypaconitine[130,131]
   Mesaconitine[130,132]
Aconitum saposhnikovii B. Fedtsch
   14-Acetyltalatizamine[81]
   14-Dehydrotalatizamine[81]
   Talatizamine[81]
Aconitum sczukini
   Mesaconitine[4,5]
   Neoline[4,5]
Aconitum senanense Nakai
   Aconitine[128]
   Hypaconitine[11]
Aconitum septentrionale Koelle (*A. lycoctonum* L.)
   N-Deacetyllappaconitine[133]
   Lapaconidine[134]
   Lappaconine[135]
   Lappaconitine[133,135,136]
   Septentriodine[137]
   Septentrionine[137]
Aconitum sinomontanum Nakai
   Lappaconitine[138,139]
   Ranaconitine[138,139]

Table 3.   (*Continued*)

Aconitum soongaricum Stapf.
    Aconitine[42]
    Neoline[140]

Aconitum spathulatum
    Yunaconitine[141]

Aconitum spicatum Stapf.
    Pseudaconitine[142]

Aconitum spicatum Donn
    Bikhaconitine[143]

Aconitum spp. ("Chuanwu" in Chinese)
    14-Acetyltalatizamine[14]
    Aconitine[14]
    Hypaconitine[14]
    Isotalatizidine[14]
    Karakoline[14]
    Lipoaconitine[14]
    Lipo-3-deoxyaconitine[14]
    Lipohypaconitine[14]
    Lipomesaconitine[14]
    Mesaconitine[14]
    Neoline[14]
    Talatizamine[14]

Aconitum stapfianum Hand Mazz. var. *pubisceps* Wang
    Aconosine[144]
    Dolaconine[144]

Aconitum stoerckianum Reichb. (*A. cammarum* L.)
    Aconitine[13]
    Neoline[13]

Aconitum subcuneatum Nakai
    Aconitine[46,127]
    Chasmanine[145]
    3-Deoxyaconitine[146]
    3-Deoxyjesaconitine[146]
    Hypaconitine[146]
    Jesaconitine[127,128,146]
    Mesaconitine[146]

Aconitum talassicum Popov
    Isotalatizidine[147,148]
    Talatizamine[122,147]
    Talatizidine[147,148]

Aconitum tasiromontanum Nakai
    Aconitine[149]
    Hypaconitine[149]
    Mesaconitine[149]

Aconitum tianschanicum Rupr. (*A. napellus* L.)
    Aconitine[147]

Table 3. (*Continued*)

Aconitum tortuosum Willd.
    Aconitine[9,10]
    Hypaconitine[9,11]
    Mesaconitine[9]
Aconitum tranzschelii
    Isotalatizidine[3]
    Talatizamine[3]
Aconitum umbrosum
    Ajacine[150]
    Anthranoyllycoctonine[150]
    Lycaconitine[4,5,150]
    Umbrosine[150]
Aconitum variegatum
    Cammaconine[151–153]
    Talatizamine[151,153]
Aconitum vilmorrianum Kom.
    Vilmorrianine A[154]
    Vilmorrianine B[154]
    Vilmorrianine C[154]
    Vilmorrianine D[84]
    Yunaconitine[154]
Aconitum violaceum
    Bikhaconitine[155]
    Indaconitine[156]
Aconitum yesoense Nakai
    14-Acetylneoline[157]
    Aconitine[98,158]
    Anisoezochasmaconitine[157,159]
    Chasmanine[98,157,159]
    Ezochasmaconitine[157,159]
    Ezochasmanine[157,159]
    Hypaconitine[98]
    Jesaconitine[98,157–159]
    Mesaconitine[98,157]
    Neoline[98,157]
    Pyrochasmanine[157]
Aconitum zuccarini Nakai
    Aconitine[9,10]
    Hypaconitine[9,11]
    Mesaconitine[9,46]
Atragne sibirica L.
    Aconitine[160]
    Delphinine[160]
Consolida ambigua L. P. W. Ball and V. H. Heywood (*Delphinium ajacis* L.)
    14-Acetylbrowniine[161]
    14-Acetyldelcosine[161,162]
    Ajacine[161]

Table 3. (*Continued*)

Consolida ambigua (*Continued*)
  Ajacusine[161]
  Ajadine[161]
  Ambiguine[161]
  Anthranoyllycoctonine[161]
  Browniine[161]
  Delcosine[161,162]
  Delphatine[161]
  Delsoline[161,162]
  Lycoctonine[161]
  Methyllycaconitine[161]

Consolida orientalis Gay, ssp. *Orientalis* (*Delphinium orientalis* Gay and
D. hispanicum Wilk.)
  Delcosine[163]
  Delsoline[163]
  18-Hydroxy-14-*O*-methylgadesine[163]
  Gigactonine[163]
  18-Methoxygadesine[164]

Delphinium ajacis L. (*Consolida ambigua*)
  14-Acetyldelcosine[165–167]
  Ajacine[165,166,168]
  Delcosine[165,166]
  Delcoline[169]
  Delosine[169]
  Delsoline[169]
  Dimethylacetyldelcosine[169]
  Lycoctonine[171]
  Trimethylacetyldelcosine[169]

Delphinium araraticum
  Methyllycaconitine[172]

Delphinium barbeyi Huth (*D. glaucum* S. Wats)
  Anthranoyllycoctonine[173]
  Delpheline[174]
  Deltaline[174]
  Dictyocarpine[255]
  Lycoctonine[173,174]

Delphinium belladonna Kelw.
  14-Acetyldelcosine[165,166]
  Ajacine[165]
  Delcosine[165]

Delphinium bicolor Nutt.
  Alkaloid A[175–179]
  Alkaloid B[175–179]
  Delcosine[175–179]
  Isotalatizidine[175–179]
  Lycoctonine[175–179]

Table 3.   (*Continued*)

Delphinium biternatum Huth.
  Anthranoyllycoctonine[180]
  14-Benzoylbrowniine[180]
  14-Benzoyldelcosine[105,180,181]
  Browniine[180]
  14-Dehydrobrowniine[180]
  14-Dehydrodelcosine (Shimoburo Base II)[105,180]
  Delbiterine[180]
  Delcosine[105,180-182]
  Delphatine[180,183-185]
  Delsoline[180]
Delphinium brownii Rydb.
  14-Acetylbrowniine[186]
  Browniine[186,187]
  Lycoctonine[188]
  Methyllycaconitine[186,187]
Delphinium buschainum
  Methyllycaconitine[189]
Delphinium cardinale Hook.
  14-Dehydrobrowniine[190]
  Browniine[190]
  Lycoctonine[190]
Delhinium cardiopetalum DC (*D. verdunense* Balbis)
  Cardiopetalidine[191]
  Cardiopetaline[191]
Delphinium carolinianum Walter
  Browniine[192,193]
  Delcaroline[193]
Delphinium cashmirianum Royle
  Anthranoyllycoctonine[194]
  Avadharidine[194]
  Cashmiradelphine[194]
  N-Deacetyllappaconitine[194]
  Lappaconitine[194]
  Lycaconitine[194]
Delphinium confusum Popov
  Anthranoyllycoctonine[195]
  Condelphine[196]
  Methyllycaconitine[195]
Delphinium consolida L.
  Anthranoyllycoctonine[197]
  Delcosine[198,199]
  Delsoline[199]
  Lycoctonine[197]
Delphinium corumbosum Rgr. (*D. corymbosum* Regel)
  Delcorine[200]
  6-Deoxydelcorine[201]
  Methyllycaconitine[200]

Table 3. (*Continued*)

Delphinium crassifolium Schrad by Mats
    Methyllycaconitine[202]
Delphinium denudatum Wall.
    Condelphine[203]
    Isotalatizidine[203]
Delphinium dictyocarpum DC.
    14-Acetyldelectine[204]
    N-Acetyldelectine[205]
    Anthranoyllycoctonine[205]
    14-Benzoyldictyocarpine[206]
    Delectine[207]
    Delectinine[95]
    Deltaline[208]
    Deltamine[208]
    Demethyleneeldelidine[205]
    Dictyocarpine[208]
    7,18-Di-*O*-methyllycoctonine[207]
    Lycoctonine[200,208]
    Methyllycaconitine[200,205,208,209]
Delphinium elatum L.
    Delpheline[210]
    Deltaline[211-213]
    Elatine[211,214]
    Methyllycaconitine[210]
Delphinium elisabethae
    Anthranoyllycoctonine[215]
    Lycoctonine[215]
    Methyllycaconitine[215]
Delphinium flexuosum
    Methyllycaconitine[216]
Delphinium formosum
    Delsemine[217]
    Lycoctonine[217]
Delphinium glaucescens Rydb.
    Anthranoyllycoctonine[218]
    Browniine[218]
    14-Dehydrobrowniine[218]
    Delcosine[218]
    Deltaline[218]
    Dictyocarpine[218]
    Dictyocarpinine[218]
    Glaucedine[218]
    Glaucenine[218]
    Glaucephine[218]
    Glaucerine[218]
    Glaudelsine[218]
    Lycoctonine[218]
    Methyllycaconitine[218]

Table 3. (*Continued*)

Delphinium grandiflora
   Anthranoyllycoctonine[219]
   Methyllycaconitine[219]
Delphinium grandiflorum L.
   Methyllycaconitine[202]
Delphinium iliense Huth.
   Browniine[220]
   6-Dehydrodelcorine[220]
   Delcoridine[220]
   Delcorine[220]
   Deltaline[220]
   Dictyocarpine[220]
   Dictyocarpinine[220]
   Ilidine[220]
   Lycoctonine[220]
Delphinium linearifolium
   Methyllycaconitine[221]
Delphinium occidentale (S. Wats.) S. Wats.
   Deltaline[174,222]
Delphinium oreophilum Huth
   14-Acetylbrowniine[223]
   Delsemine[224]
   Lycoctonine[224,225]
   Methyllycaconitine[225,226]
Delphinium orientale S. Wats
   Ajacine[227]
   Delcorine[227]
   Delsoline[227]
Delphinium pentagynum Lam.
   14-Acetyldihydrogadesine[228]
   Dihydrogadesine[228]
   Dihydropentagynine[228]
   Gadenine[229]
   Gadesine[230]
   Karakoline[231]
   Pentagydine[231]
   Pentagyline[229]
   Pentagynine[228]
Delphinium regalis (*Consolida regalis*)
   Delcosine[232]
   Delsoline[232]
   Lycoctonine[232]
Delphinium rotundifolium (Afan.) Sosk. et Fachieva
   Browniine[233]
   Delsemine[233]
   Methyllycaconitine[233]
Delphinium schmalhausenii
   Methyllycaconitine[216]

Table 3. (*Continued*)

Delphinium semibarbatum Bein ex Boiss.
    Delsemine[234-236]
    Lycoctonine[236]
    Methyllycaconitine[237]
Delphinium staphisagria L.
    Delphidine[238]
    Delphinine[9,239,240]
    Delphirine[165,241]
    Delphisine[242,243]
    Neoline[254]
Delphinium tamarae Kem. Nath.
    Anthranoyllycoctonine[244]
    Lycoctonine[244]
    Methyllycaconitine[244]
Delphinium tatsienense Franch
    Browniine[245]
    Deacetylambiguine[245]
    Delcosine[245]
    Lycoctonine[245]
    Tatsiensine[245]
Delphinium ternatum Huth
    Methyllycaconitine[195]
Delphinium tricorne Michaux
    Delsemine[246]
    Lycoctonine[246]
    Methyllycaconitine[246]
    Tricornine[246-247]
Delphinium triste Fisch.
    Methyllycaconitine[202]
Delphinium virescens Nutt.
    14-Acetylvirescenine[248]
    Browniine[192,248]
    Virescenine[248]
Inula royleana
    Anthranoyllycoctonine[249,250]
    Lycoctonine[249,251]
    Methyllycaconitine[249,251]

4.1. References for Table 3

1. T. E. Monakhova, T. F. Platonova, A. D. Kuzovkov, and A. I. Shreter, *Khim. Prir. Soedin.*, 113 (1965); *Chem. Abs.* **63**, 7347 (1965).

2. A. Katz and E. Staehelin, *Pharm. Acta Helv.* **54**, 253 (1979).

3. V. A. Te'lnov, M. S. Yunusov, and S. Yu. Yunusov, *Khim. Prir. Soedin.*, 383 (1971).

4. N. M. Golubev, V. A. Tel'nov. M. S. Yunusov, N. K. Fruentov, and S. Yu. Yunusov, *Vopr. Farm. Dal'nem. Vostoke* **2**, 10 (1977); *Chem. Abs.* **90**, 164 757 (1979).

5. N. M. Golubev, *Vopr. Farm. Dal'nem Vostoke* **2**, 37 (1977); *Chem. Abs.* **90**, 164 758 (1979).

6. T. A. Henry and T. M. Sharp, *J. Chem. Soc.,* 1105 (1928).

7. D. Yu, *Acta Pharmaceutica Sinica* **17**, 301 (1982); *Chem. Abs.* **97**, 212633 (1982).

8. J. H. Chu and S. T. Fang, *Acta Chimica Sinica* **31**, 222 (1965); *Chem. Abs.* **63**, 16400a (1965).

9. R. Majima, H. Suginome, and S. Morio, *Chem. Ber.* **57**, 1456 (1924).

10. R. Majima and S. Morio, *Ann.* **476**, 194 (1929).

11. R. Majima and S. Morio, *Ann.* **476**, 171 (1929).

12. S. Morio, *Ann.* **476**, 181 (1929).

13. H. Schulze and G. Berger, *Arch. der Pharm.* **265**, 524 (1927).

14. I. Kitagawa, M. Yoshikawa, Z. L. Chen, and K. Kobayashi, *Chem. Pharm. Bull.* Japan **30**, 758 (1982).

15. C. Konno, M. Shirasaka, and H. Hikino, *J. Nat. Prod.* **45**, 128 (1982).

16. Y. Chen, Y. L. Chu, and J. H. Chu, *Acta Pharmaceutica Sinica* **12**, 435 (1965); *Chem. Abs.* **63**, 16400a (1965).

17. L. Zhu and R. H. Zhu, *Heterocycles* **17**, 607 (1982).

18. S. W. Pelletier, N. V. Mody, K. I. Varughese, and S. Y. Chen, *Heterocycles* **18**, 47 (1982).

19. S. Y. Chen, Y. Q. Liu, and J. C. Wang, *Acta Botanica Yunnanica* **4**, 73 (1982).

20. D. H. Chen, H. Y. Li, and W. L. Sung, *Chinese Traditional and Herbal Drugs* **13**, 1 (1982).

21. H. Hikino, Y. Kuroiwa, and C. Konno, *J. Nat. Prod.* **46**, 178 (1983).

22. J. Iwasa and S. Narato, *J. Pharm. Soc.* Japan **86**, 585 (1966); *Chem. Abs.* **65**, 10629 (1966).

23. O. Achmatowicz, Jr. and L. Marion, *Can. J. Chem.* **42**, 154 (1964).

24. L. Marion, J. P. Boca, and J. Kallos, *Tetrahedron, Suppl.* **8**, Part 1, 101 (1966).

25. O. Achmatowics, Jr., Yu. Tsuda, L. Marion, T. Okamoto, M. Natsume, H. H. Chang, and K. Kajima, *Can. J. Chem.* **43**, 825 (1965).

26. O. E. Edwards, L. Fonzes, and L. Marion, *Can. J. Chem.* **44**, 583 (1966).

27. O. Achmatowicz, Jr. and L. Marion, *Can. J. Chem.* **43**, 1093 (1965); S. W. Pelletier, Z. Djarmati, S. Lajsic, and W. H. De Camp, *J. Amer. Chem. Soc.* **98**, 2617 (1976).

28. W. R. Dunstan and A. E. Andrews, *J. Chem. Soc.,* **87**, 1620 (1905).

29. U. Muramatsu, M. Takahashi, H. Shibata, and K. Watanabe, *Tochigi-ken Kachiku Eisei Kenkyusho Nempo* **12**, 31 (1977); *Chem. Abs.* **90**, 183148 (1979).

30. B. Sh. Ibragimov, G. M. Mamedov, and H. M. Ismailov, *Doklady Akad. Nauk Azerb. SSR* **32**, 58 (1976).

31. J. H. Liu, A. C. Wang, Y. L. Gao, and R. H. Chu, *Chinese Traditional and Herbal Drugs* **12**, 1 (1981).

32. F. P. Wang and Q. C. Fang, *Planta Med.* **42**, 375 (1981).

33. F. P. Wang and Q. C. Fang, *Acta Pharmaceutica Sinica* **16**, 49 (1981).

34. F. P. Wang and Q. C. Fang, *Acta Pharmaceutica Sinica* **17**, 300 (1982); *Planta Medica* **47**, 39 (1983).

35. J. H. Chu, *Acta Chimica Sinica* **21**, 332, (1955); *Chem. Zen.,* 12851 (1956).

36. J. H. Chu, S. H. Hung, and J. L. Chou, *Acta Chimica Sinica* **23**, 130(1957); *Chem. Abs.,* **52**, 14,632 (1958).

37. V. N. Aiyar, P. W. Codding, K. A. Kerr, M. H. Benn, and A. J. Jones, *Tetrahedron* **22**, 483 (1981).

38. F. P. Wang and Q. C. Fang, *Acta Pharmaceutica Sinica* **18**, 514 (1983).

39. C. Yang, D. Wang, D. Wu, X. Hao, and J. Zhou, *Acta Chimica Sinica* **39**, 445 (1981).

40. V. A. Te'lnov, M. S. Yunusov, and S. Yu. Yunusov, *Khim. Prir. Soedin.,* 129 (1973).

41. T. F. Platonova, A. D. Kuzovkov, and P. S. Massagetov, *J. Gen. Chem. USSR* (Engl. Transl.) **28**, 259 (1958).

42. S. Yu. Yunusov, *J. Gen. Chem. USSR* **18**, 515 (1948); *Chem. Abs.*, **42**, 7941 (1948).

43. N. Singh, G. S. Bajwa, and M. G. Singh, *Indian J. Chem.* **4**, 39 (1966).

44. S. W. Pelletier, N. V. Mody, and H. S. Puri, *Chem. Commun.*, 12 (1977).

45. S. W. Pelletier, N. V. Mody, and H. S. Puri, *Phytochemistry* **16**, 623 (1977).

46. R. Majima and S. Morio, *Ann.* **476**, 203 (1929); *Chem. Zen.* **1**, 388 (1930).

47. A. Klasek, V. Simanek, and F. Santavy, *Lloydia* **35**, 55 (1972).

48. K. K. Purushothaman and S. Chandrasekharan, *Phytochemistry* **13**, 1975 (1974).

49. C. R. A. Wright and A. P. Luff, *J. Chem. Soc.* **31**, 143 (1877).

50. C. R. A. Wright and A. P. Luff, *J. Chem. Soc.*, **33**, 151 (1878).

51. B. R. Chen, Y. F. Yang, R. M. Tian, C. P. Zhang, Y. Z. Xiao, and M. X. Lieu, *Acta Pharmaceutica Sinica* **16**, 70 (1981).

52. S. H. Jiang, Y. L. Zhu, Z. Y. Zhao, and R. H. Zhu, *Acta Pharmaceutica Sinica* **18**, 440 (1983).

53. S. H. Jiang, Y. L. Zhu, and R. H. Zhu, *Acta Pharmaceutica Sinica* **16**, 55 (1981).

54. S. H. Jiang, Y. L. Zhu, and R. H. Zhu, *Acta Pharmaceutica Sinica* **17**, 282 (1982).

55. L. Krowczynski, *Polska Akad. Umiej. Prace Kom. Nauk Farm., Dissert. Pharm.* **3**, 183 (1951); *Chem. Abs.*, **48**, 5877 (1954).

56. K. Makoshi, *Arch. der Pharm.* **247**, 243 (1909).

57. Z. G. Zhu, T. L. Chen, Y. Q. Lin, and K. T. Zhang, *Lan-Chou Ta Hsueh Hsueh Pao, Tzu Jan K'o Hsueh Pan,* 135 (1980).

58. X. Chang, H. Wang, L. Lu, Y. Zhu, and R. Zhu, *Acta Pharmaceutica Sinica* **16**, 474 (1981); *Chem. Abs.* **97**, 141,679f (1982).

59. Y. Liu and G. Chang, *Acta Pharmaceutica Sinica* **17**, 243 (1982).

60. C. H. Wang, D. H. Chen, and W. L. Sung, *Chinese Traditional and Herbal Drugs* **14**, 5 (1983).

61. S. W. Pelletier, S. Y. Chen, B. S. Joshi, and H. K. Desai, *J. Nat. Prod.*, in press.

62. C. H. Wang, D. H. Chen, and W. Sung, *Planta Medica* **48**, 55 (1983).

63. W. S. Chen and E. Breitmaier, *Chem. Ber.* **114**, 394 (1981).

64. D. H. Chen and W. L. Sung, *Chinese Traditional and Herbal Drugs* **13**, 8 (1982); *Chem. Abs.* **97**, 88655 (1982).

65. Y. P. Chen and H. Y. Hsu, *Chung-Xuo Nung Yeh Hua Hsueh Hui Chih,* 17 (1979).

66. S. Y. Chen, *Acta Chimica Sinica* **37**, 15 (1979).

67. S. Sakai, N. Shinma, S. Hasegawa, and T. Okamoto, *J. Pharm. Soc.* Japan **98**, 1376 (1978).

68. R. Aneja, D. M. Locke, and S. W. Pelletier, *Tetrahedron* **29**, 3297 (1973).

69. S. W. Pelletier and R. Aneja, *Tetrahedron* **6**, 557 (1967).

70. W. A. Jacobs and L. C. Craig, *J. Biol. Chem.* **143**, 605 (1942).

71. H. Takayama, S. Hasegawa, S. Sakai, J. Haginiwa, and T. Okamoto, *Chem. Pharm. Bull.* Japan **29**, 3078 (1981).

72. H. Takayama, S. Hasegawa, S. Sakai, J. Haginiwa, and T. Okamoto, *J. Pharm. Soc.* Japan **102**, 525 (1982).

73. S. Sakai, H. Takayama, and T. Okamoto, *J. Pharm. Soc.* Japan **99**, 647 (1979).

74. E. Ochiai, T. Okamoto, S. I. Sakai, and S. H. Inoue, *J. Pharm. Soc., Japan* **75**, 638 (1955); *Chem. Abs.*, **50**, 3477 (1956).

75. E. Ochiai, T. Okamoto, S. Sakai, M. Kaneko, K. Fiyisowa, U. Nagai, and H. Tani, *J. Pharm. Soc.* Japan **76**, 550 (1956); S. Morio, *Ann.* **476**, 181 (1929).

76. E. Ochiai and M. Kaneko, *J. Pharm. Soc.* Japan **76**, 469 (1956); *Chem. Abs.*, **50**, 13970 (1956).

77.   D. H. Chen and W. L. Song, *Acta Pharmaceutica Sinica* **16**, 748 (1981).

78.   M. N. Sultankhodzhaev, M. S. Yunusov, and S. Yu. Yunusov, *Khim. Prir. Soedin.,* 127 (1963); *Chem. Abs.* **78**, 159,946 (1973).

79.   M. N. Sultankhodzhaev, L. V. Beshitaishvili, M. S. Yunusov, and S. Yu. Yunusov, *Khim. Prir. Soedin.,* 826 (1979).

80.   M. N. Sultankhodzhaev, L. V. Beshitaishivili, M. S. Yunusov, M. R. Yagudov, and S. Yu. Yunusov, *Khim. Prir. Soedin.,* 665 (1980).

81.   M. N. Sultankhodzhaev, M. S. Yunusov, and S. Yu. Yunusov, *Khim. Prir. Soedin.,* 265 (1982); *Chem. Abs.* **97**, 1,416,974 (1982).

82.   M. N. Sultankhodzhaev, M. S. Yunusov, and S. Yu. Yunusov, *Khim. Prir. Soedin.,* 199 (1973).

83.   M. N. Sultankhodzhaev, M. S. Yunusov, and S. Yu. Yunusov, *Khim. Prir. Soedin.,* 399 (1972).

84.   T. R. Yang, X. J. Hao, and J. Chow, *Yun-nan Chih Wu Yen Chin.* **1**, 41 (1979); *Chem. Abs.,* **93**, 46,909 (1980).

85.   C. Wang and Q. Fang, *Acta Pharmaceutica Sinica* **17**, 395 (1982).

86.   K. Kao, F. Yo, and J. Chu, *Yao Hsueh Hsueh Pao* **13**, 186 (1966); *Chem. Abs.* **65**, 3922 (1966).

87.   J. H. Liu, H. C. Wang, Y. L. Kao, and J. H. Chu, *Chung Ts'ao Yao* **12**, 1 (1981).

88.   Y. G. Wang, Y. L. Zhu, and R. H. Zhu, *Acta Pharmaceutica Sinica* **15**, 526 (1980).

89.   V. N. Plugar, Y. V. Rashkes, M. G. Zhamierashvili, V. A. Tel'nov, M. S. Yunusov, and S. Yu. Yunusov, *Khim. Prir. Soedin.,* 80 (1982).

90.   V. A. Tel'nov, M. S. Yunusov, and S. Yu. Yunusov, *Khim. Prir. Soedin.,* 639 (1970).

91.   V. A. Tel'nov, M. S. Yunusov, Y. V. Rashkes, and S. Yu. Yunusov, *Khim. Prir. Soedin.,* 622 (1971).

92.   M. G. Zhamierashivili, V. A. Tel'nov, M. S. Yunusov, S. Yu. Yunusov, A. Nigmatullaev, and K. Taizhanov, *Khim. Prir. Soedin.,* 235 (1978).

93.   H. Guinaudeau, M. Leboeuf, and A. Cave, *Lloydia* **38**, 257 (1975).

94.   T. Kosuge and M. Yokota, *Chem. Pharm. Bull.* Japan **24**, 176 (1976).

95.   B. T. Salimov, N. D. Abdullaev, M. S. Yunusov, and S. Yu. Yunusov, *Khim. Prir. Soedin.,* 235 (1978).

96.   H. Suginome, S. Kakimoto, J. Sonoda, and S. Noguchi, *Proc. Jap. Acad.* **22**, 122 (1946); *Chem. Abs.,* **45**, 9222 (1951).

97.   M. M. Dragendorff, and Spohn, *Pharm. Z. Russl.* **23**, 131 (1884); *Pharm. J.* [3] **15**, 104 (1884).

98.   F. Kurosaki, T. Yatsunami, T. Okamoto, and Y. Ichinohe, *J. Pharm. Soc.,* Japan **98**, 1267 (1978).

99.   E. Ochiai, T. Okamoto, and S. Sakai, *J. Pharm. Soc. Japan* **75**, 545 (1955); *Chem. Abs.,* **50**, 5695 (1956).

100. H. Suginome, N. Katsui, and G. Hasegawa, *Bull. Chem. Soc.* Japan **32**, 604 (1959).

101. N. Katsui, *Bull. Chem. Soc.* Japan **32**, 774 (1959).

102. N. Katsui and G. Hasegawa, *Bull. Chem. Soc.* Japan **33**, 1037 (1960).

103. V. E. Nezhevenko, M. S. Yunusov, and S. Yu. Yunusov, *Khim. Prir. Soedin.,* 409 (1974).

104. V. Nezhevenko, M. S. Yunusov, and S. Yu. Yunusov, *Khim. Prir. Soedin.,* 389 (1975).

105. S. W. Pelletier and N. V. Mody, *Tetrahedron* **22**, 207 (1981).

106. E. F. Ametova, M. S. Yunusov, and V. A. Tel'nov, *Khim. Prir. Soedin.,* 504 (1982).

107. E. F. Ametova, M. S. Yunusov, V. E. Bannikova, N. D. Abdullaev, and V. A. Tel'nov, *Khim. Prir. Soedin.,* 466 (1981).

108. N. V. Mody, S. W. Pelletier, and S. Y. Chen, *Heterocycles* **17**, 91 (1982).

109. C. Wang, Y. L. Gao, R. S. Xu, and R. H. Zhu, *Acta Chimica Sinica* **39**, 869 (1981).

110. F. O. Wang, *Acta Pharmaceutica Sinica* **16**, 950 (1981).

111. S. W. Pelletier, N. V. Mody, and S. -Y. Chen, *Heterocycles* **19**, 1523 (1982).

112. H. C. Wang, D. Z. Zhu, Z. Y. Zhao, and R. H. Zhu, *Acta Chimica Sinica* **38**, 475 (1980).

113. J. H. Chu, S. T. Fang, and W. -K. Huang, *Acta Chimica Sinica* **30**, 139 (1964).

114. P. L. Gaiger, *Ann.* **7**, 269 (1833).

115. Peschier, *Tromsdorff's N.J. Pharm.* **5**, 1. St., 93 (1821).

116. T. R. N. Morson, *Poggend. Ann.* **42**, 175 (1839); *Berz. Jber.*, 318 (1839).

117. W. R. Dunstan and E. F. Harrison, *J. Chem. Soc.* **63**, 443 (1893); *Chem. Zen.* **I**, 654 (1893).

118. Y. Tsuda, O. Achmatowicz, Jr. and L. Marion, *Ann.* **680**, 88 (1964).

119. W. Freudenberg and E. F. Rogers, *J. Amer. Chem. Soc.* **59**, 2572 (1937).

120. H. Schulze and G. Berger, *Arch. der Pharm.* **262**, 553 (1924); *Chem. Zen.*, 2000 (1925).

121. D. A. Muravjeva, T. I. Plekhanova, and M. S. Yunusov, *Khim. Prir. Soedin.*, 128 (1972).

122. T. F. Platonova, A. D. Kuzovhov, and P. S. Massagetov, *J. Gen. Chem. USSR* (Engl. Transl.) **28**, 3157 (1958); *Zh. Obshch. Khim.* **28**, 3126 (1958); *Chem. Abs.*, **53**, 9285 (1959).

123. V. A. Tel'nov, M. S. Yunusov, S. Yu. Yunusov, and B. Sh. Ibragimov, *Khim. Prir. Soedin.*, 814 (1975).

124. A. D. Kuzovkov and P. S. Massagetov, *J. Gen. Chem.* USSR (Engl. Transl.) **25**, 161 (1955); *Chem. Zen.*, 8627 (1955).

125. L. M. Liu, H. C. Wang, and Y. L. Zhu, *Acta Pharmaceutica Sinica* **18**, 39 (1983).

126. N. Mollov, M. Haimova, P. Tscherneva, N. Pecigargova, I. Ognjanov, and P. Panov, *C.R. Acad. Bulg. Sci.* **17**, 251 (1964).

127. R. Majima, H. Suginome, and S. Morio, *Chem. Ber.* **57**, 1472 (1924); *Chem. Zen.*, 2048 (1924).

128. H. Suginome and S. Imato, *J. Fac. Sci., Hokkaido Univ., Ser.* **34**, 33 (1950).

129. Y. Ichinohe, M. Hasegawa, H. Fujimoto, and M. Tamura, *Bull Dept. Educ. Coll. Sci. Technol. Nihon Univ.* **25**, 17 (1979).

130. G. Ochiai, T. Okamoto, J. Sugasawa, H. Tani, and H. Hai, *J. Pharm. Soc.* Japan **72**, 816 (1952); *Chem. Abs.,* **48**, 2728 (1954).

131. E. Ochiai, T. Okamoto, S. Sakai, and A. Saito, *J. Pharm. Soc.* Japan, **76**, 1414 (1956); *Chem. Abs.* **51**, 6661 (1957).

132. E. Ochiai, T. Okamoto, S. I. Sakai, and A. Saito, *J. Pharm. Soc.* Japan **76**, 1436 (1956); *Chem. Abs.,* **51**, 6662 (1957).

133. L. Marion, L. Fonzes, C. K. Wilkins, Jr., J. P. Boca, F. Sandberg, R. Thorsen, and E. Linden, *Can. J. Chem.* **45**, 969 (1967).

134. S. W. Pelletier, N. V. Mody, and R. S. Sawhney, *Can. J. Chem.* **57**, 1652 (1979).

135. H. V. Rosendahl, *Arb. Pharmakol. Inst. Dorpat* **11**, 1 (1895); *J. Pharm.* **4**, 262 (1896).

136. H. Schulze and F. Ulfert, *Arch. der Pharm.* **260**, 230 (1922); G. Weidemann, *Arch. Exp. Pathol. Pharmakol.* **95**, 166 (1922).

137. S. W. Pelletier, R. S. Sawhney, and A. J. Aasen, *Heterocycles* **12**, 377 (1979).

138. B. Y. Wei, H. W. Kung, C. Y. Chao, H. C. Wang, and J. H. Chu, *Chung Yao T'ung Pao* **6**, 26 (1981).

139. S. Y. Chen, Y. Q. Liu, and T. R. Yang, *Acta Botanica Yunnanica* **2**, 473 (1980).

140. M. G. Zhamierashvili, V. A. Tel'nov, M. S. Yunusov, and S. Yu. Yunusov, *Khim. Prir. Soedin.*, 733 (1980).

141. S. W. Pelletier and H. P. Chokshi, Unpublished results.

142. Y. Tsuda and L. Marion, *Can. J. Chem.* **41**, 1485 (1963).

143. W. R. Dunstan and A. E. Andrews, *J. Chem. Soc.* **87**, 1636 (1905).

144. S. Luo and W. Chen, *Acta Chimica Sinica* **39**, 808 (1981).

145. J. S. Glasby, *Encyclopedia of the Alkaloids,* Plenum Press, New York, 1975, Vol. 1, p. 282.

146. H. Bando, Y. Kanaiwa, K. Wada, T. Mori, and T. Amiya, *Heterocycles* **16**, 1723 (1981).

147. R. A. Konovalova and A. P. Orekhov, *Bull. Soc. Chim. Fr., Mem.* **7**, 95 (1940); *Chem. Abs.* **34**, 5450 (1940).

148. R. A. Konovalova and A. P. Orekhov, *Zh. Obshch. Khim.* **10**, 745 (1940).

149. E. Ochiai, T. Okamoto, T. Sugasawa, and H. Tani, *J. Pharm. Soc.* Japan **72**, 1605 (1952); *Chem. Abs.,* **47**, 2936 (1953).

150. V. A. Tel'nov, N. M. Golubev, and M. S. Yunusov, *Khim. Prir. Soedin.,* 675 (1976).

151. M. A. Khaimova, M. D. Palamareva, L. G. Grozdanova, N. M. Mollov, and P. P. Panov, *C. R. Acad. Bulg. Sci.* **20**, 193 (1967); *Chem. Abs.* **67**, 54,296c (1967).

152. N. V. Mody, S. W. Pelletier, and N. M. Mollov, *Heterocycles* **14**, 1751 (1920).

153. M. A. Khaimova, M. O. Palamareva, N. M. Mollov, and V. P. Krestev, *Tetrahedron* **27**, 819 (1971).

154. C. R. Yang, X. J. Hao, D. Z. Wang, and J. Zhou, *Acta Chimica Sinica* **39**, 147 (1981).

155. K. P. Tiwari and M. Masood, *J. Indian Chem. Soc.* **54**, 924 (1977).

156. G. A. Miana, M. Ikram, M. I. Kahn, and F. Sultana, *Phytochemistry* **10**, 3320 (1971).

157. H. Takayama, A. Tokita, M. Ito, S. Sakai, F. Kurosaki, and T. Okamoto, *J. Pharm. Soc.* Japan **102**, 245 (1982).

158. R. Majima and S. I. Morio, *Proc. Imp. Acad. Tokyo* **7**, 351 (1931); *Chem. Ber.* **65**, 599 (1932); *Chem Zen.* **I,** 1539 and 3067 (1932).

159. H. Takayama, M. Ito, M. Koga, S. Sakai, and T. Okamoto, *Heterocycles* **15**, 403 (1981).

160. E. A. Krasnov and V. S. Bokova, *Khim. Prir. Soedin.,* 806 (1981).

161. S. W. Pelletier, R. S. Sawhney, H. K. Desai, and N. V. Mody, *J. Nat. Prod.* **43**, 395 (1980).

162. G. R. Waller, S. D. Sastry, and K. F. Kinneberg, *Proc. Okla. Acad. Sci.* **53**, 92 (1973).

163. A. G. Gonzalez, G. de la Fuente, O. Munguia, and K. Henrick, *Tetrahedron* **22**, 4843 (1981).

164. A. G. Gonzalez, G. de la Fuente, and O. Munguia, *Heterocycles* **20**, 409 (1983).

165. F. M. Hashem, S. M. Abdel-Wahab, M. Salah Ahmed, and A. A. Seida, *Bull. Fac. Pharm., Cairo Univ.* **13**, 205 (1974); *Chem. Abs.* **86**, 40,214 (1977).

166. F. M. Hashem, S. M. Abdel-Wahab, M. Salah Ahmed, and A. A. Seida, *Bull. Fac. Pharm., Cairo Univ.* **13**, 217 (1974); *Chem. Abs.* **86**, 40,215 (1977).

167. J. A. Goodson, *J. Chem. Soc.,* 245 (1945); *Chem. Abs.,* **39**, 3292 (1945).

168. O. Keller and O. Voelker, *Archives de Chimie* **251**, 207 (1913); *Chem. Zen.* **I,** 2143 (1913).

169. G. R. Waller and R. H. Lawrence, Jr., in *Recent Developments in Mass Spectrometry in Biochemistry and Medicine,* A. Figerio, Ed., Plenum Press, New York, 1978, p. 429.

170. S. W. Pelletier, N. V. Mody, R. S. Sawhney, and J. Bhattacharyya, *Heterocycles* **7**, 327 (1977).

171. M. V. Hunter, *Pharm. J.* **150**, 82–95 (1943); *Quart. J. Pharm.* **17**, 302 (1944); *Chem. Zen.* **I,** 359 (1944); *Chem. Abs.,* **39**, 2621 (1945).

172. G. M. Mamedov, *Aptechn. Deo.* **15**, 34 (1966); *Chem. Abs.* **65**, 5502 (1966).

173. W. B. Cook and O. A. Beath, *J. Am. Chem. Soc.* **74**, 1411 (1952).

174. M. Carmack, J. P. Ferris, J. Harvey, P. L. Magat, E. Martin, and D. W. Mayo, *J. Am. Chem. Soc.* **80**, 497 (1958).

175. A. J. Jones and M. H. Benn, *Tetrahedron Lett.,* 4351 (1972).

176. A. J. Jones and M. H. Benn, *Can. J. Chem.* **51**, 486 (1973).

177. P. W. Codding, K. A. Kerr, M. H. Benn, A. J. Jones, S. W. Pelletier, and N. V. Mody, *Tetrahedron* **21**, 127 (1980).

178. S. W. Pelletier, N. V. Mody, A. J. Jones, and M. H. Benn, *Tetrahedron Lett.,* 3025 (1976).

179. P. W. Codding and K. A. Kerr, *Acta Crystallogr., Sect. B.* **37,** 379 (1981).

180. B. T. Salimov, M. S. Yunusov, and S. Yu. Yunusov, *Khim. Prir. Soedin.,* 106 (1977).

181. M. S. Yunusov, V. E. Nezhevenko, and S. Yu. Yunusov, *Khim. Prir. Soedin.,* 770 (1975).

182. M. S. Yunusov, V. E. Nezhevenko, and S. Yu. Yunusov, *Khim. Prir. Soedin.,* 107 (1975).

183. M. S. Yunusov and S. Yu. Yunusov, *Dokl. Akad. Nauk SSR,* **188,** 1077 (1969).

184. M. S. Yunusov and S. Yu. Yunusov, *Khim. Prir. Soedin.,* 334 (1970).

185. S. Yunusov and N. K. Abubakirov, *J. Allg. Chem.* **19,** 869 (1948); *Chem. Abs.,* 1118 (1950).

186. V. N. Aiyar, M. H. Benn, Y. Y. Huang, J. M. Jacyno, and A. J. Jones, *Phytochemistry* **17,** 1453 (1978).

187. M. H. Benn, M. A. M. Cameron, and O. E. Edwards, *Can. J. Chem.* **41,** 477 (1963).

188. R. H. F. Manske, *Canad. J. Res. B.* **16,** 57 (1938); L. Marion and R. H. F. Manske, *Canad. J. Res. B* **24,** 1 (1946); *Chem. Abs.,* **32,** 5841 (1938); **40,** 4850 (1946).

189. G. M. Mamedov, N. M. Ismailov, and R. M. Abbasov, *Dokl. Akad. Nauk. Azerb. SSR* **20,** 61 (1964); *Chem. Abs.* **62,** 12,975 (1965).

190. M. H. Benn, *Can. J. Chem.* **44,** 1 (1966).

191. A. G. Gonzalez, G. de la Fuente, M. Reina, V. Zabel, and W. H. Watson, *Tetrahedron* **21,** 1155 (1980).

192. S. W. Pelletier and N. V. Mody, Unpublished results.

193. S. W. Pelletier, N. V. Mody, and R. C. Desai, *Heterocycles* **16,** 747 (1981).

194. M. Shamma, P. Chinnasamy, G. A. Miana, A. Khan, M. Bashir, M. Salazar, P. Patil, and J. L. Beal, *J. Nat. Prod.* **42,** 615 (1979).

195. A. S. Narzullaev, Yu. D. Sadykov, and M. Khodzhimatov, *Izv. Akad. Nauk Tadzh. SSR, Otd. Fiz.-Mat. Geol.-Khim Nauk,* 87 (1978).

196. M. S. Rabinovich and R. A. Konovalova, *J. Gen. Chem. USSR* **12,** 321 (1942).

197. L. Marion and O. E. Edwards, *J. Amer. Chem. Soc.* **69,** 2010 (1947).

198. O. Keller, *Ar.* **248,** (1910); *Chem. Abs.* **5,** 145 (1911).

199. L. N. Markwood, *J. Amer. Pharm. Assoc.* **13,** 696 (1924).

200. A. S. Narzullaev, M. S. Yunusov, and S. Yu. Yunusov, *Khim. Prir. Soedin.,* 497 (1973).

201. A. S. Narzullaev, M. S. Yunusov, and S. Yu. Yunusov, *Khim. Prir. Soedin.,* 412 (1974).

202. M. N. Mats, *Rastit. Resur.* **8,** 249 (1972).

203. S. W. Pelletier, L. H. Keith, and P. C. Parthasarathy, *J. Amer. Chem. Soc.* **89,** 4146 (1967).

204. B. T. Salimov, M. S. Yunusov, and S. Yu. Yunusov, *Khim. Prir. Soedin.,* 716 (1977).

205. B. T. Salimov, M. S. Yunusov, and S. Yu. Yunusov, *Khim. Prir. Soedin.,* 128 (1977).

206. B. T. Salimov and M. S. Yunusov, *Khim. Prir. Soedin.,* 530 (1981).

207. B. T. Salimov, M. S. Yunusov, S. Yu. Yunusov, and A. S. Narzullaev, *Khim. Prir. Soedin.,* 665 (1975).

208. A. S. Narzullaev, M. S. Yunusov, and S. Yu. Yunusov, *Khim. Prir. Soedin.,* 498 (1972).

209. A. D. Kuzovkov, P. S. Massagetov, and M. S. Rabinovich, *J. Gen. Chem. USSR* **25,** 157 (1955); *Chem. Zen.,* 8627 (1955).

210. J. A. Goodson, *J. Chem. Soc.,* 139 (1942).

211. M. S. Rabinovich, *J. Gen. Chem. USSR (English transl.)* **24,** 2211 (1954).

212. W. W. Feofilaktov and L. D. Alexejewa, *J. Allg. Chem.* **24,** 738 (1954); *Chem. Zen.,* 1977 (1955).

213. A. D. Kuzovkov, *J. Gen. Chem. USSR* **26,** 2063 (1956); *Chem. Abs.,* 5099 (1957).

214. M. S. Rabinovich, *J. Gen. Chem. USSR* **24,** 2242 (1954); *Chem. Zen.,* 10,030 (1955).

215. L. V. Beshitaishvili, M. H. Sultankhodzhaev, M. S. Yunusov, and K. S. Mudzhiri, *Soobshch. Akad. Nauk Gruz. SSR* **79**, 617 (1975); *Chem. Abs.* **84**, 86,723 (1976).

216. Ya. S. Savchenko, *Farmatsiya* **16**, 30 (1967); *Chem. Abs.* **68**, 57,368 (1968).

217. M. Tanker and S. Ozden, *Ankara Univ. Eczacilik Fak. Mecm.* **5**, 113 (1975); *Chem. Abs.* **85**, 198,082 (1976).

218. S. W. Pelletier, O. D. Dailey, Jr., N. V. Mody, and J. D. Olsen, *J. Org. Chem.* **46**, 3284 (1981).

219. Ya. S. Savchenko, *Aktual. Vopr. Farm.* **2**, 22 (1974); *Chem. Abs.* **84**, 147,613 (1976).

220. M. G. Zhamierashvili, V. A. Tel'nov, M. S. Yunusov, and S. Yu. Yunusov, *Khim. Prir. Soedin.,* 663 (1980).

221. S. Ya. Zolotnitskaya, G. O. Akopyan, and I. S. Melkumyan, *Dokl. Akad. Nauk. Arm. SSR* **43**, 246 (1966); *Chem. Abs.* **67**, 41,095 (1967).

222. J. F. Couch, *J. Amer. Chem. Soc.* **58**, 684 (1936).

223. V. G. Kazlikhin, V. A. Tel'nov, M. S. Yunusov, and S. Yu. Yunusov, *Khim. Prir. Soedin.,* 869 (1977).

224. S. Yu. Yunusov and N. K. Abubakirov, *J. Gen. Chem. USSR* **21**, 967 (1951); *Chem. Zen.* 2843 (1952).

225. A. S. Narzullaev and S. S. Sabirov, *Khim. Prir. Soedin.,* 250 (1981).

226. A. V. Boscharnikowa and E. I. Andrejewa, *J. Gen. Chem. USSR* **28**, 2892 (1958); *Chem. Abs.,* **53**, 9266 (1959).

227. T. F. Platonova and A. D. Kuzovkov, *Med. Prom. SSR* **17**, 19 (1963); *Chem. Abs.* **59**, 6723 (1963).

228. A. G. Gonzalez, G. de la Fuente, and R. Diaz, *Phytochemistry* **21**, 1781 (1982).

229. A. G. Gonzalez, G. de la Fuente, and R. D. Acosta, *Heterocycles,* **22**, 17 (1983).

230. A. G. Gonzalez, G. de la Fuente, and R. Diaz, *Tetrahedron Lett.,* 79 (1979).

231. A. G. Gonzalez, G. de la Fuente, R. Diaz, P. G. Jones and G. M. Sheldrich, *Tetrahedron Lett.,* 959 (1983).

232. M. Przyborowska, *Diss. Pharmacol.* **18**, 89 (1966); *Chem. Abs.* **65**, 7618 (1966).

233. N. K. Abubakirov and S. Yu. Yunusov, *J. Gen Chem. USSR* **26**, 1798 (1956); *Chem. Abs.,* **51**, 1994 (1957).

234. S. Yu. Yunusov and N. K. Abubakirov, *Dokl. Akad. Nauk Uzb. SSR* **8**, 21 (1949); A. D. Kuzovkov and T. F. Platonova, *J. Gen. Chem. USSR* (Engl. Transl.) **29**, 2746 (1959).

235. A. D. Kuzovkov and T. F. Platonova, *J. Gen. Chem. USSR* **29**, 2746 (1959).

236. S. Yu. Yunusov and N. K. Abubakirov, *J. Gen. Chem. USSR* **21**, 174 (1951); *Chem. Zen.,* 2843 (1952).

237. S. Yu. Yunusov and N. K. Abubakirov, *J. Gen. Chem. USSR* **22**, 1461 (1952); *Chem. Zen.,* 4797 (1953).

238. S. W. Pelletier, J. K. Thakkar, N. V. Mody, Z. Djarmati, and J. Bhattacharyya, *Phytochemistry* **16**, 404 (1977).

239. J. L. Lassaigne and H. Feneulle, *Ann. Chim. Phys.* **12**, 358 (1819).

240. R. Brandes, *Schweigger's J. Chem. Phys.* **25**, 369 (1819).

241. S. W. Pelletier and J. Bhattacharyya, *Tetrahedron Lett.,* 4679 (1976).

242. S. W. Pelletier, W. H. De Camp, S. Lajsic, Z. Djarmati, and A. H. Kapadi, *J. Amer. Chem. Soc.* **96**, 7815 (1974).

243. S. W. Pelletier, Z. Djarmati, S. Lajsic, and W. H. De Camp, *J. Amer. Chem. Soc.* **98**, 2617 (1976).

244. L. V. Beshitaishvili, M. N. Sultankhodzhaev, K. S. Mudzhiri, and M. S. Yunusov, *Khim. Prir. Soedin.,* 199 (1981).

245. S. W. Pelletier, J. A. Glinski, B. S. Joshi, and S. Y. Chen, *Heterocycles* **20**, 1347 (1983).

246. S. W. Pelletier and J. Bhattacharyya, *Tetrahedron Lett.,* 2735 (1977).

247. S. W. Pelletier and J. Bhattacharyya, *Phytochemistry* **16,** 1464 (1977).

248. S. W. Pelletier, N. V. Mody, A. P. Venkov, and S. B. Jones, Jr., *Heterocycles* **12,** 779 (1979).

249. O. E. Edwards and M. N. Rodger, *Can. J. Chem.* **37,** 1187 (1959).

250. S. K. Talapatra and A. Chatterjee, *J. Indian Chem. Soc.* **36,** 437 (1959).

251. I. C. Chopra, J. O. Kohli, and K. L. Handa, *Indian J. Med. Res.* **33,** 139 (1945); *Chem. Abs.,* **40,** 6681 (1946).

252. A. Khaleque, S. Papadopoulos, I. Wright, and Z. Valenta, *Chem. and Ind.,* 513 (1959).

253. M. N. Sultankhodzhaev, M. S. Yunosov, and S. Yu Yunosov, *Khim. Pri. Soedin* **1982** (5), 660.

254. S. W. Pelletier and M. Badawi, unpublished work.

255. S. W. Pelletier and J. A. Maddry, unpublished work.

256. D. Yu and B. C. Das, *Planta Medica* **49,** 85 (1983).

5. HIGH-RESOLUTION MASS VALUES AND FORMULA INDICES

5.1. Molecular Formulae and Calculated High-Resolution Mass Values of C_{19}-Diterpenoid Alkaloids

Table 4A.

$C_{21}H_{31}NO_4$: Heterophylline	MW: 361.22531	$C_{22}H_{35}NO_5$: Alkaloid B Dihydromonticamine Episcopalisinine Karakolidine	MW: 393.25152
$C_{21}H_{31}NO_5$: Heterophyllidine	MW: 377.22022		
$C_{21}H_{33}NO_3$: Cardiopetaline	MW: 347.24602	$C_{22}H_{35}NO_6$: Dihydromonticoline Lapaconidine	MW: 409.24644
$C_{21}H_{33}NO_4$: Cardiopetalidine Scopaline	MW: 363.24096	$C_{23}H_{31}NO_7$: 14-Dehydrooxocammaconic acid	MW: 433.21005
$C_{22}H_{31}NO_5$: 14-Dehydrooxoaconine	MW: 389.22021	$C_{23}H_{33}NO_6$: 14-Dehydrooxocammaconine	MW: 419.23078
$C_{22}H_{33}NO_4$: Heterophyllisine	MW: 375.24096	$C_{23}H_{35}NO_5$: Demethoxyisopyrodelphonine Pentagynine	MW: 405.25152
$C_{22}H_{33}NO_5$: Heteratisine Hokbusine B Monticamine Oxoaconosine Pentagydine	MW: 391.23587	$C_{23}H_{35}NO_6$: Gadesine Oxocammaconine	MW: 421.24644
$C_{22}H_{33}NO_6$: Excelsine Monticoline	MW: 407.23079	$C_{23}H_{37}NO_4$: Karasamine Sachaconitine	MW: 391.27226
$C_{22}H_{35}NO_4$: Aconosine Karakoline	MW: 377.25661	$C_{23}H_{37}NO_5$: Cammaconine Dihydropentagynine Isotalatizidine Talatizidine	MW: 407.26717

Table 4A. (*Continued*)

$C_{23}H_{37}NO_6$:   MW: 423.26209
    Dihydrogadesine
    Lappaconine
    Senbusine A
    Senbusine B
    Virescenine

$C_{23}H_{37}NO_7$:   MW: 439.25701
    Delphinifoline
    Ranaconine
    Takaosamine

$C_{24}H_{33}NO_7$:   MW: 447.22571
    6,14-Didehydrodictyocarpinine

$C_{24}H_{35}NO_6$:   MW: 433.24644
    1,14-Didehydroneoline

$C_{24}H_{37}NO_5$:   MW: 419.2671
    14-Dehydrotalatizamine
    Dolaconine
    Episcopalitine

$C_{24}H_{35}NO_7$:   MW: 449.24136
    6-Dehydrodictyocarpinine
    14-Dehydrodictyocarpinine
    Takaonine

$C_{24}H_{37}NO_6$:   MW: 435.26209
    1-Dehydroneoline
    Pyrodelphonine

$C_{24}H_{37}NO_7$:   MW: 451.25701
    14-Dehydrodelcosine
    Dictyocarpinine
    18-Hydroxy-14-*O*-methylgadesine
    18-Methoxygadesine

$C_{24}H_{39}NO_5$:   MW: 421.28282
    Talatizamine

$C_{24}H_{39}NO_6$:   MW: 437.27773
    Delphirine
    Deoxymethylenelycoctonine
    Foresticine
    Neoline
    Umbrosine

$C_{24}H_{39}NO_7$:   MW: 453.27266
    Delcosine
    Delectinine
    Delphonine
    Demethyleneeldelidine
    Gigactonine
    15-α-Hydroxyneoline
    15-β-Hydroxyneoline

$C_{24}H_{39}NO_8$:   MW: 469.26756
    Crassicaulidine
    15-*Epi*-hypaconine
    Hypaconine

$C_{25}H_{37}NO_6$:   MW: 447.26208
    Deacetyltatsiensine

$C_{25}H_{37}NO_7$:   MW: 463.25701
    6-Dehydrodeltamine
    Ilidine

$C_{25}H_{39}NO_5$:   MW: 433.28281
    Pyrochasmanine

$C_{25}H_{39}NO_6$:   MW: 449.27774
    Alkaloid A
    Condelphine
    Delpheline

$C_{25}H_{39}NO_7$:   MW: 465.27266
    14-Acetyldihydrogadesine
    14-Acetylvirescenine
    14-Dehydrobrowniine
    Delcoridine
    Deltamine
    Lycoctonal

$C_{25}H_{41}NO_6$:   MW: 451.29339
    Chasmanine
    1-Deoxydelsoline
    Deoxylycoctonine

$C_{25}H_{41}NO_7$:   MW: 467.28831
    Browniine
    Delbiterine
    Delsoline
    Ezochasmanine
    Lycoctonine

$C_{25}H_{41}NO_8$:   MW: 483.28322
    Delcaroline
    3-Deoxyaconine
    Pseudaconine

$C_{25}H_{41}NO_9$:   MW: 499.27814
    Aconine

$C_{26}H_{37}NO_8$:   MW: 491.25192
    14-Dehydrodictyocarpine

$C_{26}H_{39}NO_7$:   MW: 477.27266
    6-Dehydrodelcorine

$C_{26}H_{39}NO_8$:   MW: 493.26757
    Dictyocarpine

$C_{26}H_{41}NO_6$:   MW: 463.29339
    14-Acetyltalatizamine
    6-Deoxydelcorine

Table 4A. (*Continued*)

$C_{26}H_{41}NO_7$:	MW: 479.28831
Bullatine C	
Delcorine	
Delphidine	
$C_{26}H_{41}NO_8$:	MW: 495.28322
14-Acetyldelcosine	
$C_{26}H_{43}NO_6$:	MW: 465.30904
Homochasmanine	
$C_{26}H_{43}NO_7$:	MW: 481.30396
Deacetylambiguine	
Delphatine	
$C_{27}H_{39}NO_7$:	MW: 489.27266
Tatsiensine	
$C_{27}H_{41}NO_8$:	MW: 507.28322
Deltaline	
$C_{27}H_{43}NO_8$:	MW: 509.29887
14-Acetylbrowniine	
Tricornine	
$C_{27}H_{45}NO_6$:	MW: 479.32469
1,8,14-Tri-*O*-methylneoline	
$C_{27}H_{45}NO_7$:	MW: 495.31961
7,18-Di-*O*-methyllycoctonine	
$C_{28}H_{41}NO_8$:	MW: 519.28322
1-Dehydrodelphisine	
$C_{28}H_{41}NO_8$:	MW: 535.27814
14-Acetyldictyocarpine	
$C_{28}H_{43}NO_8$:	MW: 521.29887
6-Acetyldelcorine	
Delphisine	
1-*Epi*-Delphisine	
$C_{28}H_{45}NO_8$:	MW: 523.31452
Ambiguine	
$C_{29}H_{37}NO_6$:	MW: 495.26209
6-Benzoylheteratisine	
$C_{29}H_{39}NO_6$:	MW: 497.27773
Delavaconitine	
Episcopalisine	
$C_{30}H_{41}NO_5$:	MW: 495.29845
1-Benzoylkarasamine	
$C_{30}H_{41}NO_7$:	MW: 527.2883
Pentagyline	
$C_{30}H_{41}NO_8$:	MW: 543.28321
Gadenine	
$C_{30}H_{42}N_2O_7$:	MW: 542.29921
N-Deacetyllappaconitine	
$C_{30}H_{42}N_2O_8$:	MW: 558.29412
N-Deacetylranaconitine	

$C_{30}H_{42}N_2O_9$:	MW: 574.28903
N-Deacetylfinaconitine	
$C_{30}H_{45}NO_9$:	MW: 563.30944
1-Acetyldelphisine	
1-Acetyl-1-*epi*-delphisine	
Glaucerine	
$C_{30}H_{49}NO_8$:	MW: 551.34582
Glaucedine	
$C_{31}H_{41}NO_7$:	MW: 539.28831
Pyrodelphinine	
$C_{31}H_{41}NO_8$:	MW: 555.28322
8,15-Epoxypyrodelphinine	
$C_{31}H_{41}NO_{11}$:	MW: 603.26796
Flavaconitine	
$C_{31}H_{43}NO_8$:	MW: 557.29887
14-Benzolydelcosine	
$C_{31}H_{43}NO_{10}$:	MW: 589.28869
14-Benzoylmesaconine	
$C_{31}H_{44}N_2O_8$:	MW: 572.30977
Delectine	
$C_{31}H_{47}NO_9$:	MW: 577.32509
Glaucenine	
$C_{32}H_{41}NO_8$:	MW: 567.28322
8-Acetyl-16-demethoxyisopyro-	
delphinine	
$C_{32}H_{43}NO_8$:	MW: 569.29887
8-Acetyl-14-benzolyneoline	
Mithaconitine	
$C_{32}H_{44}N_2O_7$:	MW: 568.31486
Aconorine	
$C_{32}H_{44}N_2O_8$:	MW: 584.30977
Lappaconitine	
$C_{32}H_{44}N_2O_9$:	MW: 600.30469
Puberanine	
Ranaconitine	
$C_{32}H_{44}N_2O_{10}$:	MW: 616.29960
Finaconitine	
$C_{32}H_{45}NO_8$:	MW: 571.31452
14-Benzoylbrowniine	
$C_{32}H_{45}NO_9$:	MW: 587.30944
Ludaconitine	
$C_{32}H_{45}NO_{10}$:	MW: 603.30435
Hokbusine A	
$C_{32}H_{46}N_2O_8$:	MW: 586.32542
Anthranoyllycoctonine	

Table 4A. (*Continued*)

$C_{33}H_{43}NO_9$:   MW: 597.29379
    13-Acetyl-8,15-epoxypyro-
      delphinine
    Glaucephine

$C_{33}H_{45}NO_9$:   MW: 599.30944
    Delphinine
    Isodelphinine

$C_{33}H_{45}NO_{10}$:   MW: 615.30435
    Hypaconitine

$C_{33}H_{45}NO_{11}$:   MW: 631.29927
    Mesaconitine

$C_{33}H_{45}NO_{12}$:   MW: 647.28418
    Beiwutine

$C_{33}H_{46}N_2O_9$:   MW: 614.32034
    14-Acetyldelectine
    N-Acetyldelectine

$C_{33}H_{47}NO_9$:   MW: 601.32508
    Forestine

$C_{34}H_{45}NO_{10}$:   MW: 627.30435
    Anhydroaconitine

$C_{34}H_{47}NO_8$:   MW: 597.33017
    Ezochasmaconitine

$C_{34}H_{47}NO_9$:   MW: 613.32509
    Chasmaconitine
    Penduline

$C_{34}H_{47}NO_{10}$:   MW: 629.32000
    3-Deoxyaconitine
    Falaconitine
    Indaconitine

$C_{34}H_{47}NO_{11}$:   MW: 645.31492
    Aconitine

$C_{34}H_{47}NO_{12}$:   MW: 661.30983
    Aconifine

$C_{34}H_{48}N_2O_9$:   MW: 628.33599
    Ajacine

$C_{34}H_{49}NO_{11}$:   MW: 647.33057
    Veratroylpseudaconine

$C_{35}H_{47}NO_{10}$:   MW: 641.32000
    13-Acetyldelphinine

$C_{35}H_{48}N_2O_{10}$:   MW: 656.33090
    Ajadine

$C_{35}H_{49}NO_9$:   MW: 627.34074
    Anisoezochasmaconitine
    Foresaconitine
    Vilmorrianine C

$C_{35}H_{49}NO_{10}$:   MW: 643.33565
    Crassicauline A
    Vilmorrianine A

$C_{35}H_{49}NO_{11}$:   MW: 659.32957
    3-Deoxyjesaconitine
    Yunaconitine

$C_{35}H_{49}NO_{12}$:   MW: 675.32548
    Jesaconitine

$C_{36}H_{48}N_2O_{10}$:   MW: 668.33090
    Glaudelsine
    Lycaconitine

$C_{36}H_{49}NO_9$:   MW: 639.34074
    Chasmanthinine

$C_{36}H_{49}NO_{12}$:   MW: 687.32548
    3-Acetylaconitine

$C_{36}H_{50}N_2O_{11}$:   MW: 686.34147
    N-(Succinyl)anthranoyllycoctonine

$C_{36}H_{51}N_3O_{10}$:   MW: 685.35745
    Avadharidine

$C_{36}H_{51}NO_{11}$:   MW: 673.34622
    Bikhaconitine

$C_{36}H_{51}NO_{12}$:   MW: 689.34113
    Pseudaconitine

$C_{37}H_{50}N_2O_{10}$:   MW: 682.34655
    Methyllycaconitine

$C_{37}H_{51}NO_{11}$:   MW: 685.34621
    3-Acetylvilmorrianine A

$C_{37}H_{51}NO_{12}$:   MW: 701.34112
    3-Acetylyunaconitine

$C_{37}H_{52}N_2O_{11}$:   MW: 700.35712
    Puberaconitidine
    Septentriodine

$C_{37}H_{53}N_3O_{10}$:   MW: 699.37310
    Delsemine

$C_{38}H_{50}N_2O_{10}$:   MW: 694.34655
    Elatine

$C_{38}H_{54}N_2O_{11}$:   MW: 714.34277
    Septentrionine

$C_{40}H_{55}NO_{14}$:   MW: 773.36226
    Diacetylpseudaconitine

$C_{41}H_{53}NO_{11}$:   MW: 735.36185
    Liwaconitine

$C_{43}H_{52}N_2O_{11}$:   MW: 772.35712
    Ajacusine

$C_{49}H_{73}NO_{10}$:   MW: 835.52344
    8-O-Linoleoyl-14-benzoyl-
      hypaconine

$C_{49}H_{73}NO_{11}$:   MW: 851.51835
    8-O-Linoleoyl-14-benzoyl-
      mesaconine

$C_{50}H_{75}NO_{11}$:   MW: 865.53401
    8-O-Linoleoyl-14-benzoyl-aconine

5.2. Calculated High-Resolution Mass Values and Molecular Formulae of C_{19}-Diterpenoid Alkaloids

Table 4B.

MW: 347.24602: $C_{21}H_{33}NO_3$
Cardiopetaline

MW: 361.22531: $C_{21}H_{31}NO_4$
Heterophylline

MW: 363.24096: $C_{21}H_{33}NO_4$
Cardiopetalidine
Scopaline

MW: 375.24096: $C_{22}H_{33}NO_4$
Heterophyllisine

MW: 377.22022: $C_{21}H_{31}NO_5$
Heterophyllidine

MW: 377.25661: $C_{22}H_{35}NO_4$
Aconosine
Karakoline

MW: 389.22021: $C_{22}H_{31}NO_5$
14-Dehydrooxoaconine

MW: 391.23587: $C_{22}H_{33}NO_5$
Heteratisine
Hokbusine B
Monticamine
Oxoaconosine
Pentagydine

MW: 391.27226: $C_{23}H_{37}NO_4$
Karasamine
Sachaconitine

MW: 393.25152: $C_{22}H_{35}NO_5$
Alkaloid B
Dihydromonticamine
Episcopalisinine
Karakolidine

MW: 405.25152: $C_{23}H_{35}NO_5$
Demethoxyisopyrodelphonine
Pentagynine

MW: 407.23079: $C_{22}H_{33}NO_6$
Excelsine
Monticoline

MW: 407.26717: $C_{23}H_{37}NO_5$
Cammaconine
Dihydropentagynine
Isotalatizidine
Talatizidine

MW: 409.24644 $C_{22}H_{35}NO_6$
Dihydromonticoline
Lapaconidine

MW: 419.23078: $C_{23}H_{33}NO_6$
14-Dehydrooxocammaconine

MW: 419.2671: $C_{24}H_{37}NO_5$
14-Dehydrotalatizamine
Dolaconine
Episcopalitine

MW: 421.24644: $C_{23}H_{35}NO_6$
Gadesine
Oxocammaconine

MW: 421.28282: $C_{24}H_{39}NO_5$
Talatizamine

MW: 423.26209: $C_{23}H_{37}NO_6$
Dihydrogadesine
Lappaconine
Senbusine A
Senbusine B
Virescenine

MW: 433.21005: $C_{23}H_{31}NO_7$
14-Dehydrooxocammaconic acid

MW: 433.24644: $C_{24}H_{35}NO_6$
1,14-Didehydroneoline

MW: 433.28281: $C_{25}H_{39}NO_5$
Pyrochasmanine

MW: 435.26209: $C_{24}H_{37}NO_6$
1-Dehydroneoline
Pyrodelphonine

MW: 437.27773: $C_{24}H_{39}NO_6$
Delphirine
Deoxymethylenelycoctonine
Foresticine
Neoline
Umbrosine

MW: 439.25701: $C_{23}H_{37}NO_7$
Delphinifoline
Ranaconine
Takaosamine

MW: 447.22571: $C_{24}H_{33}NO_7$
6,14-Didehydrodictoycarpinine

MW: 447.26208: $C_{25}H_{37}NO_6$
Deacetyltatsiensine

MW: 449.24136: $C_{25}H_{35}NO_7$
6-Dehydrodictyocarpinine
14-Dehydrodictyocarpinine
Takaonine

MW: 449.27774: $C_{25}H_{39}NO_6$
Alkaloid A
Condelphine
Delpheline

Table 4B.   (*Continued*)

MW: 451.25701:	$C_{24}H_{37}NO_7$
14-Dehydrodelcosine	
Dictyocarpinine	
18-Hydroxy-14-*O*-methylgadesine	
18-Methoxygadesine	
MW: 451.29339:	$C_{25}H_{41}NO_6$
Chasmanine	
1-Deoxydelsoline	
Deoxylycoctonine	
MW: 453.27266:	$C_{24}H_{39}NO_7$
Delcosine	
Delectinine	
Delphonine	
Demethyleneeldelidine	
Gigactonine	
15-α-Hydroxyneoline	
15-β-Hydroxyneoline	
MW: 463.24701:	$C_{25}H_{37}NO_7$
6-Dehydrodeltamine	
Ilidine	
MW: 463.29339:	$C_{26}H_{41}NO_6$
14-Acetyltalatizamine	
6-Deoxydelcorine	
MW: 465.27266:	$C_{25}H_{39}NO_7$
14-Acetyldihydrogadesine	
14-Acetylvirescenine	
14-Dehydrobrowniine	
Delcoridine	
Deltamine	
Lycoctonal	
MW: 465.30904:	$C_{26}H_{43}NO_6$
Homochasmanine	
MW: 467.28831:	$C_{25}H_{41}NO_7$
Browniine	
Delbiterine	
Delsoline	
Ezochasmanine	
Lycoctonine	
MW: 469.26756:	$C_{24}H_{39}NO_8$
Crassicaulidine	
15-*epi*-Hypaconine	
Hypaconine	
MW: 477.27266:	$C_{26}H_{39}NO_7$
6-Dehydrodelcorine	
MW: 477.28831:	$C_{26}H_{41}NO_7$
Bullatine C	
Delcorine	
Delphidine	

MW: 479.32469:	$C_{27}H_{45}NO_6$
1,8,14-Tri-*O*-methylneoline	
MW: 481.30396:	$C_{26}H_{43}NO_7$
Deacetylambiguine	
Delphatine	
MW: 483.28322:	$C_{25}H_{41}NO_8$
Delcaroline	
3-Deoxyaconine	
Pseudaconine	
MW: 489.27266:	$C_{27}H_{39}NO_7$
Tatsiensine	
MW: 491.25192:	$C_{26}H_{37}NO_8$
14-Dehydrodictyocarpine	
MW: 493.26757:	$C_{26}H_{39}NO_8$
Dictyocarpine	
MW: 495.26209:	$C_{29}H_{37}NO_6$
6-Benzoylheteratisine	
MW: 495.28322:	$C_{26}H_{41}NO_8$
14-Acetyldelcosine	
MW: 495.29845:	$C_{30}H_{41}NO_5$
1-Benzoylkarasamine	
MW: 495.31961:	$C_{27}H_{45}NO_7$
7,18-Di-*O*-methyllycoctonine	
MW: 497.27773:	$C_{29}H_{39}NO_6$
Delavaconitine	
Episcopalisine	
MW: 499.27814:	$C_{25}H_{51}NO_9$
Aconine	
MW: 507.28322:	$C_{27}H_{41}NO_8$
Deltaline	
MW: 509.29887:	$C_{27}H_{43}NO_8$
14-Acetylbrowniine	
Tricornine	
MW: 519.28322:	$C_{28}H_{41}NO_8$
1-Dehydrodelphisine	
MW: 521.29887:	$C_{28}H_{43}NO_8$
6-Acetyldelcosine	
Delphisine	
1-*Epi*-delphisine	
MW: 523.31452:	$C_{28}H_{45}NO_8$
Ambiguine	
MW: 527.2883:	$C_{30}H_{41}NO_7$
Pentagyline	
MW: 535.27814:	$C_{28}H_{41}NO_9$
14-Acetyldictyocarpine	
MW: 539.28831:	$C_{21}H_{41}NO_7$
Pyrodelphinine	

Table 4B. (*Continued*)

MW: 542.29921: $C_{30}H_{42}N_2O_7$
N-Deacetyllappaconitine
MW: 543.28321: $C_{30}H_{41}NO_8$
Gadenine
MW: 551.34582: $C_{30}H_{49}NO_8$
Glaucedine
MW: 555.28322: $C_{31}H_{41}NO_8$
8,15-Epoxypyrodelphinine
MW: 557.29887: $C_{31}H_{43}NO_8$
14-Benzoyldelcosine
MW: 558.29412: $C_{30}H_{42}N_2O_8$
N-Deacetylranaconitine
MW: 563.30944: $C_{30}H_{45}NO_9$
1-Acetyldelphisine
1-Acetyl-1-*epi*-delphisine
Glaucerine
MW: 567.28322: $C_{32}H_{41}NO_8$
8-Acetyl-16-demethoxyisopyro-
delphinine
MW: 568.31486: $C_{32}H_{44}N_2O_7$
Aconorine
MW: 569.29887: $C_{32}H_{43}NO_8$
8-Acetyl-14-benzoylneoline
Mithaconitine
MW: 571.31452: $C_{32}H_{45}NO_8$
14-Benzoylbrowniine
MW: 572.30977: $C_{31}H_{44}N_2O_8$
Delectine
MW: 574.28903: $C_{30}H_{42}N_2O_9$
N-Deacetylfinaconitine
MW: 577.32509: $C_{31}H_{47}NO_9$
Glaucenine
MW: 584.30977: $C_{32}H_{44}N_2O_8$
Lappaconitine
MW: 586.32542: $C_{32}H_{46}N_2O_8$
Anthranoyllycoctonine
MW: 587.30944: $C_{32}H_{45}NO_9$
Ludaconitine
MW: 589.28869: $C_{31}H_{43}NO_{10}$
14-Benzoylmesaconine
MW: 597.29379: $C_{33}H_{43}NO_9$
Glaucephine
13-Acetyl-8,15-epoxypyro-
delphinine
MW: 597.33017: $C_{34}H_{47}NO_8$
Ezochasmaconitine

MW: 599.30944: $C_{33}H_{45}NO_9$
Delphinine
Isodelphinine
MW: 600.30469: $C_{32}H_{44}N_2O_9$
Puberanine
Ranaconitine
MW: 601.32508: $C_{33}H_{47}NO_9$
Forestine
MW: 603.26796: $C_{31}H_{41}NO_{11}$
Flavaconitine
MW: 603.30435: $C_{32}H_{45}NO_{10}$
Hokbusine A
MW: 613.32509: $C_{34}H_{47}NO_9$
Chasmaconitine
Penduline
MW: 614.32034: $C_{33}H_{46}N_2O_9$
14-Acetyldelectine
N-Acetyldelectine
MW: 615.30435: $C_{33}H_{45}NO_{10}$
Hypaconitine
MW: 616.29960: $C_{34}H_{44}N_2O_{10}$
Finaconitine
MW: 627.30435: $C_{34}H_{45}NO_{10}$
Anhydroaconitine
MW: 627.34074: $C_{35}H_{49}NO_9$
Anisoezochasmaconitine
Foresaconitine
Vilmorrianine C
MW: 628.33599: $C_{34}H_{48}N_2O_9$
Ajacine
MW: 629.32000: $C_{34}H_{47}NO_{10}$
3-Deoxyaconitine
Falaconitine
Indaconitine
MW: 631.29927: $C_{33}H_{45}NO_{11}$
Mesaconitine
MW: 639.34074: $C_{36}H_{49}NO_9$
Chasmanthinine
MW: 641.32000: $C_{35}H_{47}NO_{10}$
13-Acetyldelphinine
MW: 643.33565: $C_{35}H_{49}NO_{10}$
Crassicauline A
Vilmorrianine A
MW: 645.31492: $C_{34}H_{47}NO_{11}$
Aconitine
MW: 647.28418: $C_{33}H_{45}NO_{12}$
Beiwutine

Table 4B. (*Continued*)

MW: 647.33057:	$C_{34}H_{49}NO_{11}$	MW: 694.34655:	$C_{38}H_{50}N_2O_{10}$
Veratroylpseudaconine		Elatine	
MW: 656.33090:	$C_{35}H_{48}N_2O_{10}$	MW: 699.37310:	$C_{37}H_{53}N_3O_{10}$
Ajadine		Delsemine	
MW: 659.32957:	$C_{35}H_{49}NO_{11}$	MW: 700.35712:	$C_{37}H_{52}N_2O_{11}$
3-Deoxyjesaconitine		Puberaconitidine	
Yunaconitine		Septentriodine	
MW: 661.30983:	$C_{34}H_{47}NO_{12}$	MW: 701.34112:	$C_{37}H_{51}NO_{12}$
Aconifine		3-Acetylyunaconitine	
MW: 668.33090:	$C_{36}H_{48}N_2O_{10}$	MW: 714.34277:	$C_{38}H_{54}N_2O_{11}$
Glaudelsine		Septentrionine	
Lycaconitine		MW: 735.36185:	$C_{41}H_{53}NO_{11}$
MW: 673.34622:	$C_{36}H_{51}NO_{11}$	Liwaconitine	
Bikhaconitine		MW: 772.35712:	$C_{41}H_{52}N_2O_{11}$
MW: 675.32548:	$C_{35}H_{49}NO_{12}$	Ajacusine	
Jesaconitine		MW: 773.36226:	$C_{40}H_{55}N_2O_{11}$
MW: 682.34655:	$C_{37}H_{50}N_2O_{10}$	Diacetylpseudaconitine	
Methyllycaconitine		MW: 835.52344:	$C_{49}H_{73}NO_{10}$
MW: 685.34621:	$C_{37}H_{51}NO_{11}$	8-*O*-Linoleoyl-14-benzoyl-	
3-Acetylvilmorrianine A		hypaconine	
MW: 685.35745:	$C_{36}H_{51}N_3O_{10}$	MW: 851.51835:	$C_{49}H_{73}NO_{11}$
Avadharidine		8-*O*-Linoleoyl-14-benzoyl-	
MW: 686.34147:	$C_{36}H_{50}N_2O_{11}$	mesaconine	
N-(Succinyl)anthranoyllycoctonine		MW: 865.53401:	$C_{50}H_{75}NO_{11}$
MW: 687.32548:	$C_{36}H_{49}NO_{12}$	8-*O*-Linoleoyl-14-benzoyl-	
3-Acetylaconitine		aconine	
MW: 689.34113:	$C_{36}H_{51}NO_{12}$		
Pseudaconitine			

6. INDEX OF NATURALLY OCCURRING C$_{19}$-DITERPENOID ALKALOIDS AND THEIR DERIVATIVES

Table 5.

C$_{19}$-Diterpenoid Alkaloid	^{1}H Spectrum	^{13}C Spectrum
3-Acetylaconitine	x	x
8-Acetyl-14-benzoylneoline	x	x
14-Acetylbrowniine	x	x
6-Acetyldelcorine	x	x
14-Acetyldelcosine	x	x
14-Acetyldelectine	x	—
N-Acetyldelectine	x	—
13-Acetyldelphinine	x	x
1-Acetyldelphisine	x	x

Table 5. (*Continued*)

C$_{19}$-Diterpenoid Alkaloid	^{1}H Spectrum	^{13}C Spectrum
1-Acetyl-1-*epi*-delphisine	x	x
8-Acetyl-16-demethoxyisopyrodelphinine	x	x
14-Acetyldictyocarpine	x	x
14-Acetyldihydrogadesine	x	—
13-Acetyl-8,15-epoxypyrodelphinine	x	x
14-Acetylneoline (see Bullatine C)	x	—
14-Acetyltalatizamine	x	x
3-Acetylvilmorrianine A	x	x
14-Acetylvirescenine	x	x
3-Acetylyunaconitine	x	x
Acomonine (see Delsoline)	x	x
Aconifine	x	x
Aconine	—	x
Aconitine	x	x
Aconorine	x	—
Aconosine	x	x
Ajacine	x	x
Ajacusine	x	x
Ajadine	x	x
Alkaloid A	x	x
Alkaloid B	—	x
Ambiguine	x	x
Anhydroaconitine	x	x
Anisoezochasmaconitine	x	x
Anthranoyllycoctonine	x	x
Avadharidine (Awadcharidine)	x	—
Beiwutine	x	x
14-Benzoylbrowniine	x	x
14-Benzoyldelcosine	x	x
6-Benzoylheteratisine	—	x
14-Benzoyliliensine	x	x
(see 14-Benzoyldelcosine)		
1-Benzoylkarasamine	x	—
14-Benzoylmesaconine	x	—
Bikhaconitine	x	—
Browniine	x	x
Bullatine B (see Neoline)	x	x
Bullatine C	x	—
Cammaconine	x	x
Cardiopetalidine	x	—
Cardiopetaline	x	x
Cashmiradelphine (see Septentriodine)	x	x
Chasmaconitine	x	—
Chasmanine	x	x
Chasmanthinine	x	—

Table 5. (*Continued*)

C₁₉-Diterpenoid Alkaloid	¹H Spectrum	¹³C Spectrum
Condelphine	x	x
Crassicaulidine	x	—
Crassicauline A	x	—
Crassicaulisine (see 15-β-Hydroxyneoline)	x	x
Deacetylambiguine	x	x
N-Deacetylfinaconitine	x	—
N-Deacetyllappaconitine	x	x
N-Deacetylranaconitine	x	—
Deacetyltatsiensine	x	x
14-Dehydrobrowniine	x	x
6-Dehydrodelcorine	x	x
14-Dehydrodelcosine	x	x
1-Dehydrodelphisine	x	x
6-Dehydrodeltamine	x	x
14-Dehydrodictyocarpine	x	x
6-Dehydrodictyocarpinine	x	x
14-Dehydrodictyocarpinine	x	x
6-Dehydroeldelidine (see 6-Dehydrodeltamine)	x	x
14-Dehydroiliensine (see 14-Dehydrodelcosine)	x	x
1-Dehydroneoline	x	x
14-Dehydrooxoaconine	x	x
14-Dehydrooxocammaconic acid	x	x
14-Dehydrooxocammaconine	—	x
14-Dehydrotalatizamine	x	—
Delavaconitine	—	—
Delartine (see Methyllycaconitine)	x	x
Delbiterine	x	—
Delcaroline	x	x
Delcoridine	x	—
Delcorine	x	x
Delcosine	x	x
Delectine	x	—
Delectinine	x	x
Delphamine (see Delcosine)	x	x
Delphatine	x	x
Delphelatine (see Deltaline)	x	x
Delpheline	x	x
Delphidine	x	x
Delphinifoline	x	x
Delphinine	x	x
Delphirine	x	x
Delphisine	x	x

Table 5. (*Continued*)

C₁₉-Diterpenoid Alkaloid	¹H Spectrum	¹³C Spectrum
1-*Epi*-delphisine	x	x
Delphonine	x	x
Delsemidine (see Methyllycaconitine)	x	x
Delsemine	—	x
Delsine (see Lycoctonine)	x	x
Delsoline	x	x
Deltaline	x	x
Deltamine	x	x
Demethoxyisopryodelphonine	x	x
Demethylenedeltamine (see Demethyleneeldelidine)	x	—
Demethyleneeldelidine	x	—
3-Deoxyaconine	—	x
3-Deoxyaconitine	x	x
6-Deoxydelcorine	x	—
1-Deoxydelsoline	x	—
3-Deoxyjesaconitine	x	x
Deoxylycoctonine	—	x
Diacetylpseudaconitine	x	—
Dictyocarpine	x	x
Dictyocarpinine	x	x
6,14-Didehydrodictyocarpinine	x	x
1,14-Didehydroneoline	x	x
Dihydrogadesine	x	—
Dihydromonticamine	x	x
Dihydromonticoline	x	x
Dihydropentagynine	x	—
7,18-Di-*O*-methyllycoctonine	x	x
Dolaconine	—	—
Elatine	—	—
Eldelidine (see Deltamine)	x	x
Eldeline (see Deltaline)	x	x
Episcopalisine	x	x
Episcopalisinine	x	x
Episcopalitine	x	x
8,15-Epoxypyrodelphinine	x	x
Excelsine	x	—
Ezochasmaconitine	x	x
Ezochasmanine	x	x
Falaconitine	x	x
Finaconitine	x	x
Flavaconitine	—	—
Foresaconitine	x	x
Foresticine	x	x

Table 5.   (*Continued*)

C$_{19}$-Diterpenoid Alkaloid	[1]H Spectrum	[13]C Spectrum
Forestine	x	x
Fuziline (see 15-α-Hydroxyneoline)	x	x
Gadenine	x	x
Gadesine	x	x
Gigactonine	x	x
Glaucedine	x	x
Glaucenine	x	x
Glaucephine	x	x
Glaucerine	x	x
Glaudelsine	x	x
Heteratisine	x	x
Heterophyllidine	x	—
Heterophylline	x	—
Heterophyllisine	x	—
Hokbusine A	x	x
Hokbusine B	x	x
Homochasmanine	x	—
10-β-Hydroxyaconitine (see Aconifine)	x	x
18-Hydroxy-14-*O*-methylgadesine	x	—
15-α-Hydroxyneoline	x	x
15-β-Hydroxyneoline	x	x
10-β-Hydroxyranaconitine (see Finaconitine)	x	x
Hypaconine	—	x
15-*Epi*-Hypaconine	x	x
Hypaconitine	x	x
Ilidine	x	—
Iliensine (see Delcosine)	x	x
Indaconitine	x	x
Inuline (see Anthranoyllycoctonine)	x	x
Isoaconitine (see Yunaconitine)	x	x
Isodelphinine	x	x
Isodelphonine	x	x
15-*Epi*-isodelphonine	—	x
Isotalatizidine	x	x
Jesaconitine	x	x
Karacolidine (see Karakolidine)	x	—
Karakolidine	x	—
Karacoline (see Karakoline)	x	x
Karakoline	x	x
Karasamine	x	—
Lapaconidine	x	x
Lappaconine	x	x
Lappaconitine	x	x

Table 5. (*Continued*)

C$_{19}$-Diterpenoid Alkaloid	^{1}H Spectrum	^{13}C Spectrum
8-*O*-Linoleoyl-14-benzoylaconine	—	x
8-*O*-Linoleoyl-14-benzoylhypaconine	—	x
8-*O*-Linoleoyl-14-benzoylmesaconine	—	x
Lipoaconitine	—	x
Lipo-3-deoxyaconitine	—	x
Lipohypaconitine	—	x
Lipomesaconitine	—	x
Liwaconitine	x	—
Lucaconine (see Delcosine)	x	x
Ludaconitine	x	—
Lycaconitine	x	—
Lycoctonal	x	x
Lycoctonine	x	x
Mesaconitine	x	x
18-Methoxygadesine	x	x
Methyllycaconitine	x	x
Mithaconitine	x	x
Monticamine	x	x
Monticoline	x	x
Nagarine (see 15-β-Hydroxyneoline)	x	x
Neoline	x	x
1-*Epi*-Neoline (see Delphirine)	x	x
Neoline-8-acetate (see Delphidine)	x	x
Neoline-8,14-diacetate (see Delphisine)	x	x
Neopelline (see 8-Acetyl-14-benzoylneoline)	x	x
Oxoaconosine	x	x
Oxocammaconine	x	x
Penduline	x	—
Pentagydine	x	—
Pentagyline	x	x
Pentagynine	x	x
Pseudaconine	—	x
Pseudaconitine	x	x
Puberaconitidine	x	—
Puberaconitine (see *N*-(Succinyl)anthranoyllycoctonine)	x	x
Puberanidine (see *N*-Deacetyllappaconitine)	x	x
Puberanine	x	x
Pyrochasmanine	x	—
Pyrodelphinine	x	x
Pyrodelphonine	x	x
Ranaconine	—	x

Table 5.   (*Continued*)

C$_{19}$-Diterpenoid Alkaloid	^{1}H Spectrum	^{13}C Spectrum
Ranaconitine	x	x
Royaline (see Lycoctonine)	x	x
Sachaconitine	x	x
Scopaline	x	x
Senbusine A	x	x
Senbusine B	x	x
Senbusine C (see 15-α-Hydroxyneoline)	x	x
Septentriodine	x	x
Septentrionine	x	x
Shimoburo base II	x	x
(see 14-Dehydrodelcosine)		
N-(Succinyl)anthranoyllycoctonine	x	—
Takao base I (see Delcosine)	x	x
Takaonine	x	x
Takaosamine	x	—
Talatizamine	x	x
Talatizidine	—	—
Tatsiensine	x	x
Toroko base II (see Chasmanine)	x	x
Tricornine	x	x
1,8,14-Tri-O-methylneoline	x	x
Umbrosine	x	—
Veratroylpseudaconine	x	x
Vilmorrianine A	x	x
Vilmorrianine B (see Karakoline)	x	x
Vilmorrianine C (see Foresaconitine)	x	x
Vilmorrianine D (see Sachaconitine)	x	x
Virescenine	x	x
Wang's amd Zhu's "Nagarine"	x	x
(see Aconifine)		
Xuan-Wu 3 (see Neoline)	x	x
Xuan-Wu 4 (see Bullatine C)	x	—
Yunaconitine	x	x

7. CATALOG OF SPECTRAL DATA AND PHYSICAL CONSTANTS FOR NATURALLY OCCURRING C_{19}-DITERPENOID ALKALOIDS AND THEIR DERIVATIVES

3-ACETYLACONITINE

$C_{36}H_{49}NO_{12}$; mp 196-197°C;
$[\alpha]_D$ + 18.6° $(CHCl_3)$
Aconitum flavum, A. pendulum.
^{1}H NMR: δ 1.08 (3H, *t*, J=7Hz, N-CH$_2$-CH$_3$), 1.28 (3H, *s*, C(8)-OCOCH$_3$), 1.94 (3H, *s*, C(3)-OCOCH$_3$), 3.12, 3.16 and 3.66 (12H, each *s*, OCH$_3$), 4.64 (1H, *d*, J=4.5Hz, C(14)-β-H), 4.70 (1H, *dd*, J$_1$=7Hz, J$_2$=11Hz, C(3)-β-H), 7.34-7.97 (5H, *m*, aromatic protons).[1]

[13]C *Chemical Shift Assignments* [2]

C-1	83.8	C-16	90.4	1 (C=O)	166.4
C-2	32.1	C-17	61.1	1	129.4
C-3	71.7	C-18	71.7	2	128.8
C-4	42.6	C-19	49.1	3	128.7
C-5	46.2	N-CH$_2$	47.6	4	133.2
C-6	82.2	CH$_3$	13.3	5	128.7
C-7	45.6	C-1'	56.3	6	128.8
C-8	92.1	C-6'	58.3		
C-9	44.8	C-7'	-		
C-10	40.8	C-8'	-		
C-11	49.9	C-14'	-		
C-12	36.6	C-16'	60.7		
C-13	74.3	C-18'	58.8		
C-14	79.1	C=O	169.2,172.2		
C-15	79.1	CH$_3$	21.3,21.5		

1. X. Chang, H. Wang, L. Lu, Y. Zhu, and R. Zhu, *Acta Pharmaceutica Sinica*, 16, 474 (1981).
2. L.M. Liu, H.C. Wang and Y.L. Zhu, *Acta Pharmaceutica Sinica*, 18, 39 (1983).

8-ACETYL-14-BENZOYLNEOLINE

$C_{32}H_{43}NO_8$; mp 196-197°;
$[\alpha]_D$ + 9.7° (abs EtOH)
Prepared from delphisine
¹H NMR: δ 1.17 (3H, t, NCH$_2$-CH$_3$),
1.46 (3H, s, OCOCH$_3$), 3.24, 3.35 and
3.42 (each 3H, s, OCH$_3$), 5.11 (1H,
dd, C(14)-β-H) and 7.40-8.12 (aromatic protons).[1] *Aconitum napellus.*

¹³C Chemical Shift Assignments[2]

C-1	72.2	C-16	82.8	C=O	166.0
C-2	29.5	C-17	63.0	1	130.3
C-3	30.1	C-18	79.9	2	129.7
C-4	38.9	C-19	56.6	3	128.4
C-5	44.4	N-CH$_2$	48.3	4	133.0
C-6	84.0	CH$_3$	13.0	5	128.4
C-7	48.2	C-1'	–	6	129.7
C-8	85.9	C-6'	57.9		
C-9	43.5	C-7'	–		
C-10	38.2	C-8'	–		
C-11	49.9	C-14'	–		
C-12	29.5	C-16'	56.9		
C-13	43.2	C-18'	59.1		
C-14	75.7	C=O	169.6		
C-15	38.9	CH$_3$	21.6		

1. S.W. Pelletier, J. Bhattacharyya and N.V. Mody, *Heterocycles*, 6, 463 (1977).

2. S.W. Pelletier, N.V. Mody and N. Katsui, *Tetrahedron Letters*, 4027 (1977).

14-ACETYLBROWNIINE

$C_{27}H_{43}NO_8$; mp 123-124°;
$[\alpha]_D$ + 27.8° (CHCl$_3$)
*Consolida ambigua, Delphinium brownii,
D. oreophilum.*
^{1}H NMR: δ 0.99 (3H, t, N-CH$_2$-CH_3),
2.01 (3H, t, OCOCH_3), 3.16, 3.21, 3.25,
3.34 (each 3H, s, OCH_3) and 4.72 (1H,
dd, C(14)-β-H).[1,2,5,6]

^{13}C *Chemical Shift Assignments*[3,4]

C-1	84.2	C-11	49.5	C-1'	55.8
C-2	26.2	C-12	28.2	C-6'	57.3
C-3	32.4	C-13	45.7	C-7'	-
C-4	38.1	C-14	76.0	C-8'	-
C-5	42.6	C-15	33.7	C-14'	-
C-6	90.3	C-16	82.4	C-16'	56.2
C-7	88.3	C-17	64.8	C-18'	59.0
C-8	77.1	C-18	78.0	C=O	171.9
C-9	51.2	C-19	52.7	CH$_3$	21.5
C-10	38.1	N-CH$_2$	48.8		
		CH$_3$	14.2		

1. V.G. Kazlikhin, V.A. Tel'nov, M.S. Yunusov and S.Yu. Yunusov, *Khim. Prirod. Soedin.*, 13, 869 (1977).

2. V.N. Aiyar, M. Benn, Y.Y. Huang, J.M. Jaycyno and A.J. Jones, *Phytochemistry*, 17, 1453 (1978).

3. S.W. Pelletier, N.V. Mody, R.S. Sawhney and J. Bhattacharyya, *Heterocycles*, 7, 327 (1977).

4. S.W. Pelletier, O.D. Dailey, Jr., and N.V. Mody, *J. Org. Chem.*, 46, 3284 (1981).

5. S.W. Pelletier, R.S. Sawhney, H.K. Desai, and N.V. Mody, *J. Nat. Prod.*, 43, 395 (1980).

6. S.W. Pelletier, N.V. Mody, K.I. Varughese, J.A. Maddry, and H.K. Desai, *J. Am. Chem. Soc.*, 1981, 103, 6536.

6-ACETYLDELCORINE

$C_{28}H_{43}NO_8$; Amorphous,
Prepared from delcorine
^{1}H NMR: δ 2.01 (3H, s, OCOCH$_3$).[1]

^{13}C *Chemical Shift Assignments*[2]

C-1	83.4	C-11	50.1	C-1'	55.3
C-2	26.6	C-12	28.0	C-6'	–
C-3	31.7	C-13	38.7	C-7-O	
C-4	38.0	C-14	81.9[a]	C-8-O CH$_2$	93.5
C-5	51.1	C-15	33.9	C-14'	57.7
C-6	78.4	C-16	81.7[a]	C-16'	56.2
C-7	92.0	C-17	64.5	C-18'	59.3
C-8	83.3	C-18	78.0	C=O	170.0
C-9	48.4	C-19	53.3	CH$_3$	21.6
C-10	40.0	N-CH$_2$	50.5		
		CH$_3$	13.9		

1.   A.S. Narzallaev, M.S. Yunusov and S.Yu. Yunusov, *Khim. Prir. Soedin.*, 9, 497 (1973).

2.   S.W. Pelletier, N.V. Mody, and O.D. Dailey, Jr., *Can. J. Chem.*, 58, 1875 (1980).

14-ACETYLDELCOSINE

$C_{26}H_{41}NO_8$; mp 193-195°;
$[\alpha]_D$ + 34° (EtOH)
Consolida ambigua, Delphinium ajacis,
D. belladonna.
^{1}H NMR: δ 1.10 (3H, *t*, NCH$_2$C*H$_3$*), 2.07,
(3H, *s*, OCOC*H$_3$*), 3.29, 3.33, 3.34
(each 3H, *s*, OC*H$_3$*) and 4.79 (1H, *dd*,
C(14)-β-*H*).[1]

[13]C *Chemical Shift Assignments*[2]

C-1	72.6	C-11	49.2	C-1'	-
C-2	27.2	C-12	29.4[a]	C-6'	57.2
C-3	29.9[a]	C-13	42.6[b]	C-7'	-
C-4	37.5	C-14	76.3	C-8'	-
C-5	43.5[b]	C-15	33.8	C-14'	-
C-6	90.2	C-16	82.7	C-16'	56.3
C-7	87.6	C-17	66.1	C-18'	59.1
C-8	78.4	C-18	77.3	C=O	171.4
C-9	44.9	C-19	57.2	CH$_3$	21.4
C-10	38.0	N-CH$_2$	50.3		
		CH$_3$	13.6		

1. S.W. Pelletier, R.S. Sawhney, H.K. Desai, and N.V. Mody, *J. Nat. Prod.*, 43, 395 (1980).
2. S.W. Pelletier, N.V. Mody, R.S. Sawhney and J. Bhattacharyya, *Heterocycles*, 7, 327 (1977).

14-ACETYLDELECTINE

$C_{33}H_{46}N_2O_9$; mp 118-120°;
$[\alpha]_D$ + 42° (CHCl$_3$)
Delphinium dictyocarpum
^{1}H NMR: δ 1.04 (3H, *t*, NCH$_2$-*CH$_3$*),
2.02 (3H, *s*, OCOC*H$_3$*), 3.18, 3.26 and
3.29 (each 3H, *s*, OC*H$_3$*), 4.73 (1H,
dd, J=5Hz, C(14)-β-*H*) and 6.61-7.76
(aromatic protons).[1,2,3]

^{13}C Chemical Shift Assignments

C-1	C-16	C=O
C-2	C-17	
C-3	C-18	
C-4	C-19	
C-5	N-CH$_2$	
C-6	CH$_3$	
C-7	C-1'	
C-8	C-6'	
C-9	C-7'	
C-10	C-8'	
C-11	C-14'	
C-12	C-16'	
C-13	C-18'	
C-14	C=O	
C-15	CH$_3$	

1. B.T. Salimov, M.S. Yunusov and S.Yu. Yunusov, *Khim. Prir. Soedin.*, 716 (1977).

2. B.T. Salimov, N.D. Abdullaev, M.S. Yunusov and S.Yu. Yunusov, *Khim. Prir. Soedin.*, 235 (1978).

3. S.W. Pelletier, N.V. Mody, K.I. Varughese, J.A. Maddry, and H.K. Desai, *J. Am. Chem. Soc.*, 1981, 103, 6536.

N-ACETYLDELECTINE

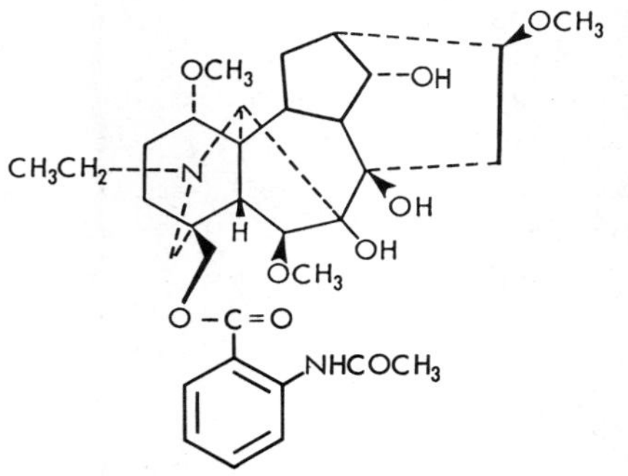

$C_{33}H_{46}N_2O_9$; mp 116-118;
$[\alpha]_D$ + 30° (CHCl$_3$)
Delphinium dictyocarpum
^{1}H NMR: δ 1.03 (3H, *s*, NCH$_2$-CH$_3$),
2.20 (3H, *s*, NHCOCH$_3$), 3.21, 3.31
and 3.33 (each 3H, *s*, OCH$_3$), 7.07-
8.66 (aromatic protons) and 10.93
(1H, *bs*, NHCOCH$_3$).[1,2,3]

[13]*C Chemical Shift Assignments*

C-1	C-16
C-2	C-17
C-3	C-18
C-4	C-19
C-5	N-CH$_2$
C-6	CH$_3$
C-7	C-1'
C-8	C-6'
C-9	C-7'
C-10	C-8'
C-11	C-14'
C-12	C-16'
C-13	C-18'
C-14	C=O
C-15	CH$_3$

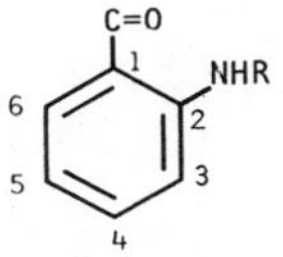

R = COCH$_3$

	1
	2
	3
	4
	5
	6

1.   B.T. Salimov, M.S. Yunusov and S.Yu. Yunusov, *Khim. Prir. Soedin.*, 128
     (1977).

2.   B.T. Salimov, N.D. Abdullaev, M.S. Yunusov and S.Yu. Yunusov, *Khim. Prir.
     Soedin.*, 235 (1978).

3.   S.W. Pelletier, N.V. Mody, K.I. Varughese, J.A. Maddry, and H.K. Desai,
     J. Am. Chem. Soc., 1981, 103, 6536.

13-ACETYLDELPHININE

$C_{35}H_{47}NO_{10}$; Amorphous;
Prepared from delphinine
^{1}H NMR: δ 1.13 (3H, s, C(8)-OCOCH_3),
2.13 (3H, s, C(13)-OCOCH_3), 2.43 (3H,
s, NCH_3), 3.29 and 3.54 (each 3H, s,
OCH_3), 3.42 (6H, s, OCH_3), 5.36 (1H,
d, C(14)-β-H), and 7.76-8.50 (aro-
matic protons).

[13]C Chemical Shift Assignments [1]

C-1	84.7	C-16	80.0	C=O		166.0
C-2	26.3	C-17	63.2		1	130.1
C-3	35.3	C-18	80.1		2	129.8
C-4	39.2	C-19	56.1		3	128.4
C-5	48.5	N-CH3	42.4		4	133.0
C-6	83.1				5	128.4
C-7	48.0	C-1'	56.1		6	129.8
C-8	85.3	C-6'	57.7			
C-9	43.6	C-7'	-			
C-10	41.7	C-8'	-			
C-11	50.3	C-14'	-			
C-12	34.8	C-16'	58.0			
C-13	82.0	C-18'	59.0			
C-14	77.5	C=O	169.4,170.0			
C-15	39.2	CH3	21.2,21.4			

1.   S.W. Pelletier and Z. Djarmati, *J. Am. Chem. Soc.*, 98, 2626 (1976).

1-ACETYLDELPHISINE

$C_{30}H_{45}NO_9$; mp 149-151°;

Prepared from delphisine

^{1}H NMR: δ 1.09 (3H, *t*, NCH$_2$-C*H*$_3$), 1.97 (3H, *s*, OCOC*H*$_3$), 2.02 (6H, *s*, OCOC*H*$_3$), 3.26, 3.28 and 3.31 (each 3H, *s*, OC*H*$_3$), 4.06 (1H, *dd*, C(6)-β-*H*), and 4.73 (1H, *dd*, C(14)-β-*H*).[1]

^{13}C *Chemical Shift Assignments*[2]

C-1	77.5	C-11	49.4	C-1'	–
C-2	27.9	C-12	29.5	C-6'	58.1
C-3	34.6	C-13	44.1	C-7'	–
C-4	39.0	C-14	75.0	C-8'	–
C-5	49.3	C-15	37.7	C-14'	–
C-6	83.5	C-16	83.1	C-16'	56.5
C-7	49.4	C-17	60.7	C-18'	59.1
C-8	85.6	C-18	80.1	C=O	169.3,170.1
C-9	44.3	C-19	54.3	CH$_3$	21.2,22.0
C-10	38.5	N-CH$_2$	48.6	C=O	170.5
		CH$_3$	13.5	CH$_3$	22.4

1. S.W. Pelletier, Z. Djarmati, S. Lajsic and W.H. De Camp, *J. Am. Chem. Soc.*, 98, 2617 (1976).

2. S.W. Pelletier and Z. Djarmati, *J. Am. Chem. Soc.*, 98, 2626 (1976).

1-ACETYL-1-*EPI*-DELPHISINE

$C_{30}H_{45}NO_9$;   Amorphous;
Prepared from delphisine
^{1}H NMR: δ 1.07 (3H, *t*, NCH_2-CH_3),
1.96, 2.00 and 2.03 (each 3H, *s*,
$OCOCH_3$), 3.26 (3H, *s*, OCH_3), 3.30
(6H, *s*, OCH_3), 4.04 (1H, *dd*, C(6)-β-*H*),
4.80 (1H, *dd*, C(14)-β-*H*) and 5.16 (1H,
m, C(1)-α-*H*).[1]

[13]*C Chemical Shift Assignments* [2]

C-1	72.6	C-11	49.3	C-1'	-
C-2	27.2	C-12	29.3	C-6'	57.9
C-3	32.0	C-13	45.4	C-7'	-
C-4	39.1	C-14	75.2	C-8'	-
C-5	39.3	C-15	38.1	C-14'	-
C-6	83.6	C-16	82.8	C-16'	56.5
C-7	47.8	C-17	62.2	C-18'	59.1
C-8	85.5	C-18	79.9	C=O CH_3	169.2,169.7 21.3,22.3
C-9	43.5	C-19	53.6		
C-10	38.4	$N-CH_2$ CH_3	48.8 13.1	C=O CH_3	170.4 22.3

1.   S.W. Pelletier, Z. Djarmati, S. Lajsic and W.H. De Camp, *J. Am. Chem. Soc.*, 98, 2617 (1976).

2.   S.W. Pelletier and Z. Djarmati, *J. Am. Chem. Soc.*, 98, 2626 (1976).

8-ACETYL-16-DEMETHOXYISOPYRODELPHININE

$C_{32}H_{41}NO_8$; Amorphous;
Prepared from delphinine
^{1}H NMR: δ 1.30 (3H, s, $OCOCH_3$), 2.33
(3H, s, NCH_3), 3.23, 3.32 and 3.38
(each 3H, s, OCH_3), 4.18 (1H, s,
C(6)-β-H), 5.12 (1H, d, C(14)-β-H),
6.27 and 6.74 (each 1H, s, CH=CH-)
and 7.60-8.28 (aromatic protons).

[13]C *Chemical Shift Assignments* [1]

C-1	85.1	C-16	125.2	C=O		166.7
C-2	26.2	C-17	64.4		1	130.1
C-3	34.9	C-18	80.5		2	129.7
C-4	39.6	C-19	56.1		3	128.4
C-5	48.8	N-CH₃	42.5		4	133.1
C-6	82.3				5	128.4
C-7	44.6	C-1'	56.2		6	129.7
C-8	83.7	C-6'	57.2			
C-9	44.2	C-7'	-			
C-10	42.2	C-8'	-			
C-11	50.1	C-14'	-			
C-12	39.2	C-16'	-			
C-13	76.0	C-18'	59.1			
C-14	78.2	C=O	169.5			
C-15	137.4	CH₃	21.6			

1. S.W. Pelletier and Z. Djarmati, *J. Am. Chem. Soc.*, 98, 2626 (1976).

14-ACETYLDICTYOCARPINE

$C_{28}H_{41}NO_9$;   mp 64-69.5° (amorphous);
$[\alpha]_D$ - 46.6° $(CHCl_3)$
Prepared from dictyocarpine
^{1}H NMR: δ 0.90 (3H, *s*, C(4)-CH_3),
1.08 (3H, *t*, NCH$_2$-CH_3), 2.08 (6H, *s*,
OCOCH_3), 3.32 and 3.35 (each 3H, *s*,
OCH_3), 4.97 and 5.03 (each 1H, *s*,
OCH_2O), 5.36 (1H, *dd*, J=5.5Hz, C(14)-
β-*H*) and 5.53 (1H, br *s*, C(6)-α-*H*).[1]

[13]C Chemical Shift Assignments [1,2]

C-1	79.0	C-11	55.8	C-1'	55.4
C-2	27.0	C-12	36.6	C-6'	-
C-3	37.3	C-13	38.9	C-7-O CH$_2$	93.8
C-4	33.7	C-14	74.7	C-8-O	
C-5	50.4[a]	C-15	35.0	C-14'	-
C-6	77.3	C-16	81.3	C-16'	56.1
C-7	91.7	C-17	63.9	C-18'	-
C-8	83.3	C-18	25.6	C=O (6)	170.0
C-9	49.9[a]	C-19	56.9	CH$_3$	21.7
C-10	81.3	N-CH$_2$	50.4	(14)	171.7
		CH$_3$	13.9		21.4

1. S.W. Pelletier, O.D. Dailey, Jr., N.V. Mody, and J.D. Olsen, *J. Org. Chem.*, 46, 3284 (1981).
2. S.W. Pelletier, N.V. Mody, and O.D. Dailey, Jr., *Can. J. Chem.*, 58, 1875 (1980).

14-ACETYLDIHYDROGADESINE

$C_{25}H_{39}NO_7$; MW: 465.2733; Resin;
Delphinium pentagynum
^{1}H NMR: δ 2.08 (3H, *s*, COC*H₃*) and
4.82 (1H, *dd*, $J_1=J_2=4.5$Hz, C(14)-β-*H*).[1]

^{13}C *Chemical Shift Assignments*

C-1	C-11	C-1'
C-2	C-12	C-6'
C-3	C-13	C-7'
C-4	C-14	C-8'
C-5	C-15	C-14'
C-6	C-16	C-16'
C-7	C-17	C-18'
C-8	C-18	C=O
C-9	C-19	CH₃
C-10	N-CH₂	
	CH₃	

1.   A.G. Gonzalez, G. de la Fuente, and R. Diaz, *Phytochemistry*, 21, 1781 (1982).

13-ACETYL-8,15-EPOXYPYRODELPHININE

$C_{33}H_{43}NO_9$; mp 125-128°;
Prepared from delphinine
^{1}H NMR: δ 2.09 (3H, *s*, OCOC*H$_3$*), 2.41
(3H, *s*, NC*H$_3$*), 3.42 and 3.80 (each
3H, *s*, OC*H$_3$*), 3.48 (6H, *s*, OC*H$_3$*),
5.30 (1H, *d*, C(14)-β-*H*), and 7.60-
8.22 (aromatic protons).

^{13}C *Chemical Shift Assignments* [1]

C-1	85.8	C-16	79.5	C=O 1	166.3
C-2	25.6	C-17	69.0	1	130.5
C-3	35.1	C-18	80.1	2	130.2
C-4	39.8	C-19	56.1	3	128.1
C-5	50.0	N-CH$_3$	42.5	4	132.6
C-6	82.5			5	128.1
C-7	53.5	C-1'	56.1	6	130.2
C-8	62.8	C-6'	57.9		
C-9	44.8	C-7'	-		
C-10	39.5	C-8'	-		
C-11	51.8	C-14'	-		
C-12	36.7	C-16'	57.5		
C-13	82.4	C-18'	59.1		
C-14	77.4	C=O	170.2		
C-15	45.3	CH$_3$	21.5		

1.   S.W. Pelletier, N.V. Mody and Y. Ohtsuka, unpublished results.

14-ACETYLTALATIZAMINE (14-ACETYLTALATISAMINE)

$C_{26}H_{41}NO_6$; mp 95-97°;
$[\alpha]_D$ + 19.7° (CHCl$_3$)
Aconitum carmichaeli, A. confertiflorum, A. japonicum, A. nemorum, A. saposhnikovii.

^{1}H NMR: δ 1.05 (3H, *t*, NCH$_2$-C*H$_3$*), 2.03 (3H, *s*, OCOC*H$_3$*), 3.21, 3.26 and 3.29 (each 3H, *s*, OC*H$_3$*), and 4.80 (1H, *t*, J= 4.5Hz, C(14)-β-*H*).[1]

[13]C *Chemical Shift Assignments*[1,2]

C-1	85.8d	C-11	48.8s	C-1'	56.1q
C-2	26.2t	C-12	28.5t	C-6'	-
C-3	32.7t	C-13	45.0d[a]	C-7'	-
C-4	38.6s	C-14	77.0d	C-8'	-
C-5	35.4d	C-15	41.0t	C-14'	-
C-6	25.0t	C-16	81.7d	C-16'	56.1q
C-7	45.4d[a]	C-17	62.2d	C-18'	59.5q
C-8	73.7s	C-18	79.7t	C=O	170.7s
C-9	46.3d[a]	C-19	53.1t	CH$_3$	21.4q
C-10	45.0d[a]	N-CH$_2$	49.4t		
		CH$_3$	13.6q		

1. S. Sakai, H. Takayama, and T. Okamoto, *J. Pharm. Soc.* Japan, 99, 647 (1979).
2. C. Konno, M. Shirasaka and H. Hikino, *J. Nat. Prod.*, 45, 128 (1982).

3-ACETYLVILMORRIANINE A

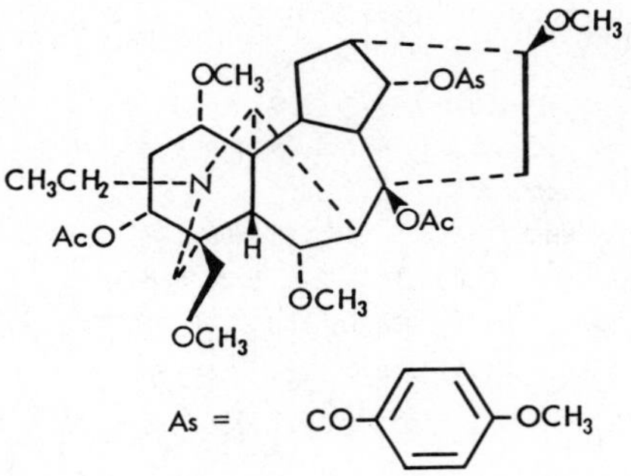

As =

$C_{37}H_{51}NO_{11}$;   mp 189-194°;

Prepared from vilmorrianine A

^{1}H NMR: δ 1.09 (3H, t, J=7Hz, NCH$_2$
CH$_3$), 1.40, 2.06 (each 3H, s, OAc),
3.2 (6H, s, OCH$_3$), 3.25, 3.38 (6H,
s, OCH$_3$), 3.85 (3H, s, aromatic
OCH$_3$), 4.2 (1H, d, J=7Hz, C(6)-β-H),
4.93 (1H, t, J=9Hz, C(3)-β-H), 5.06
(1H, t, J=4.5Hz, C(14)-β-H), 6.97,
8.09 (4H, AB q, J=9Hz, aromatic).[1]

^{13}C Chemical Shift Assignments[1]

C-1	83.8	C-16	77.3	C=O		165.2
C-2	32.1	C-17	61.1	1		123.7
C-3	71.8	C-18	71.8	2		131.8
C-4	42.4	C-19	49.0	3		113.8
C-5	46.5	N-CH$_2$	47.8	4		168.3
C-6	82.0	CH$_3$	13.5	5		113.8
C-7	44.5	C-1'	56.6	6		131.8
C-8	85.8	C-6'	58.2			
C-9	49.0	C-7'	-	OCH$_3$		55.5
C-10	44.1	C-8'	-			
C-11	49.5	C-14'	-			
C-12	28.8	C-16'	54.9			
C-13	37.9	C-18'	58.8			
C-14	75.7	C=O	171.1			
C-15	39.2	CH$_3$	21.3			

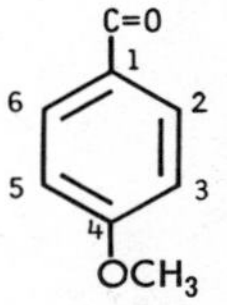

1.   C.-R. Yang, D.-Z. Wang, D.-G. Wu, X.-J. Hao and J. Zhou, *Acta Chimica
     Sinica*, 39, 445 (1981).

14-ACETYLVIRESCENINE

$C_{25}H_{39}NO_7$; mp 157-159°;
$[\alpha]_D$ + 31.8° ($CHCl_3$)
Delphinium virescens
^{1}H NMR: δ 1.10 (3H, t, NCH_2-CH_3),
2.07 (3H, s, $OCOCH_3$), 3.29 and 3.33
(each 3H, s, OCH_3) and 4.88 (1H, dd,
$C(14)-\beta-H$).[1]

[13]*C Chemical Shift Assignments*[1]

C-1	72,4	C-11	50.0	C-1'	–
C-2	29.0	C-12	26.8	C-6'	–
C-3	29.4	C-13	42.9	C-7'	–
C-4	37.7	C-14	77.1	C-8'	–
C-5	41.7	C-15	35.9	C-14'	–
C-6	33.7	C-16	82.1	C-16'	56.3
C-7	85.9	C-17	64.9	C-18'	59.4
C-8	76.9	C-18	78.8	C=O	170.9
C-9	45.9	C-19	56.1	CH3	21.3
C-10	37.7	N-CH2	50.6		
		CH3	13.9		

1. S.W. Pelletier, N.V. Mody, A.P. Venkov and S.B. Jones, Jr., *Heterocycles*, 12, 779 (1979).

3-ACETYLUNACONITINE

As =

$C_{37}H_{51}NO_{12}$; mp 190.8-192.5°
Prepared from yunaconitine
^{1}H NMR: δ 1.11 (3H, *t*, J=7Hz, N-CH$_2$ CH$_3$), 1.34, 2.06 (each 3H, *s*, O Ac), 3.19 (6H, *s*, OCH$_3$), 3.24, 3.52 (each 3H, *s*, OCH$_3$), 3.87 (3H, *s*, aromatic OCH$_3$), 4.09 (1H, *d*, J=7Hz, C(6)-β-*H*), 4.91 (1H, *t*, J=9Hz, C(3)-β-*H*), 4.86 (1H, *d*, J=4.5Hz, C(14)-β-*H*), 6.98, 8.07 (4H, AB *q*, J=9Hz, aromatic).[1]

[13]C Chemical Shift Assignments [1]

C-1	83.5	C-16	83.9	C=O		167.3
C-2	32.0	C-17	61.4		1	123.7
C-3	71.8	C-18	71.7		2	131.8
C-4	42.5	C-19	49.0		3	113.8
C-5	46.6	N-CH$_2$	47.7		4	164.8
C-6	81.9	CH$_3$	13.4		5	113.8
C-7	45.3	C-1'	56.3		6	131.8
C-8	85.5	C-6'	58.9			
C-9	46.6	C-7'	-		OCH$_3$	55.4
C-10	40.6	C-8'	-			
C-11	50.0	C-14'	-			
C-12	35.7	C-16'	58.2			
C-13	74.9	C-18'	58.8			
C-14	78.6	C=O	171.4, 176.7			
C-15	39.4	CH$_3$	21.2, 21.6			

1. C.-R. Yang, D.-Z. Wang, D.-G. Wu, X.-J. Hao and J. Zhou, *Acta Chimica Sinica*, **39**, 445 (1981).

ACONIFINE (Wang's and Zhu's NAGARINE) (10-β-HYDROXYACONITINE)

$C_{34}H_{47}NO_{12}$; mp 195-197°;
$[\alpha]_D$ + 14.8° (CH_3OH); + 30.6°($CHCl_3$)
Aconitum karakolicum, *A. nagarum*
Stapf, var. *lasiandrum* W.T. Wang.
^{1}H NMR: δ 0.96 (3H, *t*, J=6Hz, NCH_2-
CH_3), 1.26 (3H, *s*, $OCOCH_3$), 3.00,
3.14, 3.14 and 3.61 (each 3H, *s*,
OCH_3), 5.24 (1H, *d*, J=5Hz, C(14)-β
-*H*), and 7.39-7.87 (5H, *m*, aromatic
protons).[1,2,3,4]

[13]*C Chemical Shift Assignments*[1]

	$CDCl_3$	C_5D_5N		$CDCl_3$	C_5D_5N			$CDCl_3$	C_5D_5N
C-1	79.8	80.4	C-16	90.1	90.7	C=O		166.1	166.1
C-2	33.5	35.9	C-17	61.2	61.7		1	130.2	130.5
C-3	71.6	68.4	C-18	74.9	73.5		2	129.7	129.7
C-4	43.1	43.9	C-19	47.7	49.0		3	128.6	128.6
C-5	42.8	42.9	$N-CH_2$	47.2	47.4		4	133.2	133.2
C-6	83.6	84.3	CH_3	13.3	13.6		5	128.6	128.6
C-7	44.7	45.5	C-1'	55.4	55.5		6	129.7	129.7
C-8	89.7	90.5	C-6'	58.2	58.1				
C-9	54.0	55.0	C-7'	-	-				
C-10	78.6	78.5	C-8'	-	-				
C-11	55.9	56.1	C-14'	-	-				
C-12	48.9	49.4	C-16'	61.2	61.4				
C-13	77.0	75.6	C-18'	59.1	58.4				
C-14	77.3	77.4	C=O	172.1	172.2				
C-15	78.7	79.8	CH_3	21.5	21.2				

1. M.N. Sultankhodzhaev, L.V. Beshitaishvili, M.S. Yunusov, M.R. Yagudov,
   and S.Yu. Yunusov, *Khim. Prir. Soedin.*, 16, 665 (1980).
2. S.W. Pelletier, N.V. Mody and S.-Y. Chen, *Heterocycles*, 19, 1523 (1982).
3. Wang Fengpeng, *Acta Pharmaceutica Sinica*, 16, 950 (1981).
4. Zhu Yuanlong (Chu Yuan-Lung) and Zhu Renhong (Chu Jen-Hung), *Heterocycles*,
   17, 607 (1982).

ACONINE

$C_{25}H_{41}NO_9$; mp $132°$; $[\alpha]_D$ + $23°$
Hydrolysis product of aconitine

^{13}C Chemical Shift Assignments[1]

	CDCl$_3$	C$_5$D$_5$N		CDCl$_3$	C$_5$D$_5$N		CDCl$_3$	C$_5$D$_5$N
C-1	84.1	84.4	C-11	50.5	50.3	C-1'	55.7	55.6
C-2	35.5	38.3	C-12	37.4	39.1	C-6'	58.0	57.8
C-3	71.9	69.3	C-13	78.8	78.9	C-7'	-	-
C-4	43.2	44.3	C-14	80.6	81.3	C-8'	-	-
C-5	49.0	48.9	C-15	78.5	79.6	C-14'	-	-
C-6	83.0	83.6	C-16	91.8	93.4	C-16'	61.9	61.3
C-7	51.3	48.9	C-17	60.8	61.5	C-18'	59.1	58.8
C-8	76.4	77.2	C-18	77.4	77.1	C=O		
C-9	50.1	50.3	C-19	48.3	47.9	CH$_3$		
C-10	42.4	42.6	N-CH$_2$	46.2	47.1			
			CH$_3$	13.4	13.8			

1.   S.W. Pelletier, N.V. Mody, and R.S. Sawhney, *Can. J. Chem.*, 57, 1652
      (1979).

ACONITINE

$C_{34}H_{47}NO_{11}$; mp 202-205°;
$[\alpha]_D + 19°$ (CHCl$_3$)

*Aconitum callianthum, A. carmichaeli,
A. chinense, A. fauriei, A. flavum,
A. fukatomei, A. grossedentatum, A.
hakusanense, A. ibukiense, A. japoni-
cum, A. karakolicum, A. kusnezoffii,
A. majima, A. mitakense, A. mokchang-
ense, A. nagarum* var. *heterotrichum* f. *dielsianum,
A. napellus* and ssp., *A. pendulum, A. sachalinense,
A. sanyosense, A. senanense, A. soongaricum, A.
stoerckianum, A. subcuneatum, A. tasiromontanum,
A. tianschanicum, A. tortuosum, A. yezoense, A.
zuccarini.*[*]

^{1}H NMR: δ 1.15 (3H, t, NCH$_2$-CH_3), 1.44 (3H, s,
OCOCH_3), 3.29, 3.38, 3.42, and 3.88 (each 3H, s,
OCH_3), 5.04 (1H, d, C(14)-β-H), and 7.67-8.25
(aromatic protons).

^{13}C *Chemical Shift Assignments* [1,2]

C-1	83.4		C-11	49.8		C-1'	55.7
C-2	36.0		C-12	34.0		C-6'	57.9
C-3	70.4		C-13	74.0		C-16'	60.7
C-4	43.2		C-14	78.9		C-18'	58.9
C-5	46.6		C-15	78.9		C=O	172.2
C-6	82.3		C-16	90.1		CH$_3$	21.3
C-7	44.8[a]		C-17	61.0			
C-8	92.0		C-18	75.6		C=O	165.9
C-9	44.2[a]		C-19	48.8		1	129.6
C-10	40.8		N-CH$_2$	46.9		2	128.6
			CH$_3$	13.3		3	129.8
						4	133.2
						5	129.8
						6	128.6

1. S.W. Pelletier and Z. Djarmati, *J. Am. Chem. Soc.*, 98,
    2626 (1976).

2. C.-R. Yang, D.-Z. Wang, D.-G. Wu, X.-J. Hao and J. Zhou, *Acta Chimica
    Sinica*, 39, 445 (1981).

ACONORINE

$C_{32}H_{44}N_2O_7$;   mp 237° (perchlorate);

Aconitum orientale.
^{1}H NMR: δ 1.03 (3H, *t*, NCH$_2$-C*H$_3$*),
2.19 (3H, *s*, NHCOC*H$_3$*), 3.24 and 3.31
(each 3H, *s*, OC*H$_3$*) and 6.9-8.8 (aromatic protons).[1]

^{13}C *Chemical Shift Assignments*

C-1	C-11	C-1'
C-2	C-12	C-6'
C-3	C-13	C-7'
C-4	C-14	C-8'
C-5	C-15	C-14'
C-6	C-16	C-16'
C-7	C-17	C-18'
C-8	C-18	C=O
C-9	C-19	CH$_3$
C-10	N-CH$_2$	
	CH$_3$	

1. V.A. Tel'nov, M.S. Yunusov, S.Yu. Yunusov and B.Sh. Ibragimov, *Khim. Prir. Soedin.*, 11, 814 (1975).

ACONOSINE

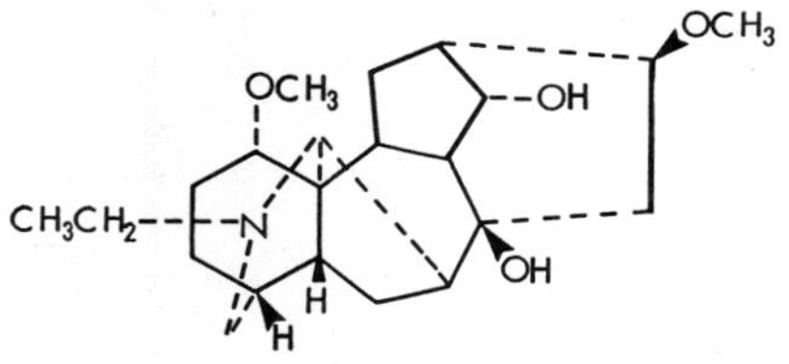

$C_{22}H_{35}NO_4$; mp 148°;
$[\alpha]_D$ - 21° (MeOH)
Prepared from cammaconine
Aconitum arenatum, A. fischeri, A. forestii, A. nasutum, A. stapfianum
Hand Mazz. var. *pubiceps* Wang.
^{1}H NMR: δ 1.05 (3H, *t*, J=7Hz, CH_2-CH_3), 3.26, 3.34 (each 3H, *s*, OCH_3), 3.56 (1H, *s*, exch. with D_2O, OH), 4.08 (1H, *dd*, J=4Hz, C(14)-β-*H*), 4.56 (1H, *d*, J=4Hz, exch. with D_2O, OH).[1,2]

[13]*C Chemical Shift Assignments*[2]

C-1	86.3d	C-11	49.0s	C-1'	55.8q
C-2	26.6t	C-12	28.2t	C-6'	-
C-3	30.3t	C-13	45.7d	C-7'	-
C-4	46.5d[a]	C-14	76.0d	C-8'	-
C-5	37.0d	C-15	40.0t	C-14'	-
C-6	29.8t	C-16	82.7d	C-16'	56.0q
C-7	46.2d[a]	C-17	63.0d	C-18'	-
C-8	73.2s	C-18	-		
C-9	47.8d	C-19	50.8t		
C-10	39.1d	N-CH_2	49.8t		
		CH_3	13.8q		

1. D.A. Muravjeva, T.I. Plekhanova and M.S. Yunusov, *Khim. Prir. Soedin.*, 8, 128 (1972).

2. O.E. Edwards, R.J. Kolt and K.K. Purushothaman, *Can. J. Chem.*, 61, 1194 (1983).

AJACINE

$C_{34}H_{48}N_2O_9$; mp 154° (Hydrate);
$[\alpha]_D$ + 50° (EtOH)

Aconitum umbrosum, Consolida ambigua, Delphinium ajacis, D. belladonna, D. orientale.

^{1}H NMR: δ 1.08 (3H, *t*, NCH$_2$-C*H$_3$*),
2.24 (3H, *s*, NHCOC*H$_3$*), 3.27, 3.35,
3.38 and 3.41 (each 3H, *s*, OC*H$_3$*),
and 7.12-8.76 (aromatic protons).[1,3]

[13]C Chemical Shift Assignments [2]

C-1	83.9	C-16	82.6	C=O	168.1
C-2	26.1	C-17	64.5	1	114.5
C-3	32.2	C-18	69.8	2	141.9
C-4	38.2	C-19	52.5	3	120.6
C-5	43.3	N-CH$_2$	51.0	4	135.0
C-6	91.0	CH$_3$	14.0	5	122.5
C-7	88.6	C-1'	55.8	6	130.3
C-8	77.5	C-6'	57.8		
C-9	50.5	C-7'	-	X=NHCO	169.0
C-10	37.6	C-8'	-	CH$_3$	25.5
C-11	49.1	C-14'	58.1		
C-12	28.6	C-16'	56.3		
C-13	46.1	C-18'	-		
C-14	83.9				
C-15	33.8				

1.   S.W. Pelletier, H.K. Desai, R.S. Sawhney and N.V. Mody, *J. Nat. Prod.*, 43, 395 (1980).

2.   S.W. Pelletier, N.V. Mody, R.S. Sawhney, and J. Bhattacharyya, *Heterocycles*, 7, 327 (1977).

3.   S.W. Pelletier, N.V. Mody, K.I. Varughese, J.A. Maddry, and H.K. Desai, *J. Am. Chem. Soc.*, 1981, 103, 6536.

AJACUSINE

$C_{43}H_{52}N_2O_{11}$; mp 158-161°;
$[\alpha]_D$ + 65.2° (abs. EtOH)
Consolida ambigua (Delphinium ajacis)
^{1}H NMR: δ 1.07 (3H, *t*, NCH$_2$-*CH$_3$*),
1.45 (3H, *d*, CH-*CH$_3$*), 3.27, 3.29 and
3.32 (each 3H, *s*, O*CH$_3$*), 5.03 (1H,
dd, C(14)-β-*H*) and 7.22-8.16 (aromatic protons).[1,3]

[13]C *Chemical Shift Assignments*[2]

C-1	84.0	C-16	82.1	C=O			167.0
C-2	26.0	C-17	64.5			1	129.9
C-3	32.1	C-18	69.4			2	128.3
C-4	37.6	C-19	52.2			3	129.9
C-5	43.1	N-CH$_2$	51.1			4	132.5
C-6	90.6	CH$_3$	14.1			5	129.9
C-7	88.5	C-1'	55.9			6	128.3
C-8	77.3	C-6'	58.1				
C-9	50.0	C-7'	-	C=O			164.1
C-10	37.6	C-8'	-			1'	127.0
C-11	48.9	C-14'	-			2'	133.1
C-12	28.1	C-16'	56.1			3'	129.4
C-13	45.7	C-18'	-			4'	133.7
C-14	75.9					5'	130.9
C-15	34.0					6'	129.9

1. S.W. Pelletier, R.S. Sawhney,
    H.K. Desai, and N.V. Mody, *J.
    Nat. Prod.*, 43, 395 (1980).
2. S.W. Pelletier and R.S. Sawhney,
    Heterocycles, 9, 463 (1978).
3. S.W. Pelletier, N.V. Mody, K.I.
    Varughese, J.A. Maddry, and H.K. Desai,
    J. Am. Chem. Soc., 1981, 103, 6536.

1"	179.8
2"	37.0
3"	35.3
4"	175.8
5"	16.8

AJADINE

$C_{35}H_{48}N_2O_{10}$; mp 134-136° (d);
$[\alpha]_D$ + 43.9° (abs. EtOH)
Consolida ambigua
^{1}H NMR: δ 1.07 (3H, *t*, NCH$_2$-C*H$_3$*),
2.07 (3H, *s*, OCOC*H$_3$*), 2.24 (3H, *s*,
NHCOC*H$_3$*), 3.28, 3.34 and 3.38 (each
3H, *s*, OC*H$_3$*), 4.77 (1H, *dd*, C(14)-β
-*H*), 7.13-7.82 (aromatic protons)
and 11.0 (1H, *bs*, N*H*CO).[1,3]

^{13}*C Chemical Shift Assignments*[2]

C-1	83.7	C-16	82.4	C=O		168.1
C-2	26.0	C-17	64.5	1		114.5
C-3	32.2	C-18	69.6	2		141.8
C-4	37.6	C-19	52.2	3		120.6[a]
C-5	42.6	N-CH$_2$	51.1	4		135.1[b]
C-6	90.7	CH$_3$	14.1	5		122.7[a]
C-7	88.4	C-1'	55.8	6		130.3[b]
C-8	77.4	C-6'	58.1			
C-9	50.1	C-7'	-	X=	NHC=O	169.2
C-10	38.1	C-8'	-		CH$_3$	25.5
C-11	49.0	C-14'	-			
C-12	28.1	C-16'	56.3			
C-13	45.7	C-18'	-			
C-14	75.9	C=O	172.1			
C-15	33.7	CH$_3$	21.5			

1. S.W. Pelletier, R.S. Sawhney, H.K. Desai, and N.V. Mody, *J. Nat. Prod.*, 43, 395 (1980).
2. S.W. Pelletier and R.S. Sawhney, *Heterocycles*, 9, 463 (1978).
3. S.W. Pelletier, N.V. Mody, K.I. Varughese, J.A. Maddry, and H.K. Desai, *J. Am. Chem. Soc.*, 1981, 103, 6536.

ALKALOID A

$C_{25}H_{39}NO_6$; Amorphous; MW: 449.2797
$[\alpha]_D + 10^\circ$ (CHCl$_3$)
Delphinium bicolor
^{1}H NMR: δ 1.04 (3H, *s*, C(4)-*CH$_3$*), 1.09
(3H, *t*, NCH$_2$-*CH$_3$*), 2.10 (3H, *s*, OCO*CH$_3$*),
3.18 and 3.46 (each 3H, *s*, O*CH$_3$*), and
6.90 (2 OH exch. with D$_2$O).[1,2]

[13]C Chemical Shift Assignments[3,4]

C-1	72.6	C-11	48.9	C-1'	–
C-2	29.7	C-12	30.6	C-6'	–
C-3	31.6	C-13	44.4	C-7'	–
C-4	32.8	C-14	75.9	C-8'	52.8
C-5	42.3	C-15	37.8	C-14'	–
C-6	72.3	C-16	83.3	C-16'	56.4
C-7	48.0	C-17	65.5	C-18'	–
C-8	79.9	C-18	27.3	C=O	170.9
C-9	44.4	C-19	61.6	CH$_3$	21.5
C-10	40.0	N-CH$_2$	48.3		
		CH$_3$	12.9		

1. A.J. Jones and M. Benn, *Tetrahedron Letters*, 4351 (1972).
2. A.J. Jones and M. Benn, *Can. J. Chem.*, 51, 486 (1973).
3. S.W. Pelletier, N.V. Mody, A.J. Jones and M.H. Benn, *Tetrahedron Letters*, 3025 (1976).
4. P.W. Codding, K.A. Kerr, M.H. Benn, A.J. Jones, S.W. Pelletier, and N.V. Mody, *Tetrahedron Letters*, 127 (1980).

ALKALOID B

$C_{22}H_{35}NO_5$;   mp 190-191°; MW: 393.2523
$[\alpha]_D$ + 16° (CHCl$_3$)
Delphinium bicolor

^{13}C *Chemical Shift Assignments*[1,2]

C-1	72.9	C-11	48.4	C-1'	–
C-2	29.7	C-12	29.7	C-6'	–
C-3	32.2	C-13	44.4	C-7'	–
C-4	32.8	C-14	76.0	C-8'	–
C-5	46.1	C-15	42.4	C-14'	–
C-6	72.0	C-16	82.4	C-16'	56.3
C-7	54.8	C-17	64.9	C-18'	–
C-8	76.0	C-18	27.4		
C-9	50.2	C-19	61.8		
C-10	40.0	N-CH$_2$	48.4		
		CH$_3$	13.0		

1.   S.W. Pelletier, N.V. Mody, A.J. Jones and M.H. Benn, *Tetrahedron Letters*, 3025 (1976).

2.   P.W. Codding, K.A. Kerr, M.H. Benn, A.J. Jones, S.W. Pelletier, and N.V. Mody, *Tetrahedron Letters*, 127 (1980).

AMBIGUINE

$C_{28}H_{45}NO_8$; mp 106-108°;
$[\alpha]_D$ + 38° (CHCl$_3$)
Consolida ambigua
^{1}H NMR: δ 1.03 (3H, *t*, NCH$_2$-*CH$_3$*),
2.05 (3H, *s*, OCOC*H$_3$*), 3.28, 3.35, 3.38,
3.48 and 3.55 (each 3H, *s*, OC*H$_3$*) and
4.72 (1H, *dd*, C(14)-β-*H*).[1,4]

^{13}C *Chemical Shift Assignments*[2,3]

C-1	83.4	C-11	47.2	C-1'	55.6
C-2	25.3	C-12	27.4	C-6'	59.4
C-3	31.5	C-13	40.9	C-7'	-
C-4	38.3	C-14	75.0	C-8'	53.7
C-5	46.1	C-15	28.5	C-14'	-
C-6	91.1	C-16	81.9	C-16'	56.3
C-7	90.0	C-17	66.7	C-18'	59.4
C-8	80.4	C-18	79.3	C=O	171.3
C-9	52.0	C-19	53.6	CH$_3$	21.4
C-10	35.9	N-CH$_2$	52.6		
		CH$_3$	14.9		

1. S.W. Pelletier, R.S. Sawhney, H.K. Desai, and N.V. Mody, *J. Nat. Prod.*, 43, 395 (1980).
2. S.W. Pelletier, R.S. Sawhney, and N.V. Mody, *Heterocycles*, 9, 1241 (1978).
3. S.W. Pelletier, R.S. Sawhney, and A.J. Aasen, *Heterocycles*, 12, 377 (1979).
4. S.W. Pelletier, N.V. Mody, K.I. Varughese, J.A. Maddry, and H.K. Desai, *J. Am. Chem. Soc.*, 1981, 103, 6536

ANHYDROACONITINE

$C_{34}H_{45}NO_{10}$; Amorphous;
Prepared from aconitine[2]
^{1}H NMR: δ 1.10 (3H, t, NCH$_2$-CH$_3$),
1.33 (3H, s, OCOCH$_3$), 3.03 and 3.60
(each 3H, s, OCH$_3$), 3.20 (6H, s,
OCH$_3$), 4.71 (1H, d, C(14)-β-H), 5.60
(1H, d,C(3)-H), 5.73 (1H, d,C(2)-H)
and 7.10-7.82 (aromatic protons).[2]

^{13}C *Chemical Shift Assignments*[1]

C-1	83.9	C-16	89.9	C=O			165.9
C-2	125.3	C-17	59.2			1	130.0
C-3	137.6	C-18	78.5			2	129.6
C-4	40.9	C-19	52.2			3	128.6
C-5	47.5	N-CH$_2$	48.1			4	133.2
C-6	81.3	CH$_3$	12.6			5	128.6
C-7	42.6[a]	C-1'	56.0			6	129.6
C-8	92.5	C-6'	57.9				
C-9	44.1[a]	C-7'	-				
C-10	41.2	C-8'	-				
C-11	48.7	C-14'	-				
C-12	34.2	C-16'	61.2				
C-13	74.3	C-18'	59.0				
C-14	79.1	C=O	172.2				
C-15	79.1	CH$_3$	21.4				

1.   S.W. Pelletier and Z. Djarmati, *J. Am. Chem. Soc.*, 98, 2626 (1976).
2.   R.E. Gilman and L. Marion, *Can. J. Chem.*, 40, 1713 (1962).

ANISOEZOCHASMACONITINE

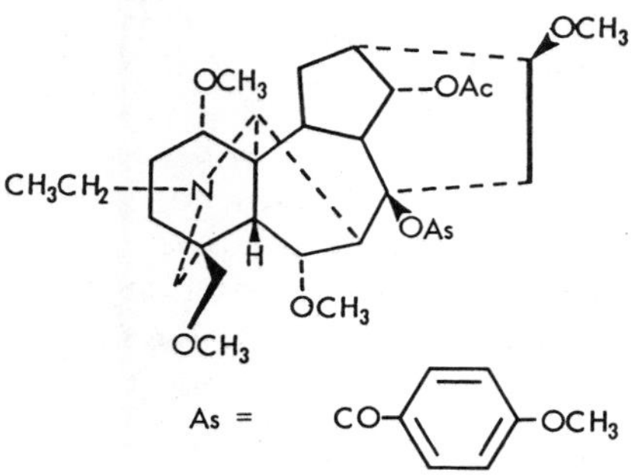

$C_{35}H_{49}NO_9$; mp 136-138.5°;
$[\alpha]_D$ + 14.1° $(CHCl_3)$
Aconitum yezoense

^{1}H NMR: δ 1.16 (3H, *t*, N-CH$_2$CH$_3$),
1.78 (3H, *s*, OCOCH$_3$), 2.96, 3.24,
3.28, 3.32 (each 3H, *s*, aliph. OCH$_3$),
3.84 (3H, *s*, arom. OCH$_3$), 4.80 (1H,
t, J=4.5Hz, C(14)-β-H), 6.86 and 7.93
(each 4H, AB *q*, J=9Hz, aromatic protons).[1,2]

As =

[13]C *Chemical Shift Assignments*[1,2,3]

C-1	84.8	C-16	82.7	C=O	164.4
C-2	26.4	C-17	61.3	1	123.5
C-3	34.8	C-18	80.3	2	131.1
C-4	39.0	C-19	53.7	3	113.3
C-5	49.1	N-CH$_2$	48.9	4	162.9
C-6	83.4	CH$_3$	13.4	5	113.3
C-7	49.2	C-1'	56.5	6	131.1
C-8	85.9	C-6'	57.8		
C-9	44.9	C-7'	-	OCH$_3$	55.3
C-10	39.3	C-8'	-		
C-11	50.2	C-14'	-		
C-12	29.0	C-16'	55.9		
C-13	43.9	C-18'	59.0		
C-14	75.6	C=O	171.1		
C-15	37.6	CH$_3$	21.4		

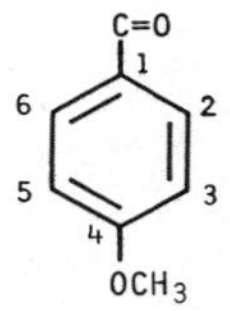

1. H. Takayama, S. Sakai and T. Okamoto, *The Chemistry of Natural Products Symposium Papers*, Nagoya, 1980, No. 52, pp. 396-403.

2. H. Takayama, M. Ito, M. Koga, S. Sakai and T. Okamoto, *Heterocycles*, 15, 403 (1981).

3. H. Takayame, A. Tokita, M. Ito, S. Sakai, F. Kurosakian and T. Okamoto, *J. Pharm. Soc.* Japan, 102, 245 (1982).

ANTHRANOYLLYCOCTONINE (INULINE)

$C_{32}H_{46}N_2O_8$; mp 165°; $[\alpha]_D$ + 51°(EtOH)
Aconitum umbrosum, Consolida ambigua, Delphinium barbeyi, D. biternatum, D. cashmirianum, D. confusum, D. consolida, D. dictyocarpum, D. elisabethœ, D. glaucescens, D. grandiflora, D. tamarae, Inula royleana.
^{1}H NMR: δ 1.07 (3H, *t*, NCH$_2$-C*H$_3$*), 3.26, 3.34, 3.37 and 3.42 (each 3H, *s*, OC*H$_3$*), and 6.72-7.83 (aromatic protons).[1,3]

^{13}C *Chemical Shift Assignments* [2]

C-1	84.0	C-16	82.6
C-2	26.2	C-17	64.6
C-3	32.3	C-18	68.7
C-4	37.6	C-19	52.6
C-5	43.3	N-CH$_2$	51.0
C-6	91.0	CH$_3$	14.1
C-7	88.6	C-1'	55.8
C-8	77.6	C-6'	57.9
C-9	50.4	C-7'	-
C-10	38.3	C-8'	-
C-11	49.1	C-14'	58.0
C-12	28.8	C-16'	56.3
C-13	46.2	C-18'	-
C-14	84.0	C=O	
C-15	33.7	CH$_3$	

C=O	167.9
1	110.4
2	150.9
3	116.9[a]
4	134.4[b]
5	116.4[a]
6	130.8[b]

1.   S.W. Pelletier, H.K. Desai, R.S. Sawhney and N.V. Mody, *J. Nat. Prod.*, 43, 395 (1980).

2.   S.W. Pelletier, N.V. Mody, R.S. Sawhney and J. Bhattacharyya, *Heterocycles*, 7, 327 (1977).

3.   M. Shamma, P. Chinnasamy, G.A. Miana, A. Khan, M. Bashir, M. Salazar, P. Patil, and J.L. Beal, *J. Nat. Prod.*, 42, 615 (1980).

AVADHARIDINE (AWADCHARIDINE)

$C_{36}H_{51}N_3O_{10}$; mp 110-125° (hydrate); $[\alpha]_D$ + 45° (EtOH)

Aconitum orientale, Delphinium cashmirianum

^{1}H NMR: δ 1.05 (3H, *t*, NCH$_2$-C*H$_3$*), 3.22, 3.30, 3.33 and 3.37 (each 3H, *s*, OC*H$_3$*), 7.0-8.57 (aromatic protons) and 10.87 (1H, *bs*, *NH*CO).[1,2,3,4]

[13]C *Chemical Shift Assignments*

C-1	C-11	C-1'
C-2	C-12	C-6'
C-3	C-13	C-7'
C-4	C-14	C-8'
C-5	C-15	C-14'
C-6	C-16	C-16'
C-7	C-17	C-18'
C-8	C-18	C=O | CH$_3$
C-9	C-19	
C-10	N-CH$_2$ | CH$_3$	

1. A.D. Kuzovkov and T.F. Platonova, *J. Gen. Chem. USSR* (English Translation), 29, 2746 (1959).

2. M. Shamma, P. Chinnasamy, G.A. Miana, A. Khan, M. Bashir, M. Salazar, P. Patil, and J.L. Beal, *J. Nat. Prod.*, 42, 615 (1979).

3. S.-H. Jiang, Y.-L. Zhu, Z.-Y. Zhao and R.-H. Zhu, *Acta Pharmaceutica Sinica*, 18, 440 (1983).

4. S.W. Pelletier, N.V. Mody, K.I. Varughese, J.A. Maddry, and H.K. Desai, *J. Am. Chem. Soc.*, 1981, 103, 6536.

BEIWUTINE

$C_{33}H_{45}NO_{12}$; mp 196-198°;
$[\alpha]_D$ + 26.9° (CHCl$_3$)
Aconitum kusnezoffii
^{1}H NMR: δ 1.34 (3H, *s*, OCOC*H₃*), 2.38
(3H, *s*, N-C*H₃*), 3.06, 3.24, 3.24 and
3.64 (each 3H, *s*, OC*H₃*), 5.28 (1H,
dd, C(14)-β-*H*), and 7.22-8.00 (aromatic protons).[1]

[13]C *Chemical Shift Assignments*[1]

C-1	83.3	C-16	89.6	C=O 1	166.3
C-2	33.7	C-17	62.8		128.8
C-3	71.3	C-18	75.7		129.7
C-4	43.2	C-19	49.7		129.9
C-5	46.8	N-CH₃	42.5		
C-6	79.8		-		
C-7	43.4	C-1'	55.8		
C-8	89.8	C-6'	58.2		
C-9	53.9	C-7'	-		
C-10	74.8	C-8'	-		
C-11	56.0	C-14'	-		
C-12	42.6	C-16'	61.2		
C-13	76.5	C-18'	59.1		
C-14	78.4	C=O	172.4		
C-15	77.1	CH₃	21.4		

1. Y.-G. Wang, Y.-L. Zhu and R.-H. Zhu, *Acta Pharmaceutica Sinica*, 15, 526 (1980).

Note: Though the C(1)-methoxyl group in the original paper is shown as
in the β-configuration, it is most likely α. The 10-hydroxyl is
shown as α-, but is most certainly in the β-configuration.

14-BENZOYLBROWNIINE

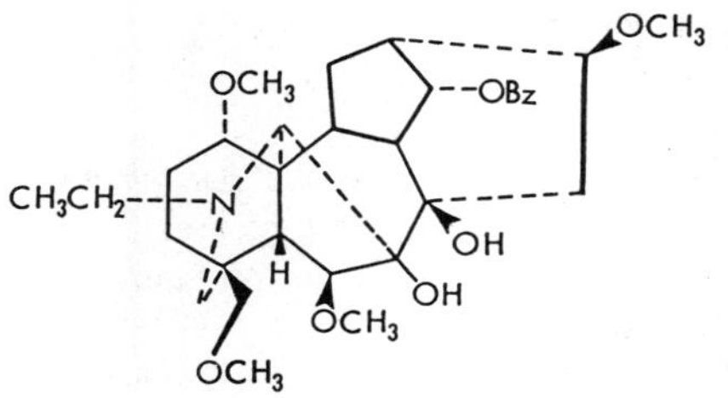

$C_{32}H_{45}NO_8$; mp 114-116°;
$[\alpha]_D$ + 53°

Delphinium biternatum

^{1}H NMR: δ 1.01 (3H, *t*, NCH_2-CH_3),
3.22 (9H, *s*, OCH_3), 3.32 (3H, *s*,
OCH_3), 5.00 (1H, *dd*, J=5Hz, C(14)-β
-*H*) and 7.42-8.10 (aromatic protons).[1,3]

[13]C *Chemical Shift Assignments* [2]

C-1	84.2	C-16	82.2	C=O	167.0
C-2	25.9	C-17	64.8		132.5
C-3	32.0	C-18	77.9		129.9
C-4	38.1	C-19	52.9		128.3
C-5	43.1	$N-CH_2$	51.3		
C-6	90.2	CH_3	14.0		
C-7	88.3	C-1'	55.9		
C-8	77.5	C-6'	57.4		
C-9	51.3	C-7'	-		
C-10	37.6	C-8'	-		
C-11	49.2	C-14'	-		
C-12	28.3	C-16'	56.1		
C-13	45.5	C-18'	59.1		
C-14	76.0				
C-15	34.0				

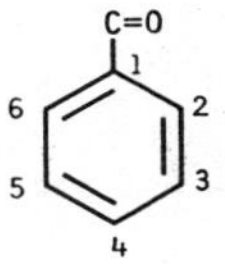

1. B.T. Salimov, M.S. Yunusov and S.Yu. Yunusov, *Khim. Prir. Soedin.*, 14, 106 (1978).
2. S.W. Pelletier and R.S. Sawhney, *Heterocycles*, 9, 463 (1978).
3. S.W. Pelletier, N.V. Mody, K.I. Varughese, J.A. Maddry, and H.K. Desai, *J. Am. Chem. Soc.*, 1981, 103, 6536.

14-BENZOYLDELCOSINE (14-BENZOYLILIENSINE)

$C_{31}H_{43}NO_8$;   mp 148-150°;
$[\alpha]_D$ + 63.8° (CHCl$_3$)
Delphinium biternatum
^{1}H NMR: δ 1.08 (3H, *t*, NCH$_2$-*CH$_3$*),
3.27 (9H, *s*, O*CH$_3$*), 5.01 (1H, *dd*,
C(14)-β-*H*), and 7.43-8.06 (aromatic
protons).[1,2]

[13]C *Chemical Shift Assignments*

C-1	72.6	C-16	82.6				166.5
C-2	27.3	C-17	66.2			1	130.7
C-3	29.8	C-18	77.1			2	129.8
C-4	37.4	C-19	57.2			3	128.3
C-5	43.6	N-CH$_2$	50.3			4	132.6
C-6	90.1	CH$_3$	13.6			5	128.3
C-7	87.7	C-1'	-			6	129.8
C-8	78.4	C-6'	57.2				
C-9	44.9	C-7'	-				
C-10	37.9	C-8'	-				
C-11	49.2	C-14'	-				
C-12	29.4	C-16'	56.1				
C-13	43.1	C-18'	59.0				
C-14	76.5						
C-15	34.2						

1.   B.T. Salimov, M.S. Yunusov and S.Yu. Yunusov, *Khim. Prir. Soedin.*, 106
     (1978).

2.   S.W. Pelletier and N.V. Mody, *Tetrahedron Letters*, 22, 207 (1981).

6-BENZOYLHETERATISINE

$C_{29}H_{37}NO_6$; mp 212-215°;
$[\alpha]_D$ + 73° (EtOH)
Aconitum heterophyllum [1,2]

[13]C *Chemical Shift Assignments* [3]

C-1	82.3	C-15	28.8	C=O		166.6
C-2	26.7	C-16	29.4		1	130.2
C-3	36.4	C-17	62.6		2	129.6
C-4	34.9	C-18	26.0		3	128.5
C-5	49.9	C-19	54.9		4	132.8
C-6	74.5	N-CH₂	48.9		5	128.5
C-7	48.9	CH₃	13.4		6	129.6
C-8	75.2	C-1'	55.8			
C-9	57.5	C-6'	-			
C-10	42.8	C-7'	-			
C-11	49.9	C-8'	-			
C-12	35.3	C-14'	-			
C-13	75.8	C-16'	-			
C-14	173.5	C-18'	-			

1. W.A. Jacobs and L.C. Craig, *J. Biol. Chem.*, 147, 571 (1943).
2. R. Aneja, D.M. Locke, and S.W. Pelletier, *Tetrahedron*, 29, 3297 (1973).
3. N.V. Mody and S.W. Pelletier, unpublished results.

1-BENZOYLKARASAMINE

$C_{30}H_{41}NO_5$; mp 206-208°
Aconitum karakolicum
^{1}H NMR: δ 0.74 (3H, s, C(4)-CH_3),
1.19 (3H, t, NCH$_2$-CH_3), 3.21, 3.24
(each 3H, s, OCH_3), 7.38-7.95 (5H,
m, aromatic protons).[1]

^{13}C *Chemical Shift Assignments*

C-1	C-11	C-1'
C-2	C-12	C-6'
C-3	C-13	C-7'
C-4	C-14	C-8'
C-5	C-15	C-14'
C-6	C-16	C-16'
C-7	C-17	C-18'
C-8	C-18	C=O
C-9	C-19	CH$_3$
C-10	N-CH$_2$	
	CH$_3$	

1.   M.N. Sultankhodzhaev, M.S. Yunusov and S.Yu. Yunusov, *Khim. Prir.*
Soedin., 660 (1982).

14-BENZOYLMESACONINE

$C_{31}H_{43}NO_{10}$;
$[\alpha]_D$ + 11.85° $(CHCl_3)$
Aconitum carmichaeli, A. chinense

^1H NMR (250MHz): δ 2.36 (3H, *s*, N-CH_3), 3.27, 3.29, 3.29, 3.67 (each 3H, *s*, OCH_3), 4.06 (1H, *d*, J=6Hz, C(6)-β-*H*), 4.52 (1H, *d*, J=6Hz, C(15)-β-*H*), 4.98 (1H, *d*, J=5Hz, C(14)-β-*H*), 7.36-8.02 (5H, *m*, aromatic protons).[1]

^{13}C *Chemical Shift Assignments*

C-1	C-16		1
C-2	C-17		2
C-3	C-18		3
C-4	C-19		4
C-5	N-CH_2		5
C-6	CH_3		6
C-7	C-1'		
C-8	C-6'		
C-9	C-7'		
C-10	C-8'		
C-11	C-14'		
C-12	C-16'		
C-13	C-18'		
C-14	C=0		
C-15	CH_3		

1.   D.-H. Chen, H.-Y. Li and W.-L. Sung, *Chinese Traditional and Herbal Drugs*, 13, 1 (1982).

BIKHACONITINE

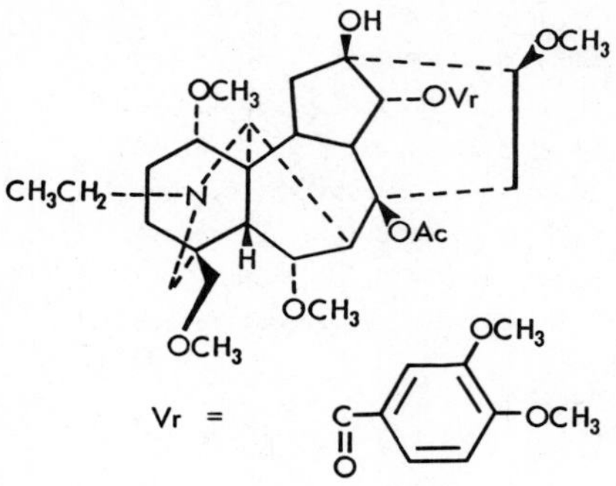

$C_{36}H_{51}NO_{11}$; MW: 673.3456
mp 105-110° (hydrate);

$[\alpha]_D$ + 16° (EtOH)

Aconitum ferox, A. spicatum, A. violaceum.

^{1}H NMR: δ 1.15 (3H, *t*, NCH$_2$-C*H$_3$*), 1.28 (3H, *s*, OCOC*H$_3$*), 3.13, 3.25, 3.27, and 3.52 (each 3H, *s*, aliphatic OC*H$_3$*), 3.88 and 3.92 (each 3H, *s*, aromatic OC*H$_3$*), 4.85 (1H, *d*, C(14)-β -*H*) and 6.80-7.78 (aromatic protons).[1,2,3]

Vr =

[13]C Chemical Shift Assignments

C-1	C-16
C-2	C-17
C-3	C-18
C-4	C-19
C-5	N-CH$_2$
C-6	CH$_3$
C-7	C-1'
C-8	C-6'
C-9	C-7'
C-10	C-8'
C-11	C-14'
C-12	C-16'
C-13	C-18'
C-14	C≟O
C-15	CH$_3$

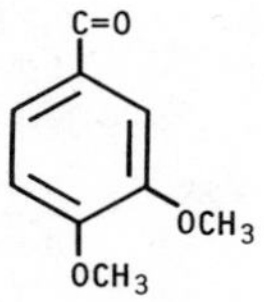

1. A. Klásek, V. Šimanek, and F. Šantavy, *Lloydia*, 35, 55 (1972).

2. K.K. Purushothaman and S. Chandrasekharan, *Phytochemistry*, 13, 1975 (1974).

3. Y. Tsuda and L. Marion, *Can. J. Chem.*, 41, 3055 (1963).

BROWNIINE

$C_{25}H_{41}NO_7$; Amorphous; $[\alpha]_D$ + 38.7 (EtOH). Perchlorate: mp 211.5-213.0°; $[\alpha]_D$ + 25° (H_2O); 29.8° (EtOH). *Consolida ambigua, Delphinium biternatum, D. brownii, D. carolinianum, D. glaucescens, D. iliense, D. rotundifolium, D. tatsienense, D. virescens.* [*]

^{1}H NMR: δ 1.04 (3H, t, NCH_2CH_3), 3.24, 3.30, 3.36 and 3.40 (each 3H, s, OCH_3).[1,3-5]

^{13}C *Chemical Shift Assignments*[2]

C-1	85.2	C-11	48.2	C-1'	56.0
C-2	25.5	C-12	27.5	C-6'	57.5
C-3	32.5	C-13	46.1	C-7'	-
C-4	38.4	C-14	75.3	C-8'	-
C-5	45.1	C-15	33.1	C-14'	-
C-6	90.1	C-16	81.7	C-16'	56.5
C-7	89.1	C-17	65.4	C-18'	59.1
C-8	76.3	C-18	78.0	C=O CH₃	
C-9	49.6	C-19	52.7		
C-10	36.4	N-CH₂ CH₃	51.3		
			14.3		

1. S.W. Pelletier, R.S. Sawhney, H.K. Desai, and N.V. Mody, *J. Nat. Prod.*, 43, 395 (1980).

2. S.W. Pelletier, N.V. Mody, R.S. Sawhney, and J. Bhattacharyya, *Heterocycles*, 7, 327 (1977).

3. M.H. Benn, M.A.M. Cameron and O.E. Edwards, *Can. J. Chem.*, 41, 477 (1963).

4. B.T. Salimov, M.S. Yunusov and S.Yu. Yunusov, *Khim. Prir. Soedin.*, 14, 106 (1978).

5. S.W. Pelletier, N.V. Mody, K.I. Varughese, J.A. Maddry, and H.K. Desai, *J. Am. Chem. Soc.*, 1981, 103, 6536.

BULLATINE C (XUAN-WU 4, 14-ACETYLNEOLINE)

$C_{26}H_{41}NO_7$;   mp 198-202°;
$[\alpha]_D$ + 42.9° ($CHCl_3$)
Aconitum jinyangense, A. yesoense.[*]
^{1}H NMR: δ 1.10 (3H, *t*, NCH_2-CH_3),
1.94 (3H, *s*, $OCOCH_3$), 3.18, 3.18, and
3.24 (each 3H, *s*, OCH_3), and 4.69
(1H, *dd*, C(14)-β-*H*).[1,2]

[13]*C Chemical Shift Assignments*

C-1	C-11	C-1'
C-2	C-12	C-6'
C-3	C-13	C-7'
C-4	C-14	C-8'
C-5	C-15	C-14'
C-6	C-16	C-16'
C-7	C-17	C-18'
C-8	C-18	C=O
C-9	C-19	CH_3
C-10	N-CH_2	
	CH_3	

1.   H.-C. Wang, D.-Z. Zhu, Z.-Y. Zhao, and R.-H. Zhu, *Acta Chimica Sinica,* 38, 475 (1980).

2.   H. Takayama, A. Tokita, M. Ito, S. Sakai, F. Kurosaki and T. Okamoto, *J. Pharm. Soc.* Japan, 102, 245 (1982).

CAMMACONINE

$C_{23}H_{37}NO_5$; mp 135-137°; $[\alpha]_D \pm 0°$
Aconitum forestii, A. variegatum.
^1H NMR: δ 1.03 (3H, t, NCH_2CH_3),
3.26 and 3.32 (each 3H, s, OCH_3) and
4.11 (1H, t, C(14)-β-H).[3]

^{13}C *Chemical Shift Assignments*[1,2]

		C_6D_6			C_6D_6			C_6D_6
C-1	86.3	86.1d	C-11	48.8	49.0s	C-1'	56.5[a]	55.7q
C-2	25.8	26.3t	C-12	27.7	28.2t	C-6'	-	-
C-3	33.2	32.7t	C-13	45.6	45.9d	C-7'	-	-
C-4	39.1	39.4s	C-14	75.6	75.9d	C-8'	-	-
C-5	46.0	46.3d[a]	C-15	38.3	39.1t	C-14'	-	-
C-6	24.6	25.4t	C-16	82.3	82.6d	C-16'	56.3[a]	56.1q
C-7	45.9	46.2d[a]	C-17	63.0	62.8d	C-18'	-	-
C-8	73.7	73.2s	C-18	68.8	68.7t	C=O		
C-9	47.0	47.5d	C-19	53.1	53.7t	CH_3		
C-10	37.6	38.5d	N-CH_2	49.5	49.6t			
			CH_3	13.7	13.8q			

1. N.V. Mody, S.W. Pelletier and N.M. Mollov, *Heterocycles*, 14, 1751 (1980).

2. O.E. Edwards, R.J. Kolt and K.K. Purushothaman, *Can. J. Chem.*, 61, 1194 (1983).

3. M.A. Khaimova, M.D. Palamareva, N.M. Mollov and V.P. Krestev, *Tetrahedron*, 27, 819 (1971).

CARDIOPETALIDINE

$C_{21}H_{33}NO_4$;   mp 223-227°;
$[\alpha]_D$ + 1.1° (EtOH)
Delphinium cardiopetalum
^{1}H NMR: δ 0.92 (3H, s, C(4)-CH_3),
1.10 (3H, t, NCH_2-CH_3), 3.08 (1H, s,
C(17)-H), 3.68 (1H, m, C(1)-β-H), and
4.18 (1H, dd, C(14)-β-H).[1]

[13]C *Chemical Shift Assignments*[2]

C-1	72.7	C-11	50.2	C-1'	
C-2	29.6	C-12	32.0	C-6'	
C-3	31.9	C-13	34.8	C-7'	
C-4	33.9	C-14	75.8	C-8'	
C-5	47.5	C-15	26.7	C-14'	
C-6	34.0	C-16	24.9	C-16'	
C-7	87.2	C-17	64.0	C-18'	
C-8	78.4	C-18	27.4	C=O	
C-9	48.0	C-19	59.4	CH₃	
C-10	43.8	N-CH₂	50.5		
		CH₃	13.5		

1.   A. Gonzalez, G. de la Fuente, M. Reina, V. Zabel, and W.H. Watson, *Tetrahedron Letters*, 21, 1155 (1980).
2.   Personal communication from Dr. G. de la Fuente, Nov. 8, 1983.

CARDIOPETALINE

$C_{21}H_{33}NO_3$; mp 179-181°;
$[\alpha]_D$ - 16° (EtOH)
Delphinium cardiopetalum
^{1}H NMR: δ 0.88 (3H, *s*, C(4)-CH*₃*),
1.12 (3H, *t*, NCH$_2$-CH*₃*), 3.07 (1H, *s*,
C(17)-*H*), 3.75 (1H, *m*, C(1)-β-*H*), and
4.12 (1H, *dd*, C(14)-β-*H*).[1]

[13]C *Chemical Shift Assignments*[2]

C-1	72.3	C-11	49.0	C-1'	
C-2	29.7	C-12	32.8	C-6'	
C-3	31.3	C-13	34.9	C-7'	
C-4	32.9	C-14	76.3	C-8'	
C-5	46.3	C-15	32.6	C-14'	
C-6	25.6	C-16	25.1	C-16'	
C-7	46.7	C-17	63.0	C-18'	
C-8	77.2	C-18	27.6	C=O	
C-9	46.9	C-19	60.4	CH$_3$	
C-10	44.1	N-CH$_2$	48.3		
		CH$_3$	13.0		

1. A.G. Gonzalez, G. de la Fuente, M. Reina, V. Zabel, and W.H. Watson, *Tetrahedron Letters*, 21, 1155 (1980).
2. Personal communication from Dr. G. de la Fuente, Nov. 8, 1983.

CHASMACONITINE

$C_{34}H_{47}NO_9$; MW: 613.3240
mp 181-182° (Hexane); 165-167°
(Ether); $[\alpha]_D$ + 10.3° (EtOH)
Aconitum chasmanthum, A. ferox, A.
forestii, A. franchetii.
^{1}H NMR: δ 1.08 (3H, *t*, NCH_2CH_3),
1.27 (3H, *s*, $OCOCH_3$), 3.16, 3.27,
3.28 and 3.53 (each 3H, *s*, OCH_3),
and 7.35-8.20 (aromatic protons).[1,2,3]

^{13}C *Chemical Shift Assignments*

C-1	C-16	
C-2	C-17	
C-3	C-18	
C-4	C-19	
C-5	N-CH$_2$	
C-6	CH$_3$	
C-7	C-1'	
C-8	C-6'	
C-9	C-7'	
C-10	C-8'	
C-11	C-14'	
C-12	C-16'	
C-13	C-18'	
C-14	C≐O	
C-15	CH$_3$	

1
2
3
4
5
6

1.   O. Achmatowicz and L. Marion, *Can. J. Chem.*, 42, 154 (1964).
2.   A. Klasek, V. Simanek and F. Santavy, *Lloydia*, 35, 55 (1972).
3.   K.B. Birnbaum, K. Wiesner, E.W.K. Jay and L. Jay, *Tetrahedron Letters*, 867 (1971).

CHASMANINE (TOROKO BASE II)

$C_{25}H_{41}NO_6$; mp 90-91°;
$[\alpha]_D$ + 23.6° (EtOH)

*Aconitum chasmanthum, A. crassicaule,
A. franchetii, A. mitakense, A. sacha-
liense* var. *compactum, A. subcuneatum,
A. yezoense, A. forestii.*

^{1}H NMR: δ 1.10 (3H, *t*, NCH_2-CH_3),
3.37 and 3.46 (each 3H, *s*, OCH_3) and
3.42 (6H, *s*, OCH_3).[2]

[13]C *Chemical Shift Assignments* [1]

C-1	86.1	C-11	50.4	C-1'	56.3
C-2	26.0	C-12	28.6	C-6'	57.2
C-3	35.2	C-13	45.7	C-7'	-
C-4	39.5	C-14	75.5	C-8'	-
C-5	48.8	C-15	39.2	C-14'	-
C-6	82.5[a]	C-16	82.2[a]	C-16'	55.9
C-7	52.8	C-17	62.4	C-18'	59.2
C-8	72.6	C-18	80.8	C=O	
C-9	50.3	C-19	54.0	CH_3	
C-10	38.4	N-CH_2	49.3		
		CH_3	13.6		

1. S.W. Pelletier and Z. Djarmati, *J. Am. Chem. Soc.*, 98, 2626 (1976).
2. S.W. Pelletier, W.H. De Camp, and Z. Djarmati, *J. Chem. Soc. Chem.
    Commun.*, 253 (1976).

CHASMANTHININE

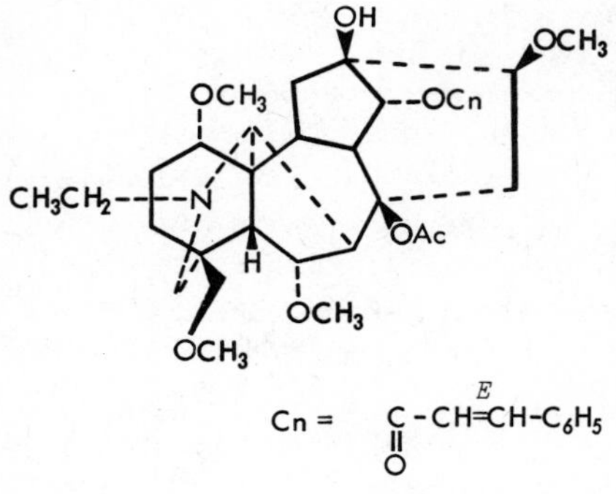

$C_{36}H_{49}NO_9$; mp 160-161°;
[α]$_D$ + 9.6° (EtOH)
Aconitum chasmanthum
^{1}H NMR: δ 1.08 (3H, *t*, NCH$_2$-*CH$_3$*),
1.77 (3H, *s*, OCOC*H$_3$*), 3.19, 3.24,
3.27, and 3.50 (each 3H, *s*, OC*H$_3$*),
4.80 (1H, *d*, J=4.5Hz, C(14)-β-*H*),
6.42 and 7.72 (2H, AB *q*, J=16Hz,
double bond) and 7.43 (5H, aromatic
protons).[1]

Cn =   C-CH=CH-C$_6$H$_5$
       ‖   *E*
       O

^{13}C Chemical Shift Assignments

C-1	C-11	C-1'
C-2	C-12	C-6'
C-3	C-13	C-7'
C-4	C-14	C-8'
C-5	C-15	C-14'
C-6	C-16	C-16'
C-7	C-17	C-18'
C-8	C-18	C=O
C-9	C-19	CH$_3$
C-10	N-CH$_2$	
	CH$_3$	

1.   O. Achmatowicz, Jr. and L. Marion, *Can. J. Chem.*, 42, 154 (1964).

CONDELPHINE

$C_{25}H_{39}NO_6$; mp 158-159°;
$[\alpha]_D$ + 21.3° (CHCl$_3$)
Aconitum anthoroideum, Delphinium
confusum, D. denudatum, A. japonicum.
^{1}H NMR: δ 1.10 (3H, *t*, NCH$_2$-C*H$_3$*),
2.02 (3H, *s*, OCOC*H$_3$*), 3.22 and 3.27
(each 3H, *s*, OC*H$_3$*) and 4.80 (1H, *dd*,
J=4.5Hz, C(14)-β-*H*).[1]

^{13}C *Chemical Shift Assignments*[2]

C-1	72.1	C-11	49.0	C-1'	–
C-2	29.1	C-12	26.7	C-6'	–
C-3	29.7	C-13	43.5	C-7'	–
C-4	37.3	C-14	76.9	C-8'	–
C-5	41.4	C-15	42.4	C-14'	–
C-6	25.1	C-16	82.2	C-16'	55.9
C-7	45.8	C-17	63.5	C-18'	59.3
C-8	74.5	C-18	79.0	C=O	170.3
C-9	44.6	C-19	56.6	CH$_3$	21.2
C-10	37.0	N-CH$_2$	48.4		
		CH$_3$	13.0		

1. S.W. Pelletier, L.H. Keith and P.C. Parthasarathy, *J. Am. Chem. Soc.*, 89, 4146 (1967).
2. S.W. Pelletier and Z. Djarmati, *J. Am. Chem. Soc.*, 98, 2626 (1976).

CRASSICAULIDINE

$C_{24}H_{39}NO_8$; MW: 469.2641; mp 206-209°;
Aconitum crassicaule

¹H NMR (200 MHz): δ 1.14 (3H, *t*, J=
7Hz, NCH₂C*H₃*), 3.34, 3.40, 3.52 (each
3H, *s*, OC*H₃*), 3.74 (1H, *m*, C(1)-β-*H*),
3.86 (1H, *d*, J=8Hz, C(15)-β-*H*), 4.09
(1H, *t*, J=4.5Hz, C(14)-β-*H*), 4.28 (1H,
dd, J₁=1Hz,J₂=6Hz, C(6)-β-*H*), 4.01
(1H, *dd*, C(3)-β-*H*), 3.14 (2H, AB type,
J=12Hz, C(19)-*H*), 4.01 (1H, br s,
exch. D₂0, OH).[1,2]

¹³C Chemical Shift Assignments

C-1	C-11	C-1'
C-2	C-12	C-6'
C-3	C-13	C-7'
C-4	C-14	C-8'
C-5	C-15	C-14'
C-6	C-16	C-16'
C-7	C-17	C-18'
C-8	C-18	C=0
C-9	C-19	CH₃
C-10	N-CH₂	
	CH₃	

1.   F.-P. Wang and Q.-C. Fang, *Planta Medica*, 47, 39 (1983).

2.   F.-P. Wang and Q.-C. Fang, *Acta Pharmaceutica Sinica*, 17, 300 (1982).

CRASSICAULINE A

$C_{35}H_{49}NO_{10}$; mp 162.5-164.5°;
$[\alpha]_D$ + 31.5° (CHCl$_3$)
Aconitum crassicaule; A. forrestii
^{1}H NMR: δ 1.09 (3H, *t*, J=7Hz, N-CH$_2$-
CH$_3$), 1.33 (3H, *s*, OAc), 1.5-1.75
(2H, *m*, C(3)-*H*), 3.13, 3.25, 3.26,
3.5 (each 3H, *s*, OCH$_3$), 3.85 (3H, *s*,
aromatic OCH$_3$), 3.75 (1H, *s*, OH), 3.9
(1H, *d*, J=8Hz, C(6)-*H*), 4.83 (1H, *d*,
J=4.5Hz, C(14)-α-*H*), 6.89, 7.98 (each
2H, A$_2$B$_2$ aromatic protons).[1]

^{13}C *Chemical Shift Assignments** *

C-1	84.8	C-16	83.9	C=O 1	166.3
C-2	25.9	C-17	61.8	2	123.1
C-3	35.4	C-18	80.4	3	131.9
C-4	39.1	C-19	53.8	4	114.1
C-5	48.9	N-CH$_2$	49.1	5	163.8
C-6	83.3	CH$_3$	13.1	6	59.7
C-7	50.4	C-1'	55.6		
C-8	85.3	C-6'	56.2		
C-9	40.9	C-7'	—		
C-10	44.9	C-8'	—		
C-11	49.8	C-14'	—		
C-12	34.5	C-16'	59.3		
C-13	75.1	C-18'	57.8		
C-14	78.9	C=O	169.8		
C-15	38.9	CH$_3$	21.3		

1. F.—P. Wang and Q.—C. Fang, *Planta Medica*, 42, 375 (1981).

*Private communication from Feng—peng Wang, Beijing, China, June 20, 1984

DEACETYLAMBIGUINE

$C_{26}H_{43}NO_7$; Amorphous;
$[\alpha]_D$ + 36.6° (EtOH)
Delphinium tatsienense
^{1}H NMR: δ 1.09 (3H, *t*, J=7Hz, NCH$_2$-
C*H*$_3$), 3.31, 3.41, 3.48, 3.53 and
3.61 (each 3H, *s*, OC*H*$_3$), 3.86 (1H,
m, C(14)-β-*H*), 4.20 (1H, *d*, J=1.5Hz,
C(6)-α-*H*).[1]

^{13}C *Chemical Shift Assignments*[1,2]

C-1	83.9	C-11	46.4	C-1'	55.7
C-2	24.9	C-12	27.6	C-6'	59.2
C-3	31.7	C-13	46.7[a]	C-7'	-
C-4	38.6	C-14	74.9	C-8'	54.0
C-5	42.7[a]	C-15	26.8	C-14'	-
C-6	91.0	C-16	82.3	C-16'	56.5
C-7	89.9	C-17	68.3	C-18'	59.5
C-8	79.6	C-18	79.1	C=O	
C-9	52.6	C-19	53.9	CH$_3$	
C-10	36.6	N-CH$_2$	52.9		
		CH$_3$	15.3		

1.   S.W. Pelletier, J.A. Glinski, B.S. Joshi and S.-Y. Chen, *Heterocycles*, 20, 1347 (1983).
2.   S.W. Pelletier, R.S. Sawhney and N.V. Mody, *Heterocycles*, 9, 1241 (1978).

N-DEACETYLFINACONITINE

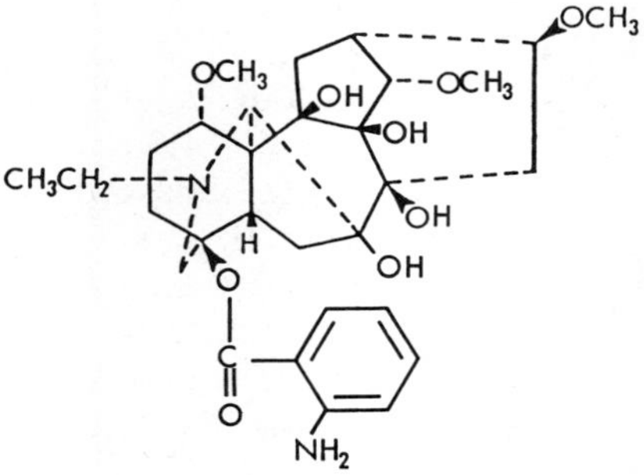

$C_{30}H_{42}N_2O_9$; mp 121-123°;
$[\alpha]_D$ + 34.9°

Aconitum finetianum

^{1}H NMR: δ 1.04 (3H, *t*, J=7Hz, NCH$_2$-
CH$_3$), 3.22, 3.24, 3.36 (each 3H, *s*,
OCH$_3$), 3.74 (1H, *d*, J=4.5Hz, C(14)-β
-*H*), 5.64 (2H, *s*, NH$_2$), 6.5-7.64
(4H, *m*, aromatic protons).[1]

^{13}C Chemical Shift Assignments

C-1	C-16	
C-2	C-17	
C-3	C-18	
C-4	C-19	
C-5	N-CH$_2$	
C-6	CH$_3$	
C-7	C-1'	
C-8	C-6'	
C-9	C-7'	
C-10	C-8'	
C-11	C-14'	
C-12	C-16'	
C-13	C-18'	
C-14	C=O	
C-15	CH$_3$	

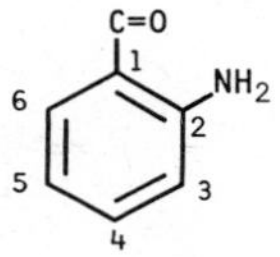

1.   S.-H. Jiang, Y.-L. Zhu, Z.-Y. Zhao and R.-H. Zhu, *Acta Pharmaceutica Sinica*, 18, 440 (1983).

N-DEACETYLLAPPACONITINE (PUBERANIDINE)

$C_{30}H_{42}N_2O_7$;   MW: 542.2997; mp 120-121°;
$[\alpha]_D$ + 42° (EtOH)[1]; + 23.2° (CHCl$_3$)[3]
Aconitum finetianum, A. ranunculae-
folium, A. septentrionale, Delphinium
cashmirianum. *

^{1}H NMR: (400 MHz) δ 1.12 (3H, t, N-CH$_2$-CH$_3$), 3.0
(1H, s, C(17)-H), 3.3, 3.32, 3.42 (each 3H, s,
OCH_3), 3.2 (1H, dd, J=9,6Hz, C(16)-α-H), 3.46
(1H, d, J=5Hz, C(14)-β-H), 2.55, 3.62 (each 1H,
d, J=11Hz, C(19)-H), 6.62, 6.65 (each 1H, t, J=
7Hz, Ar-H$_5$,-H$_4$), 7.25, 7.78 (each 1H, d, J=7Hz,
Ar-H$_4$,-H$_6$).[1-3]

^{13}C *Chemical Shift Assignments*[3]

C-1	83.2	C-16	83.1				167.7
C-2	26.3	C-17	61.7			1	112.2
C-3	32.1	C-18	–			2	150.7
C-4	84.5	C-19	55.8			3	116.8
C-5	48.9	N-CH$_2$	50.0			4	134.0
C-6	26.9	CH$_3$	13.6			5	116.4
C-7	47.7	C-1'	56.5			6	131.8
C-8	75.8	C-6'	–				
C-9	78.7	C-7'	–				
C-10	36.5	C-8'	–				
C-11	51.0	C-14'	58.0				
C-12	24.1	C-16'	56.2				
C-13	49.1	C-18'	–				
C-14	90.4	C=O	–				
C-15	44.9	CH$_3$	–				

1.   M. Shamma, P. Chinnasamy, G.A. Miana, A. Khan, M. Bashir, M. Salazar,
     P. Patil, and J.L. Beal, *J. Nat. Prod.*, 42, 615 (1979).

2.   N. Mollov, H. Haimova, P. Tscherneva, N. Pecigarova, I. Ognianov, and
     P. Panov, *Compt. Rend. Acad. Bulgare Sci.*, 17, 251 (1964).

3.   D. Yu and B.C. Das, *Planta Medica*, 49, 85 (1983).

N-DEACETYLRANACONITINE

$C_{30}H_{42}N_2O_8$; mp 125-127°;
$[\alpha]_D$ + 43.7°

Aconitum barbatum var. *puberulum*, *A. finetianum*.

^{1}H NMR: δ 1.08 (3H, *t*, J=7Hz, NCH$_2$-
CH$_3$), 3.22, 3.26, 3.36 (each 3H, *s*,
OCH$_3$), 5.60 (2H, *s*, NH$_2$), 6.5-7.66
(4H, *m*, aromatic protons).[1]

[13]*C Chemical Shift Assignments*

C-1	C-16	C=0	1
C-2	C-17		2
C-3	C-18		3
C-4	C-19		4
C-5	N-CH$_2$		5
C-6	CH$_3$		6
C-7	C-1'		
C-8	C-6'		
C-9	C-7'		
C-10	C-8'		
C-11	C-14'		
C-12	C-16'		
C-13	C-18'		
C-14	C=0		
C-15	CH$_3$		

1.   S.-H. Jiang, Y.-L. Zhu, Z.-Y. Zhao and R.-H. Zhu, *Acta Pharmaceutica Sinica*, 18, 440 (1983).

DEACETYLTATSIENSINE

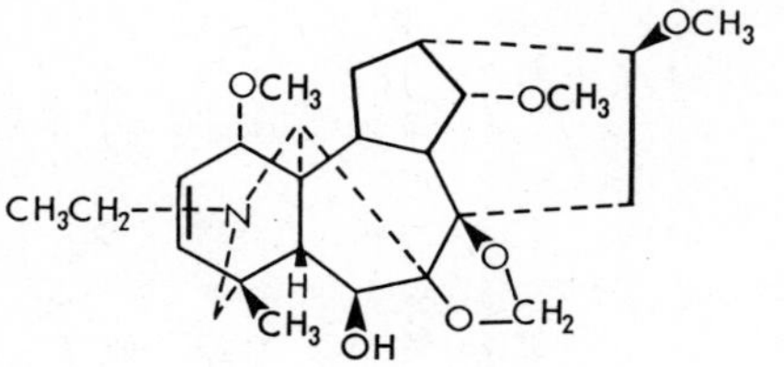

$C_{25}H_{37}NO_6$; mp 236-238°;

$[\alpha]_D$ + 28.8° (EtOH);

Prepared from tatsiensine

^{1}H NMR: δ 1.04 (3H, *s*, C(18)C*H$_3$*),
1.09 (3H, *t*, J=7Hz, N-CH$_2$C*H$_3$*), 3.10
(1H, *d*, J=2Hz, C(1)-β-*H*), 3.32, 3.41,
3.47 (each 3H, *s*, OC*H$_3$*), 3.69 (2H, *m*,
C(14)-*H*, C(16)-*H*), 4.25 (1H, *s*, C(6)-
α-*H*), 5.10, 5.13 (each 2H, *s*, O-C*H$_2$*O),
5.68 (1H, *d*, J$_{2,3}$=10Hz, C(3)-*H*), 5.94
(1H, *dd*, J$_{2,3}$=10Hz, J$_{1,2}$=4Hz, C(2)-*H*).[1]

^{13}C Chemical Shift Assignments [1]

C-1	83.6	C-11	50.8	C-1'	55.9
C-2	124.6	C-12	27.9	C-6'	-
C-3	137.8	C-13	38.4	C-7-O, C-8-O CH$_2$	93.3
C-4	34.6	C-14	80.7		
C-5	55.9	C-15	34.0	C-14'	57.7
C-6	79.5	C-16	82.3	C-16'	56.2
C-7	92.3	C-17	61.1	C-18'	-
C-8	84.9	C-18	23.1		
C-9	47.4	C-19	58.3		
C-10	40.2	N-CH$_2$ CH$_3$	48.9 12.9		

1. S.W. Pelletier, J.A. Glinski, B.S. Joshi and S.-Y. Chen, *Heterocycles*, 20, 1347 (1983).

14-DEHYDROBROWNIINE

$C_{25}H_{39}NO_7$; mp 161-163°, 172-174°.
$[\alpha]_D$ + 19° (EtOH), + 31.3° (CHCl$_3$).
Delphinium biternatum, D. cardinale,
D. glaucescens.
^{1}H NMR: δ 1.10 (3H, t, NCH$_2$-CH$_3$),
3.30 (6H, s, OCH$_3$), 3.35 and 3.42
(each 3H, s, OCH$_3$).[1,3,4]

^{13}C *Chemical Shift Assignments*[2]

C-1	85.5	C-11	49.0	C-1'	56.1
C-2	25.5	C-12	25.3	C-6'	57.6
C-3	32.5	C-13	49.5	C-7'	-
C-4	38.5	C-14	216.3	C-8'	-
C-5	46.1	C-15	33.1	C-14'	-
C-6	89.8	C-16	85.5	C-16'	56.3
C-7	88.9	C-17	65.9	C-18'	59.2
C-8	85.5	C-18	77.9	C=O	
C-9	53.8	C-19	52.7	CH$_3$	
C-10	43.9	N-CH$_2$	51.4		
		CH$_3$	14.3		

1. M.H. Benn, *Can. J. Chem.*, 44, 1 (1966).
2. S.W. Pelletier, N.V. Mody and R.S. Sawhney, *Can. J. Chem.*, 57, 1652 (1979).
3. B.T. Salimov, M.S. Yunusov and S.Yu. Yunusov, *Khim. Prir. Soedin.*, 106 (1978).
4. S.W. Pelletier, N.V. Mody, K.I. Varughese, J.A. Maddry, and H.K. Desai, *J. Am. Chem. Soc.*, 1981, 103, 6536.

6-DEHYDRODELCORINE

$C_{26}H_{39}NO_7$; mp 141-143°;
$[\alpha]_D$ - 64° (MeOH)
Delphinium iliense
^{1}H NMR: δ 1.03 (3H, *t*, NCH$_2$-C*H$_3$),
3.23, 3.26, 3.30 and 3.35 (each 3H, *s*,
OC*H$_3$*), and 5.03 and 5.46 (each 1H, J=
1.5Hz, *d*, OC*H$_2$*O).[1,3]

[13]C *Chemical Shift Assignments*[2]

C-1	82.7	C-11	46.1	C-1'	55.9
C-2	26.5	C-12	27.7	C-6'	-
C-3	32.2	C-13	38.7	C-7-O$\diagdown$CH$_2$	95.3
C-4	41.8	C-14	82.4	C-8-O$\diagup$	
C-5	56.5	C-15	32.9	C-14'	58.1
C-6	216.7	C-16	82.3	C-16'	56.5
C-7	90.4	C-17	63.0	C-18'	59.2
C-8	81.5	C-18	76.8		
C-9	47.8	C-19	53.4		
C-10	38.6	N-CH$_2$	50.2		
		CH$_3$	13.7		

1.   M.G. Zhamierashvili, V.A. Tel'nov, M.S. Yunusov, and S.Yu. Yunusov,
     Khim. Prir. Soedin., 836 (1977).

2.   S.W. Pelletier, N.V. Mody and O.D. Dailey, Jr., *Can. J. Chem.*, 58,
     1875 (1980).

3.   S.W. Pelletier, N.V. Mody, K.I. Varughese, J.A. Maddry, and H.K. Desai,
     J. Am. Chem. Soc., 1981, 103, 6536.

14-DEHYDRODELCOSINE (SHIMOBURO BASE II, 14-DEHYDROILIENSINE)

$C_{24}H_{37}NO_7$; mp 212.5-213.5°;
$[\alpha]_D$ + 25.2° $(CHCl_3)$
Aconitum japonicum, Delphinium biternatum.
^{1}H NMR: δ 1.10 (3H, *t*, J=7Hz, NCH$_2$-
CH$_3$), 3.32 (6H, *s*, OC*H$_3$*), 3.34 (3H,
s, OC*H$_3$*), 3.76 (1H, br *s*, W½ 12 Hz,
C(1)-β-*H*), 4.03 (1H, *s*, C(6)-α-*H*),
6.96 (1H, br *s*, W½=20Hz, C(1)-OH).[1,2,3-5]

[13]C *Chemical Shift Assignments* [2]

C-1	72.1	C-11	49.7	C-1'	-
C-2	27.3	C-12	27.5	C-6'	56.6
C-3	29.6	C-13	46.8	C-7'	-
C-4	37.5	C-14	214.9	C-8'	-
C-5	45.3	C-15	34.8	C-14'	-
C-6	89.7	C-16	86.5	C-16'	56.0
C-7	87.3	C-17	66.4	C-18'	59.0
C-8	82.9	C-18	77.0		
C-9	53.1	C-19	57.3		
C-10	40.9	N-CH$_2$	50.5		
		CH$_3$	13.6		

1. E. Ochiai, T. Okamoto and M. Kaneko, *Chem. Pharm. Bull.* (Japan), 6, 730 (1958).
2. S. Sakai, H. Takayama and T. Okamoto, *J. Pharm. Soc.* Japan, 99, 647 (1979)
3. S.W. Pelletier and N.V. Mody, *Tetrahedron Letters*, 22, 207 (1981).
4. V. Skaric and L. Marion, *Can. J. Chem.*, 38, 2433 (1960).
5. B.T. Salimov, M.S. Yunusov, and S.Yu. Yunusov, *Khim. Prir. Soedin.*, 14, 106 (1978).

1-DEHYDRODELPHISINE

$C_{28}H_{41}NO_8$; mp 170-171°;
Prepared from delphisine
^{1}H NMR: δ 1.10 (3H, *t*, J=7Hz, NCH$_2$-CH$_3$), 1.97 and 2.02 (each 3H, *s*, OCOCH$_3$), 3.27 (6H, *s*, OCH$_3$), 3.28 (3H, *s*, OCH$_3$), 3.57 (2H, AB type *d*, J=9Hz, C(4)-CH$_2$-O), 4.00 (1H, *dd*, J$_1$=1Hz, J$_2$=7Hz, C(6)-β-*H*) and 4.84 (1H, *dd*, J$_1$=J$_2$=4.5Hz, C(14)-β-*H*).[1]

^{13}C Chemical Shift Assignments[2]

C-1	212.7	C-11	61.0	C-1'	-
C-2	41.5	C-12	34.2	C-6'	58.2
C-3	38.4	C-13	39.1	C-7'	-
C-4	39.4	C-14	75.6	C-8'	-
C-5	52.9	C-15	39.2	C-14'	-
C-6	83.2	C-16	82.8	C-16'	56.4
C-7	48.9	C-17	63.2	C-18'	59.1
C-8	85.9	C-18	78.7	C=O	169.4,169.4
C-9	43.5	C-19	54.7	CH$_3$	21.2,22.3
C-10	38.6	N-CH$_2$	48.6		
		CH$_3$	13.3		

1.   S.W. Pelletier, Z. Djarmati, S. Lajsic, and W.H. De Camp, *J. Am. Chem. Soc.*, 98, 2617 (1976).
2.   S.W. Pelletier and Z. Djarmati, *J. Am. Chem. Soc.*, 98, 2626 (1976).

6-DEHYDRODELTAMINE (6-DEHYDROELDELIDINE)

$C_{25}H_{37}NO_7$; mp 120-122°;
Prepared from deltamine
^{1}H NMR: δ 0.87 (3H, s, C(4)-CH_3),
1.01 (3H, t, NCH_2-CH_3), 3.30, 3.32
and 3.37 (each 3H, s, OCH_3), 5.00
and 5.44 (each 1H, s, OCH_2O).[1]

[13]C Chemical Shift Assignments[2]

C-1	77.0	C-11	51.7	C-1'	55.7
C-2	27.0	C-12	37.7	C-6'	-
C-3	39.2	C-13	38.3	C-7-O–CH₂ C-8-O	95.5
C-4	35.2	C-14	81.3		
C-5	57.2	C-15	34.0	C-14'	58.1
C-6	216.7	C-16	81.0	C-16'	56.5
C-7	90.2	C-17	62.6	C-18'	-
C-8	82.8	C-18	25.0	C=O, CH₃	
C-9	51.2	C-19	56.5		
C-10	80.3	N-CH₂, CH₃	50.0, 13.7		

1. A.S. Narzullaev, M.S. Yunusov, and S.Yu. Yunusov, *Khim. Prir. Soedin.*, 9, 497 (1973).

2. S.W. Pelletier, N.V. Mody, and O.D. Dailey, Jr., *Can. J. Chem.*, 58, 1875 (1980).

14-DEHYDRODICTYOCARPINE

$C_{26}H_{37}NO_8$; mp 129-135.5°;
$[\alpha]_D$ - 17.7° ($CHCl_3$)
Prepared from dictyocarpine
^{1}H NMR: δ 0.89 (3H, s, $C(4)-CH_3$),
1.12 (3H, t, NCH_2-CH_3), 2.11 (3H, s,
$OCOCH_3$), 3.36 (6H, s, OCH_3), 5.00 and
5.03 (each 1H, s, OCH_2O), and 5.58
(1H, d, $C(6)-\alpha-H$).[1]

[13]C Chemical Shift Assignments [1]

C-1	78.2	C-11	55.2	C-1'	55.6
C-2	26.0	C-12	36.3	C-6'	-
C-3	36.3	C-13	45.4	C-7-O\	
C-4	34.2	C-14	215.0	C-8-O/ CH₂	94.7
C-5	50.5	C-15	30.9	C-14'	-
C-6	77.1	C-16	83.9	C-16'	56.1
C-7	92.7	C-17	64.9	C-18'	-
C-8	87.1	C-18	25.4	C=O	170.2
C-9	58.1	C-19	57.0	CH₃	21.6
C-10	79.4	N-CH₂	50.2		
		CH₃	14.0		

1.   S.W. Pelletier, N.V. Mody, and O.D. Dailey, Jr., *Can. J. Chem.*, 58, 1875 (1980).

6-DEHYDRODICTYOCARPININE

$C_{24}H_{35}NO_7$; mp 201-203.5°;
$[\alpha]_D$ - 23.7° (CH_3OH)
Prepared from dictyocarpinine
^{1}H NMR: δ 0.97 (3H, *s*, C(4)-C*H₃*),
1.09 (3H, *t*, NCH₂-C*H₃*), 3.34 and 3.38
(each 3H, *s*, OC*H₃*), 4.59 (1H, *bm*,
C(14)-β-*H*) and 5.18 and 5.60 (each 1H,
s, OC*H₂*O).[1]

^{13}C *Chemical Shift Assignments* [1]

C-1	77.1	C-11	51.5	C-1'	55.9
C-2	26.4	C-12	37.3	C-6'	-
C-3	37.7	C-13	36.3	C-7-O	
C-4	35.2	C-14	71.9	C-8-O—CH₂	95.8
C-5	56.5	C-15	32.7	C-14'	-
C-6	217.7	C-16	81.0	C-16'	56.5
C-7	91.2	C-17	63.4	C-18'	-
C-8	82.2	C-18	24.8	C=O	
C-9	53.8	C-19	57.2	CH₃	
C-10	79.7	N-CH₂	50.0		
		CH₃	13.8		

1.   S.W. Pelletier, N.V. Mody, and O.D. Dailey, Jr., *Can. J. Chem.*, 58, 1875
     (1980).

14-DEHYDRODICTYOCARPININE

$C_{24}H_{35}NO_7$; Amorphous;
Prepared from dictyocarpine
^{1}H NMR: δ 0.95 (3H, *s*, C(4)-C*H₃*),
1.09 (3H, *t*, NCH₂-C*H₃*), 3.32 and 3.34
(each 3H, *s*, OC*H₃*), 4.34 (1H, *m*, C(6)-
α-*H*), and 5.13 and 5.29 (each 1H, *s*,
OC*H₂*O).[1]

[13]C *Chemical Shift Assignments* [1]

C-1	79.8	C-11	55.4	C-1'	55.6
C-2	26.1	C-12	36.5	C-6'	-
C-3	35.9	C-13	45.2	C-7-O⟍ CH₂	94.3
C-4	34.1	C-14	213.6	C-8-O⟋	
C-5	51.7	C-15	31.7	C-14'	-
C-6	77.1	C-16	84.0	C-16'	56.1
C-7	93.0	C-17	64.5	C-18'	-
C-8	87.3	C-18	25.4	C=O	
C-9	57.9	C-19	57.2	CH₃	
C-10	79.5	N-CH₂	50.6		
		CH₃	14.0		

1.   S.W. Pelletier, N.V. Mody, and O.D. Dailey, Jr., *Can. J. Chem.*, <u>58</u>, 1875 (1980).

1-DEHYDRONEOLINE

$C_{24}H_{37}NO_6$; mp 150-152°;
Prepared from 1-dehydrodelphisine
^{1}H NMR: δ 1.11 (3H, *t*, NCH$_2$-CH$_3$),
3.27, 3.28 and 3.35 (each 3H, *s*,
OCH$_3$).[1]

[13]C *Chemical Shift Assignments* [2]

C-1	213.8	C-11	60.9	C-1'	–
C-2	41.4	C-12	33.5	C-6'	57.9
C-3	38.8	C-13	40.4	C-7'	–
C-4	39.5	C-14	75.8	C-8'	–
C-5	53.5	C-15	42.3	C-14'	–
C-6	82.2	C-16	82.2	C-16'	56.3
C-7	52.9	C-17	64.1	C-18'	59.2
C-8	74.0	C-18	79.2	C=O	
C-9	48.4	C-19	55.0	CH$_3$	
C-10	39.8	N-CH$_2$	48.6		
		CH$_3$	13.4		

1.   S.W. Pelletier, Z. Djarmati, S. Lajsic, and W.H. De Camp, *J. Am. Chem.*
     Soc., 98, 2617 (1976).
2.   S.W. Pelletier and Z. Djarmati, *J. Am. Chem. Soc.*, 98, 2626 (1976).

14-DEHYDROOXOACONINE

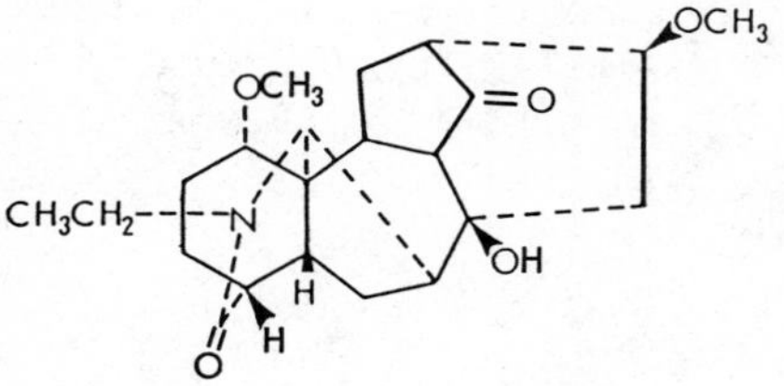

$C_{22}H_{31}NO_5$; mp 202-203°;

$[\alpha]_D$ - 44.5^{o}

Prepared from cammaconine

^{1}H NMR: δ 1.21 (3H, *t*, J=7Hz, CH$_2$ *CH$_3$*), 3.30, 3.34 (each 3H, *s*, O*CH$_3$*).[1]

^{13}C *Chemical Shift Assignments*[1]

C-1	84.1d	C-11	47.7s	C-1'	55.9q
C-2	26.5t	C-12	24.7t	C-6'	-
C-3	30.1t	C-13	46.1d	C-7'	-
C-4	52.9d	C-14	214.2s	C-8'	-
C-5	48.9d	C-15	36.4t	C-14'	-
C-6	25.4t	C-16	84.9d	C-16'	56.1q
C-7	43.2d	C-17	61.8d	C-18'	-
C-8	81.6s	C-18	-		
C-9	54.6d	C-19	171.0s		
C-10	44.1d	N-CH$_2$	41.3t		
		CH$_3$	13.2q		

1.   O.E. Edwards, R.J. Kolt and K.K. Purushothaman, *Can. J. Chem.*, 61, 1194 (1983).

14-DEHYDROOXOCAMMACONIC ACID

$C_{23}H_{31}NO_7$; mp 210° (dec.)
$[\alpha]_D$ 16°; Prepared from cammaconine

^{1}H NMR: δ 1.28 (3H, *t*, J=7Hz, CH_2
CH_3), 3.33, 3.34 (each 3H, *s*, OC*H_3*).[1]

[13]C *Chemical Shift Assignments*[1]

C-1	82.7d	C-11	47.2s	C-1'	56.2q
C-2	25.9t	C-12	24.4t	C-6'	-
C-3	33.3t	C-13	46.0d	C-7'	-
C-4	57.8s	C-14	213.5s	C-8'	-
C-5	52.5d	C-15	36.1t	C-14'	-
C-6	30.4t	C-16	84.9d	C-16'	56.2q
C-7	43.8d[a]	C-17	63.8d	C-18'	-
C-8	81.3s	C-18	172.7s		
C-9	54.2d	C-19	172.7s		
C-10	44.0d[a]	N-CH_2	42.6t		
		CH_3	12.8q		

1. O.E. Edwards, R.J. Kolt and K.K. Purushothaman, *Can. J. Chem.*, 61, 1194
   (1983).

14-DEHYDROOXOCAMMACONINE

$C_{23}H_{33}NO_6$
Prepared from cammaconine

[13]C Chemical Shift Assignments[1]

C-1	84.1d	C-11	47.8s	C-1'	55.9q
C-2	26.8t	C-12	24.7t	C-6'	-
C-3	29.9t	C-13	46.1d[a]	C-7'	-
C-4	49.7s	C-14	214.0s	C-8'	-
C-5	52.2d	C-15	36.1t	C-14'	-
C-6	25.8t	C-16	85.0d	C-16'	56.1q
C-7	45.6d[a]	C-17	62.4d	C-18'	-
C-8	81.4s	C-18	66.7t		
C-9	54.4d	C-19	174.3s		
C-10	44.2d	N-CH$_2$	41.3t		
		CH$_3$	13.1q		

1.   O.E. Edwards, R.J. Kolt and K.K. Purushothaman, *Can. J. Chem.*, 61, 1194 (1983).

14-DEHYDROTALATIZAMINE

$C_{24}H_{37}NO_5$; mp 128-130°

Aconitum saposhnikovii

^{1}H NMR: δ 1.03 (3H, *t*, NCH$_2$-*CH$_3$*),
3.19, 3.21, 3.21 (each 3H, *s*, OC*H$_3$*).[1]

[13]C Chemical Shift Assignments

C-1	C-11	C-1'
C-2	C-12	C-6'
C-3	C-13	C-7'
C-4	C-14	C-8'
C-5	C-15	C-14'
C-6	C-16	C-16'
C-7	C-17	C-18'
C-8	C-18	C=O | CH$_3$
C-9	C-19	
C-10	N-CH$_2$ | CH$_3$	

1.   M.N. Sultankhodzhaev, M.S. Yunusov and S.Yu. Yunusov, *Khim. Prir. Soedin.*, 265 (1982).

DELAVACONITINE

$C_{29}H_{39}NO_6$
Aconitum delavyi.[1,2]

[13]C Chemical Shift Assignments

C-1	C-11	C-1'
C-2	C-12	C-6'
C-3	C-13	C-7'
C-4	C-14	C-8'
C-5	C-15	C-14'
C-6	C-16	C-16'
C-7	C-17	C-18'
C-8	C-18	C=O
C-9	C-19	CH₃
C-10	N-CH₂	
	CH₃	

1.   Y.L. Zhu and R.-H. Zhu, *Heterocycles*, 17, 607 (1982).
2.   J.-H. Zhu, H.-C. Wang, S.-H. Jiang, S.-H. Hong, X.-Z. Tang, and Y.-L. Zhu, *Chem. Nat. Prod. Proc. Sino Am. Symp.*, 306 (1980).

DELBITERINE

$C_{25}H_{41}NO_7$; mp 137-138°
Delphinium biternatum
^{1}H NMR: δ 1.02 (3H, *t*, NCH_2-CH_3),
3.17 and 3.25 (each 3H, *s*, OCH_3), and
3.38 (6H, *s*, OCH_3).[1,2]

[13]C *Chemical Shift Assignments*

C-1	C-11	C-1'
C-2	C-12	C-6'
C-3	C-13	C-7'
C-4	C-14	C-8'
C-5	C-15	C-14'
C-6	C-16	C-16'
C-7	C-17	C-18'
C-8	C-18	C=O / CH₃
C-9	C-19	
C-10	N-CH₂ / CH₃	

1. B.T. Salimov, M.S. Yunusov and S.Yu. Yunusov, *Khim. Prir. Soedin.*, 106
    (1978).

2. S.W. Pelletier, N.V. Mody, K.I. Varughese, J.A. Maddry, and H.K. Desai,
    J. Am. Chem. Soc., 1981, 103, 6536.

DELCAROLINE

$C_{25}H_{41}NO_8$; Amorphous;
$[\alpha]_D$ + 49.8° (MeOH)
Delphinium carolinianum
^{1}H NMR: δ 1.05 (3H, *t*, NCH_2-CH_3),
3.25, 3.30, 3.35 and 3.42 (each 3H,
s, OCH_3) and 4.06 (1H, *m*, C(14)-β-*H*).[1,2]

^{13}C Chemical Shift Assignments[1]

C-1	79.4	C-11	53.8	C-1'	55.5
C-2	25.5	C-12	37.6	C-6'	57.7
C-3	32.2	C-13	37.0	C-7'	-
C-4	38.1	C-14	73.6	C-8'	-
C-5	45.1	C-15	33.9	C-14'	-
C-6	90.8	C-16	81.3	C-16'	56.3
C-7	88.0	C-17	66.1	C-18'	59.1
C-8	75.1	C-18	77.2	C=O CH₃	
C-9	54.0	C-19	52.5		
C-10	79.9	N-CH₂ CH₃	51.3 14.3		

1.   S.W. Pelletier, N.V. Mody, and R.C. Desai, *Heterocycles*, 16, 747 (1981).
2.   S.W. Pelletier, N.V. Mody, K.I. Varughese, J.A. Maddry and H.K. Desai,
     J. Am. Chem. Soc., 103, 6536 (1981).

DELCORIDINE

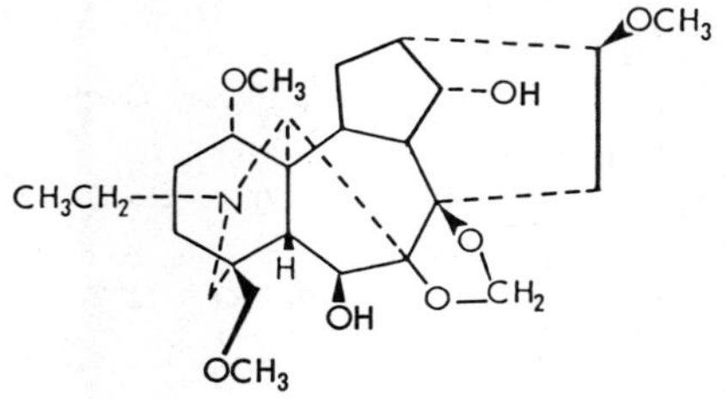

$C_{25}H_{39}NO_7$;

Delphinium iliense

^{1}H NMR: δ 1.00 (3H, t, N-CH$_2$-CH$_3$), 3.15, 3.20, and 3.25 (each 3H, s, O*CH$_3$*), and 5.00 and 5.10 (each 1H, s, O-C*H$_2$*-O).[1]

[13]C *Chemical Shift Assignments*

C-1	C-11	C-1'
C-2	C-12	C-6'
C-3	C-13	C-7'
C-4	C-14	C-8'
C-5	C-15	C-14'
C-6	C-16	C-16'
C-7	C-17	C-18'
C-8	C-18	C=O
C-9	C-19	CH$_3$
C-10	N-CH$_2$	
	CH$_3$	

1.   M.G. Zhamierashvili, V.A. Tel'nov, M.S. Yunusov, and S.Yu. Yunusov, *Khim. Prir. Soedin.*, 663 (1980).

DELCORINE

$C_{26}H_{41}NO_7$;   mp 200-202°;
$[\alpha]_D$ - 18° (CHCl$_3$)
Delphinium corumbosum (corymbosum),
D. iliense, D. orientale.
^{1}H NMR: δ 1.00(3H, *t*, NCH$_2$-C*H$_3$*), 3.22,
3.28, 3.30 and 3.40 (each 3H, *s*, OC*H$_3$*),
4.22 (1H, *s*, C(6)-α-*H*) and 5.01 and
5.07 (each 1H, *s*, OC*H$_2$*O).[1,3,4]

^{13}C Chemical Shift Assignments[2]

C-1	83.1	C-11	50.2	C-1'	55.5
C-2	26.4	C-12	28.1	C-6'	-
C-3	31.8	C-13	37.9	C-7-O$\diagdown$CH$_2$	92.9
C-4	38.1	C-14	82.5	C-8-O$\diagup$	
C-5	52.6	C-15	33.3	C-14'	57.8
C-6	78.9	C-16	81.8	C-16'	56.3
C-7	92.7	C-17	63.9	C-18'	59.6
C-8	83.9	C-18	78.9	C=O	
C-9	48.1	C-19	53.7	CH$_3$	
C-10	40.3	N-CH$_2$	50.7		
		CH$_3$	14.0		

1.   A.S. Narzullaev, M.S. Yunusov and S.Yu. Yunusov, *Khim. Prir. Soedin.*, 9, 497 (1973).

2.   S.W. Pelletier, N.V. Mody, and O.D. Dailey, Jr., *Can. J. Chem.*, 58, 1875 (1980).

3.   M.G. Zhamierashvili, V.A. Tel'nov, M.S. Yunusov and S.Yu. Yunusov, *Khim. Prir. Soedin.*, 13, 836 (1977).

4.   S.W. Pelletier, N.V. Mody, K.I. Varughese, J.A. Maddry, and H.K. Desai, *J. Am. Chem. Soc.*, 1981, 103, 6536.

DELCOSINE (DELPHAMINE, ILIENSINE, LUCACONINE, TAKAO BASE I)

$C_{24}H_{39}NO_7$; mp 203-204°;

$[\alpha]_D$ + 57° (CHCl$_3$)

Aconitum japonicum, Consolida ambigua,
C. orientalis, Delphinium Ajacis, D.
belladonna, D. biternatum, D. consolida,
D. glaucescens, D. regalis.*

^{1}H NMR: δ 1.10 (3H, *s*, NCH$_2$-CH$_3$),3.34,
3.36 and 3.39 (each 3H, *s*, O*CH*$_3$), 4.02
(1H, *s*, C(6)-β-*H*), and 4.10 (1H, *dd*,
J=4.5Hz, C(14)-β-*H*).[1,2,4]

[13]*C Chemical Shift Assignments*[3]

C-1	72.7	C-11	48.9	C-1'	-
C-2	27.5	C-12	29.4	C-6'	57.4
C-3	29.4	C-13	45.3	C-7'	-
C-4	37.6	C-14	75.8	C-8'	-
C-5	44.0	C-15	34.5	C-14'	-
C-6	90.1	C-16	82.0	C-16'	56.4
C-7	87.9	C-17	66.3	C-18'	59.1
C-8	78.1	C-18	77.4	C=O CH$_3$	
C-9	45.3	C-19	57.1		
C-10	39.4	N-CH$_2$ CH$_3$	50.4 13.7		

1. S. Sakai, H. Takayama, and T. Okamoto, *Yakugaku Zasshi,* 99(6), 647 (1979).
2. S.W. Pelletier and N.V. Mody, *Tetrahedron Letters,* 22, 207 (1981).
3. S.W. Pelletier, N.V. Mody, R.S. Sawhney and J. Bhattacharyya, *Hetero-cycles,* 7, 327 (1977).
4. S.W. Pelletier, R.S. Sawhney, H.K. Desai, and N.V. Mody, *J. Nat. Prod.,* 43, 395 (1980).

DELECTINE

$C_{31}H_{44}N_2O_8$; mp 107-109°
Delphinium dictyocarpum
^{1}H NMR: δ 1.02 (3H, *t*, NCH_2-CH_3),
3.20, 3.32 and 3.34 (each 3H, *s*,
OCH_3), 5.71 (2H, *bs*, NH_2), and 6.65-
7.80 (aromatic protons).[1]

^{13}C *Chemical Shift Assignments*

C-1	C-16
C-2	C-17
C-3	C-18
C-4	C-19
C-5	N-CH₂
C-6	CH₃
C-7	C-1'
C-8	C-6'
C-9	C-7'
C-10	C-8'
C-11	C-14'
C-12	C-16'
C-13	C-18'
C-14	C=O
C-15	CH₃

1.   B.T. Salimov, M.S. Yunusov, S.Yu. Yunusov and A.S. Narzullaev, *Khim.*
     Prir. Soedin., 11, 665 (1975).
2.   S.W. Pelletier, N.V. Mody, K.I. Varughese, J.A. Maddry, and H.K. Desai,
     J. Am. Chem. Soc., 1981, 103, 6536.

DELECTININE

$C_{24}H_{39}NO_7$; mp 167-169°;
$[\alpha]_D$ + 42° (CHCl$_3$)
Delphinium dictyocarpum
^{1}H NMR: δ 0.99 (3H, *s*, NCH$_2$-CH$_3$),
3.20, 3.29 and 3.37 (each 3H, *s*,
OCH$_3$).[1,3]

[13]C *Chemical Shift Assignments*[2]

C-1	85.1	C-11	48.2	C-1'	56.0
C-2	25.3	C-12	27.5	C-6'	58.1
C-3	31.6	C-13	46.1	C-7'	-
C-4	38.8	C-14	75.3	C-8'	-
C-5	45.1	C-15	33.1	C-14'	-
C-6	90.1	C-16	81.8	C-16'	56.5
C-7	89.0	C-17	65.4	C-18'	-
C-8	76.3	C-18	67.6	C=O	
C-9	49.5	C-19	52.8	CH$_3$	
C-10	36.5	N-CH$_2$	51.3		
		CH$_3$	14.2		

1. B.T. Salimov, N.D. Abdullaev, M.S. Yunusov and S.Yu. Yunusov, *Khim. Prir. Soedin.*, 14, 235 (1978).

2. S.W. Pelletier and R.S. Sawhney, *Heterocycles*, 9, 463 (1978).

3. S.W. Pelletier, N.V. Mody, K.I. Varughese, J.A. Maddry, and H.K. Desai, *J. Am. Chem. Soc.*, 1981, 103, 6536.

DELPHATINE

$C_{26}H_{43}NO_7$;
mp 106°; 205-207° (perchlorate);
$[\alpha]_D$ + 38.2° (CHCl$_3$)
Consolida ambigua, Delphinium biternatum.
^{1}H NMR: δ 1.07 (3H, *t*, NCH$_2$-C*H$_3$*),
3.28, 3.33 and 3.40 (each 3H, *s*, O*H$_3$*)
and 3.46 (6H, *s*, OC*H$_3$*).[1,3,4]

^{13}C *Chemical Shift Assignments*[2]

C-1	83.9	C-11	48.9	C-1'	55.7
C-2	26.2	C-12	28.7	C-6'	57.3
C-3	32.4	C-13	46.1	C-7'	-
C-4	38.1	C-14	84.3	C-8'	-
C-5	43.3	C-15	33.5	C-14'	57.8
C-6	90.6	C-16	82.6	C-16'	56.3
C-7	88.4	C-17	64.8	C-18'	59.0
C-8	77.5	C-18	78.1	C=O	
C-9	49.8	C-19	52.8	CH$_3$	
C-10	38.1	N-CH$_2$	51.1		
		CH$_3$	14.2		

1. M.S. Yunusov and S.Yu. Yunusov, *Khim. Prir. Soedin.*, 334 (1970).
2. S.W. Pelletier, N.V. Mody, R.S. Sawhney and J. Bhattacharyya, *Heterocycles*, 7, 327 (1977).
3. S.W. Pelletier, R.S. Sawhney, H.K. Desai, and N.V. Mody, *J. Nat. Prod.*, 43, 395 (1980).
4. S.W. Pelletier, N.V. Mody, K.I. Varughese, J.A. Maddry, and H.K. Desai, *J. Am. Chem. Soc.*, 1981, 103, 6536.

DELPHELINE

$C_{25}H_{39}NO_6$; mp 227°;
$[\alpha]_D$ - 26° (CHCl$_3$)
Delphinium elatum, D. barbeyi.
^{1}H NMR: δ 0.86 (3H, *s*, C(4)-CH$_3$),
3.62 (1H, *dd*, C(14)-β-H), 4.14 (1H,
s, C(6)-α-H), and 5.01 and 5.08 (each
1H, *s*, OCH$_2$O).[1,3,4,5]

^{13}C *Chemical Shift Assignments* [2]

C-1	83.1	C-11	50.4	C-1'	56.2
C-2	26.9	C-12	28.1	C-6'	-
C-3	36.9	C-13	37.9	C-7-O⟍	
C-4	33.9	C-14	81.9	C-8-O⟋ CH$_2$	92.9
C-5	55.5	C-15	33.5	C-14'	57.8
C-6	79.3	C-16	82.9	C-16'	56.8
C-7	92.8	C-17	63.7	C-18'	-
C-8	84.6	C-18	25.3		
C-9	47.9	C-19	57.3		
C-10	40.3	N-CH$_2$	50.2		
		CH$_3$	13.9		

1. A.S. Narzullaev, M.S. Yunusov and S.Yu. Yunusov, *Khim. Prir. Soedin.*,
    443 (1973).

2. S.W. Pelletier, J.A. Glinski, B.S. Joshi, and S.-Y. Chen, *Heterocycles*,
    20, 1347 (1983).

3. O.E. Edwards, L. Marion, and K.H. Palmer, *Can. J. Chem.*, 36, 1097 (1958).

4. M. Carmack, J.P. Ferris, J. Harvey, P.L. Magart, E.W. Martin and D.W.
    Mayo, *J. Am. Chem. Soc.*, 80, 497 (1958).

5. M.S. Yunusov, V.E. Nezhevenko and S.Yu. Yunusov, *Khim. Prir. Soedin.*,
    11, 107 (1975).

DELPHIDINE (NEOLINE-8-ACETATE)

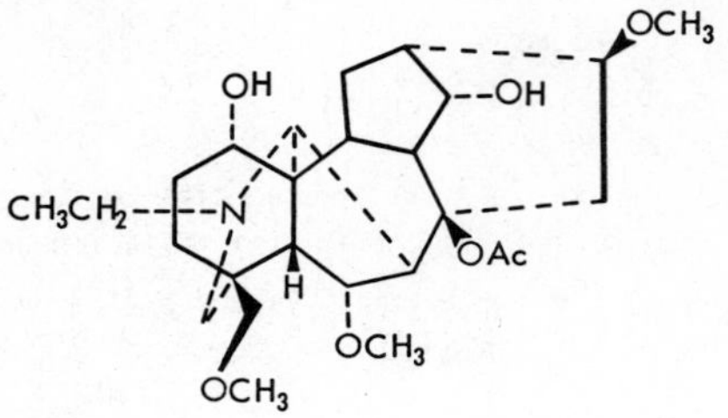

$C_{26}H_{41}NO_7$; mp 98-100° (d);

$[\alpha]_D$ + 16.6° (EtOH)

Delphinium staphisagria

^{1}H NMR: δ 1.13 (3H, *t*, NCH$_2$-CH$_3$),

2.00 (3H, *s*, OCOCH$_3$), and 3.26, 3.31

and 3.34 (each 3H, *s*, OCH$_3$).[1,2]

^{13}C *Chemical Shift Assignments* [3]

C-1	72.0	C-11	49.9	C-1'	-
C-2	29.5^a	C-12	29.5^a	C-6'	58.1
C-3	29.9^a	C-13	44.0	C-7'	-
C-4	38.2	C-14	75.0	C-8'	-
C-5	44.0	C-15	38.4	C-14'	-
C-6	84.1	C-16	82.4	C-16'	56.6
C-7	48.2	C-17	63.0	C-18'	59.1
C-8	85.4	C-18	79.8	C=O	169.9
C-9	46.1	C-19	56.8	CH$_3$	22.5
C-10	40.8	N-CH$_2$	48.4		
		CH$_3$	12.7		

1.   S.W. Pelletier, J.K. Thakkar, N.V. Mody, Z. Djarmati, and J. Battachary-
     ya, *Phytochemistry*, 16, 404 (1977).

2.   S.W. Pelletier, Z. Djarmati, S. Lajsic and W.H. De Camp, *J. Am. Chem.
     Soc.*, 98, 2617 (1976).

3.   S.W. Pelletier and Z. Djarmati, *J. Am. Chem. Soc.*, 98, 2626 (1976).

DELPHINIFOLINE

$C_{23}H_{37}NO_7$; MW: 439.2555;
mp 218-220°; $[\alpha]_D$ + 30° (EtOH)
Aconitum delphinifolium
^{1}H NMR (200 MHz): δ 1.08 (3H, *t*,
$NCH_2\text{-}CH_3$), 3.34 and 3.36 (each 3H, *s*,
OCH_3), 3.64 (1H, *m*, C(1)-β-*H*), 4.22
(1H, *s*, C(6)-α-*H*), and 4.56 (1H, *dd*,
C(14)-β-*H*).[1]

[13]C *Chemical Shift Assignments*[2]

C-1	72.8	C-11	50.3	C-1'	-
C-2	28.1	C-12	28.7	C-6'	-
C-3	29.4	C-13	44.1[a]	C-7'	-
C-4	37.5	C-14	76.4	C-8'	-
C-5	45.1[a]	C-15	36.6	C-14'	-
C-6	79.1	C-16	81.8	C-16'	56.2
C-7	-	C-17	66.9	C-18'	59.6
C-8	-	C-18	80.6		
C-9	45.1[a]	C-19	57.5		
C-10	39.2	N-CH$_2$	50.5		
		CH$_3$	13.9		

1.   V.N. Aiyar, P.W. Codding, K.A. Kerr, M.H. Benn, and A.J. Jones, *Tetra-hedron Letters*, 22, 483 (1981).

2.   Private communication to S.W. Pelletier from Prof. M.H. Benn, Dec. 31, 1980.

DELPHININE

$C_{33}H_{45}NO_9$;   MW: 599.30944
mp 191-192°; $[\alpha]_D$ + 25° (EtOH)
Delphinium staphisagria,Atragne sibirica
^{1}H NMR: δ 1.31 (3H, *s*, OCOCH_3), 2.39
(3H, *s*, NCH_3), 3.26 and 3.65 (each
3H, *s*, OCH_3), 3.40 (6H, *s*, OCH_3),
5.08 (1H, *d*, C(14)-β-*H*) and 7.66-
8.42 (aromatic protons).^{1}

^{13}C *Chemical Shift Assignments* ^{2,3,4}

C-1	84.9	C-16	83.7^a	C=O	166.0
C-2	26.3	C-17	63.3	1	129.6
C-3	34.7	C-18	80.2	2	128.4
C-4	39.3	C-19	56.1	3	130.4
C-5	48.8	N-CH$_3$	42.3	4	132.8
C-6	83.0^a			5	130.4
C-7	48.2	C-1'	56.1	6	128.4
C-8	85.4	C-6'	57.6		
C-9	45.1	C-7'	-		
C-10	41.0	C-8'	-		
C-11	50.2	C-14'	-		
C-12	35.7	C-16'	58.6		
C-13	74.8	C-18'	58.9		
C-14	78.9	C=O	169.4		
C-15	39.3	CH$_3$	21.4		

1.   K.B. Birnbaum, K. Wiesner, E.W.K. Jay and L. Jay, *Tetrahedron Letters*,
     867 (1971); S.W. Pelletier and J. Bhattacharyya, *Tetrahedron Letters*,
     4679 (1976).
2.   S.W. Pelletier and Z. Djarmati, *J. Am. Chem. Soc.*, 98, 2626 (1976).
3.   S.W. Pelletier, N.V. Mody and N. Katsui, *Tetrahedron Letters*, 4027 (1977).
4.   S.W. Pelletier, J. Finer-Moore, R.C. Desai, N.V. Mody and H.K. Desai,
     J. Org. Chem., 47, 5290 (1982).

DELPHIRINE (1-EPI-NEOLINE)

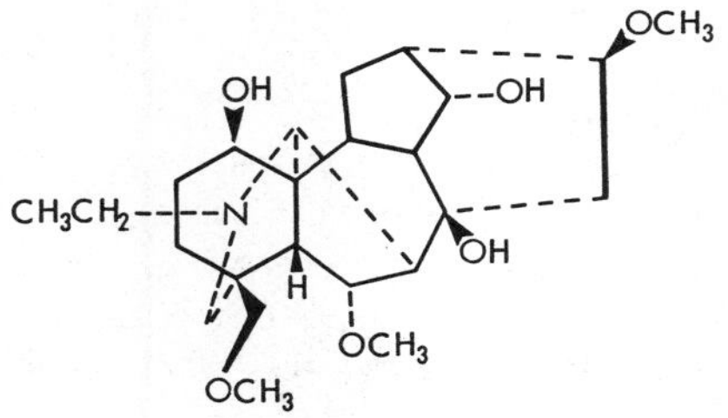

$C_{24}H_{39}NO_6$; mp 95-100°;

$[\alpha]_D$ + 3.8° (EtOH)

Delphinium staphisagria

^{1}H NMR (100 MHz): δ 1.06 (3H, *s*, NCH$_2$-C*H$_3$*), 3.32 (3H, *s*, OC*H$_3$*), 3.38 (6H, *s*, OC*H$_3$*), 3.62 (1H, br m, C(1)-α-*H*), 4.15 (1H, *dd*, J$_1$=1Hz,J$_2$=7Hz, C(6)-β-*H*), and 4.26 (1H, *dd*, J=4.5Hz, C(14)-β-*H*).[2]

[13]*C Chemical Shift Assignments*[1,2]

C-1	69.0	C-11	50.5	C-1'	-
C-2	30.3	C-12	28.9	C-6'	57.5
C-3	31.2	C-13	45.5	C-7'	-
C-4	40.1	C-14	75.7	C-8'	-
C-5	39.7	C-15	42.1	C-14'	-
C-6	82.8	C-16	82.4	C-16'	56.2
C-7	51.7	C-17	63.3	C-18'	59.2
C-8	73.9	C-18	80.4	C=O	
C-9	48.4	C-19	53.8	CH$_3$	
C-10	39.4	N-CH$_2$	48.8		
		CH$_3$	13.5		

1.   S.W. Pelletier and Z. Djarmati, *J. Am. Chem. Soc.*, 98, 2626 (1976).

2.   S.W. Pelletier and J. Bhattacharyya, *Tetrahedron Letters*, 4679 (1976).

DELPHISINE (NEOLINE-8,14-DIACETATE)

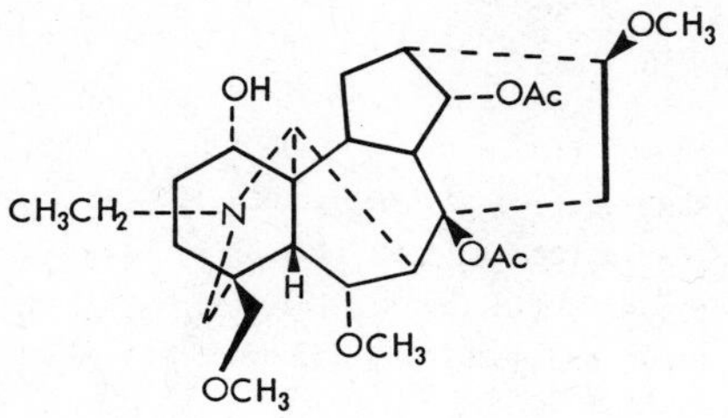

$C_{28}H_{43}NO_8$;   mp 122-123°;

$[\alpha]_D$ + 7.1° (EtOH)

Delphinium staphisagria

^{1}H NMR: δ 1.12 (3H, *t*, NCH_2-*CH₃*),

1.96 and 2.02 (each 3H, *s*, $OCOCH_3$),

3.25, 3.30 and 3.31 (each 3H, *s*, OCH_3),

4.04 (1H, *dd*, J_1=1Hz, J_2=7Hz, C(1)-β-*H*),

and 4.80 (1H, *dd*, J=4.5Hz, C(14)-β-*H*).[1]

[13]C *Chemical Shift Assignments* [2]

C-1	72.1	C-11	49.8	C-1'	-
C-2	29.5[a]	C-12	29.5[a]	C-6'	58.0
C-3	30.1[a]	C-13	43.3	C-7'	-
C-4	38.1	C-14	75.5	C-8'	-
C-5	44.1	C-15	38.5	C-14'	-
C-6	84.2	C-16	82.7	C-16'	56.5
C-7	48.3	C-17	62.7	C-18'	59.0
C-8	85.8	C-18	79.8	C=O	169.3,170.4
C-9	43.3	C-19	56.8	CH₃	21.2,22.2
C-10	38.5	N-CH₂	48.0		
		CH₃	12.9		

1.   S.W. Pelletier, Z. Djarmati, S. Lasjic and W.H. De Camp, *J. Am. Chem. Soc.*, 98, 2617 (1976).

2.   S.W. Pelletier and Z. Djarmati, *J. Am. Chem. Soc.*, 98, 2626 (1976).

1-*EPI*-DELPHISINE

$C_{28}H_{43}NO_8$;   Amorphous;
Prepared from 1-Ketodelphisine
^{1}H NMR: δ 1.08 (3H, t, J=7Hz, NCH_2-
CH_3), 1.96 and 2.02 (each 3H, s,
$OCOCH_3$), 3.25, 3.27 and 3.30 (each 3H,
s, OCH_3), 3.52 (1H, AB type d, C(4)-
CH_2-O), 3.96 (1H, t, w½ 6Hz, C(1)-α-H),
4.03 (1H, dd, C(6)-α-H) and 4.86 (1H,
dd, J_1=J_2=4.5Hz, C(14)-β-H).[1]

[13]C *Chemical Shift Assignments*[2]

C-1	68.6	C-11	50.9	C-1'	-
C-2	30.3	C-12	29.3	C-6'	57.8
C-3	31.2	C-13	44.7	C-7'	-
C-4	39.4	C-14	75.5	C-8'	-
C-5	39.2	C-15	38.2	C-14'	-
C-6	83.7	C-16	83.1	C-16'	56.4
C-7	47.9	C-17	62.3	C-18'	59.0
C-8	86.0	C-18	80.0	C=O	169.4,170.4
C-9	43.6	C-19	53.6	$\overset{\mid}{CH_3}$	21.2,22.3
C-10	38.4	$\overset{N-CH_2}{\underset{CH_3}{\mid}}$	48.9 13.2		

1.   S.W. Pelletier, Z. Djarmati, S. Lasjic and W.H. De Camp, *J. Am. Chem. Soc.*, 98, 2617 (1976).

2.   S.W. Pelletier and Z. Djarmati, *J. Am. Chem. Soc.*, 98, 2626 (1976).

DELPHONINE

$C_{24}H_{39}NO_7$; Amorphous;
$[\alpha]_D$ + 37.5° (EtOH)
Hydrolysis product of delphinine
¹H NMR: δ 2.41 (3H, s, NCH₃), 3.38
and 3.55 (each 3H, s, OCH₃) and 3.42
(6H, s, OCH₃).[1,2]

¹³C *Chemical Shift Assignments* [2]

C-1	85.7	C-11	50.2	C-1'	56.2
C-2	25.9	C-12	36.4	C-6'	57.8
C-3	34.9	C-13	76.7	C-7'	–
C-4	39.5	C-14	79.3	C-8'	–
C-5	49.4	C-15	40.3	C-14'	–
C-6	82.3	C-16	84.4	C-16'	57.2
C-7	51.5	C-17	63.5	C-18'	59.1
C-8	72.8	C-18	80.6		
C-9	50.4	C-19	56.2		
C-10	42.3	N-CH₃	42.3		

1.   L.C. Craig, L. Michaells, S. Gronick and W.A. Jacobs, *J. Biol. Chem.*,
     154, 293 (1949).
2.   S.W. Pelletier and Z. Djarmati, *J. Am. Chem. Soc.*, 98, 2626 (1976).

DELSEMINE

$C_{37}H_{53}N_3O_{10}$; mp 125° (hydrate);
$[\alpha]_D$ + 43° (EtOH)

Delphinium formosum, D. oreophilum,
D. rotundifolium, D. semibarbatum,
D. tricorne.[3]

R = Mixture of $NH-\overset{\overset{O}{\|}}{C}-\overset{\overset{CH_3}{|}}{CH}-CH_2-\overset{\overset{O}{\|}}{C}-NH_2$ and $NH-\overset{\overset{O}{\|}}{C}-CH_2-\overset{\overset{CH_3}{|}}{CH}-\overset{\overset{O}{\|}}{C}-NH_2$

^{13}C *Chemical Shift Assignments* [1,2]

C-1	83.9	C-16	82.6	C=O	168.1
C-2	26.1	C-17	64.5	1	114.7;114.8
C-3	32.2	C-18	69.8	2	141.9;141.7
C-4	37.6	C-19	52.4	3	120.7[a]
C-5	43.3	N-CH₂	50.9	4	134.9[b]
C-6	91.0	CH₃	14.0	5	122.5[a]
C-7	88.6	C-1'	55.7	6	130.3[b]
C-8	77.5	C-6'	57.8		
C-9	50.5	C-7'	-	NHC=O	174.1
C-10	38.2	C-8'	-	CH-CH₃	51.9;17.1
C-11	49.1	C-14'	58.1	CH₂	39.0
C-12	28.7	C-16'	56.3	H₂N-C=O	172.4
C-13	46.1	C-18'	-	NH-C=O	170.0
C-14	83.9	C=O		CH₂	41.4
C-15	33.8	CH₃		CH-CH₃	51.7;17.9
				H₂N-C=O	176.0

1. S.W. Pelletier, N.V. Mody, R.S. Sawhney and J. Bhattacharyya, *Hetero-cycles*, 7, 327 (1977).

2. S.W. Pelletier and J. Bhattacharyya, *Tetrahedron Letters*, 2735 (1977).

3. S.W. Pelletier, N.V. Mody, K.I. Varughese, J.A. Maddry, and H.K. Desai, *J. Am. Chem. Soc.*, 1981, 103, 6536.

DELSOLINE (ACOMONINE)

$C_{25}H_{41}NO_7$; mp 215-216°;
$[\alpha]_D$ + 53.4° (CHCl$_3$)

*Aconitum monticola, Consolida ambi-
gua, C. orientalis, Delphinium biternatum, D. consolida, D. orientale, D.
regalis.*

^{1}H NMR: δ 1.09 (3H, *t*, NCH$_2$-C*H$_3$*),
3.33 and 3.41 (each 3H, *s*, OC*H$_3$*) and
3.38 (6H, *s*, OC*H$_3$*).[1,2,3,4,5]

^{13}C *Chemical Shift Assignments*[3]

C-1	72.6	C-11	49.3	C-1'	-
C-2	27.2	C-12	30.5	C-6'	57.2
C-3	29.3	C-13	43.3	C-7'	-
C-4	37.4	C-14	84.5	C-8'	-
C-5	43.9	C-15	33.5	C-14'	57.9
C-6	90.4	C-16	82.9	C-16'	56.3
C-7	87.8	C-17	66.0	C-18'	59.1
C-8	78.5	C-18	77.3	C=O	
C-9	44.9	C-19	57.2	CH$_3$	
C-10	37.7	N-CH$_2$	50.3		
		CH$_3$	13.5		

1. S.W. Pelletier, R.S. Sawhney, H.K. Desai, and N.V. Mody, *J. Nat. Prod.*, 43, 395 (1980).

2. S.W. Pelletier and N.V. Mody, *Tetrahedron Letters*, 22, 207 (1981).

3. S.W. Pelletier, N.V. Mody, R.S. Sawhney and J. Bhattacharyya, *Hetero-cycles*, 7, 327 (1977).

4. V. Skaric and L. Marion, *Can. J. Chem.*, 38, 2433 (1960).

5. S.W. Pelletier, N.V. Mody, K.I. Varughese, J.A. Maddry, and H.K. Desai, *J. Am. Chem. Soc.*, 1981, 103, 6536.

DELTALINE (DELPHELATINE, ELDELINE)

$C_{27}H_{41}NO_8$; mp 182-184°, 186.5-188°.

$[\alpha]_D$ - 30° (MeOH), - 27.8° (EtOH).

Delphinium barbeyi, D. dictyocarpum, D. elatum, D. glaucescens, D. occidentale, D. iliense.

^{1}H NMR: δ 0.90 (3H, *s*, C(4)-CH_3), 1.07 (3H, *t*, NCH_2-CH_3), 2.07 (3H, *s*, $OCOCH_3$), 3.29, 3.36 and 3.49 (each 3H, *s*, OCH_3), 4.12 (1H, *dd*, J=6Hz, C(14)-β-*H*), 4.97 and 5.01 (each 1H, *s*, OCH_2O), and 5.53 (1H, *d*, J=2Hz, C(6)-α-*H*).[1,3,4,5]

[13]C *Chemical Shift Assignments* [2]

C-1	79.2	C-11	56.0	C-1'	55.3
C-2	27.1	C-12	36.5	C-6'	-
C-3	39.4	C-13	38.5	C-7-O	
C-4	33.7	C-14	81.7[a]	C-8-O CH_2	93.9
C-5	50.4	C-15	34.8	C-14'	57.7
C-6	77.3	C-16	81.5[a]	C-16'	56.2
C-7	91.6	C-17	63.5	C-18'	-
C-8	83.8	C-18	25.7	C=O	169.9
C-9	50.4	C-19	56.9	CH_3	21.8
C-10	81.6	N-CH_2	50.2		
		CH_3	13.8		

1. S.W. Pelletier, O.D. Dailey, Jr., N.V. Mody, and J.D. Olsen, *J. Org. Chem.*, 46, 3284 (1981).

2. S.W. Pelletier, N.V. Mody, and O.D. Dailey, Jr., *Can. J. Chem.*, 58, 1875 (1980).

3. M. Carmack, J.P. Ferris, J. Harvey, P.L. Magat, E.W. Martin and D.W. Mayo, *J. Am. Chem. Soc.*, 80, 497 (1958).

4. A.S. Narzullaev, M.S. Yunusov and S.Yu. Yunusov, *Khim. Prir. Soedin.*, 9, 443 (1973).

5. S.W. Pelletier, N.V. Mody, K.I. Varughese, J.A. Maddry, and H.K. Desai, *J. Am. Chem. Soc.*, 1981, 103, 6536.

DELTAMINE (ELDELIDINE)

$C_{25}H_{39}NO_7$; mp 226-228°;
$[\alpha]_D$ - 17° (MeOH)
Delphinium dictyocarpum.
^{1}H NMR: δ 0.90 (3H, *s*, C(4)-C*H*_3),
1.07 (3H, *t*, NCH_2-C*H*_3), 4.10 (1H, *dd*,
C(14)-β-*H*), 4.22 (1H, *s*, C(6)-α-*H*)
and 4.99 and 5.06 (each 1H, *s*, OC*H*_2O).[1,3-5]

[13]C Chemical Shift Assignments[2]

C-1	80.2	C-11	56.2	C-1'	55.5
C-2	27.0	C-12	36.8	C-6'	-
C-3	38.7	C-13	37.6	C-7-O⟍CH_2	93.3
C-4	33.6	C-14	81.6	C-8-O⟋	
C-5	51.0	C-15	34.3	C-14'	57.9
C-6	77.4	C-16	81.6	C-16'	56.2
C-7	92.4	C-17	63.2	C-18'	-
C-8	83.5	C-18	25.6	C=O	
C-9	51.5	C-19	57.3	CH_3	
C-10	82.4	N-CH_2	50.4		
		CH_3	13.9		

1. A.S. Narzullaev, M.S. Yunusov, and S.Yu. Yunusov, *Khim. Prir. Soedin.*, 443 (1973).

2. S.W. Pelletier, N.V. Mody, and O.D. Dailey, Jr., *Can. J. Chem.*, 58, 1875 (1980).

3. M. Carmack, J.P. Ferris, J. Harvey, P.L. Magat, E.W. Martin, and D.W. Mayo, *J. Am. Chem. Soc.*, 80, 497 (1958).

4. S.W. Pelletier, O.D. Dailey, Jr., N.V. Mody, and J.D. Olsen, *J. Org. Chem.*, 46, 3284 (1981).

5. S.W. Pelletier, N.V. Mody, K.I. Varughese, J.A. Maddry, and H.K. Desai, *J. Am. Chem. Soc.*, 1981, 103, 6536.

DEMETHOXYISOPYRODELPHONINE

$C_{23}H_{35}NO_5$;
Prepared from pyrodelphinine
^{1}H NMR: δ 2.30 (3H, s, NCH_3), 3.35
(3H, s, OCH_3), 3.40 (6H, s, OCH_3),
5.77 and 5.85 (each 1H, s, -$CH=CH$-).

^{13}C *Chemical Shift Assignments* [1]

C-1	86.6	C-11	50.9	C-1'	56.4
C-2	26.4	C-12	42.9	C-6'	57.8
C-3	35.1	C-13	77.2	C-7'	–
C-4	39.6	C-14	77.7	C-8'	–
C-5	48.7	C-15	134.3	C-14'	–
C-6	84.9	C-16	128.7	C-16'	–
C-7	44.6	C-17	62.1	C-18'	59.2
C-8	40.1	C-18	80.6		
C-9	42.0	C-19	56.4		
C-10	40.8	N-CH3	42.4		

1.   S.W. Pelletier and Z. Djarmati, *J. Am. Chem. Soc.*, 98, 2626 (1976).

DEMETHYLENEELDELIDINE (DEMETHYLENEDELTAMINE)

$C_{24}H_{39}NO_7$; mp 98-100°;
$[\alpha]_D$ + 30° (CHCl$_3$)
Delphinium dictyocarpum
^{1}H NMR: δ 0.91 (3H, *s*, C(4)-CH$_3$),
0.98 (3H, *t*, NCH$_2$-CH$_3$), 3.17, 3.25
and 3.35 (each 3H, *s*, OCH$_3$).[1,2,3]

^{13}C *Chemical Shift Assignments*

C-1	C-11	C-1'
C-2	C-12	C-6'
C-3	C-13	C-7'
C-4	C-14	C-8'
C-5	C-15	C-14'
C-6	C-16	C-16'
C-7	C-17	C-18'
C-8	C-18	
C-9	C-19	
C-10	N-CH$_2$	
	CH$_3$	

1.   B.T. Salimov, M.S. Yunusov and S.Yu. Yunusov, *Khim. Prir. Soedin.*, 13,
     128 (1977).
2.   B.T. Salimov, N.D. Abdullaev, M.S. Yunusov and S.Yu. Yunusov, *Khim. Prir.*
     Soedin., 14, 235 (1978).
3.   S.W. Pelletier, N.V. Mody, K.I. Varughese, J.A. Maddry, and H.K. Desai,
     J. Am. Chem. Soc., 1981, 103, 6536.

3 - DEOXYACONINE

$C_{25}H_{41}NO_8$; Amorphous;
Hydrolysis product of deoxyaconitine

^{13}C Chemical Shift Assignments[1]

C-1	85.4	C-11	48.9	C-1'	56.2
C-2	26.5	C-12	38.0	C-6'	58.0
C-3	35.3	C-13	79.2	C-7'	-
C-4	39.1	C-14	80.6	C-8'	-
C-5	49.3	C-15	78.8	C-14'	-
C-6	83.7	C-16	91.1	C-16'	61.4
C-7	49.3	C-17	61.4	C-18'	59.1
C-8	76.5	C-18	81.5	C=0	
C-9	48.2	C-19	53.8	CH₃	
C-10	41.8	N-CH₂	48.9		
		CH₃	13.5		

1.   S.W. Pelletier, N.V. Mody, R.S. Sawhney, *Can. J. Chem.*, 57, 1652 (1979).

3-DEOXYACONITINE

$C_{34}H_{47}NO_{10}$; mp 177-180;
$[\alpha]_D$ + 15.5° (EtOH)
*Aconitum japonicum, A. kusnezoffii, A.
nagarum, A. napellus, A. subcuneatum.*[*]
^{1}H NMR: δ 1.06 (3H, *t*, NCH$_2$-CH$_3$),
1.36 (3H, *s*, OCOCH$_3$), 3.14, 3.24,
3.24 and 3.72 (each 3H, *s*, OCH$_3$),
4.86 (1H, *d*, J=4.5Hz, C(14)-β-*H*) and
7.4-8.18 (aromatic protons).[1,2,4]

^{13}C Chemical Shift Assignments[3]

C-1	85.2	C-16	90.2	C=O		165.9
C-2	26.3	C-17	61.2		1	129.9
C-3	35.2	C-18	80.2		2	129.5
C-4	39.0	C-19	53.3		3	128.5
C-5	49.1	N-CH$_2$	49.1		4	133.1
C-6	83.3	CH$_3$	13.4		5	128.5
C-7	45.1[a]	C-1'	56.0		6	129.5
C-8	92.0	C-6'	57.9			
C-9	44.6[a]	C-7'	-			
C-10	41.0	C-8'	-			
C-11	49.9	C-14'	-			
C-12	36.7	C-16'	60.9			
C-13	74.0	C-18'	59.0			
C-14	78.8[b]	C=O	172.2			
C-15	79.0[b]	CH$_3$	21.3			

1. Y.-G. Wang, Y.-L. Zhu and R.-H. Zhu, *Acta Pharmaceutica Sinica,* 15, 526 (1980).

2. Y. Tsuda, O. Achmatowicz, Jr. and L. Maion, *Ann.,* 680, 88 (1964).

3. S.W. Pelletier and Z. Djarmati, *J. Am. Chem. Soc.,* 98, 2626 (1976).

4. H.-C. Wang, Y.-L. Zao, R.-S. Xu, and R.-H. Zhu, *Acta Chimica Sinica,* 39, 869 (1981).

6- DEOXYDELCORINE

$C_{26}H_{41}NO_6$; mp 93-95°;
Delphinium corumbosum.
^{1}H NMR: δ 1.01 (3H, *t*, NCH$_2$-C*H$_3$*),
3.22, 3.25, 3.30, and 3.36 (each 3H,
s, OC*H$_3$*), 3.60 (1H, *t*, C(14)-β-*H*)
4.86 and 4.96 (each 1H, *s*, -OC*H$_2$*O-).[1,2]

[13]*C Chemical Shift Assignments*

C-1	C-11	C-1'
C-2	C-12	C-6'
C-3	C-13	C-7'
C-4	C-14	C-8'
C-5	C-15	C-14'
C-6	C-16	C-16'
C-7	C-17	·C-18'
C-8	C-18	C=O
C-9	C-19	|
		CH$_3$
C-10	N-CH$_2$	
	|	
	CH$_3$	

1. A.S. Narzullaev, M.S. Yunusov, and S.Yu. Yunusov, *Khim. Prir. Soedin.*, 10, 412 (1974).
2. S.W. Pelletier, N.V. Mody, K.I. Varughese, J.A. Maddry, and H.K. Desai, *J. Am. Chem. Soc.*, 1981, 103, 6536.

1-DEOXYDELSOLINE

$C_{25}H_{41}NO_6$;　mp 134-135°
Aconitum monticola
^{1}H NMR:　δ 1.41 (3H, *t*, NCH$_2$-C*H*$_3$),
3.1, 3.12, 3.3, 3.16 (each 3H, *s*,
OC*H$_3$*), 3.48 (1H, *d*, J=5Hz, C(14)-β
-*H*).[1]

^{13}C *Chemical Shift Assignments*

C-1	C-11	C-1'
C-2	C-12	C-6'
C-3	C-13	C-7'
C-4	C-14	C-8'
C-5	C-15	C-14'
C-6	C-16	C-16'
C-7	C-17	C-18'
C-8	C-18	C=O
C-9	C-19	CH$_3$
C-10	N-CH$_2$	
	CH$_3$	

1.　E.F. Ametova, M.S. Yunusov and V.A. Tel'nov, *Khim. Prir. Soedin.*, 504 (1982).

3-DEOXYJESACONITINE

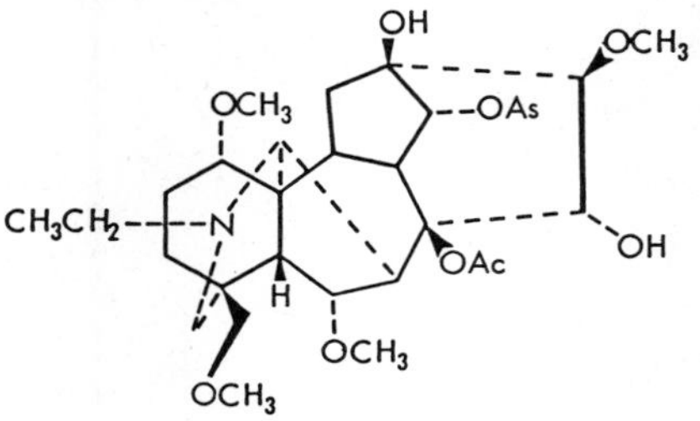

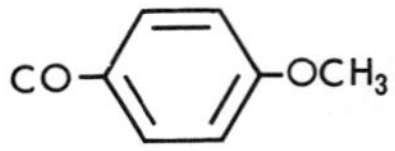

$C_{35}H_{49}NO_{11}$;   mp 174-176°;

$[\alpha]_D$ + 52° (MeOH)

Aconitum subcuneatum Nakai

^1H NMR: δ 1.08 (3H, t, J=7Hz, N-CH$_2$-CH$_3$), 1.43 (3H, s, COCH_3), 3.17, 3.28, 3.30 and 3.74 (each 3H, s, OCH_3), 3.88 (3H, s, aromatic OCH_3), 4.83 (1H, d, J=5Hz, C(14)-β-H), 6.92 and 7.97 (each 2H, A_2B_2, J=8Hz, aromatic protons).[1,2]

As =

^{13}C *Chemical Shift Assignments*[1]

C-1	85.2	C-16	90.2	C=O		165.8
C-2	26.4	C-17	61.4	1	122.3
C-3	35.2	C-18	80.3	2	131.6
C-4	39.0	C-19	53.2	3	113.8
C-5	49.2	N-CH$_2$	49.1	4	163.4
C-6	83.3	CH$_3$	13.4	5	113.8
C-7	45.2[a]	C-1'	56.2	6	131.6
C-8	92.1	C-6'	58.0		
C-9	44.1[a]	C-7'	-	OCH$_3$	55.5
C-10	41.0	C-8'	-		
C-11	49.9	C-14'	-		
C-12	36.6	C-16'	61.1		
C-13	74.2	C-18'	59.1		
C-14	78.7[b]	C=O	172.4		
C-15	78.8[b]	CH$_3$	21.5		

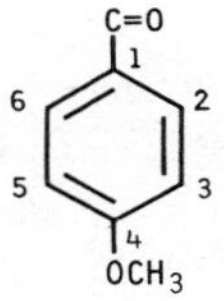

1. H. Bando, Y. Kanaiwa, K. Wada, T. Mori, and T. Amiya, *Heterocycles*, 16, 1723 (1981).

DEOXYLYCOCTONINE

$C_{25}H_{41}NO_6$;   mp 81.5-84.5°;
$[\alpha]_D$ + 50° (EtOH)
Prepared from lycoctonine.[1]

^{13}C Chemical Shift Assignments[2]

C-1	82.8	C-11	49.3	C-1'	57.7
C-2	33.8[a]	C-12	28.9	C-6'	55.6
C-3	37.3	C-13	38.2	C-7'	–
C-4	34.1	C-14	84.5	C-8'	–
C-5	43.4[b]	C-15	26.8[a]	C-14'	58.2
C-6	91.5	C-16	84.1	C-16'	56.2
C-7	88.6	C-17	64.3	C-18'	–
C-8	77.6	C-18	26.8		
C-9	55.2[b]	C-19	56.8		
C-10	46.2[b]	N-CH$_2$	50.9		
		CH$_3$	14.0		

1.   O.E. Edwards and L. Marion, *Can. J. Chem.*, 30, 627 (1952).

2.   A.J. Jones and M.H. Benn, *Can. J. Chem.*, 51, 486 (1973).

3.   S.W. Pelletier, N.V. Mody, K.I. Varughese, J.A. Maddry, and H.K. Desai, *J. Am. Chem. Soc.*, 1981, 103, 6536.

DEOXYMETHYLENELYCOCTONINE

$C_{24}H_{39}NO_6$; mp 146° (perchlorate); $[\alpha]_D$ + 27.8° (EtOH) Prepared from lycoctonine.[1,3]

^{13}C Chemical Shift Assignments[2]

C-1	82.8	C-11	48.7	C-1'	57.6
C-2	33.4[a]	C-12	29.0[a]	C-6'	55.8
C-3	29.6	C-13	38.2[b]	C-7'	-
C-4	36.9[b]	C-14	84.8	C-8'	-
C-5	43.8[c]	C-15	26.1[a]	C-14'	58.3
C-6	94.5	C-16	84.0	C-16'	56.1
C-7	88.5	C-17	65.0	C-18'	-
C-8	77.0	C-18	-		
C-9	49.0[c]	C-19	51.0		
C-10	46.2[c]	N-CH$_2$	50.2		
		CH$_3$	14.0		

1. O.E. Edwards and L. Marion, *Can. J. Chem.*, 30, 627 (1952).

2. A.J. Jones and M.H. Benn, *Can. J. Chem.*, 51, 486 (1973).

3 S.W. Pelletier, N.V. Mody, K.I. Varughese, J.A. Maddry, and H.K. Desai, *J. Am. Chem. Soc.*, 1981, 103, 6536.

DIACETYLPSEUDACONITINE

$Vr = $

$C_{40}H_{55}NO_{14}$; mp 229°
$[\alpha]_D$ + 24° (EtOH)[1]
Aconitum ferox
^{1}H NMR: δ 1.08 (3H, *t*, NCH$_2$-C*H$_3$*),
1.29 (3H, *s*, C(8)-OCOC*H$_3$*), 2.03 (6H,
s, OCOC*H$_3$*), 3.21, 3.24 and 3.39 (each
12H, *s*, OC*H$_3$*), 3.85 (6H, *s*, aromatic
OC*H$_3$*), 4.81 (2H, *m*, C(3)-β-*H* and
C(14)-β-*H*) and 6.58-7.72 (aromatic
protons).[1,2]

^{13}C *Chemical Shift Assignments*

C-1	C-16
C-2	C-17
C-3	C-18
C-4	C-19
C-5	N-CH$_2$
C-6	CH$_3$
C-7	C-1'
C-8	C-6'
C-9	C-7'
C-10	C-8'
C-11	C-14'
C-12	C-16'
C-13	C-18'
C-14	C=O
C-15	CH$_3$

1. T.M. Sharp, *J. Chem. Soc.*, 3094 (1928); Y. Tsuda and L. Marion, *Can. J. Chem.*, 41, 1485 (1963).

2. K.K. Purushothaman and S. Chandrasekharan, *Phytochemistry*, 13, 1975 (1974).

DICTYOCARPINE

$C_{26}H_{39}NO_8$; mp 214.5-216.5° (200-202°).
$[\alpha]_D$ - 14.7° (CHCl$_3$), - 12.8° (MeOH).
Delphinium dictyocarpum, D. glauces-
cens, D. barbeyi, D. iliense.
^{1}H NMR: δ 0.88 (3H, *s*, C(4)-CH$_3$),
1.08 (3H, *t*, NCH$_2$-CH$_3$), 2.12 (3H, *s*,
OCOCH$_3$), 3.30 and 3.39 (each 3H, *s*,
OCH$_3$), 4.70 (1H, *dd*, J=6.6Hz, C(14)-
β-*H*), 5.00 and 5.05 (each 1H, *s*,
OCH$_2$O) and 5.55 (1H, *d*, J=1.5Hz, C(6)-
α-*H*).[1,3,4,5]

[13]C *Chemical Shift Assignments* [2]

C-1	78.7	C-11	55.1s	C-1'	55.6
C-2	26.4t	C-12	36.5t	C-6'	-
C-3	37.6t	C-13	36.6d	C-7-O	
C-4	34.0s	C-14	72.8	C-8-O CH$_2$	94.0
C-5	51.8d	C-15	32.9t	C-14'	-
C-6	77.2	C-16	81.2	C-16'	56.3
C-7	93.0s	C-17	64.4	C-18'	-
C-8	82.9s	C-18	25.5	C=O	170.2s
C-9	50.4d(t)	C-19	56.9t	CH$_3$	21.8
C-10	79.9s	N-CH$_2$ CH$_3$	50.4t(d) 14.0		

1. S.W. Pelletier, O.D. Dailey, Jr., N.V. Mody, and J.D. Olsen, *J. Org. Chem.*, 46, 3284 (1981).

2. S.W. Pelletier, N.V. Mody, and O.D. Dailey, Jr., *Can. J. Chem.*, 58, 1875 (1980).

3. A.S. Narzullaev, M.S. Yunusov, and S.Yu. Yunusov, *Khim. Prir. Soedin.*, 8, 498 (1972), 9, 443 (1973).

4. B.T. Salimov, N.D. Abdullaev, M.S. Yunusov, and S.Yu. Yunusov, *Khim. Prir. Soedin.*, 14, 235 (1978).

5. S.W. Pelletier, N.V. Mody, K.I. Varughese, J.A. Maddry, and H.K. Desai, *J. Am. Chem. Soc.*, 1981, 103, 6536.

DICTYOCARPININE

$C_{24}H_{37}NO_7$; mp 205-206.5°;

$[\alpha]_D$ - 5° (CH$_3$OH)

Delphinium glaucescens, D. iliense.

^{1}H NMR: δ 0.93 (3H, *s*, C(4)-*CH*$_3$),
1.06 (3H, *t*, NCH$_2$-*CH*$_3$), 3.26 and 3.38
(each 3H, *s*, O*CH*$_3$), 4.28 (1H, *d*, J=
1Hz, C(6)-α-*H*), 4.72 (1H, *dd*, J=6.5Hz,
C(14)-β-*H*), 5.12 and 5.24 (each 1H, *s*,
O*CH*$_2$O).[1,2]

[13]C Chemical Shift Assignments[3]

C-1	79.9	C-11	55.4	C-1'	55.6
C-2	26.4	C-12	36.7	C-6'	-
C-3	36.9	C-13	36.5	C-7-O CH$_2$	93.4
C-4	33.9	C-14	72.6	C-8-O	
C-5	51.9	C-15	33.2	C-14'	-
C-6	77.3	C-16	81.2	C-16'	56.3
C-7	93.4	C-17	64.0	C-18'	-
C-8	82.8	C-18	25.4	C=O CH$_3$	
C-9	51.6	C-19	57.2		
C-10	80.5	N-CH$_2$ CH$_3$	50.5 14.0		

1. S.W. Pelletier, O.D. Dailey, Jr., N.V. Mody, and J.D. Olsen, *J. Org. Chem.*,
    46, 3284 (1981).

2. A.S. Narzullaev, M.S. Yunusov and S.Yu. Yunusov, *Khim. Prir. Soedin.*, 9,
    443 (1973).

3. S.W. Pelletier, N.V. Mody, and O.D. Dailey, Jr., *Can. J. Chem.*, 58, 1875
    (1980).

6,14-DIDEHYDRODICTYOCARPININE

$C_{24}H_{33}NO_7$; mp 239-244° (dec);
$[\alpha]_D$ - 26.8° (CH_3OH)
Prepared from dictyocarpinine
^{1}H NMR: δ 0.98 (3H, s, C(4)-CH_3),
1.13 (3H, t, NCH$_2$-CH_3), 3.37 and 3.38
(each 3H, s, OCH_3), 5.17 and 5.63
(each 1H, s, OCH_2O).

^{13}C *Chemical Shift Assignments* [1]

C-1	77.1	C-11	51.6	C-1'	55.7
C-2	26.2	C-12	36.1	C-6'	-
C-3	37.3	C-13	45.2	C-7-O$\!$CH$_2$	96.4
C-4	35.2	C-14	211.2	C-8-O	
C-5	56.5	C-15	31.1	C-14'	-
C-6	215.6	C-16	83.4	C-16'	56.2
C-7	90.6	C-17	63.8	C-18'	-
C-8	86.2	C-18	24.6	C=O	
C-9	59.6	C-19	57.0	CH$_3$	
C-10	79.0	N-CH$_2$	50.2		
		CH$_3$	13.8		

1. S.W. Pelletier, N.V. Mody, and O.D. Dailey, Jr., *Can. J. Chem.*, 58, 1875 (1980).

1,14-DIDEHYDRONEOLINE

$C_{24}H_{35}NO_6$; mp 170-172°;
Prepared from neoline
¹H NMR: δ 1.15 (3H, t, J=7Hz, NCH₂-
CH_3), 3.25, 3.27 and 3.35 (each 3H, s,
OCH_3).[1]

¹³C Chemical Shift Assignments [2]

C-1	213.2	C-11	61.0	C-1'	-
C-2	41.5	C-12	31.9	C-6'	58.1
C-3	38.9	C-13	47.2	C-7'	-
C-4	39.5	C-14	216.5	C-8'	-
C-5	53.0	C-15	41.8	C-14'	-
C-6	82.8	C-16	86.8	C-16'	56.0
C-7	53.0	C-17	64.0	C-18'	59.2
C-8	80.3	C-18	79.0	C=O	
C-9	54.8	C-19	54.8	CH₃	
C-10	37.6	N-CH₂	48.6		
		CH₃	13.4		

1.   S.W. Pelletier, Z. Djarmati, S. Lajsic, and W.H. De Camp, *J. Am. Chem.
     Soc.*, 98, 2617 (1976).
2.   S.W. Pelletier and Z. Djarmati, *J. Am. Chem. Soc.*, 98, 2626 (1976).

DIHYDROGADESINE

$C_{23}H_{37}NO_6$; mp 136-138;
$[\alpha]_D$ + 54° (EtOH)
Delphinium pentagynum
^{1}H NMR: δ 1.10 (3H, s, C(4)-CH_3),
1.10 (3H, t, J=7Hz, N-CH_2-CH_3), 3.37
(6H, s, O-CH_3), 3.65 (1H, m, w½=6Hz,
C(1)-β-H), 3.98 (1H, s, C(6)-α-H),
and 4.12 (1H, dd, J_1=J_2=4.5Hz, C(14)-
β-H).^{1}

^{13}C *Chemical Shift Assignments*

C-1	C-11	C-1'
C-2	C-12	C-6'
C-3	C-13	C-7'
C-4	C-14	C-8'
C-5	C-15	C-14'
C-6	C-16	C-16'
C-7	C-17	C-18'
C-8	C-18	C=O CH₃
C-9	C-19	
C-10	N-CH₂ CH₃	

1. A.G. Gonzalez, G. de la Fuente, and R. Diaz, *Phytochemistry*, 21, 1781 (1982).

DIHYDROMONTICAMINE

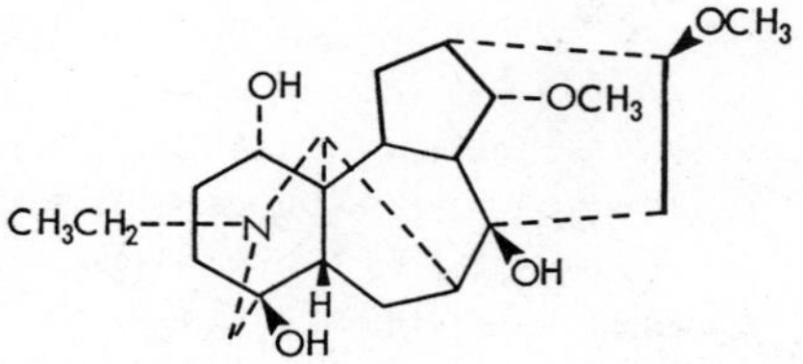

$C_{22}H_{35}NO_5$; mp 156-157°

Aconitum monticola

^{1}H NMR (Py.): δ 0.92 (3H, *t*, NCH$_2$-*CH$_3$*), 3.14 and 3.28 (each 3H, *s*, O*CH$_3$*) and 3.75 (1H, *t*, C(14)-β-*H*).[1]

[13]C Chemical Shift Assignments[2]

C-1	72.0	C-11	50.0	C-1'	–
C-2	29.9	C-12	30.3	C-6'	–
C-3	33.4	C-13	43.4	C-7'	–
C-4	70.3	C-14	84.8	C-8'	–
C-5	48.0	C-15	42.9	C-14'	57.6
C-6	24.6	C-16	82.8	C-16'	56.1
C-7	45.3	C-17	62.7	C-18'	–
C-8	75.0	C-18	–		
C-9	45.7	C-19	60.2		
C-10	37.1	N-CH$_2$	48.0		
		CH$_3$	13.0		

1.   E.F. Ametova, M.S. Yunusov and V.A. Tel'nov, *Khim. Prir. Soedin.*, 504 (1982).

2.   M.S. Yunusov, Personal Communication, May 14, 1981.

DIHYDROMONTICOLINE

$C_{22}H_{35}NO_6$; mp 211-214°; (HCl salt)
Reduction product of monticoline
^{1}H NMR: δ 1.06 (3H, t, NCH_2-CH_3),
3.29 and 3.36 (each 3H, s, OCH_3),
and 3.69 (1H, t, C(14)-β-H).[1]

[13]C *Chemical Shift Assignments*[1]

C-1	72.2	C-11	51.2	C-1'	–
C-2	29.7	C-12	30.2	C-6'	–
C-3	33.0	C-13	43.1	C-7'	–
C-4	70.9	C-14	84.8	C-8'	–
C-5	47.0	C-15	36.2	C-14'	57.7
C-6	33.2	C-16	82.6	C-16'	56.2
C-7	85.9	C-17	63.7	C-18'	–
C-8	77.1	C-18	–		
C-9	48.4	C-19	59.6		
C-10	37.1	$N-CH_2$	50.0		
		CH_3	13.7		

1.   E.F. Ametova, M.S. Yunusov, V.E. Bannikova, N.D. Abdullaev, and V.A.
Tel'nov, *Khim. Prir. Soedin.*, 466 (1981).

DIHYDROPENTAGYNINE

$C_{23}H_{37}NO_5$; mp 150-154°;
$[\alpha]_D$ + 43° (EtOH)
Delphinium pentagynum
^{1}H NMR: δ 1.08 (3H, *s*, C(4)-CH$_3$),
1.12 (3H, *t*, J=7Hz, N-CH$_2$-CH$_3$), 3.38
(6H, *s*, OCH$_3$), 3.78 (1H, *m*, w½=6Hz,
C(1)-β-H), 3.97 (1H, *d*, J=8Hz, C(6)-β
-H) and 4.12 (1H, *dd*, J$_1$=J$_2$=4.5Hz,
C(14)-β-H).[1]

^{13}C Chemical Shift Assignments

C-1	C-11	C-1'
C-2	C-12	C-6'
C-3	C-13	C-7'
C-4	C-14	C-8'
C-5	C-15	C-14'
C-6	C-16	C-16'
C-7	C-17	C-18'
C-8	C-18	C=O
C-9	C-19	CH$_3$
C-10	N-CH$_2$	
	CH$_3$	

1.   A.G. Gonzalez, G. de la Fuente and R. Diaz, *Phytochemistry*, 21, 1781
     (1982).

7,18-DI-*O*-METHYLLYCOCTONINE

$C_{27}H_{45}NO_7$; Amorphous;
Delphinium dictyocarpum
^{1}H NMR: δ 0.99 (3H, *t*, NCH_2-CH_3),
3.20, 3.28, 3.32, 3.38, 3.40 and 3.57
(each 3H, *s*, OCH_3), 3.97 (1H, *s*, OH),
and 4.00 (1H, *s*, C(6)-α-*H*).[1,2,3]

[13]*C Chemical Shift Assignments*[1]

C-1	82.9	C-11	49.0	C-1'	55.8
C-2	30.0	C-12	26.2	C-6'	56.3[a]
C-3	32.3	C-13	38.9	C-7'	55.5
C-4	38.0	C-14	84.5	C-8'	–
C-5	50.4	C-15	33.5	C-14'	57.7
C-6	85.5	C-16	84.2	C-16'	56.0[a]
C-7	92.8	C-17	66.4	C-18'	59.0
C-8	80.0	C-18	78.1		
C-9	44.5	C-19	52.4		
C-10	45.6	$N-CH_2$	51.4		
		CH_3	14.0		

1. S. Sakai, N. Shinma, S. Hasegawa, and T. Okamoto, *Yakugaku Zasshi*, 98, 1376 (1978).

2. B.T. Salimov, M.S. Yunusov, S.Yu. Yunusov and A.S. Narzullaev, *Khim. Prir. Soedin.*, 11, 665 (1975).

3. S.W. Pelletier, N.V. Mody, K.I. Varughese, J.A. Maddry, and H.K. Desai, *J. Am. Chem. Soc.*, 1981, 103, 6536.

DOLACONINE (14-ACETYLACONOSINE)

$C_{24}H_{37}NO_5$;   mp 44-46°
Aconitum stapfianum Hand Mazz var.
pubisceps Wang[1]

[13]C Chemical Shift Assignments

C-1	C-11	C-1'
C-2	C-12	C-6'
C-3	C-13	C-7'
C-4	C-14	C-8'
C-5	C-15	C-14'
C-6	C-16	C-16'
C-7	C-17	C-18'
C-8	C-18	C=O
C-9	C-19	CH₃
C-10	N-CH₂	
	CH₃	

1.   S. Luo and W. Chen, *Huaxue Xuebao.*, 39, 808 (1981); (CA, 97, 10707 (1982)).

ELATINE

$C_{38}H_{50}N_2O_{10}$; mp 222-225°;
$[\alpha]_D$ + 3.4° (CHCl$_3$)
Delphinium elatum.[1,2,3,4]

^{13}C *Chemical Shift Assignments*

C-1	C-11	C-1'
C-2	C-12	C-6'
C-3	C-13	C-7'
C-4	C-14	C-8'
C-5	C-15	C-14'
C-6	C-16	C-16'
C-7	C-17	C-18'
C-8	C-18	C=O
C-9	C-19	CH$_3$
C-10	N-CH$_2$	
	CH$_3$	

1. M.S. Rabinovich, *J. Gen. Chem.* USSR (English Transl.), 24, 2211 (1954).

2. A.D. Kuzovkov, *J. Gen. Chem.* USSR (English Transl.), 25, 399 (1955).

3. A.D. Kuzovkov and A.V. Bocharnikora, *J. Gen. Chem.* USSR (English Transl.) 28, 546 (1958).

4. S.W. Pelletier, N.V. Mody, K.I. Varughese, J.A. Maddry and H.K. Desai, *J. Am. Chem. Soc.*, 1981, 103, 6536.

EPISCOPALISINE

$C_{29}H_{39}NO_6$; MW: 497.2779;
$[\alpha]_D$ - 11.7°
Aconitum episcopale
¹H NMR (200 MHz): δ 1.05 (3H, *t*, J=
7Hz, NCH$_2$-*CH$_3$*), 1.10-1.50 (4H, *m*, C(3),
C(6)-*H*), 3.28, 3.40 (each 3H, *s*, O*CH$_3$*),
5.10 (1H, *d*, J=4.5Hz, C(14)-β-*H*), 7.42,
7.54, 8.03 (5H, *m*, aromatic protons).[1]

¹³C Chemical Shift Assignments [1]

C-1	86.1	C-16	83.5	C=O		167.3
C-2	25.9	C-17	63.1		1	130.4
C-3	35.7	C-18	-		2	130.0
C-4	36.1	C-19	50.1		3	128.7
C-5	46.9	N-CH$_2$	49.7		4	133.3
C-6	29.1^a	CH$_3$	13.1		5	128.7
C-7	44.9	C-1'	56.7		6	130.0
C-8	76.7	C-6'	-			
C-9	74.3	C-7'	-			
C-10	36.1	C-8'	-			
C-11	48.9	C-14'	-			
C-12	29.7^a	C-16'	58.3			
C-13	42.1	C-18'	-			
C-14	80.8	C=O				
C-15	41.3	CH$_3$				

1.   F.-P. Wang and Q.-C. Fang, *Acta Pharmaceutica Sinica*, 18, 514 (1983).

EPISCOPALISININE

$C_{22}H_{35}NO_5$; MW: 393.2492;
mp 152-154°; $[\alpha]_D$ - 3.8°
Aconitum episcopale
^{1}H NMR (200 MHz): δ 1.06 (3H, *t*, J=
7Hz, NCH_2-*CH_3*), 1.10-1.50 (4H, *m*, C(3),
C(6)-*H*), 3.26, 3.42 (each 3H, *s*,
OC*H_3*), 4.02 (1H, *d*, J=4.5Hz, C(14)-β
-*H*).[1]

[13]*C Chemical Shift Assignments*[1]

C-1	85.8	C-11	47.9	C-1'	56.5
C-2	25.5	C-12	29.6	C-6'	-
C-3	35.1	C-13	52.7	C-7'	-
C-4	35.9	C-14	79.9	C-8'	-
C-5	45.5	C-15	39.5	C-14'	-
C-6	28.5	C-16	84.1	C-16'	57.5
C-7	44.7	C-17	63.5	C-18'	-
C-8	77.3	C-18	-		
C-9	73.5	C-19	53.9		
C-10	41.9	N-CH_2	50.5		
		CH_3	13.1		

1. F.-P. Wang and Q.-C. Fang, *Acta Pharmaceutica Sinica*, 18, 514 (1983).

EPISCOPALITINE

$C_{24}H_{37}NO_5$;   MW: 419.2646;
$[\alpha]_D$ - 0.9°
Aconitum episcopale
^{1}H NMR (200 MHz):   δ 1.05 (3H, *t*, J=
7Hz, NCH_2-CH_3), 1.10-1.50 (4H, *m*, C(3),
C(6)-*H*), 2.02 (3H, *s*, OCOCH_3), 3.22,
3.26 (each 3H, *s*, OCH_3), 4.81 (1H, *t*,
J=4.5Hz, C(14)-β-*H*).[1]

[13]C *Chemical Shift Assignments*[1]

C-1	86.1	C-11	50.3	C-1'	56.0
C-2	26.3	C-12	29.1[a]	C-6'	-
C-3	36.8	C-13	44.7	C-7'	-
C-4	35.3	C-14	77.6	C-8'	-
C-5	49.5	C-15	41.4	C-14'	-
C-6	28.3[a]	C-16	81.9	C-16'	56.5
C-7	48.7	C-17	62.9	C-18'	-
C-8	73.9	C-18	-		
C-9	46.3	C-19	56.0		
C-10	35.5	N-CH_2	50.3		
		CH_3	13.1		

1.   F.-P. Wang and Q.-C. Fang, *Acta Pharmaceutica Sinica*, 18, 514 (1983).

8,15-EPOXYPYRODELPHININE

$C_{31}H_{41}NO_8$; mp 203.5-204°;
Prepared from pyrodelphinine
^{1}H NMR: δ 2.37 (3H, *s*, N*CH₃*), 3.39
and 3.42 (each 3H, *s*, O*CH₃*), 3.41
(6H, *s*, O*CH₃*), 5.01 (1H, *dd*, C(14)-β
-*H*) and 7.60-8.25 (aromatic protons).

[13]C *Chemical Shift Assignments* [1]

C-1	85.8	C-16	83.4	C=O		167.3
C-2	25.5	C-17	69.0		1	130.4
C-3	35.0	C-18	80.1		2	130.0
C-4	39.8	C-19	56.3		3	128.1
C-5	50.0	N-CH₃	42.5		4	132.6
C-6	82.5				5	128.1
C-7	53.5	C-1'	56.3		6	130.0
C-8	62.8	C-6'	57.9			
C-9	44.1	C-7'	-			
C-10	41.4	C-8'	-			
C-11	51.7	C-14'	-			
C-12	37.6	C-16'	57.6			
C-13	75.5	C-18'	59.1			
C-14	78.3	C±O				
C-15	45.4	CH₃				

1.   S.W. Pelletier, N.V. Mody and Y. Ohtsuka, unpublished results.

EXCELSINE

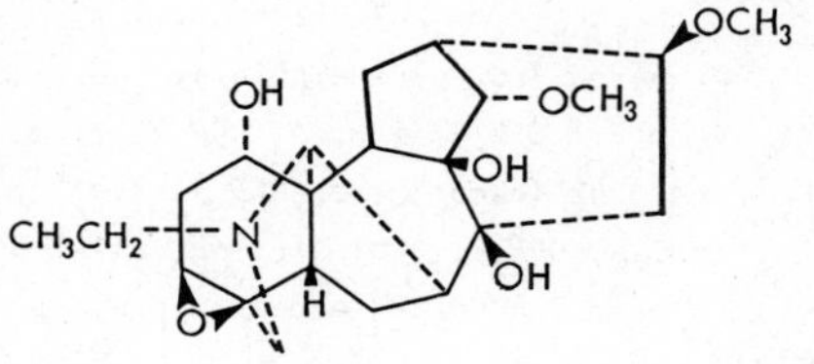

$C_{22}H_{33}NO_6$; MW: 407.2317

mp 103-105°;

Aconitum excelsum

^{1}H NMR: δ 1.01 (3H, *t*, NCH_2-CH_3),

3.13 and 3.18 (each 3H, *s*, OCH_3).[1,2,3]

^{13}C *Chemical Shift Assignments*

C-1	C-11	C-1'
C-2	C-12	C-6'
C-3	C-13	C-7'
C-4	C-14	C-8'
C-5	C-15	C-14'
C-6	C-16	C-16'
C-7	C-17	C-18'
C-8	C-18	C=O CH₃
C-9	C-19	
C-10	N-CH₂ CH₃	

1. V.A. Tel'nov, M.S. Yunusov and S.Yu. Yunusov, *Khim. Prir. Soedin.*, 9, 129 (1973).

2. S.M. Nasirov, V.G. Andrianov, Yu.T. Struchkov, V.A. Tel'nov, M.S. Yunusov, and S.Yu. Yunusov, *Khim. Prir. Soedin.*, 10, 812 (1974).

3. S.M. Nasirov, V.G. Andrianov, Yu.T. Struchkov and S.Yu. Yunusov, *Khim. Prir. Soedin.*, 12, 206 (1976).

EZOCHASMACONITINE

$C_{34}H_{47}NO_8$; mp 163-165°;
$[\alpha]_D$ + 26.1° (CHCl$_3$)
Aconitum yesoense

^{1}H NMR: δ 1.10 (3H, *t*, NCH$_2$-C*H$_3$*),
1.76 (3H, *s*, OCOC*H$_3$*), 2.94, 3.24,
3.24, 3.32 (each 3H, *s*, OC*H$_3$*), 4.82
(1H, *t*, J=4.5Hz, C(14)-β-*H*), 7.45-
8.0 (5H, aromatic protons).[1]

^{13}C *Chemical Shift Assignments* [2]

C-1	84.7	C-16	82.6^b	C=O		164.5
C-2	26.4	C-17	61.4		1	131.0
C-3	34.8	C-18	80.2		2	129.1
C-4	39.0	C-19	53.7		3	128.1
C-5	49.1^a	N-CH$_2$	48.9		4	132.5
C-6	83.4^b	CH$_3$	13.4		5	128.1
C-7	49.2^a	C-1'	56.5		6	129.1
C-8	86.4	C-6'	57.8			
C-9	44.8	C-7'	-			
C-10	39.2	C-8'	-			
C-11	50.2	C-14'	-			
C-12	29.0	C-16'	55.9			
C-13	43.9	C-18'	58.9			
C-14	75.6	C=O	171.1			
C-15	37.6	CH$_3$	21.4			

1. H. Takayama, M. Ito, M. Koga, S. Sakai and T. Okamoto, *Heterocycles*, 15, 403 (1981).

2. H. Takayama, A. Tokita, M. Ito, S. Sakai, F. Kurosaki and T. Okamoto, *J. Pharm. Soc. Japan*, 102, 245 (1982).

EZOCHASMANINE

$C_{25}H_{41}NO_7$; mp 115-118°;
$[\alpha]_D$ + 40° (CHCl$_3$)
Aconitum yesoense
^{1}H NMR: δ 1.08 (3H, *t*, J=7Hz, NCH$_2$-
CH$_3$), 3.21, 3.31, 3.31, 3.33 (each 3H,
s, OCH$_3$), and 4.10 (1H, *t*, J=5Hz, C(14)
-β-*H*).

^{13}C *Chemical Shift Assignments* [1,2]

C-1	83.2	C-11	50.2	C-1'	56.4
C-2	33.9	C-12	28.1	C-6'	57.3
C-3	72.2	C-13	45.3	C-7'	–
C-4	43.5	C-14	75.5	C-8'	–
C-5	48.5	C-15	39.1	C-14'	–
C-6	82.2[a]	C-16	82.0[a]	C-16'	56.0
C-7	52.4	C-17	62.2	C-18'	59.2
C-8	72.5	C-18	77.4	C=O	
C-9	48.8	C-19	47.4	CH$_3$	
C-10	38.1	N-CH$_2$	49.1		
		CH$_3$	13.7		

1.   H. Takayama, M. Ito, M. Koga, S. Sakai and T. Okamoto, *Heterocycles*, 15, 403 (1981).

2.   H. Takayama, A. Tokita, M. Ito, S. Sakai, F. Kurosaki and T. Okamoto, *J. Pharm. Soc.* Japan, 102, 245 (1982).

FALACONITINE

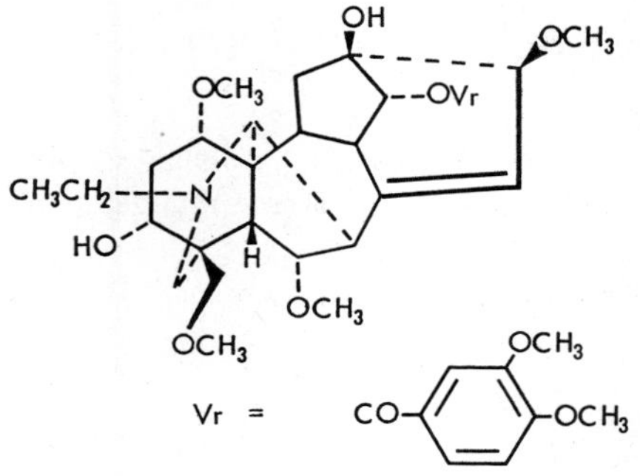

$C_{34}H_{47}NO_{10}$; Amorphous;
$[\alpha]_D$ + 111.5° (EtOH)
Aconitum falconeri
^{1}H NMR: δ 1.10 (3H, *t*, NCH$_2$-*CH$_3$*),
3.25, 3.30, 3.34 and 3.45 (each 3H, *s*,
O*CH$_3$*), 3.92 (6H, *s*, aromatic O*CH$_3$*),
4.92 (1H, *d*, C(14)-β-*H*) and 5.57 (1H,
d, J=6Hz, C(15)-*H*).[1]

Vr =

[13]*C Chemical Shift Assignments*[1,2]

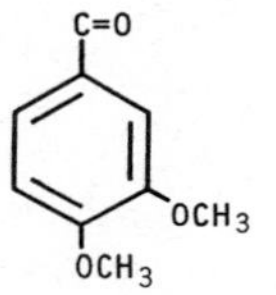

C-1	83.8[a]	C-16	83.1	C=O		167.5
C-2	38.0	C-17	77.8			122.9
C-3	71.4	C-18	76.1			110.3
C-4	44.0	C-19	49.7		OCH$_3$	153.0
C-5	48.0	N-CH$_2$	47.7	OCH$_3$		148.5
C-6	83.7[a]	CH$_3$	13.5			112.5
C-7	49.5	C-1'	56.2			
C-8	146.6	C-6'	58.0		OCH$_3$	55.9
C-9	48.2	C-7'	-			
C-10	46.2	C-8'	-			
C-11	51.6	C-14'	-			
C-12	33.4	C-16'	57.3			
C-13	77.4	C-18'	59.2			
C-14	78.1					
C-15	116.1					

1. S.W. Pelletier, N.V. Mody, and H.S. Puri, *J.C.S. Chem. Comm.*, 12 (1977).
2. S.W. Pelletier, N.V. Mody, R.S. Sawhney and J. Bhattacharyya, *Heterocycles*, **7**, 327 (1977).

FINACONITINE (10-β-HYDROXYRANACONITINE)

$C_{32}H_{44}N_2O_{10}$; mp 220-221°C;
$[\alpha]_D$ + 44.7° (CH$_3$OH)
Aconitum finetianum
^{1}H NMR: δ 1.08 (3H, *t*, J=7Hz, NCH$_2$-
CH$_3$), 2.20 (3H, *s*, NHCOCH$_3$), 3.26, 3.30,
3.40 (each 3H, *s*, OCH$_3$), 3.78 (1H, *d*,
J=4.5Hz, C(14)-β-*H*), 7.02-8.70 (4H, *m*,
aromatic protons), 11.10 (1H, *s*,
N*H*COCH$_3$).[1,2,3]

^{13}C *Chemical Shift Assignments* [2]

C-1	77.1d	C-16	82.7d		C=O		167.4s
C-2	26.5t	C-17	64.3d			1	115.8s
C-3	31.5t	C-18	-			2	141.6s
C-4	84.6s	C-19	55.1t			3	120.3s
C-5	44.0d	N-CH$_2$ CH$_3$	51.0t			4	134.4s
C-6	32.9t		14.5q			5	122.5s
C-7	76.6s	C-1'	55.9q			6	131.0s
C-8	84.9s	C-6'	-				
C-9	79.5s	C-7'	-				
C-10	78.5s	C-8'	-				
C-11	57.0s	C-14'	57.9q				
C-12	37.1t	C-16'	56.3q				
C-13	34.8d	C-18'	-				
C-14	87.7d	NH-C=O CH$_3$	169.3s 25.5q				
C-15	37.6t						

1. Zhu Yuanlong (Chu Yuan-Lung) and Zhu Renhong (Chu Jen-Hung), *Hetero-
    cycles*, 17, 607 (1982).

2. S.-H. Jiang, Y.-L. Zhu and R.-H. Zhu, *Acta Pharmaceutica Sinica*, 17,
    282 (1982).

3. Wei Biyu, Kong Xiancheng, Zhao Zhiyuan, Wang Hongcheng, and Zhu Renhong,
    Bull. Chinese Materia Medica, 6 (2), 26 (1981).

FLAVACONITINE

$C_{31}H_{41}NO_{11}$;
Aconitum flavum[1]

^{13}C *Chemical Shift Assignments*

C-1	C-16
C-2	C-17
C-3	C-18
C-4	C-19
C-5	N-CH$_2$
C-6	CH$_3$
C-7	C-1'
C-8	C-6'
C-9	C-7'
C-10	C-8'
C-11	C-14'
C-12	C-16'
C-13	C-18'
C-14	C=O
C-15	CH$_3$

1.   Y. Liu and G. Chang, *Acta Pharmaceutica Sinica*, 17, 243 (1982).

FORESACONITINE (VILMORRIANINE C)

$C_{35}H_{49}NO_9$; mp 153-154°;
$[\alpha]_D$ + 30.5° (CHCl$_3$)

Aconitum forestii Stapf var. *albo-villosum* (Chen et Liu) W.T. Wang, *A. vilmorrianum*.

^{1}H NMR: δ 1.08 (3H, *t*, J=7Hz, NCH$_2$-CH$_3$), 1.40 (3H, *s*, C(8)-OCOCH$_3$), 3.18, 3.26, 3.28, 3.37 and 3.87 (each 3H, *s*, OCH$_3$), 4.07 (1H, *d*, J=6Hz, C(6)-β-*H*), 5.04 (1H, *d*, J=4.5Hz, C(14)-β-*H*), 6.90 and 8.00 (aromatic protons).[1,2]

^{13}C *Chemical Shift Assignments*[1]

C-1	85.1d	C-16	83.5d[a]	C=0		166.2s
C-2	26.4t	C-17	61.7d	1		123.0s
C-3	34.9t	C-18	80.4t	2		131.8d
C-4	39.1s	C-19	53.8t	3		113.7d
C-5	49.2d	N-CH$_2$	49.0t	4		163.5s
C-6	82.9d[a]	CH$_3$	13.4q	5		113.7d
C-7	44.9d	C-1'	56.6q	6		131.8d
C-8	85.9s	C-6'	57.8q			
C-9	49.3d	C-7'	-	OCH$_3$		55.4q
C-10	43.9d	C-8'	-			
C-11	50.3s	C-14'	-			
C-12	29.0t	C-16'	56.0q			
C-13	39.1d	C-18'	59.1q			
C-14	75.4d	C=0	169.8s			
C-15	37.9t	CH$_3$	21.8q			

1. W.-S. Chen and E. Breitmaier, *Chem. Ber.*, 114, 394 (1981).
2. C.-R. Yang, X.-H. Hao, D.-Z. Wang and J. Zhou, *Acta Chimica Sinica*, 39(2), 147 (1981).

FORESTICINE

$C_{24}H_{39}NO_6$; MW: 437.2785;
mp 79-80°; $[\alpha]_D$ - 1.9° $(CHCl_3)$;
Aconitum forestii

^{1}H NMR: δ 1.12 (3H, *t*, J=7Hz, NCH$_2$-
C*H$_3$*), 3.32, 3.40,3.42 (each 3H, *s*,
OC*H$_3$*), 4.20 (1H, *t*, J=4.5Hz, C(14)-β-
H), 4.90 (1H, *d*, J=7Hz, C(6)-β-*H*).[1]

[13]C *Chemical Shift Assignments* [1]

C-1	85.7	C-11	50.6(s)	C-1'	56.1
C-2	25.8	C-12	28.8	C-6'	-
C-3	34.8	C-13	49.6	C-7'	-
C-4	39.1(s)	C-14	75.3	C-8'	-
C-5	49.3	C-15	39.4	C-14'	-
C-6	71.9	C-16	82.2	C-16'	56.4
C-7	54.3	C-17	62.6	C-18'	59.2
C-8	74.0(s)	C-18	80.8		
C-9	48.9	C-19	54.3		
C-10	38.7	N-CH$_2$	50.4		
		CH$_3$	13.5		

1.   S.W. Pelletier, S.-Y. Chen, B.S. Joshi, and H.K. Desai, *J. Nat. Prod.*, in press.

FORESTINE

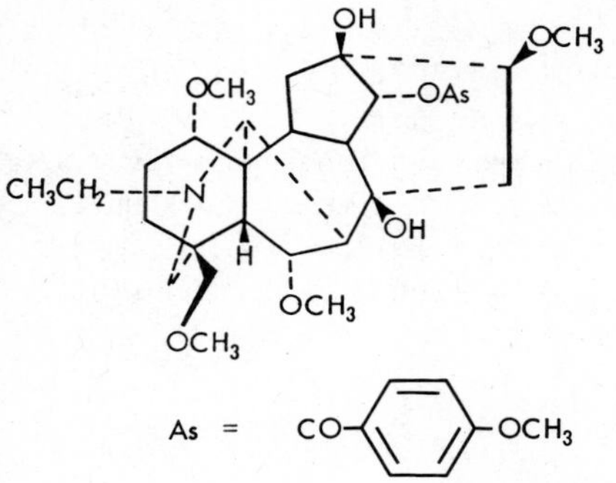

$C_{33}H_{47}NO_9$; Amorphous;
$[\alpha]_D$ + 21.6° (EtOH);
Aconitum forestii
^{1}H NMR: δ 1.10 (3H, *t*, J=7Hz, NCH$_2$-CH$_3$), 3.30, 3.32, 3.33, 3.40 (each 3H, *s*, OCH$_3$), 3.86 (3H, *s*, aromatic OCH$_3$), 4.06 (1H, *d*, J=6Hz, C(6)-β-H), 5.12 (1H, *d*, J=5Hz, C(14)-β-H), 6.95, 8.05 (each 2H, *d*, J=9Hz, A B *q*, aromatic protons).[1]

As =

^{13}C *Chemical Shift Assignments*[1]

C-1	85.4	C-16	83.3[b]	C=O	166.6(s)
C-2	26.0	C-17	62.2	1	122.4(s)
C-3	34.9	C-18	80.6	2	131.8
C-4	39.3(s)	C-19	53.6	3	113.8
C-5	49.6[a]	N-CH$_2$	48.3	4	163.4(s)
C-6	82.5[b]	CH$_3$	13.6	5	113.8
C-7	49.2[a]	C-1'	56.3	6	131.8
C-8	73.7(s)	C-6'	58.3		
C-9	53.6	C-7'	-	OCH$_3$	55.4
C-10	42.3	C-8'	-		
C-11	50.2(s)	C-14'	-		
C-12	36.4	C-16'	57.5		
C-13	76.1(s)	C-18'	59.2		
C-14	80.1	C=O			
C-15	41.9	CH$_3$			

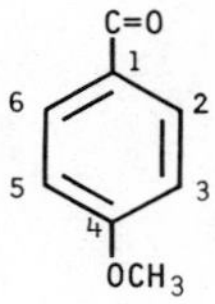

1.   S.W. Pelletier, S.-Y. Chen, B.S. Joshi and H.K. Desai, *J. Nat. Prod.*, in press.

GADENINE

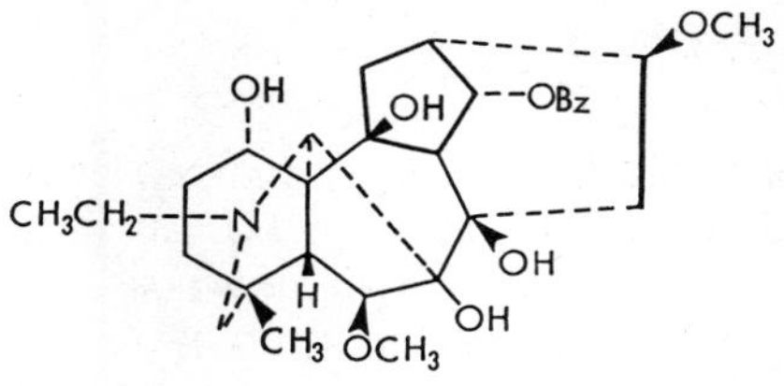

$C_{30}H_{41}NO_8$;   mp 147-150°

Delphinium pentagynum

^{1}H NMR: δ 1.11 (3H, *s*, CH_3), 1.11 (3H, *t*, J=7Hz, NCH_2-CH_3), 3.32, 3.36 (each 3H, *s*, OCH_3), 4.02 (1H, *s*, C(6)-α-*H*), 4.15 (1H, *m*, w½=7Hz, C(1)-β-*H*), 5.55 (1H, *t*, J=4.5Hz, C(14)-β-*H*), 7.5-8.15 (5H, *m*, aromatic protons).[1]

[13]*C Chemical Shift Assignments* [1]

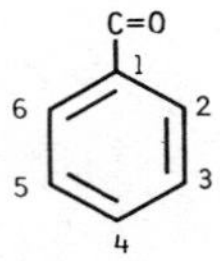

C-1	70.0	C-16	81.5	C=O	166.6
C-2	30.0	C-17	66.3	1	130.5
C-3	31.5	C-18	27.7	2	130.0
C-4	32.7	C-19	61.0	3	128.4
C-5	46.0	$N-CH_2$	50.4	4	132.8
C-6	91.5	CH_3	13.4	5	128.4
C-7	87.0	C-1'	-	6	130.0
C-8	76.8	C-6'	58.2		
C-9	53.0	C-7'	-		
C-10	81.7	C-8'	-		
C-11	54.3	C-14'	-		
C-12	40.5	C-16'	56.2		
C-13	38.2	C-18'	-		
C-14	75.5	C=O			
C-15	35.2	CH_3			

1.   A.G. Gonzalez, G. de la Fuente and R.O. Acosta, *Heterocycles*, **22**, 17 (1983).

GADESINE

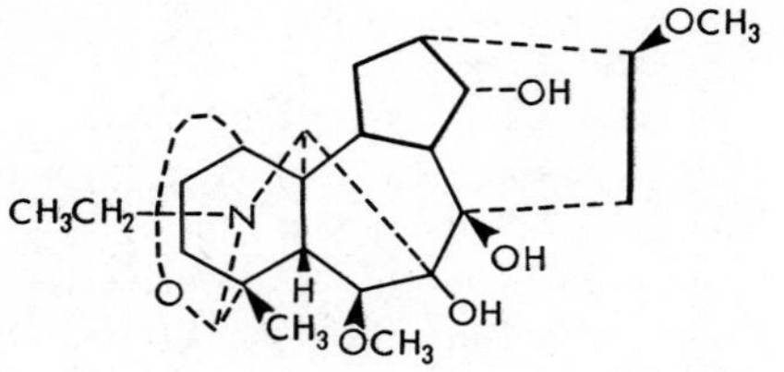

$C_{23}H_{35}NO_6$; mp 174-177°;

$[\alpha]_D$ + 76° (EtOH)

Delphinium pentagynum

^{1}H NMR: δ 1.00 (3H, *s*, C(4)-CH₃), 1.09 (3H, *t*, NCH₂-CH₃), 3.35 and 3.37 (each 3H, *s*, OCH₃), 3.69 (1H, *m*, w½=7Hz, C(1)-β-H), 3.78 (1H, *s*, C(19)-H), 3.82 (1H, *d*, J=1Hz, C(6)-α-H), and 3.97 (1H, *dd*, J₁=J₂=4.5Hz, C(14)-β-H).[1]

[13]C Chemical Shift Assignments[2]

C-1	89.0	C-11	46.4	C-1'	–
C-2	30.3*	C-12	27.7	C-6'	58.4
C-3	22.5*	C-13	45.4	C-7'	–
C-4	38.4	C-14	75.3	C-8'	–
C-5	38.3	C-15	33.8	C-14'	–
C-6	90.9	C-16	81.8	C-16'	56.4
C-7	84.9	C-17	64.1	C-18'	–
C-8	76.0	C-18	20.1	C=O	
C-9	52.2	C-19	68.3	CH₃	
C-10	36.9	N-CH₂	47.3		
		CH₃	13.7		

1. A.G. Gonzalez, G. de la Fuente, R. Diaz, J. Fayos and M. Martinez-Ripoll, *Tetrahedron Letters*, 79 (1979).

2. A.G. Gonzalez, R. Diaz, and G. de la Fuente, Private Communication, January 8, 1980.

*. G. de la Fuente, private communication, November 8, 1983.

GIGACTONINE

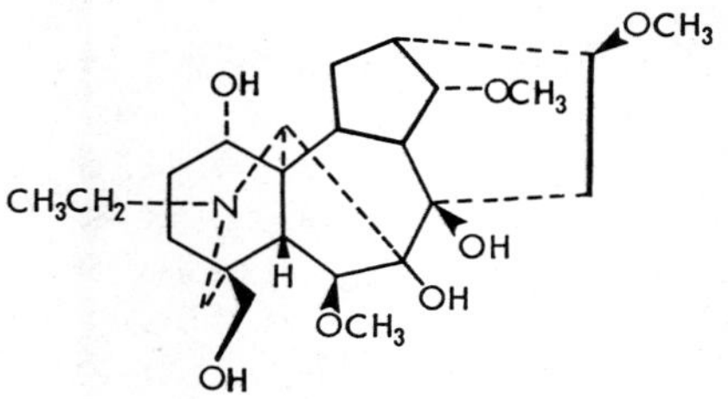

$C_{24}H_{39}NO_7$; mp 168-169° (acetone:benzene); 94-95° ($CHCl_3$:hexane);

$[\alpha]_D$ + 49° (EtOH)

Aconitum gigas, Consolida orientalis.

^{1}H NMR: δ 1.08 (3H, *t*, J=7Hz, NCH_2-CH_3), 2.92 (2H, *q*, J=7Hz, NCH_2-CH_3), 3.33 (3H, *s*, OCH_3) and 3.40 (6H, *s*, OCH_3), 3.98 (1H, *s*, C(6)-α-*H*).[1,2,3]

[13]*C Chemical Shift Assignments*[1]

C-1	72.7	C-11	49.4	C-1'	-
C-2	29.4	C-12	26.7	C-6'	57.7
C-3	30.5	C-13	37.8	C-7'	-
C-4	38.2	C-14	84.6	C-8'	-
C-5	44.7	C-15	33.5	C-14'	57.7
C-6	90.6	C-16	83.0	C-16'	56.4
C-7	87.8	C-17	66.1	C-18'	-
C-8	78.5	C-18	66.8	C=O CH3	
C-9	43.4	C-19	57.3		
C-10	44.0	N-CH2 CH3	50.4 13.6		

1. S. Sakai, N. Shinma, S. Hasegawa, and T. Okamoto, *J. Pharm. Soc.* Japan, 98, 1376 (1978).

2. S. Sakai, N. Shinma, and T. Okamoto, *Heterocycles*, 8, 207 (1977).

3. S.W. Pelletier, N.V. Mody, K.I. Varughese, J.A. Maddry, and H.K. Desai, *J. Am. Chem. Soc.*, 1981, 103, 6536.

GLAUCEDINE

$C_{30}H_{49}NO_8$; mp 117-120°;
$[\alpha]_D$ + 36.4° (CH_3OH), + 39.1 (MeOH).
Delphinium glaucescens
^{1}H NMR: δ 0.90 (3H, *t*, J=7.5Hz, CH_2-CH_3), 1.03 (3H, *t*, NCH_2-CH_3), 1.15 (3H, *d*, J=7Hz, $CHCH_3$), 3.28 and 3.43 (each 3H, *s*, OCH_3), 3.33 (6H, *s*, OCH_3), 3.88 (1H, *d*, J=3Hz, C(6)-α-*H*) and 4.82 (1H, *dd*, J=5.5Hz, C(14)-β-*H*).[1]

$$^{13}C \ \textit{Chemical Shift Assignments}^1$$

C-1	84.3	C-16	82.3	C=O	176.9
C-2	26.2	C-17	64.8	CH	41.3
C-3	32.4	C-18	78.1	CH_3	16.2
C-4	37.7	C-19	52.8	CH_2	26.2
C-5	43.2	N-CH_2	48.9	CH_3	11.6
C-6	90.5	CH_3	14.2		
C-7	88.4	C-1'	55.8		
C-8	77.4	C-6'	57.4		
C-9	51.1	C-7'	–		
C-10	38.1	C-8'	–		
C-11	49.6	C-14'	–		
C-12	28.3	C-16'	55.8		
C-13	45.7	C-18'	59.0		
C-14	75.6	C=O	–		
C-15	33.8	CH_3	–		

1.   S.W. Pelletier, O.D. Dailey, Jr., N.V. Mody, and J.D. Olsen, *J. Org. Chem.*, 46, 3284 (1981).

GLAUCENINE

$C_{31}H_{47}NO_9$; Amorphous;
$[\alpha]_D$ - 45.0° (CHCl$_3$)
Delphinium glaucescens
^{1}H NMR: δ 0.89 (3H, *s*, C(4)-CH$_3$),
0.97 (3H, *t*, J=7.5Hz, CH$_2$-CH$_3$), 1.06
(3H, *t*, NCH$_2$-CH$_3$), 1.12 (3H, *d*, J=7Hz,
CH-CH$_3$), 2.05 (3H, *s*, OCOCH$_3$), 3.32
(6H, *s*, OCH$_3$), 4.89 and 4.97 (each 1H,
s, OCH$_2$O), 5.28 (1H, *dd*, J=6.6Hz, C(14)
-β-H) and 5.46 (1H, *d*, J=1.5Hz, C(6)-α
-H).[1]

[13]*C Chemical Shift Assignments*[1]

C-1	79.1	C-16	81.2	C=O	176.9
C-2	26.9	C-17	63.8	CH	41.3
C-3	37.3	C-18	25.5	CH$_3$	16.2
C-4	33.7	C-19	56.9	CH$_2$	26.3
C-5	50.3	N-CH$_2$	50.4	CH$_3$	11.4
C-6	77.3	CH$_3$	13.8		
C-7	91.6	C-1'	55.4		
C-8	83.2	C-6'	-		
C-9	50.1	C-7-O			
C-10	81.3	C-8-O⟍CH$_2$	93.7		
C-11	55.8	C-14'	-		
C-12	36.6	C-16'	55.8		
C-13	38.9	C-18'	-		
C-14	74.1	C=O	170.0		
C-15	34.9	CH$_3$	21.6		

1. S.W. Pelletier, O.D. Dailey, Jr., N.V. Mody, and J.D. Olsen, *J. Org. Chem.*, 46, 3284 (1981).

GLAUCEPHINE

$C_{33}H_{43}NO_9$; Amorphous;
$[\alpha]_D$ - 33.6° $(CHCl_3)$
Delphinium glaucescens
^{1}H NMR: δ 0.88 (3H, *s*, C(4)-CH*z*),
1.06 (3H, *t*, NCH$_2$-CH*z*), 2.05 (3H, *s*,
OCOCH*z*), 3.31 and 3.33 (each 3H, *s*,
OCH*z*), 4.89 and 4.95 (each 1H, *s*,
OCH$_2$O), 5.48 (2H, *m*, C(6) and C(14)-
H) and 7.38-8.25 (aromatic protons).[1]

[13]C Chemical Shift Assignments [1]

C-1	79.0	C-16	81.2	C=O		166.9
C-2	26.9	C-17	64.1	1		130.7
C-3	36.9	C-18	25.6	2		129.9
C-4	33.8	C-19	56.9	3		128.3
C-5	50.2	N-CH$_2$	50.4	4		132.7
C-6	77.4	CH$_3$	13.9	5		128.3
C-7	91.7	C-1'	55.5	6		129.9
C-8	83.2	C-6'	-			
C-9	50.1	C-7-O				
C-10	81.2	C-8-O CH$_2$	93.9			
C-11	55.7	C-14'	-			
C-12	36.6	C-16'	55.9			
C-13	38.7	C-18'	-			
C-14	74.3	C=O	170.2			
C-15	35.1	CH$_3$	21.6			

1.   S.W. Pelletier, O.D. Dailey, Jr., N.V. Mody, and J.D. Olsen, *J. Org. Chem.*, 46, 3284 (1981).

GLAUCERINE

$C_{30}H_{45}NO_9$; Amorphous;
$[\alpha]_D$ - 48.5° (CHCl$_3$)
Delphinium glaucescens
^{1}H NMR: δ 0.90 (3H, *s*, C(4)-CH$_3$),
1.07 (3H, *t*, NCH$_2$-CH$_3$), 1.17 (6H, *d*,
J=7Hz, CH(CH$_3$)$_2$, 2.07 (3H, *s*, OCOCH$_3$),
3.32 and 3.35 (each 3H, *s*, OCH$_3$), 4.91
and 4.98 (each 1H, *s*, OCH$_2$O), 5.32
(1H, *dd*, J=6.5Hz, C(14)-β-H) and 5.45
(1H, *s*, C(6)-α-H).[1]

^{13}C *Chemical Shift Assignments* [1]

C-1	79.0	C-16	81.2	C=O	177.4
C-2	26.9	C-17	63.9	CH	34.2
C-3	37.3	C-18	25.6	(CH$_3$)$_2$	18.9
C-4	33.7	C-19	56.9		
C-5	50.2	N-CH$_2$	50.4		
C-6	77.3	CH$_3$	13.9		
C-7	91.6	C-1'	55.4		
C-8	83.2	C-6'	-		
C-9	49.9	C-7-O CH$_2$	93.7		
C-10	81.2	C-8-O			
C-11	55.7	C-14'	-		
C-12	36.5	C-16'	55.9		
C-13	38.8	C-18'	-		
C-14	74.3	C=O	170.1		
C-15	34.8	CH$_3$	21.6		

1. S.W. Pelletier, O.D. Dailey, Jr., N.V. Mody, and J.D. Olsen, *J. Org. Chem.*, **46**, 3284 (1981).

GLAUDELSINE

$C_{36}H_{48}N_2O_{10}$; Amorphous;
$[\alpha]_D$ + 36.1° (CHCl$_3$)
Delphinium glaucescens
^{1}H NMR: δ 1.08 (3H, *t*, NCH$_2$-C*H$_3$*),
1.48 (3H, *d*, J=6.5Hz, CH-C*H$_3$*), 3.29,
3.41 and 3.52 (each 3H, *s*, OC*H$_3$*),
3.92 (1H, *m*, C(14)-β-*H*), 4.16 (1H,
d, J=0.5Hz, C(6)-α-*H*), and 7.18-8.25
(aromatic protons).[1]

[13]C Chemical Shift Assignments [1]

C-1	84.9	C-16	81.7	C=O		164.2
C-2	25.3	C-17	65.0	1		127.0
C-3	32.2	C-18	69.5	2		133.1
C-4	37.0	C-19	52.4	3		129.5
C-5	45.8	N-CH$_2$	51.2	4		133.7
C-6	90.3	CH$_3$	14.3	5		131.0
C-7	89.2	C-1'	56.1	6		130.1
C-8	76.3	C-6'	-			
C-9	50.2	C-7'	-			
C-10	37.9	C-8'	-	1'		179.8
C-11	48.3	C-14'	58.3	2'		37.0
C-12	27.6	C-16'	56.5	3'		35.3
C-13	46.1	C-18'	-	4'		175.9
C-14	84.9	C=O		5'		16.5
C-15	33.1	CH$_3$				

1. S.W. Pelletier, O.D. Dailey, Jr., N.V. Mody, and J.D. Olsen, *J. Org. Chem.*, 46, 3284 (1981).

HETERATISINE

$C_{22}H_{33}NO_5$; mp 261-265° (dec.);
$[\alpha]_D$ + 40° (MeOH)
Aconitum heterophyllum
^{1}H NMR: δ 0.97 (3H, *s*, C(4)-*CH_3*),
1.02 (3H, *t*, J=7.5Hz, NCH_2-*CH_3*), 3.25
(3H, *s*, OCH_3), 3.49 (1H, *d*, J=2Hz,
C(17)-*H*), 4.03 (1H, C(9)-*H*), 4.5 (1H,
C(6)-α-*H*), 4.74 (1H, C(13)-*H*).[1,3]

[13]*C Chemical Shift Assignments*[2]

C-1	83.5	C-11	49.3	N-CH_2 CH_3	49.0 13.5
C-2	26.9	C-12	33.1		
C-3	36.8	C-13	75.8	C-1'	55.2
C-4	34,7	C-14	176.0	C-6'	-
C-5	50.9	C-15	29.1[a]	C-7'	-
C-6	72.9	C-16	29.2[a]	C-8'	-
C-7	49.3	C-17	62.2	C-14'	-
C-8	75.4	C-18	26.2	C-16'	-
C-9	57.8	C-19	58.3	C-18'	-
C-10	42.8				

1. R. Aneja, D.M. Locke and S.W. Pelletier, *Tetrahedron*, 29, 3297 (1973).
2. S.W. Pelletier, N.V. Mody, A.J. Jones, and M.H. Benn, *Tetrahedron Letters*, 3025 (1976).
3. S.W. Pelletier, N.V. Mody and N. Katsui, *Tetrahedron Letters*, 4027 (1977).

HETEROPHYLLIDINE

$C_{21}H_{31}NO_5$;   mp 269-272°;
$[\alpha]_D$ + 42.3° (MeOH)
Aconitum heterophyllum
^{1}H NMR: δ 1.08 (3H, *s*, C(4)CH_3),
1.08 (3H, *t*, J=6.5Hz, NCH_2-CH_3), 3.43
(1H, *d*, J=1.5Hz, C(17)-*H*), 3.81 (1H,
C(9)-*H*), 4.67 (1H, *m*, C(6)-*H*), and
4.92 (1H, *m*, C(13)*H*).[1]

[13]C *Chemical Shift Assignments*

C-1	C-11	N-CH_2
C-2	C-12	CH_3
C-3	C-13	C-1'
C-4	C-14	C-6'
C-5	C-15	C-7'
C-6	C-16	C-8'
C-7	C-17	C-14'
C-8	C-18	C-16'
C-9	C-19	C-18'
C-10		

1.   S.W. Pelletier, R. Aneja, and K.N. Gopinath, *Phytochemistry*, 7, 625
     (1968).

HETEROPHYLLINE

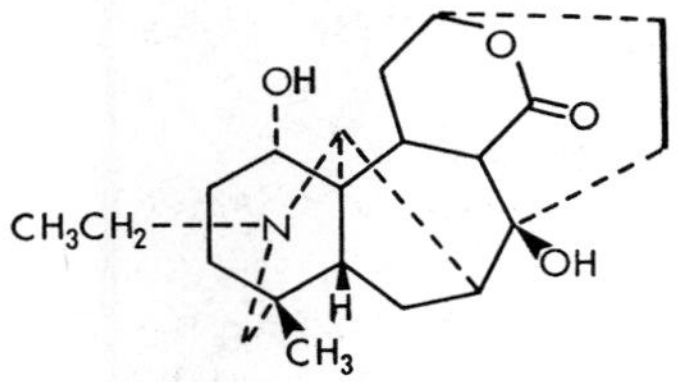

$C_{21}H_{31}NO_4$; mp 221.5-223°;

$[\alpha]_D$ + 10.5° (MeOH)

Aconitum heterophyllum

^{1}H NMR: δ 0.92 (3H, *s*, C(4)-CH$_3$),
1.13 (3H, *t*, J=7.5Hz, NCH$_2$-CH$_3$), 3.30
(1H, *d*, w½=3Hz, C(17)-*H*), 3.82 (1H, *d*,
J=4Hz, C(9)-*H*), and 4.86 (1H, *m*, C(13)
-*H*).[1]

[13]*C Chemical Shift Assignments*

C-1	C-11	N-CH$_2$ CH$_3$
C-2	C-12	
C-3	C-13	C-1'
C-4	C-14	C-6'
C-5	C-15	C-7'
C-6	C-16	C-8'
C-7	C-17	C-14'
C-8	C-18	C-16'
C-9	C-19	C-18'
C-10		

1. S.W. Pelletier, R. Aneja, and K.W. Gopinath, *Phytochemistry*, 7, 625 (1968).

HETEROPHYLLISINE

$C_{22}H_{33}NO_4$; mp 178-179°;
$[\alpha]_D$ + 15.5° (MeOH)
Aconitum heterophyllum
^{1}H NMR: δ 0.81 (3H, *s*, C (4)-CH_3),
1.06 (3H, *t*, J=7Hz, NCH_2-CH_3), 2.10
(1H, *s*, exch. with D_2O, OH), 3.28
(3H, *s*, OCH_3), 3.38 (1H, *d*, J=1Hz,
C(17)-*H*), and 4.74 (1H, *m*, C(13)-*H*).[1]

13C Chemical Shift Assignments

C-1	C-11	N-CH_2
C-2	C-12	$\quad$$CH_3$
C-3	C-13	C-1'
C-4	C-14	C-6'
C-5	C-15	C-7'
C-6	C-16	C-8'
C-7	C-17	C-14'
C-8	C-18	C-16'
C-9	C-19	C-18'
C-10		

1.   S.W. Pelletier, R. Aneja, and K.W. Gopinath, *Phytochemistry*, 7, 625
     (1968).

HOKBUSINE A

$C_{32}H_{45}NO_{10}$; 603.3069
Amorphous; $[\alpha]_D$ + 11.4° (MeOH)
Aconitum carmichaeli.
^{1}H NMR: δ 2.36 (3H, *s*, N-CH_3), 3.13
(3H, *s*, OCH_3), 3.28 (6H, *s*, OCH_3),
3.31 and 3.72 (each 3H, *s*, OCH_3),
4.04 (1H, *bd*, J=7Hz, C(16)-*H*), 4.53
(1H, *bs*, C(15)-*H*), 4.84 (1H, *d*, J=
5Hz, C(14)-*H*), 7.50 (3H, *m*, aromatic
protons) and 8.04 (2H, *dd*, J=8,2Hz,
aromatic protons).[1]

[13]*C Chemical Shift Assignments* (CD_3OD)[1]

C-1	83.9d	C-16	95.3d	C=O		167.8s
C-2	35.8t	C-17	63.2d		1	131.4s
C-3	69.9d	C-18	76.1t		2	130.6d
C-4	44.7s	C-19	50.3t		3	129.2d
C-5	46.7d	N-CH_3	42.7q		4	133.9d
C-6	84.7d				5	129.2d
C-7	43.1s	C-1'	56.3q		6	130.6d
C-8	83.7s	C-6'	59.0q			
C-9	49.9d	C-7'	-			
C-10	42.4d	C-8'	59.0q			
C-11	51.3s	C-14'	-			
C-12	38.2t	C-16'	62.2q			
C-13	74.2s	C-18'	59.0q			
C-14	80.9d	C=O				
C-15	77.6d	CH$_3$				

1. H. Hikino, Y. Kuroiwa, and C. Konno, *J. Nat. Prod.*, **46**, 178 (1983).

HOKBUSINE B

$C_{22}H_{33}NO_5$; mp 183-185°C;
Aconitum carmichaeli
^{1}H NMR: δ 0.75 (3H, *s*, C(4)-*CH₃*),
1.94 (3H, *s*, CO*CH₃*), 3.16 (3H, *s*,
O*CH₃*), 3.69 (1H, *m*, C(1)-*H*) and
4.69 (1H, *t*, J=5Hz, C(14)-*H*).[1]

[13]C Chemical Shift Assignments[1] (C_5D_5N)

C-1	72.7d	C-11	49.1s	C-1'	–
C-2	30.8t	C-12	29.6t	C-6'	–
C-3	32.1t	C-13	43.6d	C-7'	–
C-4	33.3s	C-14	77.0d	C-8'	–
C-5	47.0d	C-15	41.8t	C-14'	–
C-6	25.8t	C-16	83.1d	C-16'	55.8q
C-7	55.0d	C-17	57.7d	C-18'	–
C-8	74.0s	C-18	27.6q	C=O	171.0s
C-9	44.4d	C-19	52.4t	CH₃	21.4q
C-10	38.3d	N-CH₂	–		
		CH₃	–		

1.   H. Hikino, Y. Kuroiwa, and C. Konno, *J. Nat. Prod.*, **46**, 178 (1983).

HOMOCHASMANINE

$C_{26}H_{43}NO_6$; mp 105-107°;
$[\alpha]_D$ + 19.2° (EtOH)
Aconitum chasmanthum
^{1}H NMR: δ 1.07 (3H, t, NCH_2-CH_3),
3.23, 3.24, 3.30, 3.32 and 3.36 (each
3H, s, OCH_3).[1,2]

^{13}C *Chemical Shift Assignments*

C-1	C-11	C-1'
C-2	C-12	C-6'
C-3	C-13	C-7'
C-4	C-14	C-8'
C-5	C-15	C-14'
C-6	C-16	C-16'
C-7	C-17	C-18'
C-8	C-18	C=O
C-9	C-19	CH₃
C-10	N-CH₂	
	CH₃	

1.   O. Achmatowicz, Jr. and L. Marion, *Can. J. Chem.*, 43, 1093 (1965).
2.   S.W. Pelletier, Z. Djarmati, S. Lajsic and W.H. De Camp, *J. Am. Chem. Soc.*, 98, 2617 (1976).

18-HYDROXY-14-O-METHYLGADESINE

$C_{24}H_{37}NO_7$;   MW: 451.2567;
mp 110-114°;

Consolida orientalis

¹H NMR: δ 1.10 (3H, t, J=7Hz, N-CH$_2$-
CH$_3$), 3.35 (3H, s, OCH$_3$), 3.43 (6H, s,
OCH$_3$), 3.67 (2H, bs, CH$_2$OH), 4.00 and
4.08 (each 1H, s, C(6)-α-H & C(19)-H).[1]

¹³C Chemical Shift Assignments

C-1	C-11	C-1'
C-2	C-12	C-6'
C-3	C-13	C-7'
C-4	C-14	C-8'
C-5	C-15	C-14'
C-6	C-16	C-16'
C-7	C-17	C-18'
C-8	C-18	C=O
C-9	C-19	CH$_3$
C-10	N-CH$_2$	
	CH$_3$	

1.   A.G. Gonzales, G. de la Fuente, O. Munguia, and K. Henrich, *Tetrahedron Letters*, 22, 4843 (1981).

15-α-HYDROXYNEOLINE   (FUZILINE; SENBUSINE C)

$C_{24}H_{39}NO_7$; MW: 453.2694
mp 206.5-207°; mp 214-216°;
$[\alpha]_D$ + 11.6°; + 19.3° $(CHCl_3)$
Aconitum carmichaeli, A. japonicum
^{1}H NMR: δ 1.11 (3H, *t*, J=7Hz, NCH$_2$-
C*H$_3$*), 3.33, 3.36 and 3.45 (each 3H, *s*,
OC*H$_3$*), 3.63 (2H, *m*, C(18)-*H*), 3.71 (1H,
t, C(1)-β-*H*), 4.12 (2H, *m*, C(6)-β-*H*,
C(14)-β-*H*), and 4.48 (1H, *d*, J=7Hz,
C(15)-β-*H*), 7.60 (1H, br, exch. D$_2$O,
C(1)-α-O*H*).[1,3]

Prepared from neoline.[3]

[13]*C Chemical Shift Assignments* [2,3]

C-1	72.1d	72.1	C-11	49.3s	49.4	C-1'	-	-
C-2	29.3t[a]	29.5[a]	C-12	29.9t	30.7[a]	C-6'	58.5q	57.5
C-3	29.9t[a]	30.1[a]	C-13	43.6d	43.6d	C-7'	-	-
C-4	38.1s	38.1	C-14	75.6d	75.7	C-8'	-	-
C-5	44.1d	44.1	C-15	78.6d	79.0	C-14'	-	-
C-6	84.1d	84.3	C-16	90.5d	90.4	C-16'	57.5q	58.0
C-7	49.5d	46.6	C-17	62.6d	62.6	C-18'	59.1q	59.1
C-8	79.1s	79.0	C-18	80.0t	80.1			
C-9	46.7d	48.5	C-19	56.8t	56.7			
C-10	40.7d	40.7	N-CH$_2$	48.5t	48.5			
			CH$_3$	13.0q	13.1			

1. S.W. Pelletier, N.V. Mody, K.I. Varughese, and C. Szu-Ying, *Heterocycles*, 18, 47 (1982).

2. C. Konno, M. Shirasaka, and H. Hikino, *J. Nat. Prod.*, 45, 128 (1982).

3. H. Takayama, S. Hasegawa, S. Sakai, J. Haginawa and T. Okamoto, *Chem. Pharm. Bull.* Japan, 29, 3078 (1981); *J. Pharm. Soc.* Japan, 102, 525 (1982).

15-β-HYDROXYNEOLINE (CRASSICAULISINE, NAGARINE)

$C_{24}H_{39}NO_7$; MW: 453.2725;
mp 192-192°; $[\alpha]_D^{21°}$ + 55.4° (EtOH);[1,2]
mp 190-191°; $[\alpha]_D$ + 20.4° (CHCl$_3$);[4]
Aconitum crassicaule, A. nagarum var.
heterotrichum.
^{1}H NMR (200 MHz): δ 1.10 (3H, *t*, J=
7Hz, NCH$_2$C*H$_3$*), 1.50-1.60(2H, *m*, C(3)*H*),
3.31, 3.36, 3.48 (each 3H, *s*, OC*H$_3$*),
3.22-3.65 (2H, AB type, J=9Hz, C(18)-
H), 3.78 (1H, *d*, J=8Hz, C(15)-β-*H*),
3.97 (1H, *t*, J=4.5Hz, C(14)-β-*H*), 4.23
(1H, *dd*, J$_1$=1Hz,J$_2$=6Hz, C(6)-β-*H*),
2.68 (2H, AB type, J=12Hz, C(19)-*H*),
3.78, 4.08 (each 1H, *s*, exch.D$_2$0,
OH).[1,2]
Prepared from neoline, mp 175-177°;
$[\alpha]_D$ + 24.7°.[3]

^{13}C Chemical Shift Assignments[1]

C-1	72.0	C-11	49.5	C-1'	-
C-2	29.5[a]	C-12	29.3[a]	C-6'	57.9
C-3	29.8[a]	C-13	44.1	C-7'	-
C-4	38.1	C-14	74.5	C-8'	-
C-5	44.1	C-15	68.1	C-14'	-
C-6	83.3	C-16	83.9	C-16'	57.9
C-7	52.8	C-17	61.9	C-18'	59.1
C-8	74.4	C-18	80.0		
C-9	47.9	N-CH$_2$	48.2		
C-10	42.4	CH$_3$	12.9		

1.   F.-P. Wang and Q.-C. Fang, *Planta Medica*, 47, 39 (1983).

2.   F.-P. Wang and Q.-C. Fang, *Acta Pharmaceutica Sinica*, 17, 300 (1982).

3.   H. Takayama, S. Hasegawa, S. Sakai, J. Haginawa and T. Okamoto, *Chem. Pharm. Bull.* Japan, 29, 3078 (1981).

4.   N.V. Mody, S.W. Pelletier and S.-Y. Chen, *Heterocycles*, 17, 91 (1982).

HYPACONINE

$C_{24}H_{39}NO_8$; Amorphous;
Hydrolysis product of hypaconitine

^{13}C *Chemical Shift Assignments*[1,2]

C-1	85.2	C-11	50.1	C-1'	56.5
C-2	26.5	C-12	35.3	C-6'	57.9
C-3	34.9	C-13	76.4	C-7'	-
C-4	39.3	C-14	79.0	C-8'	-
C-5	49.2	C-15	78.9	C-14'	-
C-6	83.6	C-16	91.2	C-16'	61.5
C-7	50.0	C-17	62.4	C-18'	59.0
C-8	75.7	C-18	80.4		
C-9	48.1	C-19	56.1		
C-10	42.0	N-CH3	42.8		

1. S.W. Pelletier, N.V. Mody, and R.S. Sawhney, *Can. J. Chem.*, 57, 1652 (1979).
2. N.V. Mody, S.W. Pelletier and S.-Y. Chen, *Heterocycles*, 17, 91 (1982).

15-*EPI*-HYPACONINE

$C_{24}H_{39}NO_8$; mp 183.5-186°;
Prepared from delphinine
^{1}H NMR: δ 2.37 (3H, *s*, NCH$_3$), 3.30
and 3.73 (each 3H, *s*, OCH$_3$), and
3.37 (6H, *s*, OCH$_3$).[1]

[13]*C Chemical Shift Assignments* [1,2]

C-1	85.7	C-11	50.3	C-1'	56.3
C-2	25.5	C-12	36.6	C-6'	57.2
C-3	34.9	C-13	76.3	C-7'	-
C-4	39.6	C-14	79.2	C-8'	-
C-5	49.0	C-15	68.2	C-14'	-
C-6	82.5	C-16	88.1	C-16'	62.5
C-7	50.1	C-17	63.3	C-18'	59.2
C-8	73.2	C-18	80.4		
C-9	48.3	C-19	56.1		
C-10	42.4	N-CH$_3$	42.4		

1.   S.W. Pelletier, N.V. Mody and Y. Ohtsuka, unpublished results.
2.   N.V. Mody, S.W. Pelletier and S.-Y. Chen, *Heterocycles*, 17, 91 (1982).

HYPACONITINE

$C_{33}H_{45}NO_{10}$; mp 197-198°
$[\alpha]_D$ + 22° (CHCl$_3$)
Aconitum callianthum, A. carmichaeli,
A. coreanum, A. grossedentatum, A. ha-
kusanense, A. ibukiense, A. japonicum,
A. kamtschaticum (fischeri), A. kusne-
zoffii, A. mitakense, A. napellus, A.
pendulum, A. sanyoense, A. senanense,
A. subcuneatum, A. tasiromontanum, A.
tortuosum, A. yezoense, A. zuccarini.
^{1}H NMR: δ 1.30 (3H, *s*, OCOC*H$_3$*), 2.23
(3H, *s*, NC*H$_3$*), 3.00 and 3.50 (each 3H,
s, OC*H$_3$*), 3.12 (6H, *s*, OC*H$_3$*), 4.68
(1H, *d*, C(14)-β-*H*) and 7.20-7.78 (aromatic protons).[2,3]

^{13}C *Chemical Shift Assignments*[1]

C-1	85.0	C-11	49.9	C-1'	56.5
C-2	26.4	C-12	36.3	C-6'	57.9
C-3	34.9	C-13	74.1	C-16'	60.9
C-4	39.3	C-14	78.8	C-18'	59.0
C-5	48.2	C-15	78.8	C=O	172.3
C-6	83.1	C-16	90.1	CH$_3$	21.4
C-7	44.5	C-17	62.1		
C-8	91.9	C-18	80.1	C=O	166.1
C-9	43.8	C-19	56.0	1	129.9
C-10	41.1	N-CH$_3$	42.6	2	129.6
				3	128.6
				4	133.2
				5	128.6
				6	129.6

1.   A. Katz and E. Staehelin, *Pharm. Acta Helv.*, 54, 253 (1979).
2.   Y. Tsuda, O. Achmatowicz and L. Marion, *Ann.*, 680, 88 (1964).
3.   K.B. Birnbaum, K. Wiesner, E.W.K. Jay and L. Jay, *Tetrahedron Letters*,
     867 (1971).

ILIDINE

$C_{25}H_{37}NO_7$;   mp 141-143°;
Delphinium iliense
^{1}H NMR: δ 1.02 (3H, *t*, NCH_2-*CH_3*),
3.25, 3.27 and 3.31 (each 3H, *s*, O*CH_3*),
5.07 and 5.53 (each 1H, *s*, O*CH_2*O).[1,2]

^{13}C *Chemical Shift Assignments*

C-1	C-11	C-1'
C-2	C-12	C-6'
C-3	C-13	C-7'
C-4	C-14	C-8'
C-5	C-15	C-14'
C-6	C-16	C-16'
C-7	C-17	C-18'
C-8	C-18	C=O
C-9	C-19	CH_3
C-10	N-CH_2	
	CH_3	

1.   M.G. Zhamierashvili, V.A. Tel'nov, M.S. Yunusov and S.Yu. Yunusov, *Khim. Prir. Soedin.*, 13, 836 (1977).
2.   S.W. Pelletier, N.V. Mŏdy, K.I. Varughese, J.A. Maddry, and H.K. Desai, *J. Am. Chem. Soc.*, 1981, 103, 6536.

INDACONITINE

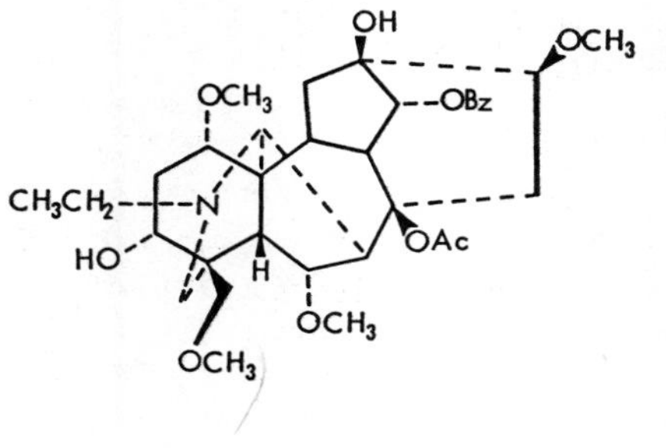

$C_{34}H_{47}NO_{10}$; MW: 629.3218;
mp 203-204°; $[\alpha]_D$ + 18° (EtOH)
Aconitum chasmanthum, A. falconeri,
A. ferox, A. franchetii, A. violaceum.
^{1}H NMR: δ 1.08 (3H, *t*, NCH$_2$-C*H$_3$*),
1.23 (3H, *s*, OCOC*H$_3$*), 3.15, 3.25,
3.28 and 3.53 (each 3H, *s*, OC*H$_3$*),
4.92 (1H, *d*, J=4.5Hz, C(14)-β-*H*) and
7.27-8.20 (aromatic protons).[1]

[13]C *Chemical Shift Assignments* [2,3]

C-1	83.2	C-16	82.8			
C-2	35.1	C-17	61.4		1	129.7
C-3	71.2	C-18	76.5		2	129.2
C-4	43.0	C-19	48.6		3	128.1
C-5	48.6	N-CH$_2$	47.2		4	132.7
C-6	82.0	CH$_3$	13.3		5	129.2
C-7	48.6	C-1'	55.6		6	128.1
C-8	85.3	C-6'	57.5			
C-9	47.2	C-7'	-			
C-10	40.7	C-8'	-			
C-11	50.0	C-14'	-			
C-12	33.5	C-16'	58.5			
C-13	74.5	C-18'	58.9			
C-14	78.5	C=O	169.2			
C-15	39.4	CH$_3$	21.5			

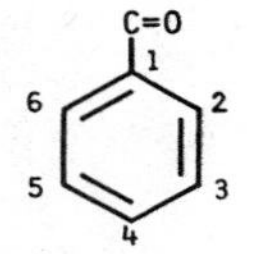

C=O 165.7

1. A. Klasek, V. Simanek and F. Santavy, *Lloydia*, 35, 55 (1972).
2. S.W. Pelletier, N.V. Mody, R.S. Sawhney and J. Bhattacharyya, *Hetero-cycles*, 7, 327 (1977).
3. S.W. Pelletier, N.V. Mody and H.S. Puri, *Phytochemistry*, 16, 623 (1977).

ISODELPHININE

$C_{33}H_{45}NO_9$; mp 167-168°;
$[\alpha]_D$ + 20.1° (EtOH)
Aconitum carmichaeli, A. miyabei.
^{1}H NMR: δ 1.45 (3H, *s*, OCOC*H$_3$*), 2.36
(3H, *s*, NC*H$_3$*), 3.20 and 3.54 (each
3H, *s*, OC*H$_3$*), 3.32 (6H, *s*, OC*H$_3$*),
5.06 (1H, *dd*, C(14)-β-*H*) and 7.40-
8.20 (aromatic protons).[1]

^{13}C *Chemical Shift Assignments* [1]

C-1	85.1	C-16	89.3	C=O	166.1
C-2	26.4	C-17	62.2		
C-3	34.9	C-18	80.2	1	130.2
C-4	39.3	C-19	56.5	2	129.7
C-5	47.9	N-CH$_3$	42.6	3	128.6
C-6	83.7			4	133.1
C-7	44.5	C-1'	56.1	5	128.6
C-8	92.1	C-6'	57.7	6	129.7
C-9	44.7	C-7'	–		
C-10	38.7	C-8'	–		
C-11	50.0	C-14'	–		
C-12	29.4	C-16'	58.0		
C-13	43.9	C-18'	59.1		
C-14	76.4	C=O	172.3		
C-15	78.8	CH$_3$	21.5		

1.   S.W. Pelletier, N.V. Mody and N. Katsui, *Tetrahedron Letters,* 4027 (1977).

ISODELPHONINE

$C_{24}H_{39}NO_7$; MW: 453.2710;
Amorphous; $[\alpha]_D$ + 13.0°
Prepared from chasmanine
^{1}H NMR (270 MHz): δ 2.35 (3H, *s*, N-*CH₃*), 3.25, 3.30, 3.36, 3.45 (each 3H, *s*, O*CH₃*), 4.09 (1H, *t*, J=4.6Hz, C(14)-β-*H*), 4.17 (1H, *d*, J=6.9Hz, C(6)-β-*H*), 4.37 (1H, *d*, J=6.4Hz, C(15)-β-*H*).[1]

[13]*C Chemical Shift Assignments* [1]

C-1	85.5	C-11	50.3	C-1'	56.4
C-2	26.3	C-12	30.4	C-6'	57.4
C-3	34.6	C-13	46.9	C-7'	-
C-4	39.4	C-14	75.9	C-8'	-
C-5	48.1	C-15	80.2	C-14'	-
C-6	83.6	C-16	90.7	C-16'	57.8
C-7	45.9	C-17	62.8	C-18'	59.2
C-8	78.4	C-18	80.6	C=O	
C-9	49.8	C-19	56.6	CH₃	
C-10	41.2	N-CH₃			

1. H. Takayama, S. Sakai, K. Yamaguchi and T. Okamoto, *Chem. Pharm. Bull. Japan*, 30, 386 (1982).

15-*EPI*-ISODELPHONINE

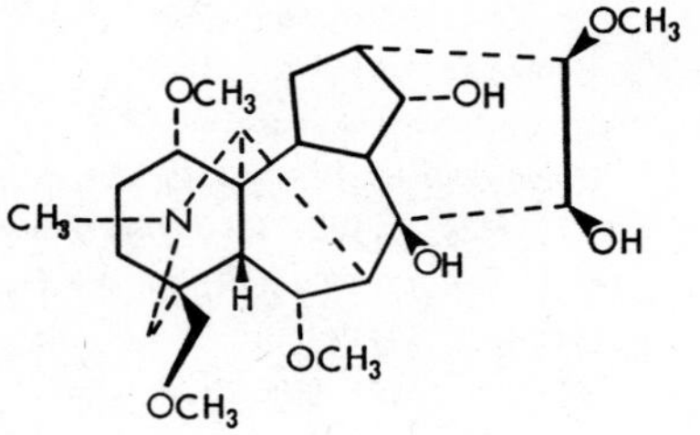

$C_{24}H_{39}NO_7$

Prepared from chasmanine.

[13]C Chemical Shift Assignments[1]

C-1	86.1	C-11	50.7	C-1'	56.4
C-2	25.6	C-12	27.7	C-6'	57.2
C-3	25.0	C-13	45.5	C-7'	-
C-4	39.8	C-14	75.5	C-8'	-
C-5	47.5	C-15	66.4	C-14'	-
C-6	82.4	C-16	83.6	C-16'	59.2
C-7	49.6	C-17	63.5	C-18'	59.3
C-8	73.7	C-18	80.5	C=O	
C-9	50.2	C-19	56.1	CH₃	
C-10	36.4	N-CH₃	42.4		

1. Personal communications from Dr. S. Sakai to Dr. S.W. Pelletier
    (April 10, 1982).

ISOTALATIZIDINE

$C_{23}H_{37}NO_5$;   mp 116-117°; $[\alpha]_D \pm 0°$
*Aconitum carmichaeli, A. japonicum,
A. talassicum, A. tranzschelii, Delphinium denudatum, D. bicolor.*
^{1}H NMR: δ 1.10 (3H, *t*, NCH$_2$-*CH$_3$*),
2.96, 3.16 (1H, *d*, J=9Hz, C(19)-*H*),
3.30 and 3.32 (each 3H, *s*, O*CH$_3$*),
3.70(1H, *m*, C(1)-β-*H*) and 4.18 (1H,
dd, J=4.5Hz, C(14)-β-*H*).[1]

[13]*C Chemical Shift Assignments*[2,3,4,5]

C-1	72.3	C-11	48.7	C-1'	–
C-2	29.2	C-12	26.8	C-6'	–
C-3	29.7	C-13	44.1	C-7'	–
C-4	37.3	C-14	75.6	C-8'	–
C-5	41.7	C-15	42.3	C-14'	–
C-6	25.0	C-16	82.4	C-16'	56.2
C-7	45.3	C-17	63.7	C-18'	59.3
C-8	74.3	C-18	79.0	C=O	
C-9	46.7	C-19	56.6	CH$_3$	
C-10	40.4	N-CH$_2$	48.4		
		CH$_3$	13.0		

1.   S.W. Pelletier, L.H. Keith and P.C. Parthasarathy, *J. Am. Chem. Soc.*,
     89, 4146 (1967).
2.   S.W. Pelletier and Z. Djarmati, *J. Am. Chem. Soc.*, 98, 2626 (1976).
3.   S.W. Pelletier, N.V. Mody and N. Katsui, *Tetrahedron Letters*, 4027 (1977).
4.   S.W. Pelletier, N.V. Mody, A.P. Venkov and S.B. Jones, Jr., *Heterocycles*,
     12, 779 (1979).
5.   N.V. Mody, S.W. Pelletier and N.M. Mollov, *Heterocycles*, 14, 1751 (1980).

JESACONITINE

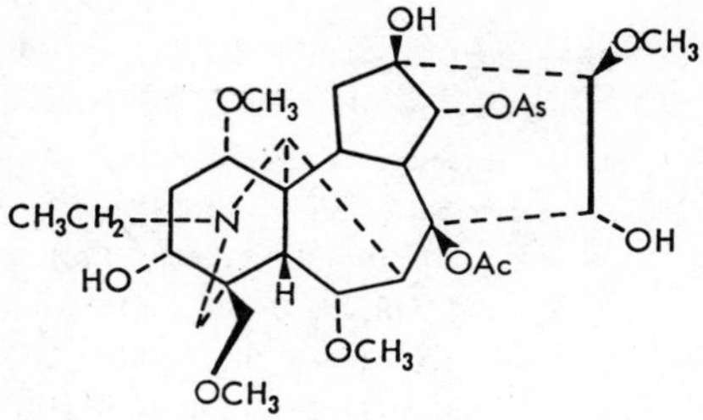

As =

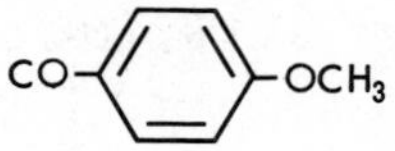

$C_{35}H_{49}NO_{12}$; MW: 675.32548;
mp 222-227° (perchlorate);
$[\alpha]_D$ - 17° for perchlorate (H_2O)
Aconitum fischeri, A. mitakense, A.
sachalinense, A. subcuneatum, A.
yezoense.
^{1}H NMR: δ 1.19 (3H, *t*, NCH$_2$-*CH$_3$*),
3.13, 3.22, 3.24 and 3.72 (each 3H,
s, OC*H$_3$*), 3.81 (3H, *s*, aromatic
OC*H$_3$*), 4.62 (1H, *d*, J=4.5Hz, C(14)-
β-*H*), 6.85 and 7.89 (4H, AB *q*, J=
9Hz, aromatic protons).[1,2]

^{13}C Chemical Shift Assignments [3]

C-1	83.3	C-16	90.0	C=O		165.7
C-2	33.6	C-17	60.9	1		122.1
C-3	70.9	C-18	75.8	2		131.6
C-4	43.1	C-19	46.9	3		113.8
C-5	46.6	N-CH$_2$	48.9	4		163.5
C-6	82.3	CH$_3$	13.3	5		113.8
C-7	44.6^a	C-1'	55.8	6		131.6
C-8	91.9	C-6'	57.9			
C-9	44.2^a	C-7'	-	OCH$_3$		55.4
C-10	40.8	C-8'	-			
C-11	49.9	C-14'	-			
C-12	35.8	C-16'	61.1			
C-13	74.0	C-18'	59.0			
C-14	78.6^b	C=O	172.4			
C-15	78.8^b	CH$_3$	21.5			

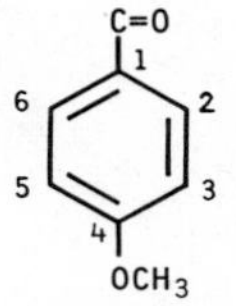

1. S.W. Pelletier, W.H. De Camp, J. Finer-Moore, and Y. Ichinohe, *Cryst.*
    Struct. Comm., 8, 299 (1979).

2. L.H. Keith and S.W. Pelletier, *J. Org. Chem.*, 33, 2497 (1968).

3. H. Bando, Y. Kanaiwa, K. Wada, T. Mori and T. Amiya, *Heterocycles*, 16,
    1723 (1981).

KARAKOLIDINE (KARACOLIDINE)

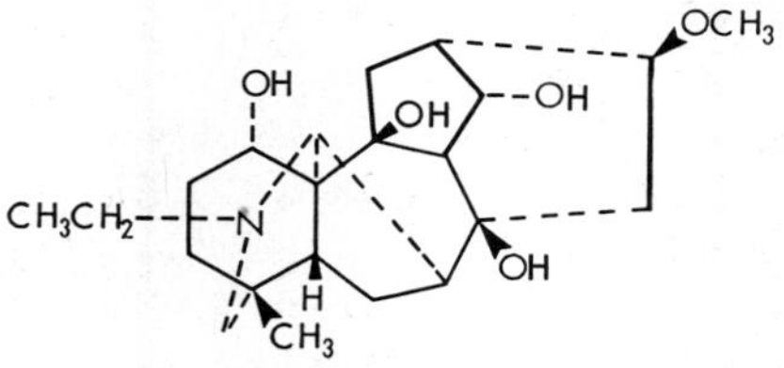

$C_{22}H_{35}NO_5$; mp 222-224°;
Aconitum karakolicum
^{1}H NMR: δ 0.87 (3H, *s*, C(4)-CH$_3$), 1.07
(3H, *t*, NCH$_2$-CH$_3$), 3.28 (3H, *s*, OCH$_3$)
and 4.65 (1H, *t*, C(14)-β-H).[1]

^{13}C *Chemical Shift Assignments*

C-1	C-11	C-1'
C-2	C-12	C-6'
C-3	C-13	C-7'
C-4	C-14	C-8'
C-5	C-15	C-14'
C-6	C-16	C-16'
C-7	C-17	C-18'
C-8	C-18	C=O
C-9	C-19	CH$_3$
C-10	N-CH$_2$	
	CH$_3$	

1.   M.N. Sultankhodzhaev, M.S. Yunusov and S.Yu. Yunusov, *Khim. Prir. Soedin.*, 11, 481 (1975), 9, 199 (1973).

KARAKOLINE (KARACOLINE, VILMORRIANINE B)

$C_{22}H_{35}NO_4$; mp 183-184°;
$[\alpha]_D$ - 10° (MeOH)
Aconitum carmichaeli, A. karakolicum,
*A. vilmorrianum, Delphinium pentagynum.**
^{1}H NMR: δ 0.84 (3H, *t*, C(4)-CH$_3$), 1.07
(3H, *t*, NCH$_2$-CH$_3$), 3.29 (3H, *s*, OCH$_3$)
and 4.16 (1H, *t*, C(14)-β-H).[1]

^{13}C *Chemical Shift Assignments*[2,3]

C-1	72.4	72.6d	C-11	48.9	49.1s	C-1'	–	
C-2	29.6	29.1t	C-12	29.3	29.9t	C-6'	–	
C-3	31.3	31.5t	C-13	44.1	44.3d	C-7'	–	
C-4	32.9	33.1s	C-14	75.6	75.9d	C-8'	–	
C-5	45.1	47.0d	C-15	42.5	42.5t	C-14'	–	
C-6	25.2	25.4t	C-16	82.4	82.5d	C-16'	56.2	56.6q
C-7	46.6	45.4d	C-17	63.0	63.5d	C-18'	–	
C-8	74.3	74.6s	C-18	27.6	27.8q			
C-9	46.7	47.0d	C-19	60.3	60.5t			
C-10	40.4	40.5d	N-CH$_2$	48.3	48.6t			
			CH$_3$	13.0	13.3q			

1.   M.N. Sultankhodzhaev, M.S. Yunusov and S.Yu. Yunusov, *Khim. Prir. Soedin.,*
     199 (1973), 399 (1972).

2.   T.-R. Yang, X.-J. Hao, and J. Chow, *Acta Botanica Yunnanica,* 1 (a), 41
     (1979).

3.   C. Konno, M. Shirasaka and H. Hikino, *J. Nat. Prod., 45,* 128 (1982).

KARASAMINE

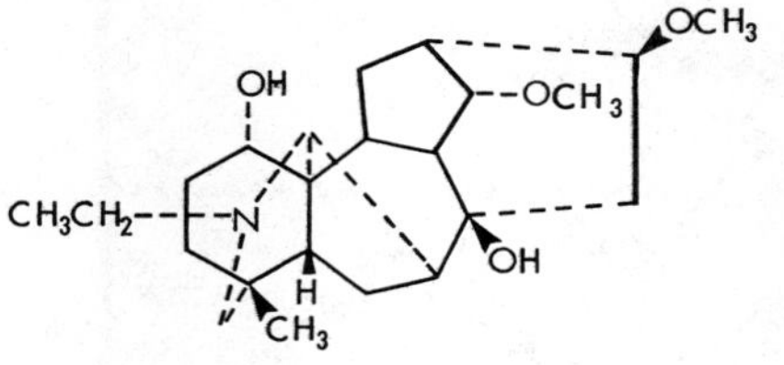

$C_{23}H_{37}NO_4$;   mp 110-112°

Aconitum karakolicum

^{1}H NMR: δ 0.81 (3H, *s*, C(4)-CH_3),
1.03 (3H, *t*, NCH$_2$-CH_3), 3.22, 3.30
(each 3H, *s*, OCH_3).[1]

^{13}C *Chemical Shift Assignments*

C-1	C-11	C-1'
C-2	C-12	C-6'
C-3	C-13	C-7'
C-4	C-14	C-8'
C-5	C-15	C-14'
C-6	C-16	C-16'
C-7	C-17	C-18'
C-8	C-18	C=O
C-9	C-19	CH$_3$
C-10	N-CH$_2$	
	CH$_3$	

1.   M.N. Sultankhodzhaev, M.S. Yunusov and S.Yu. Yunusov, *Khim. Prir. Soedin.*, 660 (1982).

LAPACONIDINE

$C_{22}H_{35}NO_6$; mp 206-207°;
$[\alpha]_D$ + 120° (CHCl$_3$)
Aconitum leucostomum (excelsum),
A. septentrionale.
^{1}H NMR: δ 1.07 (3H, *t*, NCH$_2$-C*H$_3$*),
3.26 and 3.36 (each 3H, *s*, OC*H$_3$*).[1,3]

^{13}C *Chemical Shift Assignments*[2]

	CDCl$_3$	C$_5$D$_5$N		CDCl$_3$	C$_5$D$_5$N		CDCl$_3$	C$_5$D$_5$N
C-1	72.5	73.0	C-11	50.4	51.0	C-1'	-	-
C-2	28.9	30.7	C-12	23.1	24.0	C-6'	-	-
C-3	33.5	34.6	C-13	48.4	47.9	C-7'	-	-
C-4	70.7	70.0	C-14	90.4	90.8	C-8'	-	-
C-5	48.2	48.3	C-15	45.1	44.2	C-14'	58.1	57.6
C-6	27.4	27.9	C-16	83.0	83.9	C-16'	56.3	56.0
C-7	47.0	47.6	C-17	63.1	62.9	C-18'	-	-
C-8	76.3	75.5	C-18	-	-			
C-9	77.6	78.2	C-19	60.4	61.5			
C-10	36.3	37.1	N-CH$_2$	46.5	50.2			
			CH$_3$	13.1	13.2			

1.   V.A. Tel'nov, M.S. Yunusov, and S.Yu. Yunusov, *Khim. Prirodn. Soedin.*, 6, 639 (1970).

2.   S.W. Pelletier, N.V. Mody and R.S. Sawhney, *Can. J. Chem.*, 57, 1652 (1979).

3.   V.A. Tel'nov, M.S. Yunusov, Ya.V. Rashkes and S.Yu. Yunusov, *Khim. Prir. Soedin.*, 622 (1971).

LAPPACONINE

$C_{23}H_{37}NO_6$; mp 96°; $[\alpha]_D$ + 27° (CHCl$_3$)

Hydrolysis product of lappaconitine[*]

^{1}H NMR: δ 1.07 (3H, t, NCH$_2$-CH_3) and 3.25, 3.27 and 3.37 (each 3H, s, OCH_3).[1,2]

[13]C *Chemical Shift Assignments*[3]

C-1	85.2	C-11	51.0	C-1'	56.5
C-2	26.6	C-12	23.7	C-6'	-
C-3	36.3	C-13	49.0	C-7'	-
C-4	71.1	C-14	90.3	C-8'	-
C-5	50.8	C-15	44.7	C-14'	58.0
C-6	26.9	C-16	83.1	C-16'	56.1
C-7	47.8	C-17	61.7	C-18'	-
C-8	75.7	C-18	-		
C-9	78.8	C-19	58.0		
C-10	37.4	N-CH$_2$	49.9		
		CH$_3$	13.5		

1. L Marion, L. Fonzes, C.K. Wilkins, Jr., J.P. Boca, F. Sandberg R. Thorsen, and E. Linden, *Can. J. Chem.*, 45, 969 (1967).
2. N. Mollov, M. Tada and L. Marion, *Tetrahedron Letters*, 2189 (1969).
3. S.W. Pelletier, N.V. Mody and R.S. Sawhney, *Can. J. Chem.*, 57, 1652 (1979).

LAPPACONITINE

$C_{32}H_{44}N_2O_8$; MW: 584.3073
mp 229°; $[\alpha]_D$ + 27° (CHCl₃)
Aconitum barbatum var *puberulum, A.excelsum, A.finetianum, A.leucostomum, A.orientale, A.septentrionale, A.sinomontanum, Delphinium cashmirianum.* *

¹H NMR: δ 1.11 (3H, *t*, NCH₂-C*H₃*),
2.18 (3H, *s*, NHCOC*H₃*), 3.27 (6H, *s*, OC*H₃*), 3.38 (3H, *s*, OC*H₃*), 6.93-8.58
(aromatic protons) and 10.90 (1H, *bs*, N*H*).[1,2]

¹³C *Chemical Shift Assignments* [3,4]

C-1	84.2	C-16	82.9	C=O		167.7
C-2	26.2	C-17	61.5		1	115.9
C-3	31.9	C-18	-		2	141.8
C-4	84.7	C-19	55.5		3	120.4ᵃ
C-5	48.6	N-CH₂	49.9		4	134.6ᵇ
C-6	26.8	CH₃	13.5		5	122.6ᵃ
C-7	47.6	C-1'	56.5		6	131.3ᵇ
C-8	75.6	C-6'	-			
C-9	78.6	C-7'	-	X =	NHC=O	169.5
C-10	36.4	C-8'	-		CH₃	25.6
C-11	51.0	C-14'	57.9			
C-12	24.2	C-16'	56.1			
C-13	49.0	C-18'	-			
C-14	90.2	C=O				
C-15	44.9	CH₃				

1. L. Marion, L. Fonzes, C.K. Wilkins, Jr., J.P. Boca, F. Sandberg, R. Thorsen, and E. Linden, *Can. J. Chem.*, 45, 969 (1967).

2. M. Shamma, P. Chinnasamy, G.A. Miana, A. Khan, M. Bashir, M. Salazar, P. Patil, and J.L. Beal, *J. Nat. Prod.*, 42, 615 (1979).

3. S.W. Pelletier, N.V. Mody, A.P. Venkov, and N.M. Mollov, *Tetrahedron Letters*, 5045 (1978).

4. S.W. Pelletier, N.V. Mody and R.S. Sawhney, *Can. J. Chem.*, 57, 1652 (1979).

8-o-LINOLEOYL-14-BENZOYLACONINE

$C_{50}H_{75}NO_{11}$; MW: 865
Colorless oil; $[\alpha]_D$ + 20.4° (CHCl$_3$)
Prepared from aconitine

R = $-(CH_2)_7-CH=CH-CH_2-CH=CH-(CH_2)_4-CH_3$

^{13}C Chemical Shift Assignments [1]

C-1	83.5d	C-11	50.1s	C-1'	55.8q
C-2	35.8t	C-12	34.8t	C-6'	58.2q
C-3	70.4d	C-13	74.1s	C-7'	-
C-4	43.2s	C-14	79.0d	C-8'	-
C-5	46.4d	C-15	79.0d	C-14'	-
C-6	82.4d	C-16	90.1d	C-16'	60.9q
C-7	44.8d	C-17	61.2d	C-18'	59.0q
C-8	91.7s	C-18	75.6t	O-C=O R	175.1s
C-9	44.4d	C-19	48.9t		
C-10	41.0d	N-CH$_2$ CH$_3$	47.1t 13.3q	O-C=O Ph	166.0s

1. I. Kitigawa, M. Yoshikawa, Z.L. Chen and K. Kobayashi, *Chem. Pharm. Bull. Japan,* 30, 758 (1982).

8-O-LINOLEOYL-14-BENZOYLHYPACONINE

$C_{49}H_{73}NO_{10}$; MW: 835
Colorless oil; $[\alpha]_D$ + 13.0°
Prepared from hypaconitine

$R = (CH_2)_7-CH=CH-CH_2-CH=CH-(CH_2)_4-CH_3$

[13]C Chemical Shift Assignments[1]

C-1	85.0d	C-11	50.1s	C-1'	56.1q
C-2	26.3t	C-12	34.8t	C-6'	58.1q
C-3	34.8t	C-13	74.1s	C-7'	-
C-4	39.3s	C-14	79.0d	C-8'	-
C-5	50.1d	C-15	79.0d	C-14'	-
C-6	83.3d	C-16	90.2d	C-16'	61.1q
C-7	44.7d	C-17	62.3d	C-18'	59.0q
C-8	91.6s	C-18	80.1t	O-C=O R	175.1s
C-9	44.1d	C-19	50.1t		
C-10	41.2d	N-CH3	42.6q	O-C=O Ph	166.0s

1.   I. Kitagawa, M. Yoshikawa, Z.L. Chen and K. Kobayashi, *Chem. Pharm. Bull. Japan*, 30, 758 (1982).

8-*o*-LINOLEOYL-14-BENZOYLMESACONINE

$C_{49}H_{73}NO_{11}$; MW: 851
Colorless oil; $[\alpha]_D$ + 15.0° (CHCl$_3$)
Prepared from mesaconitine

R = (CH$_2$)$_7$-CH=CH-CH$_2$-CH=CH-(CH$_2$)$_4$-CH$_3$

^{13}C *Chemical Shift Assignments*[1]

C-1	83.3d	C-11	50.2s	C-1'	56.5q
C-2	35.8t	C-12	33.8t	C-6'	58.1q
C-3	71.0d	C-13	74.1s	C-7'	-
C-4	43.5s	C-14	79.0d	C-8'	-
C-5	46.1d	C-15	79.0d	C-14'	-
C-6	82.3d	C-16	90.1d	C-16'	61.2q
C-7	44.4d	C-17	62.4d	C-18'	59.1q
C-8	91.5s	C-18	76.1t	O-C=O (R)	175.1s
C-9	43.9d	C-19	49.6t		
C-10	40.9d	N-CH$_3$	42.4q	O-C=O (Ph)	166.0s

1.   I. Kitagawa, M. Yoshikawa, Z.L. Chen and K. Kobayashi, *Chem. Pharm. Bull.
Japan*, 30, 758 (1982).

LIPOACONITINE

Colorless oil

$[\alpha]_D$ + 6.0° $(CHCl_3)$[1]

Aconitum spp. ("Chuanwu" Chinese)

$-(CH_2)_{14}-CH_3$ (Palmitoyl)
$-(CH_2)_{16}-CH_3$ (Stearoyl)
R = $-(CH_2)_7-CH=CH-CH_2-CH=(CH_2)_4-CH_3$ (Linoleoyl)
$-(CH_2)_7-CH=CH-(CH_2)_7-CH_3$ (Oleoyl)
$-(CH_2)_7-(CH_2-CH=CH)_3-CH_3$ (Linolenoyl)

[13]C Chemical Shift Assignments[1]

C-1	83.2d	C-11	49.7s	C-1'	55.5q
C-2	35.6t	C-12	34.4t	C-6'	57.8q
C-3	70.4d	C-13	73.8s	C-7'	-
C-4	42.9s	C-14	78.7d	C-8'	-
C-5	46.0d	C-15	78.7d	C-14'	-
C-6	82.1d	C-16	89.8d	C-16'	60.5q
C-7	44.5d	C-17	60.9d	C-18'	58.7q
C-8	91.4s	C-18	75.7t	O-C=O R	174.7s
C-9	44.0d	C-19	48.6t		
C-10	40.6d	N-CH_2	46.7t	O-C=O Ph	165.6s
		CH_3	13.0q		

1.   I. Kitagawa, M. Yoshikawa, Z.L. Chen and K. Kobayashi, *Chem. Pharm. Bull.*
      Japan, 30, 758 (1982).

LIPO-3-DEOXYACONITINE

Colorless oil;

$[\alpha]_D$ + 12.4° $(CHCl_3)$

Aconitum spp ("Chuanwu" Chinese)

R = -(CH_2)_{14}-CH_3 (Palmitoyl)
-(CH_2)_{16}-CH_3 (Stearoyl)
-(CH_2)_7-CH=CH-(CH_2)_7-CH_3 (Oleoyl)
-(CH_2)_7-CH=CH-CH_2-CH=CH-(CH_2)_4-CH_3 (Linoleoyl)
-(CH_2)_7-(CH_2-CH=CH)_3-CH_3 (Linolenoyl)

13C Chemical Shift Assignments [1]

C-1	85.0d	C-11	49.9s	C-1'	56.0q
C-2	26.2t	C-12	36.5t	C-6'	58.0q
C-3	35.1t	C-13	74.0s	C-7'	-
C-4	39.0s	C-14	78.8d	C-8'	-
C-5	49.0d	C-15	78.9d	C-14'	-
C-6	83.2d	C-16	90.1d	C-16'	60.9q
C-7	45.0d	C-17	61.0d	C-18'	58.8q
C-8	91.6s	C-18	80.0t	O-C=O (R)	174.8s
C-9	44.5d	C-19	53.1t		
C-10	40.9d	N-CH_2 (CH_3)	48.6t / 13.2q	O-C=O (Ph)	165.8s

1.   I. Kitagawa, M. Yoshikawa, Z.L. Chen and K. Kobayashi, *Chem. Pharm. Bull. Japan*, 30, 758 (1982).

LIPOHYPACONITINE

Colorless oil;

$[\alpha]_D$ + 13.5° (CHCl$_3$)

Aconitum spp ("Chuanwu" Chinese)

R = -(CH$_2$)$_{14}$-CH$_3$ (Palmitoyl)
-(CH$_2$)$_{16}$-CH$_3$ (Stearoyl)
-(CH$_2$)$_7$-CH=CH-(CH$_2$)$_7$-CH$_3$ (Oleoyl)
-(CH$_2$)$_7$-CH=CH-CH$_2$-CH=CH-(CH$_2$)$_4$-CH$_3$ (Linoleoyl)
-(CH$_2$)$_7$-(CH$_2$-CH=CH)$_3$-CH$_3$ (Linolenoyl)

^{13}C Chemical Shift Assignments[1]

C-1	84.8d	C-11	49.8s	C-1'	56.2q
C-2	26.1t	C-12	34.5t	C-6'	57.8q
C-3	34.7t	C-13	73.9s	C-7'	-
C-4	39.0s	C-14	78.8d	C-8'	-
C-5	49.8d	C-15	78.8d	C-14'	-
C-6	83.0d	C-16	90.0d	C-16'	60.8q
C-7	44.4d	C-17	61.9d	C-18'	58.7q
C-8	91.3s	C-18	79.8t	O-C=O R	174.7s
C-9	43.7d	C-19	49.8t		
C-10	40.9d	N-CH$_3$	42.3q	O-C=O Ph	165.6s

1.   I. Kitigawa, M. Yoshikawa, Z.L. Chen and K. Kobayashi, *Chem. Pharm. Bull.*
      Japan, 30, 758 (1982).

LIPOMESACONITINE

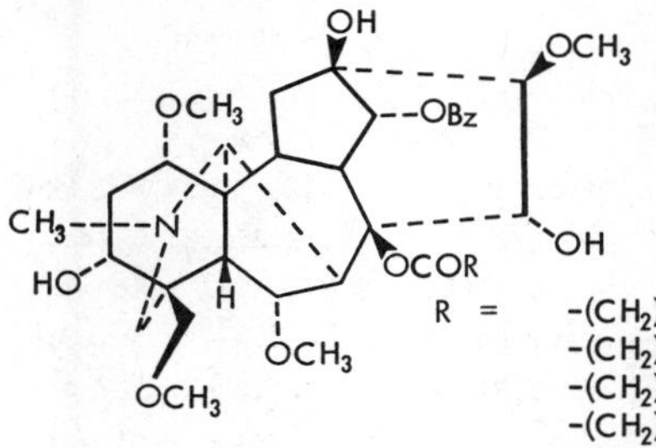

Colorless oil;

$[\alpha]_D$ + 13.8°

Aconitum spp. ("Chuanwu" Chinese)

R =   -(CH$_2$)$_{14}$-CH$_3$ (Palmitoyl)
      -(CH$_2$)$_{16}$-CH$_3$ (Stearoyl)
      -(CH$_2$)$_7$-CH=CH-(CH$_2$)$_7$-CH$_3$ (Oleoyl)
      -(CH$_2$)$_7$-CH=CH-CH$_2$-CH=CH-(CH$_2$)$_4$-CH$_3$ (Linoleoyl)
      -(CH$_2$)$_7$-(CH$_2$-CH=CH)$_3$-CH$_3$ (Linolenoyl)

^{13}C *Chemical Shift Assignments* [1]

C-1	83.1d	C-11	49.9s	C-1'	56.1q
C-2	35.6t	C-12	34.0t	C-6'	58.0q
C-3	70.5d	C-13	73.9s	C-7'	-
C-4	43.3s	C-14	78.8d	C-8'	-
C-5	46.0d	C-15	78.8d	C-14'	-
C-6	82.3d	C-16	90.0d	C-16'	60.9q
C-7	44.2d	C-17	62.0d	C-18'	58.8q
C-8	91.3s	C-18	75.5t	C=O	174.9s
C-9	43.6d	C-19	49.2t	R	
C-10	40.7d	N-CH$_3$	42.2q	O-C=O	165.8s
				Ph	

1. I. Kitigawa, M. Yoshikawa, Z.L. Chen and K. Kobayashi, *Chem. Pharm. Bull.* Japan, 30, 758 (1982).

LIWACONITINE

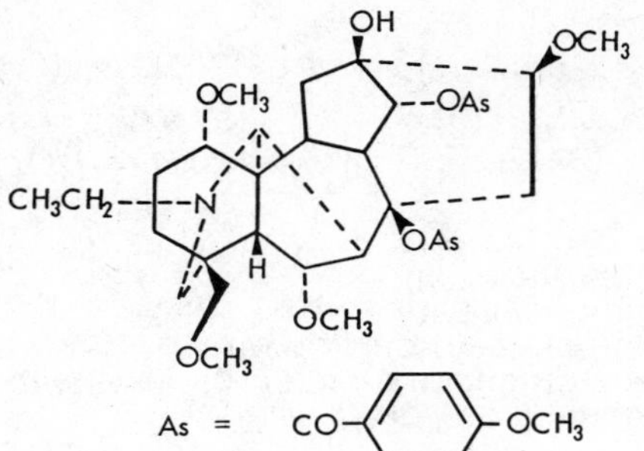

$C_{41}H_{53}NO_{11}$; MW: 735.3596;
mp 201-202.5°; $[\alpha]_D$ + 133.3° (CHCl3)
Aconitum forestii
¹H NMR: δ 1.13 (3H, *t*, J=7Hz, NCH₂-
CH₃), 2.87, 3.16, 3.27, 3.56 (each
3H, *s*, OC*H₃*), 3.67, 3.72 (each 3H,
s, aromatic OC*H₃*), 6.47, 6.50, 7.48
and 7.71 (each 2H, *d*, J=9Hz, 2 sets
of A₂B₂ system, aromatic protons).¹

As =

¹³C Chemical Shift Assignments

C-1	C-16	
C-2	C-17	
C-3	C-18	
C-4	C-19	
C-5	N-CH₂	
C-6	CH₃	
C-7	C-1'	
C-8	C-6'	
C-9	C-7'	
C-10	C-8'	
C-11	C-14'	
C-12	C-16'	
C-13	C-18'	
C-14	C=O	
C-15	CH₃	

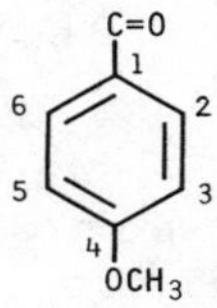

1.   C.-H. Wang, D.-H. Chen and W. Sung, *Planta Medica*, 48, 55 (1983).

LUDACONITINE

$C_{32}H_{45}NO_9$; $[\alpha]_D$ + 28° (EtOH)
Aconitum franchetii.
^{1}H NMR: δ 1.13 (3H, *t*, J=7Hz, NCH$_2$-
CH$_3$), 3.21, 3.24, 3.26, 3.36 (each
3H, *s*, OCH$_3$), 5.13 (1H, *d*, J=4.5Hz,
C(14)-β-*H*), 4.07 (1H, *d*, J=7Hz,
C(6)-β-*H*), 3.58 (1H, *d*, J=9Hz, C(18)
-*H*), 7.43-8.06 (5H, *m*, aromatic pro-
tons).[1]

^{13}C *Chemical Shift Assignments*

C-1	C-16	
C-2	C-17	
C-3	C-18	
C-4	C-19	
C-5	N-CH$_2$	
C-6	CH$_3$	
C-7	C-1'	
C-8	C-6'	
C-9	C-7'	
C-10	C-8'	
C-11	C-14'	
C-12	C-16'	
C-13	C-18'	
C-14	C=O	
C-15	CH$_3$	

1. D.-H. Chen and W.-L. Sung, *Chinese Traditional and Herbal Drugs,* 13,
   8 (1982).

LYCACONITINE

$C_{36}H_{48}N_2O_{10}$; MW: 668.3186;
Amorphous; $[\alpha]_D$ + 43° (EtOH)
Aconitum gigas, A. lycoctonum, A. umbrosum, Delphinium cashmirianum.[*]
^{1}H NMR: δ 1.08 (3H, *t*, NCH_2-CH_3), 2.88 (4H, *s*, -$\overset{O}{\overset{\|}{C}}CH_2CH_2\overset{O}{\overset{\|}{C}}$-), 3.23 and 3.38 (each 3H, *s*, OCH_3), 3.32 (6H, *s*, OCH_3), 3.58 (1H, *t*, J=4Hz, C(14)-β-*H*), and 7.20-8.02 (aromatic protons).[1,2,3,4]

^{13}C *Chemical Shift Assignments*

C-1	C-11	C-1'
C-2	C-12	C-6'
C-3	C-13	C-7'
C-4	C-14	C-8'
C-5	C-15	C-14'
C-6	C-16	C-16'
C-7	C-17	C-18'
C-8	C-18	C=O
C-9	C-19	CH₃
C-10	N-CH₂	
	CH₃	

1.   S. Sakai, N. Shinma, S. Hasegawa, and T. Okamoto, *Yakugaku Zasshi*, 98(10), 1376 (1978).

2.   M. Shamma, P. Chinnasamy, G.A. Miana, A. Khan, M. Bashir, M. Salazar, P. Patil, and J.L. Beal, *J. Nat. Prod.*, 42, 615 (1979).

3.   S.W. Pelletier and N.V. Mody, in *The Alkaloids* (R.H.F. Manske and R. Rodrigo Eds.), Vol. 17, Chapter 1, Academic Press, New York, 1979.

4.   S.W. Pelletier, N.V. Mody, K.I. Varughese, J.A. Maddry, and H.K. Desai, *J. Am. Chem. Soc.*, 1981, 103, 6536.

LYCOCTONAL

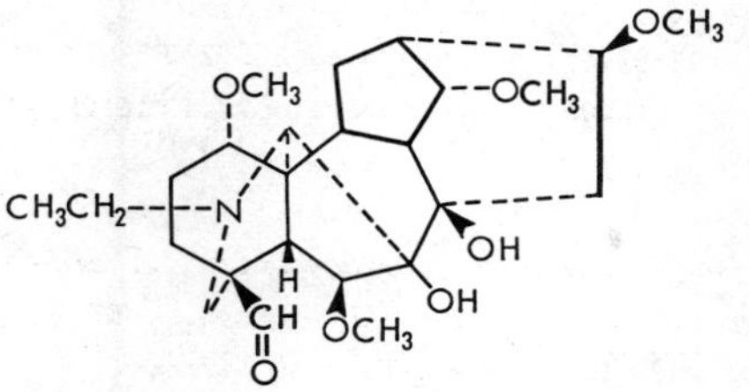

$C_{25}H_{39}NO_7$; MW: 465.2733;
mp 73°; 93-103°; (perchlorate)
213° (dec.);

$[\alpha]_D$ + 74° (EtOH)

Prepared from lycoctonine

^{1}H NMR: δ 1.06 (3H, t, J=7Hz, NCH$_2$-
CH_3), 3.23, 3.26, 3.30 and 3.37 (each
3H, s, OCH_3) and 9.38 (1H, s, CHO).[1,2]

^{13}C *Chemical Shift Assignments*[3]

C-1	82.7	C-11	49.7	C-1'	57.6
C-2	33.8[a]	C-12	29.0[a]	C-6'	55.8
C-3	29.8[a]	C-13	38.1	C-7'	-
C-4	48.5	C-14	83.9	C-8'	-
C-5	43.4[b]	C-15	25.2[a]	C-14'	59.2
C-6	92.0	C-16	83.5	C-16'	56.1
C-7	88.6	C-17	64.6	C-18'	-
C-8	77.6	C-18	203.6		
C-9	49.0[b]	C-19	50.8		
C-10	46.0[b]	N-CH$_2$	50.0		
		CH$_3$	13.9		

1. M. Shamma, P. Chinnasamy, G.A. Miana, A. Khan, M. Bashir, M. Salazar, P. Patil, and J.L. Beal, *J. Nat. Prod.*, 42, 615 (1979).
2. O.E. Edwards and L. Marion, *Can. J. Chem.*, 30, 627 (1952).
3. A.J. Jones and M.H. Benn, *Can. J. Chem.*, 51, 486 (1973).

LYCOCTONINE (DELSINE, ROYALINE)

$C_{25}H_{41}NO_7$; mp 151-153°, 112-114°, 95-97.5°; perchlorate, 206-207°; $[\alpha]_D$ + 53° (EtOH); perchlorate, + 27.1° (EtOH).

Aconitum finetianum, A. gigas, A. lycoctonum, Consolida ambigua, Delphinium ajacis, D. barbeyi, D. brownii, D. consolida, D. dictyocarpum, D. elisabethae, D. formosum, D. glaucescens, D. iliense, D. oreophilum, D. regalis, D. semibarbatum, D. tamarae, D. tricorne, Inula royleana.

^{1}H NMR: δ 1.04 (3H, *t*, NCH_2-CH_3), 3.25, 3.34, 3.41 and 3.45 (each 3H, *s*, OCH_3).[1,4,5]

[13]C *Chemical Shift Assignments* [2,3]

C-1	84.2[a]	C-11	48.9	C-1'	55.7
C-2	26.1	C-12	28.8	C-6'	57.7
C-3	31.6	C-13	46.1	C-7'	-
C-4	38.6	C-14	84.0[a]	C-8'	-
C-5	43.3	C-15	33.7	C-14'	58.0
C-6	90.6	C-16	82.7	C-16'	56.2
C-7	88.3	C-17	64.8	C-18'	-
C-8	77.5	C-18	67.6	C=0 CH_3	
C-9	49.7	C-19	52.9		
C-10	38.0	$N-CH_2$ CH_3	51.1 14.1		

1. S. Sakai, N. Shinma, S. Hasegawa, and T. Okamoto, *J. Pharm. Soc.* Japan, 98(10), 1376 (1978).

2. A.J. Jones and M.H. Benn, *Can. J. Chem.*, 51, 486 (1973).

3. S.W. Pelletier, N.V. Mody, R.S. Sawhney and J. Bhattacharyya, *Heterocycles*, 7, 327 (1977).

4. S.W. Pelletier, R.S. Sawhney, H.K. Desai, and N.V. Mody, *J. Nat. Prod.*, 43, 395 (1980).

5. S.W. Pelletier, N.V. Mody, K.I. Varughese, J.A. Maddry, and H.K. Desai, *J. Am. Chem. Soc.*, 1981, 103, 6536.

MESACONITINE

$C_{33}H_{45}NO_{11}$;   mp 208-209°;
$[\alpha]_D$ + 25° (CHCl$_3$)

Aconitum alatiacum, A. callianthum,
A. carmichaeli, A. excelsum, A. fau-
riei, A. grossedentatum, A. hakusan-
ense, A. ibukiense, A. japonicum, A.
kamtschaticum (fischeri), A. kusne-
zoffi, A. majimai, A. manshuricum, A.
mitakense, A. mokchangense, A. napel-
lus and ssp., A. sanyoense, A. sczu-
kini, A. subcuneatum, A. tasiromon-
tanum, A. tortuosum, A. yezoense, A.
zuccarini.

^{1}H NMR: δ 1.43 (3H, *s*, OCOC*H*$_3$), 2.59 (3H, *s*,
NC*H*$_3$), 3.32 and 3.88 (each 3H, *s*, OC*H*$_3$), 3.43 (6H,
s, OC*H*$_3$), 4.92 (1H, *d*, J=5Hz, C(14)-β-*H*) and 7.38-
8.20 (aromatic protons).

[13]C *Chemical Shift Assignments* [1]

C-1	83.2	C-11	50.0	C-1'	56.2
C-2	35.9	C-12	34.2	C-6'	57.9
C-3	70.8	C-13	74.1	C-16'	61.0
C-4	43.5	C-14	78.9	C-18'	59.0
C-5	46.5	C-15	78.9	C=O	172.3
C-6	82,4	C-16	90.1	CH$_3$	21.4
C-7	44.3[a]	C-17	62.2		
C-8	91.8	C-18	75.8	C=O	166.0
C-9	43,8[a]	C-19	49.4	1	129.9
C-10	40.9	N-CH$_3$	42.4	2	129.6
				3	128.6
				4	133.2

1.   S.W. Pelletier and Z. Djarmati, *J. Am. Chem. Soc.*, 98, 2626 (1976).

18-METHOXYGADESINE

$C_{24}H_{37}NO_7$; MW: 451.2566
mp 180–184°;

Consolida orientalis

^{1}H NMR: δ 1.09 (3H, *t*, NCH$_2$-C*H$_3$*),
3.30, 3.41, 3.90 (each 3H, *s*, OC*H$_3$*),
3.70 (1H, *m*, C(1)-β-*H*), 4.13 (1H, *t*,
C(14)-β-*H*), 3.88, 3.95 (each 1H, *s*,
C(16)-β-*H*) or C(19)*H*).[1]

[13]*C Chemical Shift Assignments* [1]

C-1	85.2	C-11	46.4	C-1'	–
C-2	25.5*	C-12	27.7	C-6'	58.9
C-3	21.9*	C-13	45.3	C-7'	–
C-4	43.2	C-14	75.3	C-8'	–
C-5	38.2	C-15	33.8	C-14'	–
C-6	90.2	C-16	81.7	C-16'	56.5
C-7	85.1	C-17	64.1	C-18'	59.1
C-8	76.1	C-18	73.3	C=O	
C-9	49.6	C-19	68.8	$\mid$	
C-10	36.8	N-CH$_2$	47.4	CH$_3$	
		$\mid$CH$_3$	13.7		

1.   A.G. Gonzalez, G. de la Fuente and O. Munguia, *Heterocycles*, 20, 409
      (1983).

*G. de la Fuente, Private communication, November 8, 1983.

METHYLLYCACONITINE (DELARTINE, DELSEMIDINE)

$C_{37}H_{50}N_2O_{10}$; Amorphous (130°);
$[\alpha]_D + 49°$ (EtOH)

Consolida ambigua, Delphinium arara-
ticum, D. brownii, D. buschainum, D.
confusum, D. corumbosum, D. crassi-
folium, D. dictyocarpum, D. elatum,
D. elisabethae, D. flexuosum, D. glaucescens, D.
grandiflora, D. grandiflorum, D. linearifolium,
D. oreophilum, D. rotundifolium, D. schmalhausenii,
D. semibarbatum, D. tamarae, D. ternatum, D. tri-
corne, D. triste, D. araraticum, Inula royleana.

^{1}H NMR: δ 1.08 (3H, *t*, NCH$_2$-C*H$_3$*), 1.47 (3H, *d*,
J=6Hz, CH-C*H$_3$*), 3.28, 3.38, 3.42, 3.45 (each 3H,
s, OC*H$_3$*), 3.98 (1H, *dd*, J=4.1Hz, C(14)-*H*), 4.15
(1H, *d*, J=1Hz, C(6)-α-*H*), and 7.54-8.01 (aromatic protons).[1]

[13]C *Chemical Shift Assignments*[1,2]

C-1	83.9	C-16	82.5	C=O		164.1
C-2	26.0	C-17	64.5	1		127.1
C-3	32.0	C-18	69.5	2		133.1
C-4	37.6	C-19	52.3	3		129.4
C-5	43.2	N-CH$_2$	50.9	4		133.6
C-6	90.8	CH$_3$	14.0	5		131.0
C-7	88.5	C-1'	55.7	6		130.0
C-8	77.4	C-6'	57.8			
C-9	50.3	C-7'	-			
C-10	38.0	C-8'	-	4'		179.8
C-11	49.0	C-14'	58.2	3'		37.0
C-12	28.7	C-16'	56.3	2'		35.3
C-13	46.1	C-18'	-	1'		175.8
C-14	83.9			5'		16.4
C-15	33.6					

1. S.W. Pelletier, O.D. Dailey, Jr. and N.V. Mody, *J. Org. Chem.*, 46, 3284 (1981).

2. S.W. Pelletier, N.V. Mody, R.S. Sawhney and J. Bhattacharyya, *Hetero-cycles*, 7, 327 (1977).

3. S.W. Pelletier, N.V. Mody, K.I. Varughese, J.A. Maddry, and H.K. Desai, *J. Am. Chem. Soc.*, 1981, 103, 6536.

MITHACONITINE

$C_{32}H_{43}NO_8$; Amorphous;
$[\alpha]_D$ + 94° (EtOH)
Aconitum falconeri
^{1}H NMR: δ 1.10 (3H, *t*, NCH_2-CH_3),
3.27, 3.32, 3.35 and 3.43 (each 3H,
s, OCH_3), 4.96 (1H, *d*, $C(14)-\beta-H$),
5.57 (1H, *d*, J=6Hz, C(15)H), and
7.38-8.12 (aromatic protons).[1]

^{13}C *Chemical Shift Assignments*[2]

C-1	83.6	C-16	83.1	C=O		168.0
C-2	38.2	C-17	78.5		1	130.2
C-3	71.8	C-18	76.4		2	130.0
C-4	44.1	C-19	49.9		3	128.2
C-5	48.3	N-CH$_2$	47.9		4	132.8
C-6	83.6	CH$_3$	13.5		5	128.2
C-7	49.6	C-1'	56.3		6	130.0
C-8	146.5	C-6'	58.1			
C-9	48.3	C-7'	-			
C-10	46.4	C-8'	-			
C-11	51.7	C-14'	-			
C-12	33.4	C-16'	57.2			
C-13	77.6	C-18'	59.2			
C-14	78.3	C=O				
C-15	116.4	CH$_3$				

1. S.W. Pelletier, N.V. Mody and H.S. Puri, *J.C.S. Chem. Comm.*, 12 (1977).
2. S.W. Pelletier, N.V. Mody, R.S. Sawhney and J. Bhattacharyya, *Hetero-cycles*, 7, 327 (1977).

MONTICAMINE

$C_{22}H_{33}NO_5$; mp 163-164°;
$[\alpha]_D$ + 3° (CH_3OH)
Aconitum monticola
^{1}H NMR (Py.): δ 0.84 (3H, *t*, NCH_2-CH_3), 3.14 and 3.26 (each 3H, *s*, OCH_3), 3.54 (1H, *t*, J=4.5Hz, C(14)-β-*H*), 3.92 (1H, C(1)-β-*H*), 6.36 (1H, C(1)-OH), and 4.16 (1H, C(8)-OH).[1]

[13]*C Chemical Shift Assignments*[1,2]

C-1	77.0	C-11	53.6	C-1'	-
C-2	32.3	C-12	30.6	C-6'	-
C-3	57.7	C-13	42.3	C-7'	-
C-4	58.7	C-14	84.6	C-8'	-
C-5	46.3	C-15	42.8	C-14'	57.6
C-6	25.9	C-16	82.6	C-16'	56.1
C-7	45.5	C-17	64.5	C-18'	-
C-8	74.4	C-18	-		
C-9	45.3	C-19	57.7		
C-10	37.2	N-CH$_2$ / CH$_3$	47.6 / 13.3		

1. E.F. Ametova, M.S. Yunusov, V.E. Bannikova, N.D. Abdullaev, and V.A. Tel'nov, *Khim. Prir. Soedin.*, 466 (1981).

2. M.S. Yunusov, Personal Communication, May 14, 1981.

MONTICOLINE

$C_{22}H_{33}NO_6$;   mp 166-167°;

$[\alpha]_D$ + 15° (CH_3OH)

Aconitum karakolicum, A. monticola

^{1}H NMR (Py): δ 0.92 (3H, *t*, NCH_2-CH_3), 3.13 and 3.26 (each 3H, *s*, OCH_3) and 3.56 (1H, *t*, J=5Hz, C(14)-β-*H*).[1]

[13]C *Chemical Shift Assignments*[1,2]

C-1	77.3	C-11	54.4	C-1'	-
C-2	31.8	C-12	30.5	C-6'	-
C-3	58.3	C-13	42.0	C-7'	-
C-4	59.5	C-14	84.6	C-8'	-
C-5	45.7	C-15	36.0	C-14'	57.6
C-6	34.2	C-16	82.5	C-16'	56.2
C-7	86.6	C-17	65.7	C-18'	-
C-8	76.9	C-18	-		
C-9	46.6	C-19	53.2		
C-10	37.1	N-CH_2	50.0		
		CH_3	14.1		

1.   E.F. Ametova, M.S. Yunusov, V.E. Bannikova, N.D. Abdullaev, and V.A. Tel'nov, *Khim. Prir. Soedin.*, 466 (1981).

2.   M.S. Yunusov, Personal Communication. May 14, 1981.

NEOLINE (BULLATINE B)

$C_{24}H_{39}NO_6$; mp 159-161°;
$[\alpha]_D$ + 22° (EtOH)

*Aconitum carmichaeli, A. japonicum,
A. mitakense, A. nagarum* Stapf var.
lasiandrum W.T. Wang, *A. napellus,
A. sachaliense* var. *compactum, A.
sczukini, A. soongaricum, A. stoerckianum, A. yezoense, A. bullatifolium.**[*]

^{1}H NMR: δ 1.12 (3H, *t*, NCH_2-C*H*$_3$),
3.32 (3H, *s*, OC*H*$_3$) and 3.35 (6H, *s*,
OC*H*$_3$).[1]

[13]*C Chemical Shift Assignments*[2]

C-1	72.3	C-11	49.6	C-1'	-
C-2	29.5[a]	C-12	29.8[a]	C-6'	57.8
C-3	29.9[a]	C-13	44.3	C-7'	-
C-4	38.2	C-14	75.9	C-8'	-
C-5	44.9	C-15	42.7	C-14'	-
C-6	83.3	C-16	82.3	C-16'	56.3
C-7	52.3	C-17	63.6	C-18'	59.1
C-8	74.3	C-18	80.3		
C-9	48.3	C-19	57.2		
C-10	40.7	$N-CH_2$	48.2		
		CH_3	13.0		

1. S.W. Pelletier, Z. Djarmati, S. Lajsic, and W.H. De Camp, *J. Am. Chem. Soc.*, 98, 2617 (1976).
2. S.W. Pelletier and Z. Djarmati, *J. Am. Chem. Soc.*, 98, 2626 (1976).

OXOACONOSINE

$C_{22}H_{33}NO_5$; mp 208-210°;
$[\alpha]_D$ - 4.6°
Prepared from cammaconine
^{1}H NMR: δ 1.18 (3H, *t*, J=7Hz, CH_2
CH_3), 3.24, 3.38 (each 3H, *s*,
OCH_3).

[13]*C Chemical Shift Assignments* [1]

C-1	84.1d	C-11	47.7s	C-1'	55.9q
C-2	26.6t	C-12	27.2t	C-6'	-
C-3	30.3t	C-13	46.0d	C-7'	-
C-4	53.8d	C-14	75.2d	C-8'	-
C-5	49.2d	C-15	37.6t	C-14'	-
C-6	25.5t	C-16	82.0d	C-16'	56.7q
C-7	43.4d	C-17	61.1d	C-18'	-
C-8	71.8s	C-18	-		
C-9	46.0d	C-19	171.2s		
C-10	37.4d	N-CH₂	41.2t		
		CH₃	13.2q		

1.   O.E. Edwards, R.J. Kolt and K.K. Purushothaman, *Can. J. Chem.*, 61, 1194
     (1983).

OXOCAMMACONINE

$C_{23}H_{35}NO_6$; mp 193-195°;
Prepared from cammaconine
^{1}H NMR: δ 1.18 (3H, *t*, J=7Hz,
CH$_2$*CH$_3$*), 3.30,3.42 (each 3H, *s*,
O*CH$_3$*).[1]

^{13}C *Chemical Shift Assignments* [1]

C-1	83.7d	C-11	47.4s	C-1'	55.6q
C-2	26.5t[a]	C-12	26.8t[a]	C-6'	-
C-3	29.7t	C-13	45.7d[b]	C-7'	-
C-4	49.4s	C-14	74.9d	C-8'	-
C-5	52.6d	C-15	36.8t	C-14'	-
C-6	25.7t	C-16	81.8d	C-16'	56.5q
C-7	45.4d[b]	C-17	61.6d	C-18'	-
C-8	71.4s	C-18	66.6t		
C-9	45.7d[b]	C-19	174.3s		
C-10	36.9d	N-CH$_2$	41.0t		
		CH$_3$	12.9q		

1.   O.E. Edwards, R.J. Kolt and K.K. Purushothaman, *Can. J. Chem.*, 61, 1194
      (1983).

PENDULINE

$C_{34}H_{47}NO_9$;   mp 166-167°;
Aconitum pendulum.

^{1}H NMR: δ 1.04 (3H, *t*, NCH$_2$CH$_3$),
1.39 (3H, *s*, OCOCH$_3$), 3.12, 3.24,
3.48, 3.48 (each 3H, *s*, OCH$_3$), 5.00
(1H, *t*, J=4.5Hz, C(14)-β-H), 7.36-
8.03 (5H, *m*, aromatic protons).[1,2]

[13]C Chemical Shift Assignments

C-1	C-16	C=O
C-2	C-17	
C-3	C-18	
C-4	C-19	
C-5	N-CH$_2$	
C-6	CH$_3$	
C-7	C-1'	
C-8	C-6'	
C-9	C-7'	
C-10	C-8'	
C-11	C-14'	
C-12	C-16'	
C-13	C-18'	
C-14	C=O	
C-15	CH$_3$	

1.   Y.-L. Zhu and R.-H. Zhu, *Heterocycles*, 17, 607 (1982).
2.   L.-M. Liu, H.-C. Wang and Y.-L. Zhu, *Acta Pharmaceutica Sinica*, 18, 39 (1983).

PENTAGYDINE

$C_{22}H_{33}NO_5$; MW: 391.2358; mp 130-131°;
Delphinium pentagynum
^{1}H NMR: δ 0.88 (3H, *s*, C(4)-CH_3),
1.11 (3H, *t*, J=7Hz, N-CH_2-CH_3), 3.38
(3H, *s*, OCH_3), 4.20 (1H, *t*, J=4.5Hz,
C(14)-β-*H*).[1]

^{13}C *Chemical Shift Assignments*

C-1	C-11	C-1'
C-2	C-12	C-6'
C-3	C-13	C-7'
C-4	C-14	C-8'
C-5	C-15	C-14'
C-6	C-16	C-16'
C-7	C-17	C-18'
C-8	C-18	C=O
C-9	C-19	CH₃
C-10	N-CH₂	
	CH₃	

1.   A.G. Gonzalez, G. de la Fuente and R. Diaz, *Tetrahedron Letters*, 959 (1983).

PENTAGYLINE

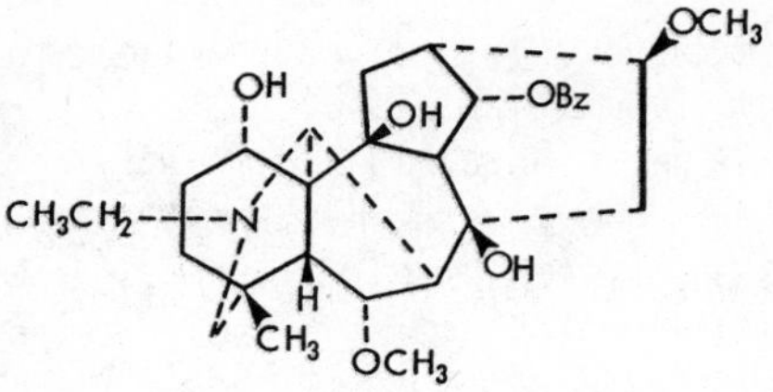

$C_{30}H_{41}NO_7$; mp 198-200°

Delphinium pentagynum

^{1}H NMR: δ 1.00(3H, *s*, CH_3), 1.03 (3H, *t*, J=7Hz, NCH_2-CH_3), 3.31, 3.38 (each 3H, *s*, OCH_3), 3.99 (1H, *d*, J= 8Hz, C(6)-β-*H*), 4.22 (1H, *m*, w½=7Hz, C(1)-β-*H*), 5.55 (1H, *t*, J=4.5Hz, C(14)-β-*H*), 7.5-8.15 (5H, *m*, aromatic protons).[1]

^{13}C *Chemical Shift Assignments*[1]

C-1	69.8	C-16	81.8	C=O		166.6
C-2	29.7	C-17	65.2	1		130.6
C-3	31.4	C-18	27.9	2		129.9
C-4	32.2	C-19	61.5	3		128.3
C-5	46.5	N-CH_2	48.7	4		132.6
C-6	82.7	CH_3	12.7	5		128.3
C-7	50.7	C-1'	–	6		129.9
C-8	73.9	C-6'	57.8			
C-9	54.4	C-7'	–			
C-10	82.4	C-8'	–			
C-11	53.8	C-14'	–			
C-12	40.5	C-16'	56.1			
C-13	38.4	C-18'	–			
C-14	75.6	C=O				
C-15	41.6	CH_3				

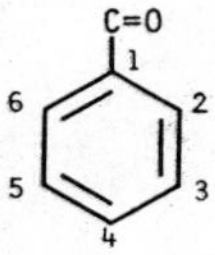

1.   A.G. Gonzalez, G. de la Fuente and R.D. Acosta, *Heterocycles*, in press.

PENTAGYNINE

$C_{23}H_{35}NO_5$; MW: 405.25152;
mp 198-201°; $[\alpha]_D$ + 72° (EtOH)
Delphinium pentagynum

^{1}H NMR: δ 0.75 (3H, *s*, C(4)-CH_3),
0.96 (3H, *t*, J=7Hz, N-CH$_2$-CH_3), 2.90
and 3.15 (each 3H, *s*, OCH_3), 3.62 (1H,
s, C(19)-*H*), 3.64 (1H, *m*, w½=7Hz, C(1)
-β-*H*), 3.94 (1H, *d*, J=7Hz, C(6)-β-*H*),
and 4.13 (1H, *dd*, J$_1$=J$_2$=4.5Hz, C(14)-
β-*H*). [1]

[13]*C Chemical Shift Assignments* [1]

C-1	91.2d	C-11	47.5s	C-1'	–
C-2	30.2t*	C-12	28.7t	C-6'	58.0q
C-3	23.0t*	C-13	45.7d	C-7'	–
C-4	38.3s	C-14	75.5d	C-8'	–
C-5	37.3d	C-15	39.0t	C-14'	–
C-6	84.3d	C-16	82.2d	C-16'	56.4q
C-7	56.9d	C-17	61.7d	C-18'	–
C-8	73.6s	C-18	20.2q		
C-9	52.6d	C-19	68.8d		
C-10	39.0d	N-CH$_2$	47.8t		
		CH$_3$	14.4q		

1. A.G. Gonzalez, G. de la Fuente, and R. Diaz, *Phytochemistry*, 21, 1781
   (1982).

*G. de la Fuente, Private communication, November 8, 1983.

PSEUDACONINE

$C_{25}H_{41}NO_8$; mp 91-92°;

$[\alpha]_D$ + 49° $(CHCl_3)$

Hydrolysis product of pseudaconitine.[1]

^{13}C Chemical Shift Assignments[2]

	CDCl$_3$	C$_5$D$_5$N		CDCl$_3$	C$_5$D$_5$N		CDCl$_3$	C$_5$D$_5$N
C-1	84.5	85.3	C-11	50.2	50.4	C-1'	56.3	55.8
C-2	35.9	38.1	C-12	33.8	35.8	C-6'	57.4	57.5
C-3	72.2	69.0	C-13	76.9	77.3	C-7'	-	-
C-4	43.5	44.6	C-14	79.6	80.0	C-8'	-	-
C-5	49.3	49.5	C-15	40.1	42.1	C-14'	-	-
C-6	82.4	83.5	C-16	83.2	83.7	C-16'	57.9	58.3
C-7	52.2	53.6	C-17	62.4	62.6	C-18'	59.4	58.9
C-8	72.9	73.5	C-18	77.5	80.0	C=O		
C-9	50.4	51.0	C-19	48.7	48.0	CH$_3$		
C-10	42.0	42.7	N-CH$_2$	47.4	47.8			
			CH$_3$	13.7	13.9			

1.   Y. Tsuda and L. Marion, *Can. J. Chem.*, 41, 1485 (1963).
2.   S.W. Pelletier, N.V. Mody and R.S. Sawhney, *Can. J. Chem.*, 57, 1652
     (1979).

PSEUDACONITINE (α-PSEUDACONITINE)

$C_{36}H_{51}NO_{12}$;   MW: 689.3430

mp 205-208°; $[\alpha]_D$ + 24° (CHCl$_3$)

Aconitum balfourii, A. deinorrhizum,
A. falconeri, A. ferox, A. genicu-
latum Flet. et Lane var. *unguicu-*
latum W.T. Wang, *A. hemsleyanum* Pritz
var. *circinatum* W.T. Wang, *A. spictatum.*

^{1}H NMR: δ 1.12 (3H, *t*, NCH$_2$-C*H*$_3$), 1.32 (3H, *s*,
OCOC*H*$_3$), 3.17, 3.27, 3.30 and 3.57 (each 3H, *s*,
OC*H*$_3$), 3.92 and 3.95 (each 3H, *s*, aromatic OC*H*$_3$),
4.88 (1H, *d*, J=4.5Hz, C(14)-β-*H*) and 6.83-7.87
(aromatic protons).[2,3]

Vr =

^{13}C *Chemical Shift Assignments*[1]

C-1	83.6	C-16	83.0	C=O		165.6
C-2	35.1	C-17	61.4	1		122.5
C-3	70.9	C-18	76.2	2		110.2
C-4	43.1	C-19	48.7	3		152.8
C-5	48.7	N-CH$_2$	47.2	4		148.4
C-6	82.1	CH$_3$	13.3	5		111.8
C-7	48.7	C-1'	55.7	6		
C-8	85.3	C-6'	57.6			
C-9	47.2	C-7'	-	R =	OCH$_3$	55.7
C-10	40.7	C-8'	-			
C-11	50.1	C-14'	-			
C-12	33.7	C-16'	58.7			
C-13	74.7	C-18'	58.9			
C-14	78.4	C=O	169.4			
C-15	39.6	CH$_3$	21.5			

1.   S.W. Pelletier, N.V. Mody, R.S. Sawhney, and J. Bhattacharyya, *Hetero-*
     cycles, 7, 327 (1977).

2.   S.W. Pelletier, N.V. Mody, and H.S. Puri, *Phytochemistry*, 16, 623 (1977).

3.   A. Klasek, V. Simanek, and F. Santavy, *Lloydia*, 35, 55 (1972).

PUBERACONITIDINE

$C_{37}H_{52}N_2O_{11}$: CI MS: 700
$[\alpha]_D$ + 22.4° (CHCl$_3$).

Aconitum barbatum var. *puberulum*[1,2]

^{1}H NMR: (400 MHz) δ 1.12 (3H, *t*, J= 7Hz, N-CH$_2$-CH$_3$), 2.8 (4H, *s*, -CH$_2$CH$_2$-), 3.0 (1H, *m*, C(17)-*H*), 3.3, 3.38, 3.38 (each 3H, *s*, OCH$_3$), 3.48 (6H, *s*, OCH$_3$), 3.3 (1H, *m*, C(16)-α-*H*), 3.5 (1H, *d*, J=5Hz, C(14)-β-*H*), 3.71 (1H, *s*, C(6)-α-*H*), 4.18, 4.22 (each 1H, *d*, J=11Hz, C(18)-*H*), 7.12 (1H, *t*, J=7Hz, Ar-H$_5$), 7.58 (1H, *t*, J=7Hz, Ar-H$_4$), 8.0 (1H, *d*, J=7Hz, Ar-H$_3$), 8.7 (1H, *d*, J=7Hz, Ar-H$_6$), 11.0 (1H, *s*, N*H*-CO).[2]

[13]C Chemical Shift Assignments

C-1	C-16		1
C-2	C-17		2
C-3	C-18		3
C-4	C-19		4
C-5	N-CH$_2$		5
C-6	CH$_3$		6
C-7	C-1'		
C-8	C-6'		
C-9	C-7'		
C-10	C-8'		
C-11	C-14'		
C-12	C-16'		
C-13	C-18'		
C-14	C=O		
C-15	CH$_3$		

1.   D. Yu, *Yaoxue Tongbao*, 17, 301 (1982); *CA*, 97,212633 (1982).

2.   D. Yu and B.C. Das, *Planta Medica*, 49, 85 (1983).

PUBERACONITINE

$C_{36}H_{50}N_2O_{11}$; M^+-H_2O 668.3368
$[\alpha]_D$ + 34° $(CHCl_3)$.

Aconitum barbatum var. *puberulum*[1,2]

^{1}H NMR: (400 MHz) δ 1.1 (3H, t, J=
7Hz, NCH_2CH_3), 2.8 (4H, s, $-CH_2CH_2-$),
3.0 (1H, m, C(17)-H), 3.3, 3.38, 3.4, 3.42
(each 3H, s, OCH_3), 3.2 (1H, m, C(16)-α-H),
3.65 (1H, d, C(14)-β-H), 3.95 (1H, s, C(6)-β
-H), 4.2, 4.22 (each 1H, d, J=11Hz), 7.3 (1H,
t, J=7Hz, Ar-H_5), 7.58 (1H, t, J=7Hz, Ar-H_4),
8.0 (1H, d, J=7Hz, Ar-H_3), 8.72 (1H, t, J=
7Hz, Ar-H_6).[2]

[13]C Chemical Shift Assignments[2]

C-1	83.8	C-16	82.5	C=O		168.0
C-2	25.7	C-17	64.5		1	114.7
C-3	31.7	C-18	69.2		2	141.6
C-4	37.5	C-19	52.5		3	120.7
C-5	43.2	N-CH₂	50.9		4	134.8
C-6	90.9	CH₃	13.7		5	122.5
C-7	88.3	C-1'	55.7		6	130.3
C-8	77.7	C-6'	57.6			
C-9	50.3	C-7'	-	X	= NHCO	170.7
C-10	37.9	C-8'	-		CH₂	29.5
C-11	49.0	C-14'	58.0		CH₂	29.8
C-12	28.7	C-16'	56.1		COOH	170.7
C-13	45.9	C-18'	-			
C-14	83.8	C=O	-			
C-15	33.6	CH₃	-			

1. D. Yu, *Yaoxue Tongbao*, 17, 301 (1982); *CA*, 97, 212633 (1982).
2. D. Yu and B.C. Das, *Planta Medica*, 49, 85 (1983).

PUBERANINE

C$_{32}$H$_{44}$N$_2$O$_9$; MW: 600.3014
Amorphous; $[\alpha]_D$ + 16.6° (CHCl$_3$)
Aconitum barbatum var. *puberulum*[1]
^{1}H NMR: (400 MHz) δ 1.15 (3H, *t*, J=
7Hz, NCH$_2$-C*H$_3$*), 2.25 (3H, *s*, CO*CH$_3$*),
2.85 (1H, *s*, C(17)-*H*), 3.28, 3.34, 3.44
(each 3H, *s*, O*CH$_3$*), 3.1 (1H, *m*, C(16)-
α-*H*), 3.5 (1H, *d*, J=5Hz, C(14)-β-*H*),
7.05, 7.5 (each 1H, *t*, J=7Hz, Ar-H$_5$,
-H$_4$), 7.9, 8.7 (each 1H, *d*, J=7Hz,
Ar-H$_5$,-H$_6$), 11.0 (1H, *s*, *NH*-CO).[1]

^{13}*C Chemical Shift Assignments*[1]

C-1	81.4	C-16	82.9	C=O		169.2
C-2	26.5	C-17	63.1			115.7
C-3	37.0	C-18	–			141.2
C-4	84.0	C-19	55.3			120.2
C-5	51.1	N-CH$_2$	48.6			134.6
C-6	32.5	CH$_3$	14.4			122.5
C-7	78.0	C-1'	56.3			131.3
C-8	85.5	C-6'	–			
C-9	78.4	C-7'	–			
C-10	36.8	C-8'	–			
C-11	51.4	C-14'	58.0			
C-12	26.0	C-16'	56.3			
C-13	49.8	C-18'	–			
C-14	90.1	C=O	169.3			
C-15	38.4	CH$_3$	25.6			

1.   D. Yu and B.C. Das, *Planta Medica*, 49, 85 (1983).

PYROCHASMANINE

$C_{25}H_{39}NO_5$; mp 126-129°;
$[\alpha]_D$ + 243.9°
Aconitum yesoense
^{1}H NMR: δ 1.06 (3H, *t*, J=7Hz, NCH$_2$-
CH$_3$), 3.92 (1H, *t*, J=4.5Hz, C(14)-β
-H), 5.44 (1H, *d*, J=6Hz, C(15)-*H*).[1,2]

[13]C *Chemical Shift Assignments*

C-1	C-11	C-1'
C-2	C-12	C-6'
C-3	C-13	C-7'
C-4	C-14	C-8'
C-5	C-15	C-14'
C-6	C-16	C-16'
C-7	C-17	C-18'
C-8	C-18	C=O | CH$_3$
C-9	C-19	
C-10	N-CH$_2$ | CH$_3$	

1.   H. Takayama, A. Tokita, M. Ito, S. Sakai, F. Kurosaki and T. Okamoto,
     J. Pharm. Soc. Japan, 102, 245 (1982).

2.   O. Achamatowicz, Jr., Y. Tsuda, L. Marion, T. Okamoto, M. Natsume,
     H. Chang and K. Kajima, *Can. J. Chem.*, 43, 825 (1965).

PYRODELPHININE

$C_{31}H_{41}NO_7$; mp 208-212°;
Prepared from delphinine
^{1}H NMR: δ 2.44 (3H, *s*, N-C*H₃*), 3.35
and 3.48 (each 3H, *s*, OC*H₃*), 3.40
(6H, *s*, OC*H₃*), 4.06 (1H, *s*, C(6)-β-*H*),
5.17 (1H, *d*, C(14)-β-*H*), 5.78 (1H, *d*,
C(15)-*H*) and 7.76-8.38 (aromatic protons).

^{13}C *Chemical Shift Assignments* [1,2,3]

C-1	86.1	C-16	83.6	C=O	168.0
C-2	25.3	C-17	78.6	1	130.5
C-3	35.3	C-18	80.3	2	130.0
C-4	40.0	C-19	56.5	3	128.1
C-5	48.5	N-CH₃	42.7	4	132.7
C-6	83.6			5	128.1
C-7	50.4	C-1'	56.5	6	130.0
C-8	146.6	C-6'	58.1		
C-9	47.6	C-7'	-		
C-10	46.7	C-8'	-		
C-11	51.9	C-14'	-		
C-12	38.4	C-16'	57.1		
C-13	77.7	C-18'	59.2		
C-14	79.1	C=O			
C-15	116.3	CH₃			

1.   S.W. Pelletier and Z. Djarmati, *J. Am. Chem. Soc.*, 98, 2626 (1976).
2.   S.W. Pelletier, N.V. Mody, R.S. Sawhney, and J. Bhattacharyya, *Hetero-cycles*, 7, 327 (1977).
3.   S.W. Pelletier, J. Finer-Moore, R.C. Desai, N.V. Mody, and H.K. Desai, *J. Org. Chem.*, 47, 5290 (1982).

PYRODELPHONINE

$C_{24}H_{37}NO_6$; Amorphous;
Prepared from pyrodelphinine
^{1}H NMR: δ 2.46 (3H, s, N-CH_3), 3.35
and 3.58 (each 3H, s, OCH_3), 3.41 (6H,
s, OCH_3), 4.20 (1H, d, C(14)-β-H) and
5.74 (1H, d, C(15)-H).

^{13}C *Chemical Shift Assignments*[1]

C-1	86.0	C-11	51.8	C-1'	56.3
C-2	25.2	C-12	37.2	C-6'	58.1
C-3	35.2	C-13	77.0	C-7'	-
C-4	40.0	C-14	79.5	C-8'	-
C-5	49.4	C-15	114.3	C-14'	-
C-6	83.5[a]	C-16	83.8[a]	C-16'	56.6
C-7	50.5	C-17	77.5	C-18'	59.1
C-8	148.7	C-18	80.3		
C-9	47.9	C-19	56.6		
C-10	47.6	N-CH$_3$	42.6		

1. S.W. Pelletier and Z. Djarmati, *J. Am. Chem. Soc.*, 98, 2626 (1976).

RANACONINE

$C_{23}H_{37}NO_7$; mp 107-109°;
Hydrolysis product of ranaconitine.[1]

[13]C Chemical Shift Assignments[2]

C-1	84.9	C-11	51.4	C-1'	56.3
C-2	27.1	C-12	26.3	C-6'	-
C-3	36.8	C-13	51.1	C-7'	-
C-4	71.1	C-14	90.2	C-8'	-
C-5	51.1	C-15	38.1	C-14'	57.9
C-6	32.4	C-16	83.0	C-16'	56.3
C-7	86.5	C-17	63.2	C-18'	-
C-8	78.0	C-18	-	C=O	
C-9	78.7	C-19	56.8	CH₃	
C-10	37.5	N-CH₂	50.0		
		CH₃	14.5		

1.   S.W. Pelletier, N.V. Mody, A.P. Venkov, and N.M. Mollov, *Tetrahedron Letters*, 5045 (1978).

2.   S.W. Pelletier, N.V. Mody and R.S. Sawhney, *Can. J. Chem.*, 57, 1652 (1979).

RANACONITINE

$C_{32}H_{44}N_2O_9$; mp 132-134°;
$[\alpha]_D$ + 33.2° (CHCl$_3$)
Aconitum barbatum var. *puberulum, A.
finetianum, A. ranunculaefolium, A.
sinomontanum.*
^{1}H NMR: δ 1.13 (3H, *t*, NCH$_2$-C*H$_3$*),
2.24 (3H, *s*, NHCOC*H$_3$*), 3.28, 3.33
and 3.43 (each 3H, *s*, OC*H$_3$*), 7.13-
8.68 (aromatic protons) and 11.07
(1H, *bs*, N*H*COCH$_3$).[1]

[13]C *Chemical Shift Assignments* [1]

C-1	83.5	C-16	82.9	C=O		167.7
C-2	26.5	C-17	63.1		1	115.9
C-3	31.6	C-18	-		2	141.8
C-4	84.4	C-19	55.2		3	120.4
C-5	51.1	N-CH$_2$	48.7		4	134.6
C-6	32.5	CH$_3$	14.4		5	122.6
C-7	85.7	C-1'	56.3		6	131.3
C-8	77.9	C-6'	-			
C-9	78.4	C-7'	-	X = NHCO		169.5
C-10	36.6	C-8'	-	CH$_3$		25.6
C-11	51.4	C-14'	58.0			
C-12	25.9	C-16'	56.3			
C-13	49.8	C-18'	-			
C-14	90.0	C=O				
C-15	37.8	CH$_3$				

1.   S.W. Pelletier, N.V. Mody, A.P. Venkov and N.M. Mollov, *Tetrahedron
Letters*, 5045 (1978).

SACHACONITINE (VILMORRIANINE D)

$C_{23}H_{37}NO_4$; MW: 391.2700;
mp 129-130°; 113-115°;

$[\alpha]_D$ - 13.1° (EtOH)

Aconitum miyabei, A. vilmorrianum

^{1}H NMR: δ 0.8 (3H, s, C(4)-CH_3),
1.06 (3H, t, J=7Hz, NCH$_2$-CH_3), 3.28
and 3.37 (each 3H, s, OCH_3) and 4.15
(1H, dd, C(14)-β-H).[1,2]

^{13}C Chemical Shift Assignments [1,2,3]

C-1	86.7	86.7	C-11	51.0	50.9	C-1'	56.3	56.4
C-2	26.3	26.3	C-12	27.8	26.3	C-6'	-	-
C-3	37.8	37.9	C-13	45.9	45.9	C-7'	-	-
C-4	34.7	34.7	C-14	75.7	75.6	C-8'	-	-
C-5	49.5	45.9	C-15	38.0	38.8	C-14'	-	-
C-6	25.2	25.2	C-16	82.3	82.4	C-16'	56.9	56.1
C-7	45.9	47.1	C-17	62.5	62.3	C-18'	-	-
C-8	72.9	72.5	C-18	26.3	27.9			
C-9	47.1	49.0	C-19	57.5	57.0			
C-10	38.5	37.9	N-CH$_2$	49.5	49.3			
			CH$_3$	13.7	13.7			

1. S.W. Pelletier, N.V. Mody, and N. Katsui, *Tetrahedron Letters*, 4027
    (1977).

2. T.-R. Yang, X.-J. Hao and J. Chow, *Acta Botanica Yunnacia*, 1(2), 41
    (1979).

3. C.-R. Yang, D.-Z. Wang, D.-G. Wu, X.-J. Hao and J. Zhou, *Acta Chimica
    Sinica*, 39, 445 (1981).

SCOPALINE

$C_{21}H_{33}NO_4$; mp 167-169°;
Aconitum episcopale
^{1}H NMR: δ 1.12 (3H, *t*, NCH$_2$-CH$_3$),
3.10 (1H, *s*, O*H*), 3.35 (3H, *s*, OCH$_3$),
3.72 (1H, *t*, J=3Hz, C(1)-β-*H*), 4.23
(1H, *dd*, J=4.5Hz, C(14)-β-*H*) and 7.40
(1H, *bs*, exch. with D$_2$O, O*H*).[1]

[13]*C Chemical Shift Assignments*[1]

C-1	72.5	C-11	47.9	C-1'	–
C-2	28.7	C-12	28.7	C-6'	–
C-3	29.7	C-13	44.1	C-7'	–
C-4	33.3	C-14	75.7	C-8'	–
C-5	41.2	C-15	42.5	C-14'	–
C-6	23.8	C-16	82.4	C-16'	56.3
C-7	45.4	C-17	64.0	C-18'	–
C-8	74.4	C-18	–		
C-9	46.6	C-19	54.0		
C-10	41.0	N-CH$_2$	48.6		
		CH$_3$	12.9		

1.   C.-R. Yang, D.-Z. Wang, D.-G. Wu, X.-J. Hao and J. Zhou, *Acta Chimica Sinica*, 39, 445 (1981).

SENBUSINE A

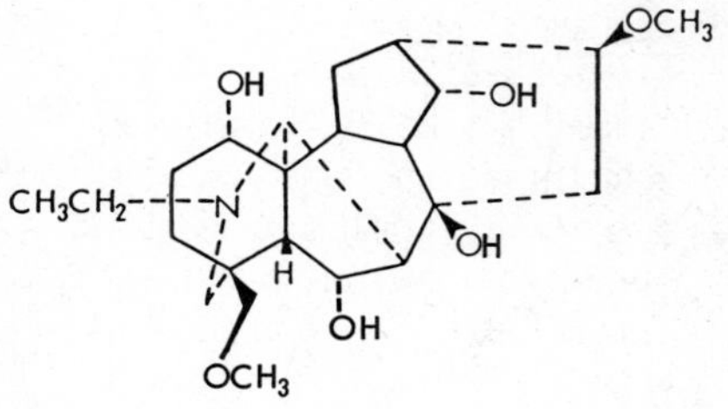

$C_{23}H_{37}NO_6$; MW: 423.2620;
Amorphous

Aconitum carmichaeli

^{1}H NMR: δ 1.14 (3H, *t*, J=7Hz, NCH_2-CH_3), 3.34 (6H, *s*, 2 OCH_3), and 4.20 (1H, *t*, J=4.5Hz, C(14)-β-*H*).[1]

[13]C Chemical Shift Assignments[1]

C-1	72.1d	C-11	48.2s	C-1'	-
C-2	29.2t[a]	C-12	29.9t	C-6'	-
C-3	29.8t[a]	C-13	44.2d	C-7'	-
C-4	37.9s	C-14	75.4d	C-8'	-
C-5	48.2d	C-15	42.2t	C-14'	-
C-6	72.6d	C-16	82.4d	C-16'	56.3q
C-7	55.4d	C-17	63.5d	C-18'	59.2q
C-8	75.6s	C-18	80.3t		
C-9	45.6d	C-19	57.1t		
C-10	40.6d	N-CH_2	49.7t		
		CH_3	12.9q		

1.   C. Konno, M. Shirasaka, and H. Hikino, *J. Nat. Prod.*, 45, 128 (1982).

SENBUSINE B

$C_{23}H_{37}NO_6$; Amorphous;
Aconitum carmichaeli
^{1}H NMR: δ 1.13 (3H, *t*, J=7Hz, NCH$_2$-
CH$_3$), 3.12 (2H, *m*, C(18)-H$_2$), 3.33
(3H, *s*, OCH$_3$), 3.45 (3H, *s*, OCH$_3$),
4.11 (1H, *t*, J=4.5Hz, C(14)-β-*H*), and
4.37 (1H, *d*, J=6Hz, C(15)-β-*H*).[1]

[13]*C Chemical Shift Assignments*[1]

C-1	72.1d	C-11	48.7s	C-1'	–
C-2	29.2t[a]	C-12	26.4t	C-6'	–
C-3	29.6t[a]	C-13	44.0d	C-7'	–
C-4	37.2s	C-14	75.4d	C-8'	–
C-5	40.9d	C-15	77.6d	C-14'	–
C-6	24.8t	C-16	90.5d	C-16'	57.4q
C-7	46.8d	C-17	63.2d	C-18'	59.4q
C-8	79.0s	C-18	79.0t		
C-9	47.9d	C-19	56.5t		
C-10	40.7d	N-CH$_2$	48.7t		
		CH$_3$	12.8q		

1.   C. Konno, M. Shirasaka, and H. Hikino, *J. Nat. Prod.*, 45, 128 (1982).

SEPTENTRIODINE (CASHMIRADELPHINE)

$C_{37}H_{52}N_2O_{11}$; MW: 700.3570;
mp 130-135°; $[\alpha]_D$ + 56° (EtOH);
Aconitum barbatum var. *puberulum, A.
gigas, A. septentrionale, Delphinium
cashmirianum.*
^{1}H NMR: δ 1.06 (3H, t, J=7Hz, NCH$_2$-
CH_3), 2.72 (4H, s, -CH_2CH_2COOCH$_3$),
3.20 and 3.28 (each 3H, s, OCH_3),
3.35 (6H, s, OCH_3), 3.62 (3H, s,
COOCH_3), 6.80-8.46 (aromatic protons)
and 10.75 (1H, bs, NHCO).[1,3,4]

^{13}C *Chemical Shift Assignments* [2]

C-1	84.0	C-16	82.7	C=O		168.3
C-2	26.1	C-17	64.6		1	114.7
C-3	31.6	C-18	69.9		2	141.9
C-4	37.6	C-19	52.4		3	120.8
C-5	43.3	N-CH$_2$	51.0		4	135.2
C-6	91.1	CH$_3$	14.1		5	122.8
C-7	88.7	C-1'	55.9		6	130.5
C-8	77.6	C-6'	57.9			
C-9	50.4	C-7'	-	X = NHCO		170.6
C-10	38.1	C-8'	-	CH$_2$		28.9
C-11	49.1	C-14'	58.1	CH$_2$		32.7
C-12	28.7	C-16'	56.4	CO		173.3
C-13	46.1	C-18'		OCH$_3$		51.9
C-14	84.0	C=O				
C-15	33.7	CH$_3$				

1. M. Shamma, P. Chinnasamy, G.A. Miana, A. Khan, M. Basir, M. Salazar,
    P. Patil, and J.L. Beal, *J. Nat. Prod.*, 42, 615 (1979).
2. S.W. Pelletier, R.S. Sawhney, and A.J. Aasen, *Heterocycles*, 12, 377
    (1979).
3. S. Sakai, N. Shinma, S. Hasegawa, and T. Okamoto, *Yakugaku Zasshi*, 98,
    1376 (1978).
4. S.W. Pelletier, N.V. Mody, K.I. Varughese, J.A. Maddry, and H.K. Desai,
    J. Am. Chem. Soc., 1981, 103, 6536.

SEPTENTRIONINE

$C_{38}H_{54}N_2O_{11}$; mp 123-125°;
$[\alpha]_D$ + 21.2° (CHCl$_3$)
Aconitum barbatum var. *puberulum*, A.
septentrionale.

^{1}H NMR: δ 1.08 (3H, *t*, J=6.8Hz,
NCH$_2$-CH$_3$), 2.80 (4H, *s*, COCH$_2$CH$_2$CO$_2$
CH$_3$), 3.25 (3H, *s*, OCH$_3$), 3.37 and
3.47 (each 6H, *s*, OCH$_3$), 3.70 (3H,
s, COOCH$_3$), 7.04-8.74 (aromatic protons) and 11.15 (1H, *bs*, N*H*CO).[1,2]

[13]C *Chemical Shift Assignments* [1]

C-1	83.1	C-16	82.8	C=O		168.4
C-2	25.6	C-17	66.2	1		115.1
C-3	31.9	C-18	70.6	2		141.8
C-4	37.7	C-19	53.2	3		120.6
C-5	46.7	N-CH$_2$	51.9	4		134.9
C-6	91.5	CH$_3$	14.8	5		122.7
C-7	90.4	C-1'	55.7	6		130.8
C-8	80.9	C-6'	60.0			
C-9	51.9	C-7'	-	X= NHCO		170.6
C-10	37.7	C-8'	54.4	CH$_2$		29.0
C-11	47.6	C-14'	57.7	CH$_2$		32.7
C-12	27.9	C-16'	56.5	C=O		173.3
C-13	40.6	C-18'	-	OCH$_3$		51.9
C-14	83.5	C=O				
C-15	27.9	CH$_3$				

1. S.W. Pelletier, R.S. Sawhney and A.J. Aasen, *Heterocycles*, 12, 377 (1979).
2. S.W. Pelletier, N.V. Mody, K.I. Varughese, J.A. Maddry, and H.K. Desai, *J. Am. Chem. Soc.*, 1981, 103, 6536.

puberulum,[2,3] A. gigas[1]

^{1}H NMR: (400 MHz) δ 1.1 (3H, *t*, J=7Hz, NCH$_2$CH$_3$),
2.8 (4H, *s*, -CH$_2$CH$_2$-), 3.0 (1H, *m*, C(17)-H), 3.3,
3.38, 3.4, 3.42 (each 3H, *s*, OCH$_3$), 3.2 (1H, *m*,
C(16)-α-H), 3.65 (1H, *d*, C(14)-β-H), 3.95 (1H, *s*,
C(6)-β-H), 4.2, 4.22 (each 1H, *d*, J=11Hz), 7.3
(1H, *t*, J=7Hz, Ar-H$_5$), 7.58 (1H, *t*, J=7Hz, Ar-
H$_4$), 8.0 (1H, *d*, J=7Hz, Ar-H$_3$), 8.72 (1H, *t*, J=
7Hz, Ar-H$_6$).[3,4]

[13]C *Chemical Shift Assignments* [3]

C-1	83.8	C-16	82.5	C=O		168.0
C-2	25.7	C-17	64.5		1	114.7
C-3	31.7	C-18	69.2		2	141.6
C-4	37.5	C-19	52.5		3	120.7
C-5	43.2	N-CH$_2$	50.9		4	134.8
C-6	90.9	CH$_3$	13.7		5	122.5
C-7	88.3	C-1'	55.7		6	130.3
C-8	77.7	C-6'	57.6			
C-9	50.3	C-7'	-	X = NHCO		170.7
C-10	37.9	C-8'	-	CH$_2$		29.5
C-11	49.0	C-14'	58.0	CH$_2$		29.8
C-12	28.7	C-16'	56.1	COOH		170.7
C-13	45.9	C-18'	-			
C-14	83.8	C=O	-			
C-15	33.6	CH$_3$				

1. S. Sakai, N. Shinma , S. Hasegawa, and T. Okamoto, *J. Pharm. Soc.* Japan, 98, 1376 (1978).

2. D. Yu, *Yaoxue Tongbao*, 17, 301 (1982); *CA*, 97, 212633 (1982).

3. D. Yu and B.C. Das, *Planta Medica*, 49, 85 (1983).

4. S.W. Pelletier, N.V. Mody, K.I. Varughese, J.A. Maddry, and H.K. Desai, *J. Am. Chem. Soc.*, 1981, 103, 6536.

TAKAONINE

$C_{24}H_{35}NO_7$; mp 186-187.5°;
$[\alpha]_D$ + 52° (CHCl$_3$)
Aconitum japonicum
^{1}H NMR: δ 1.08 (3H, *t*, J=7Hz, NCH$_2$-
C*H$_3$*), 3.34, 3.36 and 3.37 (each 3H,
s, OC*H$_3$*), 3.96 (1H, *s*, C(6)-α-*H*),
4.05 and 4.34 (each 1H, exch. with
D$_2$O, O*H*), 5.86 and 5.88 (each 1H, *s*,
-C*H*=C*H*-).[1]

[13]*C Chemical Shift Assignments*

C-1	70.8	C-11	49.6	C-1'	–
C-2	131.3	C-12	25.8	C-6'	57.3
C-3	134.5	C-13	47.1	C-7'	–
C-4	40.0	C-14	215.3	C-8'	–
C-5	49.0	C-15	35.1	C-14'	–
C-6	90.4	C-16	86.7	C-16'	56.1
C-7	86.2	C-17	65.7	C-18'	59.3
C-8	83.0	C-18	75.7		
C-9	53.2	C-19	51.6		
C-10	41.6	N-CH$_2$	50.5		
		CH$_3$	13.8		

1.   S. Sakai, H. Takayama, and T. Okamoto, *Yakugaku Zasshi*, 99, 647 (1979).

TAKAOSAMINE

$C_{23}H_{37}NO_7$; mp 174-175°;
$[\alpha]_D$ + 61.2° (CHCl$_3$)
Aconitum japonicum
^{1}H NMR: δ 1.09 (3H, *t*, J=7Hz, NCH$_2$-
CH$_3$), 3.36 and 3.40 (each 3H, *t*,
OCH$_3$), 4.00 (1H, *s*, C(6)-α-*H*) and
4.10 (1H, *t*, J=4.5Hz, C(14)-β-*H*).[1]

^{13}C *Chemical Shift Assignments*

C-1	C-11	C-1'
C-2	C-12	C-6'
C-3	C-13	C-7'
C-4	C-14	C-8'
C-5	C-15	C-14'
C-6	C-16	C-16'
C-7	C-17	C-18'
C-8	C-18	C=O
C-9	C-19	CH$_3$
C-10	N-CH$_2$	
	CH$_3$	

1.   S. Sakai, H. Takayama, and T. Okamoto, *J. Pharm. Soc.* Japan, 99, 647
     (1979).

TALATIZAMINE (TALATISAMINE)

$C_{24}H_{39}NO_5$; mp 145-146°; $[\alpha]_D \pm 0°$

Aconitum arenatum, A. carmichaeli,
A. fischeri, A. franchetii, A. nemo-
rum, A. saposhnikovii, A. talassicum,
A. tranzschelii, A. variegatum.[*]

^{1}H NMR: δ 1.08 (3H, *t*, NCH$_2$-CH$_3$),
3.00 and 3.12 (each 1H, *d*, C(18)-H$_2$),
3.28, 3.30 and 3.36 (each 3H, *s*,
OCH$_3$), and 4.13 (1H, *t*, C(14)-β-H).[1,3,4]

[13]*C Chemical Shift Assignments*[2]

C-1	86.1d	C-11	48.6s	C-1'	56.1q
C-2	25.7t	C-12	28.6t	C-6'	-
C-3	32.6t	C-13	45.7d[a]	C-7'	-
C-4	38.6s	C-14	75.7d	C-8'	-
C-5	37.7d	C-15	39.2t	C-14'	-
C-6	24.8t	C-16	82.2d	C-16'	56.3q
C-7	45.7d[a]	C-17	62.8d	C-18'	59.3q
C-8	72.7s	C-18	79.4t		
C-9	46.9d[a]	C-19	53.1t		
C-10	45.7d[a]	N-CH$_2$ CH$_3$	49.4t 13.6q		

1.   M.S. Yunusov and S.Yu. Yunusov, *Khim. Prirodn. Soedin.*, 6, 90 (1970).
2.   C. Konno, M. Shirasaka and H. Hikino, *J. Nat. Prod.*, 45, 128 (1982).
3.   K. Wiesner, T.Y.R. Tsai, K. Huber, S.E. Bolton and R. Vlahov, *J. Am. Chem. Soc.*, 96, 4990 (1974).
4.   K. Wiesner, *Pure Appl. Chem.*, 41 (1-2), 93 (1975).

TALATIZIDINE

$C_{23}H_{37}NO_5$; mp 220-221°
$[\alpha]_D$ - 20° (MeOH)
Aconitum talassicum[1,2]

^{13}C *Chemical Shift Assignments*

C-1	C-11	C-1'
C-2	C-12	C-6'
C-3	C-13	C-7'
C-4	C-14	C-8'
C-5	C-15	C-14'
C-6	C-16	C-16'
C-7	C-17	C-18'
C-8	C-18	C=O CH₃
C-9	C-19	
C-10	N-CH₂ CH₃	

1.   S.W. Pelletier, L.H. Keith, and P.C. Parasarathy, *J. Am. Chem. Soc.*, 89, 4146 (1967).

2.   S.W. Pelletier, W.H. De Camp, D.L. Herald, Jr., S.W. Page and M.G. Newton, *Acta Crystallogr. Sect. B 33*, 716 (1977).

TATSIENSINE

$C_{27}H_{39}NO_7$; Amorphous;
$[\alpha]_D$ + 17.4° (EtOH)
Delphinium tatsienense
^{1}H NMR: δ 1.00 (3H, *s*, C(18)-CH$_3$),
1.08 (3H, *t*, J=7Hz, N-CH$_2$-CH$_3$), 3.20
(1H, *d*, J=2Hz, C(1)-*H*), 2.05 (3H, *s*,
OCOCH$_3$), 3.30,3.35, 3.42 (each 3H, *s*,
OCH$_3$), 3.69 (2H, *m*, C(14),C(16)-*H*),
4.93 (2H, br *s*, O-CH$_2$-O), 5.39 (1H,
br *s*, C(6)-α-*H*), 5.62 (1H, *d*, J$_{2,3}$=
10 Hz, C(3)-*H*), 5.91 (1H, *dd*, J$_{2,3}$=
10Hz;J$_{1,2}$=4Hz, C(2)-*H*).[1]

[13]C *Chemical Shift Assignments*

C-1	83.6	C-11	50.8s	C-1'	55.9
C-2	124.6	C-12	28.3	C-6'	-
C-3	137.4	C-13	38.7	C-7-O$\diagdown$	
C-4	34.8s	C-14	80.2	C-8-O$\diagup$CH$_2$	93.4
C-5	54.8	C-15	34.2	C-14'	57.6
C-6	78.5	C-16	81.9	C-16'	56.2
C-7	91.3s	C-17	61.5	C-18'	-
C-8	84.3s	C-18	23.2	C=O	169.6s
C-9	47.4	C-19	57.6	CH$_3$	21.7
C-10	40.0	N-CH$_2$	48.7		
		CH$_3$	13.0		

1. S.W. Pelletier, J.A. Glinski, B.S. Joshi, and S.-Y. Chen, *Heterocycles*, 20, 1347 (1983).

TRICORNINE

$C_{27}H_{43}NO_8$; mp 187-189°;
$[\alpha]_D$ + 47.3° (EtOH)
Delphinium tricorne
^{1}H NMR: δ 1.02 (3H, *t*, J=7Hz, NCH$_2$-
CH$_3$), 2.02 (3H, *s*, OCOCH$_3$), 3.18,
3.26, 3.31, 3.34 (each 3H, *s*, OCH$_3$).[1,3,4]

[13]C Chemical Shift Assignments[2]

C-1	84.0	C-11	49.0	C-1'	55.7
C-2	26.1	C-12	28.7	C-6'	57.8
C-3	31.9	C-13	46.1	C-7'	-
C-4	37.2	C-14	84.0	C-8'	-
C-5	43.3	C-15	33.7	C-14'	58.0
C-6	90.9	C-16	82.6	C-16'	56.3
C-7	88.5	C-17	64.6	C-18'	-
C-8	77.5	C-18	69.1	C=O	170.9
C-9	50.4	C-19	52.4	CH$_3$	20.8
C-10	38.1	N-CH$_2$	51.0		
		CH$_3$	14.1		

1. S.W. Pelletier and J. Bhattacharyya, *Phytochemistry*, 16, 1464 (1977).
2. S.W. Pelletier, N.V. Mody, R.S. Sawhney, and J. Bhattacharyya, *Hetero-cycles*, 7, 327 (1977).
3. S.W. Pelletier, N.V. Mody, K.I. Varughese, J.A. Maddry, and H.K. Desai, *J. Am. Chem. Soc.*, 1981, 103, 6536.
4. The [1]H N.M.R. values reported in reference 1 are in error and have been corrected in the catalogue.

1,8,14-TRI-*O*-METHYLNEOLINE

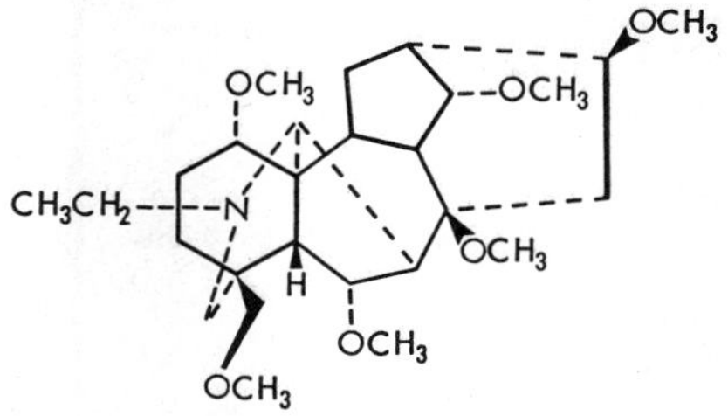

$C_{27}H_{45}NO_6$; mp 118-119°;

$[\alpha]_D$ + 4.5°

Prepared from delphisine

^{1}H NMR: δ 1.11 (3H, *t*, NCH$_2$-*CH$_3$*),
3.31 (6H, *s*, O*CH$_3$*), 3.38, 3.40, 3.43
and 3.45 (each 3H, *s*, O*CH$_3$*) and 4.20
(1H, *dd*, C(14)-β-*H*).[1,2]

^{13}C *Chemical Shift Assignments*[3]

C-1	85.4	C-11	50.9	C-1'	56.3
C-2	26.3	C-12	30.2	C-6'	57.6
C-3	35.2	C-13	46.1	C-7'	-
C-4	39.0	C-14	83.6[a]	C-8'	48.0
C-5	48.4	C-15	34.7	C-14'	58.5
C-6	84.1[a]	C-16	83.6[a]	C-16'	56.1
C-7	48.4	C-17	60.8	C-18'	59.1
C-8	78.3	C-18	80.4	C=O	
C-9	45.6	C-19	54.3	CH$_3$	
C-10	38.1	N-CH$_2$	48.9		
		CH$_3$	13.4		

1.　　S.W. Pelletier, W.H. De Camp, S.D. Lajsic, Z. Djarmati, and A.H. Kapadi, *J. Am. Chem. Soc.*, 96, 7815 (1974).

2.　　L. Marion, J.P. Boca and J. Kallos, *Tetrahedron* Suppl. 8, 101 (1966).

3.　　S.W. Pelletier and Z. Djarmati, *J. Am. Chem. Soc.*, 98, 2626 (1976).

UMBROSINE

$C_{24}H_{39}NO_6$;   MW: 437.2738;
mp 150-151°

Aconitum umbrosum

^{1}H NMR: δ 1.05 (3H, *t*, NCH$_2$-CH$_3$),
3.26, 3.28 and 3.33 (each 3H, *s*,
OCH$_3$).[1]

^{13}C Chemical Shift Assignments

C-1	C-11	C-1'
C-2	C-12	C-6'
C-3	C-13	C-7'
C-4	C-14	C-8'
C-5	C-15	C-14'
C-6	C-16	C-16'
C-7	C-17	C-18'
C-8	C-18	
C-9	C-19	
C-10	N-CH$_2$	
	CH$_3$	

1.   V.A. Tel'nov, N.M. Golubev and M.S. Yunusov, *Khim. Prir. Soedin.*, 675 (1976).

VERATROYLPSEUDACONINE

$C_{34}H_{49}NO_{11}$; mp 211-213°;
$[\alpha]_D$ + 36.8° (EtOH)
Aconitum falconeri, A. ferox.
^{1}H NMR: δ 1.11 (3H, *s*, NCH$_2$C*H$_3$*),
3.26, 3.29, 3.32 and 3.44 (each 3H,
s, OC*H$_3$*), 3.92 (6H, *s*, aromatic
OC*H$_3$*), 5.10 (1H, *d*, J=4.5Hz, C(14)-β
-*H*), and 6.86-7.68 (aromatic protons).[1,3]

Vr = CO——⟨ ⟩——OCH$_3$ OCH$_3$

[13]*C Chemical Shift Assignments*[2]

C-1	83.4	C-16	82.5	C=O		166.2
C-2	35.8	C-17	61.6			122.5
C-3	71.3	C-18	76.7			110.5
C-4	43.3	C-19	48.9		OCH$_3$	153.1
C-5	47.5	N-CH$_2$	47.5	OCH$_3$		148.6
C-6	82.5	CH$_3$	13.5			112.3
C-7	47.5	C-1'	55.8			
C-8	73.6	C-6'	57.5		OCH$_3$	55.8
C-9	53.8	C-7'	-			
C-10	41.9	C-8'	-			
C-11	50.2	C-14'	-			
C-12	33.7	C-16'	58.3			
C-13	75.8	C-18'	59.1			
C-14	79.8					
C-15	42.4					

1. S.W. Pelletier, N.V. Mody, and H.S. Puri, *Phytochemistry*, 16, 623 (1977).

2. S.W. Pelletier, N.V. Mody, R.S. Sawhney and J. Bhattacharyya, *Heterocycles*, 7, 327 (1977).

3. K.K. Purushothaman and S. Chandrasekhanan, *Phytochemistry*, 13, 1975 (1974).

VILMORRIANINE A

$C_{35}H_{49}NO_{10}$;   mp 182-184°

Aconitum kongboense, A. vilmorrianum

^{1}H NMR: δ 1.04 (3H, *t*, J=7Hz, NCH$_2$-
CH$_3$), 1.40 (3H, *s*, OCOCH$_3$), 3.18,
3.25, 3.30 and 3.38 (each 3H, *s*, OCH$_3$)
3.85 (3H, *s*, aromatic OCH$_3$), 4.11
(1H, *d*, J=6Hz, C(6)-β-*H*), 5.03 (1H,
dd, J=4.5Hz, C(14)-β-*H*) and 6.93-
8.02 (aromatic protons).[1]

As =

^{13}C Chemical Shift Assignments[2]

C-1	83.6	C-16	82.8	1	122.9
C-2	33.5	C-17	61.3	2	131.7
C-3	71.4	C-18	75.3	3	113.7
C-4	43.1	C-19	48.8	4	163.4
C-5	47.0	N-CH$_2$ CH$_3$	47.6	5	113.7
C-6	82.4		13.3	6	131.7
C-7	44.8	C-1'	55.4		
C-8	85.9	C-6'	58.0	OCH$_3$	56.5
C-9	48.6	C-7'	-		
C-10	43.7	C-8'	-		
C-11	50.5	C-14'	-		
C-12	28.4	C-16'	55.6		
C-13	38.1	C-18'	59.1		
C-14	76.8	C=O CH$_3$	169.7		
C-15	39.2		21.7		

1. C.-R. Yang, X.-J. Hao, D.-Z. Wang and J. Zhou, *Acta Chimica Sinica*,
    39, 147 (1981).
2. C.-R. Yang, D.-Z. Wang, D.-C. Wu, X.-J. Hao and J. Zhou, *Acta Chimica
    Sinica*, 39, 445 (1981).

VIRESCENINE

$C_{23}H_{37}NO_6$; mp 68-70°;
$[\alpha]_D$ + 16.9° (EtOH)
Delphinium virescens
^{1}H NMR: δ 1.10 (3H, *t*, NCH_2-CH_3),
3.36 and 3.38 (each 3H, *s*, OCH_3), and
4.30 (1H, *dd*, C(14)-β-H).[1]

[13]C *Chemical Shift Assignments* [1]

C-1	72.4	C-11	49.4	C-1'	–
C-2	28.5	C-12	26.9	C-6'	–
C-3	29.3	C-13	43.6	C-7'	–
C-4	37.7	C-14	75.5	C-8'	–
C-5	41.9	C-15	36.0	C-14'	–
C-6	33.5	C-16	81.9	C-16'	56.4
C-7	86.1	C-17	64.9	C-18'	59.4
C-8	76.2	C-18	78.7		
C-9	48.0	C-19	55.8		
C-10	39.7	N-CH_2	50.5		
		CH_3	13.9		

1.   S.W. Pelletier, N.V. Mody, A.P. Venkov and S.B. Jones, Jr., *Heterocycles*,
     12, 779 (1979).

YUNACONITINE

$C_{35}H_{49}NO_{11}$;   mp 141-143°;
$[\alpha]_D$ + 37.7° (CHCl$_3$)

Aconitum crassicaule, A. delavyi,
A. forestii, A. geniculatum Flet. et
Laue. var. *unguiculatum* W.T. Wang,
A. hemsleyanum Pritz. var. *circinatum* W.T. Wang,
A. vilmorrianum.

^{1}H NMR: δ 1.10 (3H, *t*, J=7Hz, NCH$_2$-C*H$_3$*), 1.34
(3H, *s*, OCOC*H$_3$*), 3.16, 3.25, 3.30 and 3.55
(each 3H, *s*, OC*H$_3$*), 3.87 (3H, *s*, aromatic OC*H$_3$*),
4.84 (1H, *d*, J=4.5Hz, C(14)-β-*H*) and 6.93-8.01
(4H, A$_2$B$_2$, J=9Hz, aromatic protons).[1]

[13]C *Chemical Shift Assignments* [2]

C-1	83.2	C-16	83.6	C=O		166.1
C-2	33.7	C-17	61.6	1		122.6
C-3	71.3	C-18	76.6	2		131.7
C-4	43.2	C-19	48.8	3		113.8
C-5	47.4	N-CH$_2$	47.4	4		163.5
C-6	82.3	CH$_3$	13.3	5		113.8
C-7	44.8	C-1'	56.8	6		131.7
C-8	85.6	C-6'	58.8			
C-9	48.8	C-7'	-	OCH$_3$		55.4
C-10	40.8	C-8'	-			
C-11	50.3	C-14'	-			
C-12	35.3	C-16'	57.8			
C-13	74.8	C-18'	59.1			
C-14	78.6	C=O	169.9			
C-15	39.6	CH$_3$	21.7			

1.   S.-Y. Chen, *Acta Chimica Sinica*, 37, 15 (1979).
2.   C.-R. Yang, D.-Z. Wang, D.-G. Wu, X.-J. Hao and J. Zhou, *Acta Chimica Sinica*, 39, 445 (1981).

7.1. Abbreviations Used in Spectral Catalog

The ^{13}C values are given in ppm downfield relative to Me$_4$Si. The solvent is deuteriochloroform unless stated otherwise.

a, b, c	Values within any vertical columns may be interchanged.
s	Singlet
d	Doublet
t	Triplet
m	Multiplet
*	See Section 7.2.

7.2. Addendum to Alkaloid-Bearing Plants

Aconitine
  Aconitum angustifolium
  Aconitum bullatifolium
  Aconitum cammarum
  Aconitum firmum
  Aconitum fischeri
  Aconitum fukutomei
  Aconitum nagarum var. *lasiandrum*
  Atragne sibirica
Browniine
  Delphinium cardinale
Bullatine C
  Aconitum bullatifolium var. *homotrichum*
  Aconitum nagarum var. *lasiandrum*
N-Deacetyllappaconitine
  Aconitum barbatum var. *puberulum*
Delcosine
  Aconitum finetianum
  Aconitum lucidusculum
  Delphinium bicolor
  Delphinium tatsienense
Delsoline
  Aconitum karakolicum
3-Deoxyaconitine
  Aconitum pendulum
Hypaconitine
  Aconitum angustifolium
  Aconitum bullatifolium

Hypaconitine (*Continued*)
  Aconitum koreanum
  Aconitum spp.

Karakoline
  Aconitum heterophyllum
  Aconitum spp.

Lappaconine
  Aconitum leucostomum
  Aconitum septentrionale

Lappaconitine

Lycaconitine
  Aconitum barbatum var. *puberulum*
  Aconitum septentrionale

Lycoctonine
  Delphinium bicolor
  Delphinium cardinale
  Delphinium regalis
  Delphinium tatsienense

Mesaconitine
  Aconitum angustifolium
  Aconitum chinense
  Aconitum napellus ssp. *fissurae* and *superbum*
  Aconitum spp. ("Chuanwu" in Chinese)

Neoline
  Aconitum karakolicum
  Delphinium staphisagria

Talatizamine
  Aconitum forrestii

Subject Index

Organism Index

Acacia argentea, 58
Acacia polystachia, 58
Acacia spirorbis, 58
Acalymma vittatum, 200
Acanthaceae, 90
Acanthias vulgaris, 58
Acanthina spirata, 57
Aconitum, 206, 208
Aconitum alatiacum:
   diterpenoid alkaloids from, 214
   mesaconitine and, 419
Aconitum angustifolium, 214
Aconitum anthoroideum:
   condelphine and, 297
   diterpenoid alkaloids and, 214
Aconitum arcuatum, 214
Aconitum arenatum:
   aconosine and, 271
   talatizamine and, 451
Aconitum balfouri:
   diterpenoid alkaloids and, 214
   pseudaconitine, 433
Aconitum barbatum:
   N-deacetylranaconitine and, 303
   diterpenoid alkaloids from, 214, 215
   lappaconitine and, 406
   puberaconitidine and, 434
   puberaconitine and, 435
   puberanine and, 436
   ranaconitine and, 441
   septentriodine and, 446
   septentrionine and, 447
   N-(succinyl)anthranoyllycoctonine, 448
Aconitum bullatifolium:
   diterpenoid alkaloids from, 214
   neoline and, 425
Aconitum callianthum:
   aconitine and, 269
   diterpenoid alkaloids from, 214
   hypaconitine and, 393
   mesaconitine and, 419
Aconitum cammarum, 214, 221
Aconitum carmichaeli:
   14-acetyltalatizamine and, 263

aconitine and, 269
14-benzoylmesaconine and, 287
diterpenoid alkaloids from, 214
hokbusine A, 385
hokbusine B, 386
15-alpha hydroxyneoline, 389
hypaconitine and, 393
isodelphinine and, 396
isotalatizidine and, 399
karakoline and, 402
mesaconitine and, 419
neoline and, 425
senbusine A and, 444
senbusine B and, 445
talatizamine and, 451
Aconitum chasmanthum:
   chasmaconitine and, 294
   chasmanien and, 295
   chasmanthinine and, 296
   diterpenoid alkaloids from, 215
   homochasmanine, 387
   indaconitine and, 395
Aconitum chinese:
   aconitine and, 269
   14-benzoylmesaconine and, 287
   diterpenoid alkaloids from, 215
   lipoaconitine and, 410
   lipo-3-deoxyaconitine and, 411
   lipohypaconitine and, 412
   lipomesaconitine and, 412
Aconitum confertiflorum:
   14-acetyltalatizamine, 263
   diterpenoid alkaloids from, 215
Aconitum coreanum, 393
Aconitum crassicaule:
   chasmanien and, 295
   crassicaulidine and, 298
   crassicauline A and, 299
   diterpenoid alkaloids from, 215
   15-beta hydroxyneoline, 390
   yunaconitine and, 460
Aconitum deinorrhizum:
   diterpenoid alkaloids from, 215
   pseudaconitine, 433